Lattice-Ordered Groups

T0338484

Mathematics and Its Applications

Lattice-Ordered Groups

Advances and Techniques

edited by

A. M. W. Glass

and

W. Charles Holland

*Department of Mathematics and Statistics,
Bowling Green State University, Ohio, U.S.A*

KLUWER ACADEMIC PUBLISHERS

DORDRECHT / BOSTON / LONDON

Library of Congress Cataloging in Publication Data

Lattice-ordered groups.

  (Mathematics and its applications)
  Bibliography: p.
  Includes index.
  1. Lattice ordered groups. I. Glass, A. M. W.
(Andrew Martin William), 1944-   . II. Holland,
W. Charles. III. Series: Mathematics and its
application (Kluwer Academic Publishers)
QA171.L295 1989   512'.2   89-2448

ISBN 0-7923-0116-1

Published by Kluwer Academic Publishers,
P.O. Box 17, 3300 AA Dordrecht, The Netherlands.

Kluwer Academic Publishers incorporates
the publishing programmes of
D. Reidel, Martinus Nijhoff, Dr W. Junk and MTP Press.

Sold and distributed in the U.S.A. and Canada
by Kluwer Academic Publishers,
101 Philip Drive, Norwell, MA 02061, U.S.A.

In all other countries, sold and distributed
by Kluwer Academic Publishers Group,
P.O. Box 322, 3300 AH Dordrecht, The Netherlands.

printed on acid free paper

SERIES EDITOR'S PREFACE

'Et moi, ..., si j'avait su comment en revenir, je n'y serais point allé.'

Jules Verne

The series is divergent; therefore we may be able to do something with it.

O. Heaviside

One service mathematics has rendered the human race. It has put common sense back where it belongs, on the topmost shelf next to the dusty canister labelled 'discarded nonsense'.

Eric T. Bell

Mathematics is a tool for thought. A highly necessary tool in a world where both feedback and non-linearities abound. Similarly, all kinds of parts of mathematics serve as tools for other parts and for other sciences.

Applying a simple rewriting rule to the quote on the right above one finds such statements as: 'One service topology has rendered mathematical physics ...'; 'One service logic has rendered computer science ...'; 'One service category theory has rendered mathematics ...'. All arguably true. And all statements obtainable this way form part of the raison d'être of this series.

This series, *Mathematics and Its Applications*, started in 1977. Now that over one hundred volumes have appeared it seems opportune to reexamine its scope. At the time I wrote

> "Growing specialization and diversification have brought a host of monographs and textbooks on increasingly specialized topics. However, the 'tree' of knowledge of mathematics and related fields does not grow only by putting forth new branches. It also happens, quite often in fact, that branches which were thought to be completely disparate are suddenly seen to be related. Further, the kind and level of sophistication of mathematics applied in various sciences has changed drastically in recent years: measure theory is used (non-trivially) in regional and theoretical economics; algebraic geometry interacts with physics; the Minkowsky lemma, coding theory and the structure of water meet one another in packing and covering theory; quantum fields, crystal defects and mathematical programming profit from homotopy theory; Lie algebras are relevant to filtering; and prediction and electrical engineering can use Stein spaces. And in addition to this there are such new emerging subdisciplines as 'experimental mathematics', 'CFD', 'completely integrable systems', 'chaos, synergetics and large-scale order', which are almost impossible to fit into the existing classification schemes. They draw upon widely different sections of mathematics."

By and large, all this still applies today. It is still true that at first sight mathematics seems rather fragmented and that to find, see, and exploit the deeper underlying interrelations more effort is needed and so are books that can help mathematicians and scientists do so. Accordingly MIA will continue to try to make such books available.

If anything, the description I gave in 1977 is now an understatement. To the examples of interaction areas one should add string theory where Riemann surfaces, algebraic geometry, modular functions, knots, quantum field theory, Kac-Moody algebras, monstrous moonshine (and more) all come together. And to the examples of things which can be usefully applied let me add the topic 'finite geometry'; a combination of words which sounds like it might not even exist, let alone be applicable. And yet it is being applied: to statistics via designs, to radar/sonar detection arrays (via finite projective planes), and to bus connections of VLSI chips (via difference sets). There seems to be no part of (so-called pure) mathematics that is not in immediate danger of being applied. And, accordingly, the applied mathematician needs to be aware of much more. Besides analysis and numerics, the traditional workhorses, he may need all kinds of combinatorics, algebra, probability, and so on.

In addition, the applied scientist needs to cope increasingly with the nonlinear world and the extra mathematical sophistication that this requires. For that is where the rewards are. Linear models are honest and a bit sad and depressing: proportional efforts and results. It is in the non-linear world that infinitesimal inputs may result in macroscopic outputs (or vice versa). To appreciate what I am hinting at: if electronics were linear we would have no fun with transistors and computers; we would have no TV; in fact you would not be reading these lines.

There is also no safety in ignoring such outlandish things as nonstandard analysis, superspace and anticommuting integration, p-adic and ultrametric space. All three have applications in both electrical engineering and physics. Once, complex numbers were equally outlandish, but they frequently proved the shortest path between 'real' results. Similarly, the first two topics named have already provided a number of 'wormhole' paths. There is no telling where all this is leading - fortunately.

Thus the original scope of the series, which for various (sound) reasons now comprises five sub-series: white (Japan), yellow (China), red (USSR), blue (Eastern Europe), and green (everything else), still applies. It has been enlarged a bit to include books treating of the tools from one subdiscipline which are used in others. Thus the series still aims at books dealing with:

- a central concept which plays an important role in several different mathematical and/or scientific specialization areas;
- new applications of the results and ideas from one area of scientific endeavour into another;
- influences which the results, problems and concepts of one field of enquiry have, and have had, on the development of another.

A lattice-ordered group is a mathematical structure combining a (partial) order (lattice) structure and a group structure (on a set) in a compatible way. Thus it is a composite structure, one of several which are vigorously studied today. Some others are Hopf-algebras (compatibly combined algebra and co-algebra structures), Poisson algebras (an algebra structure plus a compatible Lie algebra structure, λ-algebras, Riesz spaces (Banach space plus order structure), quantum groups, ... Much of the earlier three quarters of the 20-th century in mathematics has been devoted to the definition and study of 'simple' structures: groups, partially ordered sets, algebras, co-algebras, Lie algebras, topological spaces. Now the time has come to study, use, and apply composite structures: sets carrying two or more simple structures in a compatible way. For many of these composite structures the work has only just started. The field of lattice-ordered groups is one of the exceptions and their study has reached a certain plateau of maturity. Moreover they naturally turn up in a surprisingly wide range of (functional) analysis to universal algebra. Thus, next to a basic introductory text on the subject (M. Anderson, T. Feil, Lattice-ordered groups, Reidel, 1988) there is definitely room for a collection of some 10 authoritative surveys of the various aspects of the field, and in view of the interspecialistic nature of these studies and their wide (potential) applicability I am happy to be able to present this volume within the framework of this series.

Perusing the present volume is not guaranteed to turn you into an instant expert, but it will help, though perhaps only in the sense of the last quote on the right below.

The shortest path between two truths in the real domain passes through the complex domain. J. Hadamard	Never lend books, for no one ever returns them; the only books I have in my library are books that other folk have lent me. Anatole France
La physique ne nous donne pas seulement l'occasion de résoudre des problèmes ... elle nous fait pressentir la solution. H. Poincaré	The function of an expert is not to be more right than other people, but to be wrong for more sophisticated reasons. David Butler

Bussum, February 1989

Michiel Hazewinkel

TABLE OF CONTENTS

PREFACE

Ordered fields have been studied extensively, as have ordered Abelian groups (in areas as diverse as functional analysis and valuation theory). In contrast, the more general theory of lattice-ordered groups was relatively ignored until the last thirty years despite the fact that the theory is as natural as that of rings. Recently, lattice-ordered groups have cropped up in many other areas of mathematics. For example, in the theory of Bézout domains; in the study of unstable theories; in amalgamation and varietal questions in universal algebra; in the sudy of symmetries (where they play a certain universal role); in analogues of theorems from combinatorial group theory; and in completion theory in naturally arising classes of "semi-topological" groups. This richness and diversity has necessitated an extraordinary growth in the techniques used and their applications. In order to better understand this development and be familiar with the new techniques on both an intuitive and formal level, the editors organized a week-long conference in Bowling Green in May 1985 with eight speakers, each giving a series of talks in special areas of lattice-ordered groups.

Having decided that it would be worthwhile to make the material available to a larger audience, we asked each of the speakers (as well as some additional colleagues) to survey their particular specialty and make the material accessible not only to experts but also to other mathematicians whose disicipline overlapped with the particular topic. To ensure that our goals were achieved, every resulting chapter was refereed both by an expert and by an "interested bystander". We then wrote the necessary preliminary chapters), 0, 1, and 2 (of which only 0.1 and 1.1 are necessary background for almost all chapters) to complete the accessibility of all material and avoid too much repetition; for more complete background on lattice-ordered groups, see the list of references or the new and interesting book *"Lattice-Ordered Groups: An Introduction"* by M. Anderson and T. Feil (1988, same publisher). We have strived to keep inter-dependence between chapters to a minimum so that those interested in a particular topic can just read the necessary chapter and little else in the book. We hope that these efforts have been successful.

We are most grateful to the authors, who by supplying their own unique insight and viewpoint have produced a book which would not have been possible for any single individual. For the time and perceptive comments of the many referees, we express our heartfelt thanks. The financial support of Bowling Green State University helped make the workshop possible. Our thanks also to Cyndi Patterson for typing portions of the book, and to our families and colleagues for their support in what turned out to be a very long time-consuming task. Finally, we wish to thank Dr. David Larner of Kluwer Academic Publishers for his help and patience.

In the Black Swamp
Bowling Green, Ohio
All Souls Day 1988

A. M. W. Glass
W. C. Holland

Symbol Index

xiii

CHAPTER 0

ELEMENTARY FACTS

In this chapter and the next we provide the necessary background for the reader who is completely unversed in the subject of lattice-ordered groups. We begin with the definition and some examples and then provide the requisite elementary properties (those that concern elements). The last section of the chapter includes results which will not be needed later but which might be of interest to people outside the subject and that are of a sufficiently basic nature.

A *partially ordered group* G is a group that is also a partially ordered set such that $cad \leq cbd$ whenever $a \leq b$. We write G is a *po-group*, for short. The order of any po-group G is completely determined by the set $G^{+}=$ $\{g \in G \mid e \leq g\}$ of *positive* elements. G^{+} is sometimes called the *positive cone of* G and is a normal subsemigroup of G containing no element together with its inverse, except e. Conversely, any such subsemigroup of G serves as the set of positive elements for some partial order which makes G a po-group. If the partial order is a linear or total order (i.e., for all $a,b \in G$, $a \leq b$ or $b \leq a$), we say that G is an *o-group*; if the partial order is a lattice (for all $a,b \in G$, there is a least upper bound of a and b denoted by $a \vee b$ and a greatest lower bound denoted by $a \wedge b$) we say that G is a *lattice-ordered group*, or *ℓ-group* for short. We can view ℓ-groups as algebras under the group operations and lattice operations $\vee$ and $\wedge$. The hypotheses imply that multiplication on the right (or left) by any fixed element of G induces a lattice automorphism of G, and so the group operation distributes over the lattice operations. Throughout we will use e for the identity element of an ℓ-group (or 0 in the case of

1

A. M. W. Glass and W. C. Holland (eds.), *Lattice-Ordered Groups*, 1–10.
© 1989 by Kluwer Academic Publishers.

certain Abelian ℓ-groups when we are writing the group operation additively). If the group G is also a partially ordered set and we merely require that the order is preserved by multiplication on the right ($ad \le bd$ whenever $a \le b$) G will be called *right partially ordered*; in this case, if the partial order is linear, we say that the group is a *right ordered group*.

EXAMPLES:

1. $\mathbf{Z}$, the integers; $\mathbf{Q}$, the rationals; $\mathbf{R}$, the reals; all under the usual addition and order, are Abelian o-groups.

2. If $\{G_\lambda \mid \lambda \in \Lambda\}$ is a collection of ℓ-groups, then $\Pi\{G_\lambda \mid \lambda \in \Lambda\}$, the cartesian product of the family of ℓ-groups $\{G_\lambda \mid \lambda \in \Lambda\}$ ordered by: $f \le h$ if and only if $f_\lambda \le h_\lambda$ in G_λ for all $\lambda \in \Lambda$, is also an ℓ-group called the *cardinal product* of the family $\{G_\lambda \mid \lambda \in \Lambda\}$. This order on $\Pi\{G_\lambda \mid \lambda \in \Lambda\}$ is called the *cardinal order*. The direct sum $\{f \in \Pi G_\lambda \mid f_\lambda = e_\lambda$ for all but a finite number of $\lambda \in \Lambda\}$ is also a sublattice and hence an ℓ-group. We write it $\Sigma\{G_\lambda \mid \lambda \in \Lambda\}$ or $G_1 \boxplus \ldots \boxplus G_n$ if $\Lambda = \{1,\ldots,n\}$. Other sublattice subgroups in the case that $\Lambda = \{1,2,\ldots\}$ and all $G_\lambda \subseteq \mathbf{R}$ are the sets of eventually constant sequences and bounded sequences.

3. Let G be an ℓ-group and A and o-group. The *lexicographic product* of A and G, denoted $A \overset{\leftarrow}{\otimes} G$, is the direct product of the groups A and G ordered by: $(a_1,g_1) \le (a_2,g_2)$ if $a_1 < a_2$ or both $a_1 = a_2$ and $g_1 \le g_2$. It is an ℓ-group. The ℓ-group $G \overset{\leftarrow}{\otimes} A$ is similarly defined and, moreover, if $\{G_\lambda \mid \lambda \in \Lambda\}$ is a family of o-groups with Λ a linearly ordered set, then $\overset{\leftarrow}{\Sigma}\{G_\lambda \mid \lambda \in \Lambda\}$ is the direct sum of the groups $\{G_\lambda\}$ ordered by $f < g$ if there is $\lambda_0 \in \Lambda$ such that $f_{\lambda_0} < g_{\lambda_0}$ and $f_\lambda = g_\lambda$ for all $\lambda \in \Lambda$ with $\lambda > \lambda_0$. Note that $\overset{\leftarrow}{\Sigma}\{G_\lambda \mid \lambda \in \Lambda\}$ is an o-group. Caution: if $G \ne \{e\}$ and A is an ℓ-group, then $G \overset{\leftarrow}{\otimes} A$ *is an* ℓ-group if and only if A is an o-group: Let $a_1, a_2 \in A$ be such that $a_1 \not\le a_2$ and $a_2 \not\le a_1$. Now $a_1(a_1 \wedge a_2)^{-1} \wedge a_2(a_1 \wedge a_2)^{-1} = e$ and $b_j = a_j(a_1 \wedge a_2)^{-1} \ne e$ ($j = 1,2$). Then $(g,e) \le (e,b_1),(e,b_2)$ for all $g \in G$ so (e,b_1) and (e,b_2) have no greatest lower bound in $G \overset{\leftarrow}{\otimes} A$, for it is easily seen that the non-trivial ℓ-group G can have no largest element.

4. Examples 2 and 3 give two distinct orders on the group $\mathbf{Z} \otimes \mathbf{Z}$. Another separate family of linear orders (and hence lattice orders) is furnished by: Choose an arbitrary irrational number ξ. Define $(m_1,n_1) \le (m_2,n_2)$ if and only if $m_1 + \xi n_1 \le m_2 + \xi n_2$ in $\mathbf{R}$.

5. Any ℓ-group is torsion free (Corollary 0.1.2). For a partial converse, if G is any Abelian torsion-free group, then G can be embedded in a divisible torsion-free Abelian group V, which is, therefore, a rational vector space. If Λ is a basis for V, we can totally order the set Λ. Since V is

the direct sum of groups $\{G_\lambda \mid \lambda \in \Lambda\}$ with each G_λ isomorphic to $\mathbf{Q}$, we can totally order V as $\overset{\leftarrow}{\Sigma}\{G_\lambda \mid \lambda \in \Lambda\}$ as in example 3. Then G becomes an o-group with the inherited order. Of course, we could also make V an ℓ-group with the cardinal order as in example 2. However, G would not necessarily be a sublattice thereof.

Again, if G is an Abelian ℓ-group (written additively) then as a torsion-free Abelian group, G is contained in its divisible hull V whose elements may be described as $V = \{\frac{m}{n}g \mid g \in G, 0 \leq m, n \in \mathbf{Z}, n \neq 0\}$. The order on G may be extended to V by letting $\frac{m}{n}g \geq 0$ iff $g \geq 0$. This makes V into an ℓ-group containing G as a sublattice subgroup, that is, as an ℓ-*subgroup*.

6. We synthesize examples 2 and 3 into a more general example. Let Λ be a *root system*; i.e., a partially ordered set such that the lower bounds of each element form a totally ordered subset. Let $\{G_\lambda \mid \lambda \in \Lambda\}$ be a family of o-groups. For each $g: \Lambda \to \cup G_\lambda$ with $g(\lambda) \in G_\lambda$, let $\mathrm{supp}(g) = \{\lambda \in \Lambda \mid g(\lambda) \neq e\}$. $V(\Lambda, G_\lambda)$ is the group of all such g for which every non-empty linearly ordered subset of $\mathrm{supp}(g)$ has a largest element. We order $V(\Lambda, G_\lambda)$ by: $e < g$ if and only if $e_\lambda < g(\lambda)$ for each λ that is a maximal element of $\mathrm{supp}(g)$. Usually, each G_λ will be a subgroup of $\mathbf{R}$. In this case, $V(\Lambda, G_\lambda)$ is called a *Hahn group on* Λ.

7. $C(X, \mathbf{R})$, the additive group of continuous functions from a topological space X into $\mathbf{R}$, equipped with the pointwise order: $f \leq g$ if and only if $f(x) \leq g(x)$ for all $x \in X$. Then $C(X, \mathbf{R})$ is an ℓ-group.

8. Let $(\Omega, \leq)$ be an infinite linearly ordered set and $A(\Omega) = \mathrm{Aut}(\Omega, \leq)$. So $A(\Omega)$ is the group (under composition) of all order-preserving permutations of Ω. The pointwise order makes $A(\Omega)$ an ℓ-group. Chapter 2 is devoted to the study of such ℓ-groups. If $g \in A(\Omega)$, let $\mathrm{supp}(g) = \{\alpha \in \Omega \mid \alpha g \neq \alpha\}$. Let $B(\Omega) = \{g \in A(\Omega) \mid \exists \alpha, \beta \in \Omega$ such that $\alpha < \mathrm{supp}(g) < \beta\}$, an ℓ-subgroup of $A(\Omega)$. If whenever $\alpha_1 < \alpha_2$ and $\beta_1 < \beta_2$ in Ω, there exists $g \in A(\Omega)$ such that $\alpha_i g = \beta_i$ $(i = 1,2)$, we say that Ω is *doubly homogeneous*. Such linearly ordered sets are n-homogeneous for all $0 < n \in \mathbf{Z}$ (see Chapter 2) and for such Ω, $B(\Omega)$ is a simple group. Another example of a simple group is due to Chehata [1952] (or see [Glass, 1981b, Chapter 6].) Let $C = \{g \in B(\mathbf{R}) \mid \exists \alpha_1 < ... < \alpha_n$ such that $g \mid [\alpha_j, \alpha_{j+1}]$ is linear $(1 \leq j < n)$ and $\mathrm{supp}(g) \subseteq (\alpha_1, \alpha_n)\}$. Then C is a simple group (op. cit.) and it can be reordered to make it an o-group: g is declared to be greater than f if the left-most slope of gf^{-1} other than 1 exceeds 1.

9. Our final example involves orderings of free groups. Let F be a free group. Let $\gamma_n(F)$ be the nth term of the lower central series of F. By Witt's Theorem, $\bigcap_{n=0}^{\infty} \gamma_n(F) = \{e\}$. Moreover, each $\gamma_n(F)/\gamma_{n+1}(F)$ is a free

Abelian group and so can be linearly ordered to become an o-group (see example 5.) If a and b are distinct elements of F, there is a natural number n such that $ba^{-1} \in \gamma_n(F) \setminus \gamma_{n+1}(F)$. Define $a < b$ if $\gamma_{n+1}(F)ba^{-1} > \gamma_{n+1}(F)$ in the linear ordering on $\gamma_n(F)/\gamma_{n+1}(F)$. This makes F an o-group. It can be shown that any lattice ordering in F that makes it an ℓ-group actually makes it an o-group [Weinberg, 1963; 5.8]. In the special case that F is free on $\aleph_1$ generators, there is a linear ordering on F making it an o-group such that any other group structure on the carrier of F which makes it an o-group with respect to the given ordering generates the variety of all groups. This deep result is due to Holland, Mekler, and Shelah [1985].

The above proof that free groups can be made into o-groups yields:

Every torsion-free nilpotent group can be ordered to be an o-group.

0.1. Elementary Properties

In this section we list some important but straightforward properties concerning elements and identities in ℓ-groups. These will be used repeatedly throughout the book. In the next section we will obtain some not-so-obvious elementary facts using them.

LEMMA 0.1.1. *In any l-group*
  (a) $a \vee b = (a^{-1} \wedge b^{-1})^{-1}$ *and* $a \wedge b = (a^{-1} \vee b^{-1})^{-1}$.
  (b) $(a \vee b)^n = a^n \vee a^{n-1}b \vee ... \vee ab^{n-1} \vee b^n$ *if a and b commute.*
  (c) $(a \wedge b)^n = a^n \wedge a^{n-1}b \wedge ... \wedge ab^{n-1} \wedge b^n$ *if a and b commute.*

Proof. (a) $(a \vee b)^{-1} \leq a^{-1}, b^{-1}$ since $a, b \leq a \vee b$. If $c \leq a^{-1}, b^{-1}$, then $c^{-1} \geq a, b$; so $c^{-1} \geq a \vee b$. Thus $c \leq (a \vee b)^{-1}$ proving that $(a \vee b)^{-1} = a^{-1} \wedge b^{-1}$. The other half of (a) is similar.

(b) and (c) are immediate by induction on n.

This lemma immediately shows that there are considerable restrictions on the class of groups that are carriers of ℓ-groups.

COROLLARY 0.1.2. *In any ℓ-group, if $a^n \geq e$ for some positive integer n, then $a \geq e$. Hence any ℓ-group is torsion free.*

Proof. By (c), if $a^n \geq e$, then $(a \wedge e)^n = a^n \wedge a^{n-1} \wedge ... \wedge a \wedge e = a^{n-1} \wedge ... \wedge a \wedge e = (a \wedge e)^{n-1}$. Hence $a \wedge e = e$; so $a \geq e$. If $a^n = e$, then $a^n \geq e$ and $(a^{-1})^n \geq e$. Thus $a \geq e$ and $a^{-1} \geq e$; so $a = e$.

COROLLARY 0.1.3. *In any ℓ-group G, if $n \neq 0$ and $a^n = b^n$, then $b = c^{-1}ac$ for some $c \in G$.*

Proof. Let $c = a^{n-1} \vee a^{n-2}b \vee ... \vee ab^{n-2} \vee b^{n-1}$. Then

$$ac = a^n \vee a^{n-1}b \vee ... \vee ab^{n-1} = b^n \vee a^{n-1}b \vee ... \vee ab^{n-1} = cb.$$

In general, $a^n = b^n$ does not imply $a = b$ in an ℓ-group: Consider the ℓ-group $A(\mathbf{R})$. Let a be translation by $+1$ (so $\alpha a = \alpha + 1$ if $\alpha \in \mathbf{R}$), and let b be defined by

$$\alpha b = \begin{cases} \sqrt{\alpha - 2n} + 2n + 1 & \text{if } \alpha \in [2n, 2n + 1] \quad (n \in \mathbf{Z}) \\ (\alpha - 2n - 1)^2 + 2n + 2 & \text{if } \alpha \in [2n + 1, 2n + 2] \quad (n \in \mathbf{Z}). \end{cases}$$

Then $a, b \in A(\mathbf{R})$ and $b^2 = a^2 =$ translation by $+2$, but $a \neq b$. Indeed, $ab \neq ba$. However, if $[c,d] = c^{-1}d^{-1}cd$ denotes the commutator, then

LEMMA 0.1.4. *If G is an o-group, then*
  (a) *If $a^n = b^n$ for some positive integer n, then $a = b$;*
  (b) *If $[a^n, b] = e$ for some positive integer n, then $[a,b] = e$.*

Proof. (a) If $a \neq b$, then without loss of generality $a < b$. Hence $a^n < a^{n-1}b < a^{n-2}b^2 < ... < b^n$; so $a^n \neq b^n$.

(b) If $[a^n, b] = e$, then $(b^{-1}ab)^n = b^{-1}a^nb = a^n$; so $b^{-1}ab = a$ from part (a).

Another consequence of Lemma 0.1.1 is:

LEMMA 0.1.5. *The lattice of an ℓ-group is distributive; i.e., $a \vee (b \wedge c) = (a \vee b) \wedge (a \vee c)$ and $a \wedge (b \vee c) = (a \wedge b) \vee (a \wedge c)$.*

Proof. Any lattice that satisfies $(a \wedge b = a \wedge c$ and $a \vee b = a \vee c \Rightarrow b = c)$ is distributive. So assume $a \wedge b = a \wedge c$ and $a \vee b = a \vee c$. Then, using Lemma 0.1.1(a), $b = (a \wedge b)aa^{-1}(a \wedge b)^{-1}b = (a \wedge b)aa^{-1}(a^{-1} \vee b^{-1})b = (a \wedge b)a^{-1}(b \vee a) = (a \wedge c)a^{-1}(c \vee a) = (a \wedge c)a^{-1}a(a^{-1} \vee c^{-1})c = c$.

More can be proved (see [Bigard, Keimel, and Wolfenstein, 1977, Proposition 6.1.2]):

LEMMA 0.1.6. *In any ℓ-group, if $\vee\{a_\lambda \mid \lambda \in \Lambda\}$ exists, then so does $\vee\{a_\lambda \wedge b \mid \lambda \in \Lambda\}$ and $\vee\{a_\lambda \wedge b \mid \lambda \in \Lambda\} = (\vee\{a_\lambda \mid \lambda \in \Lambda\}) \wedge b$.*

A direct consequence of Lemma 0.1.5 is:

COROLLARY 0.1.7. *In any ℓ-group G, the sublattice subgroup generated by $g_1,...,g_n$ comprises all elements of the form $\vee_i \wedge_j w_{ij}$, where the elements w_{ij} all belong to the subgroup of G generated by $g_1,...,g_n$ and i and j range over finite sets.*

Unfortunately, this normal form is not unique as we will shortly show. Let $a^+ = a \vee e$, $a^- = a^{-1} \vee e$ and $|a| = a \vee a^{-1}$.

LEMMA 0.1.8. *In any ℓ-group*
(a) *If $a \wedge b = e$, then $ab = a \vee b = ba$ and $a \wedge bc = a \wedge c$ if $c \geq e$.*
(b) *$|a| = a^+ a^- \geq e$. Moreover, $|a| = e$ if and only if $a = e$.*
(c) *$a^+ \wedge a^- = e$.*
(d) *$a = a^+(a^-)^{-1}$.*
(e) *$(ab)^+ \leq a^+ b^+$.*
(f) *$(a^+)^n = (a^n)^+$, $(a^-)^n = (a^n)^-$ and $|a|^n = |a^n|$.*
(g) *$|a \vee b| \leq |a| \vee |b| \leq |a||b|$.*
(h) *$|ab| \leq |a||b||a|$.*

Proof. (a) If $a \wedge b = e$, then $ab = a(a \wedge b)^{-1}b = a(a^{-1} \vee b^{-1})b = b \vee a = a \vee b = ba$ using Lemma 0.1.1(a). Also $a \wedge c \leq a \wedge bc \leq (a^2 \wedge ba \wedge ac) \wedge bc = (a \wedge b)(a \wedge c) = a \wedge c$.

(b) If $b \geq a, a^{-1}$, then $b^2 \geq e$. By Corollary 0.1.2, $b \geq e$. Thus $|a| = a \vee a^{-1} \geq e$ and $a^+ a^- = (a \vee e)(a^{-1} \vee e) = e \vee a \vee a^{-1} \vee e = a \vee a^{-1} = |a|$. If $|a| = e$, then $a \vee a^{-1} = e$; so $a, a^{-1} \leq e$. Hence $a = e$.

(c) $a^+ \wedge a^- = (a \vee e) \wedge (a^{-1} \vee e) = (a \wedge a^{-1}) \vee e = (a^{-1} \vee a)^{-1} \vee e = e$ using (b), Lemma 0.1.1(a) and Lemma 0.1.5.

(d) $aa^- = a(a^{-1} \vee e) = e \vee a = a^+$, so $a = a^+(a^-)^{-1}$.

(e) $ab \vee e \leq ab \vee a \vee b \vee e = (a \vee e)(b \vee e)$.

(f) If $0 \leq k \leq n$, then $(a^{n-k} \vee a^{-k})^n = \overset{n}{\underset{j=0}{\vee}} a^{(n-k)j} a^{-k(n-j)} \geq a^{(n-k)k} a^{-k(n-k)} = e$, using Lemma 0.1.1(b). By Corollary 0.1.2, $a^{n-k} \vee a^{-k} \geq e$. Thus $a^n \vee e \geq a^k$. Hence $(a^+)^n = a^n \vee a^{n-1} \vee ... \vee a \vee e = a^n \vee e = (a^n)^+$. Similarly, $(a^-)^n = (a^n)^-$. By (a), (b), and (c), $|a|^n = (a^+ a^-)^n = (a^+)^n(a^-)^n = (a^n)^+(a^n)^- = |a^n|$.

(g) $a,a^{-1} \leq |a|$ and $b,b^{-1} \leq |b|$, so $a \vee b, a^{-1} \wedge b^{-1} \leq |a| \vee |b|$. Thus $|a \vee b| = (a \vee b) \vee (a \vee b)^{-1} \leq |a| \vee |b| \leq |a| \, |b| \vee |a| \, |b| = |a| \, |b|$ since $|x| \geq e$ for all x.

(h) $ab \leq |a| \, |b| \leq |a| \, |b| \, |a|$ and $b^{-1}a^{-1} \leq |b| \, |a| \leq |a| \, |b| \, |a|$. Therefore $|ab| \leq |a| \, |b| \, |a|$.

Caution: $|ab| \leq |a| \, |b|$ fails in general, though it does hold in Abelian ℓ-groups by the proof above. Let $a,b \in A(\mathbf{R})$ be such that $\alpha a = \alpha-1$ and $0b < 0$ and $1b = 1$. Then $0|a| \, |b| = 1|b| = 1 < 0b^{-1} + 1 = 0b^{-1}a^{-1} = 0|ab|$. Hence $|ab| \not\leq |a| \, |b|$.

By Lemma 0.1.8(b), $|a| \wedge e = e$; i.e., $(a \vee a^{-1}) \wedge e = e$. By Lemma 0.1.5, it follows that $(a \wedge e) \vee (a^{-1} \wedge e) = e$ for any $a \in G$. Thus the normal form given by Corollary 0.1.7 is not unique.

By Lemma 0.1.8(d) and Corollary 0.1.2:

COROLLARY 0.1.9. *Every ℓ-group is generated as a group by its set of positive elements, and this set is semi-isolated.*

The following fact about G^+ will prove useful:

LEMMA 0.1.10 (The Reisz Interpolation Property). *If $g_1,...g_n \in G^+$ and $e \leq f \leq g_1 g_2 \cdots g_n$, then $f = f_1 f_2 \cdots f_n$ for some $f_j \in G^+$ with $f_j \leq g_j$ $(1 \leq j \leq n)$.*

Proof. By induction on n. If $n = 1$, this is clear and if $f \leq g_1 \cdots g_{n+1}$ then $f g_{n+1}^{-1} \vee e \leq g_1 \cdots g_n$. By induction $f g_{n+1}^{-1} \vee e = f_1 \cdots f_n$ for some $f_j \in G^+$, $f_j \leq g_j$. But $f(f \wedge g_{n+1})^{-1} = e \vee f g_{n+1}^{-1}$ by Lemma 0.1.1(a), so $f = f_1 \cdots f_n (f \wedge g_{n+1})$.

0.2. Further (Less) Elementary Properties

The above might lead one to believe that the collection of groups that are carriers of ℓ-groups can be easily characterized. The following example due to Vinogradov [1971] shows this hope to be quite unfounded:

THEOREM 0.2.1. *There are metabelian groups G_1 and G_2 satisfying the same sentences of group theory such that only G_1 can be made into an ℓ-group.*

Proof. Let H be the semidirect product of $\mathbf{Q}$ by $\mathbf{Z}$, where $(0,1)^{-1}(1,0)(0,1)$ $= (-1,0)$. Now $\mathbf{Q} \otimes \mathbf{Z}$ and $\mathbf{Z}$ satisfy the same sentences of group theory (Szmielew [1955]), so the same is true of $G_1 = \mathbf{Q} \otimes \mathbf{Z} \otimes H$ and $G_2 = \mathbf{Z} \otimes H$. Define $G_1^+ = \{(a,m,b,n) \mid m > 0$ or $(m = 0$ and $n > 0)$ or $(m = 0 = n$ and $a \geq |b|)\}$. This makes G_1 into an ℓ-group, as can easily be verified. If G_2 could be made into an ℓ-group, then $|a| = (0,1,0) \vee (0,-1,0)$ would exist. Conjugation by $b = (0,0,1)$ gives $(0,-1,0) \vee (0,1,0) = a$; so $|a|$ is left fixed. But the only elements of G_2 that are fixed under conjugation by b have the form $(m,0,n)$ for $m,n \in \mathbf{Z}$. Now $|a|^{1/k} = (0,1/k,0) \vee (0,-1/k,0)$ for any positive integer k, by Lemma 0.1.8(f). Thus $(m/k,0,n/k) \in G_2$ for all positive integers k; so $m = n = 0$ and $|a| = (0,0,0)$. By Lemma 0.1.8(b), $(0,1,0) = (0,0,0)$, the desired contradiction.

This example shows that there is no set of first order sentences of group theory that are satisfied precisely by the class of groups that are carriers of ℓ-groups. Things are more propitious for o-groups. The class of groups that are carriers of o-groups is closed under ultraproducts, and if G is the group carrier of an o-group and the group H satisfies the same group theoretic sentences as G, then G and H have isomorphic ultrapowers. Thus this ultrapower of H can be made into an o-group; restriction makes H into an o-group. This means that there is a set of sentences in the language of group theory that is satisfied by a group precisely when the group is the carrier of an o-group.

Corollary 0.1.3 was used by Holland [1985b] to show:

THEOREM 0.2.2. *The class of groups that are carriers of ℓ-groups is not residual; i.e., there is a group G having a family $\{N_i \mid i \in I\}$ of normal subgroups such that $\cap \{N_i \mid i \in I\} = \{e\}$, G/N_i can be made into an ℓ-group for each $i \in I$, yet G cannot be made into an ℓ-group.*

Proof. Let A be the semidirect product of $(\mathbf{Z} \boxplus \mathbf{Z})$ by $\mathbf{Z}$, where $(m,n,0)$ conjugated by $(0,0,1)$ is $(n,m,0)$ and $(m,n,k) \geq (0,0,0)$ if and only if $k > 0$ or $(k = 0$ and $m,n \geq 0)$ (cf. example 3). For $n = 1,2,\ldots$, let $A_n \cong A$.

We denote a member g of the direct product $\prod_{n=1}^{\infty} A_n$ by $g = (g_n)$.

Let $b = (b_n)$, $c = (c_n)$, where $b_n = (1,-1,1)$ and $c_n = (-1,1,1)$ for all n.

Now let G be the subgroup of $\prod_{n=1}^{\infty} A_n$ generated by b,c, and the restricted sum $\sum_{n=1}^{\infty} A_n$. Note that $b^2 = c^2 = (0,0,2)$, but b and c are not conjugate in G. By Corollary 0.1.3, G cannot be made into an ℓ-group.

However, for each $i = 1,2,...$, if $N_i = \{g \in G \mid g_i = 0\}$, then $N_i \lhd G$, $\bigcap_{i=1}^{\infty} N_i = \{(0,0,0)\}$ and $G/N_i \cong A_i$ is an ℓ-group.

An example can be given of a group that cannot be made into an ℓ-group yet all of its finitely generated subgroups can [Vinogradov and Vinogradov, 1969]; i.e., being the carrier of an ℓ-group is not a local property for groups.

We complete this section by considering the lattice of an ℓ-group.

As we already saw, the lattice of an ℓ-group is distributive (Lemma 0.1.5); indeed, it satisfies

$$a \wedge \vee \{b_j \mid j \in J\} = \vee \{a \wedge b_j \mid j \in J\} \quad \text{whenever} \quad \vee \{b_j \mid j \in J\} \quad \text{exists (Lemma 0.1.6),}$$

and hence

$$a \vee \wedge \{b_j \mid j \in J\} = \wedge \{a \vee b_j \mid j \in J\} \quad \text{whenever} \quad \wedge \{b_j \mid j \in J\} \quad \text{exists.}$$

A lattice is called *completely distributive* if

$$\wedge_i \vee_j a_{ij} = \vee \{\wedge_i a_{if(i)} \mid f \in {}^I J\},$$

whenever all the indicated suprema and infima exist. We close this chapter with a surprising result of Richard Ball [1984] that these statements are all expressible in the first order theory of lattices, even though infinite suprema and infima are present.

Let φ_1 and φ_2 be the sentences

$$\forall x \forall y \forall z (x \leq y < z \rightarrow (\exists t)[x < t \leq z \ \& \ \forall w (w \vee y \geq z \rightarrow w \vee x \geq t)]) \text{ and}$$

$$\forall x \forall y \forall z (x < y \leq z \rightarrow (\exists t)[x \leq t < z \ \& \ \forall w (w \wedge y \leq x \rightarrow w \wedge z \leq t)])$$

respectively, and let φ be the sentence

$$\forall x \forall y (x < y \rightarrow \exists z_1 \exists z_2 \exists t_1 \exists t_2 [\overset{2}{\underset{k=1}{\&}} (z_k < t_k \ \& \ x \wedge t_k \leq z_k \ \& \ y \vee z_k \geq t_k) \ \& \ (\forall w)(z_1 \vee w \geq t_1 \text{ or } t_2 \wedge w \leq z_2)]).$$

PROPOSITION 0.2.3. *If L is a distributive lattice,*

    (1) $L \vDash \varphi_1 \Leftrightarrow a \vee \wedge_j b_j = \wedge_j (a \vee b_j) \quad$ *whenever $\wedge_j b_j$ exists*

    (2) $L \vDash \varphi_2 \Leftrightarrow a \wedge \vee_j b_j = \vee_j (a \wedge b_j) \quad$ *whenever $\vee_j b_j$ exists*

    (3) $L \vDash \varphi_1 \& \varphi_2 \& \varphi \Leftrightarrow L$ *is completely distributive.*

Proof. (1) Assume $L \vDash \varphi_1$, $b = \wedge_j b_j$ but $a \vee b < c = \wedge_j (a \vee b_j)$. We may assume that $b \leq a < c$ (replace a by $a \vee b$ if necessary). Let d be guaranteed by

φ_1 ; so $b < d \le c$ and since $b_j \lor a \ge c$, $b_j = b_j \lor b \ge d$ for all $j \in J$. Thus $b = \bigwedge_j b_j \ge d$, a contradiction.

Conversely, assume $L \models \neg\varphi_1$ and let $b \le a < c$ witness it. If $W = \{w \mid w \lor a \ge c\}$, let $\{b_j \mid j \in J\}$ be an enumeration of $\{w \lor b \mid w \in W\}$. Our assumption $L \models \neg\varphi_1$ ensures that $\bigwedge_j b_j = b$. But $a \lor \bigwedge_j b_j = a \lor b = a < c \le \bigwedge_j (a \lor b_j)$, since if $b_j = w \lor b$ and $w \lor a \ge c$, then $a \lor b_j = a \lor w \lor b = w \lor a \ge c$.

(2) is proved similarly.

(3) If L is not completely distributive, let $b = \bigwedge_i \bigvee_j a_{ij} > \bigvee_i \bigwedge_i a_{if(i)} = a$. By (1) and (2), we may assume that L is infinitely distributive; so, without loss of generality, $\bigvee_j a_{ij} = b$ for all $i \in I$ and $\bigwedge_i a_{if(i)} = a$ for all $f \in {}^I J$. Let $c < d$ with $a \land d \le c$ and $b \lor c \ge d$. If for all $i \in I$ there is $j \in J$ such that $a_{ij} \lor c \ge d$, then let $f_0(i)$ be such a $j \in J$. Hence $d \le \bigwedge_i (a_{if_0(i)} \lor c) = (\bigwedge_i a_{if_0(i)}) \lor c = a \lor c$; so $(a \land d) \lor c = (a \lor c) \land (d \lor c) \ge d > c$ contradicting $a \land d \le c$. Thus for some $i_0 \in I$, $a_{i_0 j} \lor c \not\ge d$ for all $j \in J$. Consequently, if $c_1 < d_1$ and $c_2 < d_2$ with $a \land d_k \le c_k$ and $b \lor c_k \ge d_k$ $(k = 1,2)$, then $c_1 \land a_{i_0 j} \not\ge d_1$ for all $j \in J$. If $d_2 \land a_{i_0 j} \le c_2$ for all $j \in J$, we would obtain $d_2 \land b \le c_2$; so $d_2 = d_2 \land (b \lor c_2) = (d_2 \land c_2) \le c_2$, a contradiction. Thus $L \models \neg\varphi$.

Conversely, if $L \models \neg(\varphi_1 \& \varphi_2 \& \varphi)$, then by (1) and (2) we may assume that $L \models \varphi_1 \& \varphi_2 \& \neg\varphi$. Let $a < b$ witness the failure of φ and $X = \{(c,d) \mid c < d \ \&\ a \land d \le c \ \&\ b \lor c \ge d\}$. Let I index X and for each $i \in I$, let $\{a_{ij} \mid j \in J\}$ enumerate $X_i = \{r \mid a \le r \le b \ \&\ r \lor c_i \not\ge d_i\}$. We claim that $\bigvee_j a_{ij} = b$ for all $i \in I$ and $\bigwedge_i a_{if(i)} = a$ for all $f \in {}^I J$, whence $\bigwedge_i \bigvee_j a_{ij} = b > a = \bigvee_i \bigwedge_i a_{if(i)}$ and so L is not completely distributive.

If $\bigvee_j a_{ij} \ne b$ for some $i \in I$, let $d' \in L$ be such that $a_{ij} \le d' < b$ for all $j \in J$. Now $(d',b) \in X$ and if $w \in L$ satisfies $w \lor c_i \not\ge d_i$ and $b \land w \not\le d'$ ($L \models \neg\varphi$), then an easy computation gives $(w \land b) \lor a \in X_i$. Say $(w \land b) \lor a = a_{ij}$. But then $w \land b \le (w \land b) \lor a = a_{ij} \le d'$, a contradiction. Hence $\bigvee_j a_{ij} = b$ for all $i \in I$.

If $\bigvee_i \bigwedge_i a_{if(i)} \ne a$ for some $f \in {}^I J$, let $a < c' \le a_{if(i)}$ for all $i \in I$. Now $c' \lor c_i \le a_{if(i)} \lor c_i$ and $a_{if(i)} \lor c_i \not\ge d_i$; so $c' \lor c_i \not\ge d_i$. Thus $c' \in X_i$ for all $i \in I$. But $i_0 = (a,c') \in X$ and $c' \lor a \ge c'$, so $c' \notin X_{i_0}$, the desired contradiction.

CHAPTER 1

HOMOMORPHISMS, PRIME SUBGROUPS, VALUES AND STRUCTURE THEOREMS

1.1. Homomorphisms, Prime Subgroups and Values

In this section we focus on the basic homomorphism and subalgebra picture. Clearly, we will want homomorphisms to preserve not only the group operation but also the lattice operations. So a *homomorphism* φ from an ℓ-group G into an ℓ-group H satisfies:

(1) $(fg)\varphi = (f\varphi)(g\varphi)$,

(2) $(f \vee g)\varphi = f\varphi \vee g\varphi$,

and   (3) $(f \wedge g)\varphi = f\varphi \wedge g\varphi$.

In view of Lemma 0.1.1(a), a group homomorphism that satisfies $(f \vee e)\varphi = f\varphi \vee e$ is a homomorphism for ℓ-groups, so (3) follows from (1) and (2). We adopt the category theoretic use of the term homomorphism above; some people, including several authors in this book, use ℓ-*homomorphism* for the above to distinguish the concept from group homomorphism.

One might expect the kernels of homomorphisms to be the normal sublattice subgroups. However, if $f_1\varphi = e = f_2\varphi$ and $f_1 \leq g \leq f_2$ then $g\varphi = e$. This leads to the following definitions:

A subset C of an ℓ-group is called *convex* if $c_1, c_2 \in C$ and $c_1 \leq g \leq c_2$ imply that $g \in C$. A sublattice subgroup is called an ℓ-*subgroup*. A normal

11

A. M. W. Glass and W. C. Holland (eds.), Lattice-Ordered Groups, 11–22.
© *1989 by Kluwer Academic Publishers.*

convex ℓ-subgroup is said to be an *ℓ-ideal*, or *ideal* for short, if no confusion arises. Using Lemmas 0.1.8 and 0.1.10 we obtain

LEMMA 1.1.1. *Let C be a subgroup of an ℓ-group G. Then C is a convex ℓ-subgroup of G if and only if $C = \{g \in G \mid |g| \le |c|$ for some $c \in C\}$.*

By Lemma 0.1.8, if C and D are ℓ-subgroups and $C^+ = D^+$, then $C = D$. We also have the following easily proved fundamental result:

THEOREM 1.1.2. *Let $\varphi : G \to H$ be a homomorphism between ℓ-groups. Then the kernel of φ is an ideal of G and $G\varphi$ is an ℓ-subgroup of H. For any ideal N of G, the quotient group G/N becomes an ℓ-group with $Nx \vee Ny = N(x \vee y)$ and dually, and the natural map $v : G \to G/N$ given by $gv = Ng$ is a homomorphism. If $\varphi : G \to H$ is a homomorphism with kernel N, then the map $Ng \mapsto g\varphi$ is an embedding of G/N into H.*

Let $C(G)$ denote the collection of all convex ℓ-subgroups of an ℓ-group G. Clearly $C(G)$ is closed under arbitrary intersection and so forms a complete lattice under inclusion, with meet operation equal to intersection. The join operation, which we also denote by $\vee$, is given by: $\vee\{A_j \mid j \in J\} = \cap\{C \in C(G) \mid C \supseteq A_j$ for all $j \in J\}$. In general, $\vee\{A_j \mid j \in J\} \ne \cup\{A_j \mid j \in J\}$ even when J is finite.

THEOREM 1.1.3. *For any ℓ-group G, $C(G)$ is a distributive lattice and a complete sublattice of the lattice of all subgroups of G. Moreover, $A \cap \vee\{C_j \mid j \in J\} = \vee\{A \cap C_j \mid j \in J\}$ in $C(G)$.*

Proof. If $\{C_j \mid j \in J\} \subseteq C(G)$ and C is the subgroup of G generated by $\{C_j \mid j \in J\}$, let $|g| \le |c|$ with $c \in C$. Then $c = \prod_{k=1}^{n} c_k$ with $c_k \in C_{j_k}$. By Lemmas 0.1.8(h) and 0.1.10, g^+ and g^- can be written as a product of elements from $\cup\{C_j^+ \mid j \in J\}$. Hence $g \in C$; so $C \in C(G)$ by Lemma 1.1.1. Clearly $A \cap \vee\{C_j \mid j \in J\} \supseteq \vee\{A \cap C_j \mid j \in J\}$ and if $g \in A^+ \cap \vee\{C_j \mid j \in J\}$, then $g \le \prod_{k=1}^{n} c_k$ with $c_k \in C_{j_k}^+$. By Lemma 0.1.10, $g = \prod_{k=1}^{n} g_k$ with $e \le g_k \le c_k$ $(1 \le k \le n)$. Since each $g_k \le g \in A$, each $g_k \in A \cap C_{j_k}$.

COROLLARY 1.1.4. *The set of ideals of an ℓ-group is a distributive lattice under inclusion.*

Given a convex ℓ-subgroup C of an ℓ-group G, the set of right cosets $R(C)$ of C in G is a lattice under the ordering: $Cg \leq Cf$ if $cg \leq f$ for some $c \in C$. Indeed, $Cg \vee Cf = C(g \vee f)$ and dually. In the special case that $R(C)$ becomes linearly ordered with this ordering, we call C a *prime* subgroup of G.

PROPOSITION 1.1.5. *Let G be an ℓ-group and C a convex ℓ-subgroup of G. Then the following are equivalent:*

(i) *C is a prime subgroup;*

(ii) *If D_1 and D_2 are convex ℓ-subgroups of G containing C, then $D_1 \subseteq D_2$ or $D_2 \subseteq D_1$;*

(iii) *If D_1 and D_2 are convex ℓ-subgroups of G and $D_1 \cap D_2 = C$, then either $D_1 = C$ or $D_2 = C$;*

(iv) *If $f \wedge g = e$, then $f \in C$ or $g \in C$.*

To prove the proposition, we will assume the following easily established Lemma (use Lemma 0.1.8(a), (g), and (h)).

For $g \in G$, let $G(g)$ be the convex ℓ-subgroup generated by g.

LEMMA 1.1.6. $G(g) = \{f \in G \mid |f| \leq |g|^n \text{ for some positive integer } n\}$. *Hence, if $f,g > e$, $G(f) \cap G(g) = G(f \wedge g)$ and $G(f \vee g) = G(f) \vee G(g)$.*

If $G(g) = G$ and $g \in G^+$, then g is called a *strong order unit*.

Proof of Proposition 1.1.5. (i) $\Rightarrow$ (ii). If $e < d_1 \in D_1 \setminus D_2$ and $e < d_2 \in D_2 \setminus D_1$, then without loss of generality $Cd_1 \leq Cd_2$. Hence $cd_1 \leq d_2$ for some $c \in C \subseteq D_2$. Thus $e \leq d_1 \leq c^{-1}d_2 \in D_2$; so $d_1 \in D_2$, a contradiction.

(ii) $\Rightarrow$ (iii). a fortiori.

(iii) $\Rightarrow$ (iv). By Theorem 1.1.3 and Lemma 1.1.6, $[C \vee G(f)] \cap [C \vee G(g)] = C \vee [G(f) \cap G(g)] = C \vee G(f \wedge g) = C$. So $C \vee G(f) = C$ or $C \vee G(g) = C$; i.e., $f \in C$ or $g \in C$.

(iv) $\Rightarrow$ (i). $f(f \wedge g)^{-1} \wedge g(f \wedge g)^{-1} = e$ for all $f,g \in G$. So, without loss of generality, $f(f \wedge g)^{-1} \in C$. Hence $Cf = C(f \wedge g) \leq Cg$. Therefore C is prime.

COROLLARY 1.1.7. *The set of convex ℓ-subgroups of an ℓ-group G is linearly ordered by inclusion if and only if G is an o-group.*

There is one way to obtain prime subgroups that will prove very useful. Let C be a convex ℓ-subgroup of an ℓ-group G and $g \in G \setminus C$. By Zorn's Lemma, there is a convex ℓ-subgroup V of G containing C, maximal with respect to the property that $g \notin V$. Then V is called a *value* of g and a *regular subgroup* of G. Each $g \in G \setminus \{e\}$ has a value - take $C = \{e\}$. If V^* is the intersection of all convex ℓ-subgroups of G that properly contain V, then $g \in V^* \setminus V$. Clearly there is no convex ℓ-subgroup B of G such that $V \subset B \subset V^*$; i.e., V^* *covers* V in $C(G)$. By Proposition 1.1.5, regular subgroups are prime. So regular subgroups are precisely the meet irreducible elements of the lattice $C(G)$ and the set of prime subgroups is the set of finitely meet irreducible elements of the lattice $C(G)$.

Values are useful in distinguishing the order on G.

LEMMA 1.1.8. *If G is an ℓ-group, then $f \leq g$ if and only if $Pf \leq Pg$ for all regular subgroups P in G.*

Proof. If $f \leq g$, then $Pf \leq P(f \vee g) = Pg$ for all $P \in C(G)$. Conversely, if $f \nleq g$, then $fg^{-1} \vee e \neq e$ so there is a value P of $fg^{-1} \vee e$. Now $e \leq fg^{-1} \vee e \notin P$; hence $P < P(fg^{-1} \vee e) = P(f \vee g)g^{-1}$. Thus $Pg < P(f \vee g) = \max\{Pf, Pg\}$ since P is prime. Consequently, $Pf \nleq Pg$.

If every value in an ℓ-group G is normal in its cover then G is said to be *normal valued*. This concept will prove very important in later chapters. By the previous corollary we have:

COROLLARY 1.1.9. *Every o-group is normal valued.*

Caution: It is possible for every non-identity element of an ℓ-group to have a normal value without the ℓ-group being normal valued; C of example 8 in Chapter 0 provides such an example .

We will write $\Gamma(G)$ for the set of regular subgroups of an ℓ-group G, and $\Pi(G)$ for the set of prime subgroups of G. Note that $\Pi(G)$ and $\Gamma(G)$ are root systems (see page 3) under inclusion by Proposition 1.1.5.

We call $C \in C(G)$ *closed* provided that any set of elements of C having a supremum in G has this supremum in C; i.e., if $\{c_j \mid j \in J\} \subseteq C$ and $g = \vee_G \{c_j \mid j \in J\}$, then $g \in C$. Let $\mathcal{K}(G) = \{C \in \Pi(G) \mid C$ is closed$\}$, the set of closed primes of G. The *distributive radical* $D(G)$ of an ℓ-group G is the intersection of all closed prime subgroups of G. Recall the definition of completely distributive from Chapter 0 section 2. It can be shown (see [Glass,1981b, Appendix 2, Theorem 2]):

THEOREM 1.1.10 [Byrd and Lloyd, 1967]. *An ℓ-group G is completely distributive if and only if $D(G) = \{e\}$.*

We next wish to examine the natural relationship between $C(G)$ and $C(C)$ when C is a convex ℓ-subgroup of G. This is tightest when $C(G)$ is restricted to certain subsets of $\Pi(G)$, $\Gamma(G)$ and $\mathcal{K}(G)$ as we now demonstrate.

Let L be a *Brouwerian lattice*; i.e., L satisfies the identity $a \wedge \vee \{b_j \mid j \in J\} = \vee \{a \wedge b_j \mid j \in J\}$ whenever $\vee \{b_j \mid j \in J\}$ exists. Clearly, for any $a,b \in L$, there is a largest $c \in L$ such that $a \wedge c \leq b$; we denote this element by $b:a$. We saw in Theorem 1.1.3 that $C(G)$ is Brouwerian. Hence for any $A,B \in C(G)$, there is a largest $B:A \in C(G)$ such that $A \cap (B:A) \subseteq B$. Let $b \in B^+$. Then $(G(b) \vee (B:A)) \cap A = (G(b) \cap A) \vee ((B:A) \cap A) \subseteq (B \cap A) \vee B = B$. Hence $G(b) \vee (B:A) = B:A$ and, consequently, $A \cap (B:A) = B$. If $C \in C(G)$, let $C_C(G) = \{B \in C(G) \mid B \not\supseteq C\}$, $\Pi_C(G) = \Pi(G) \cap C_C(G)$, $\Gamma_C(G) = \Gamma(G) \cap C_C(G)$, and $\mathcal{K}_C(G) = \mathcal{K}(G) \cap C_C(G)$.

THEOREM 1.1.11. *If $C \in C(G)$, then $A \mapsto A \cap C$ maps $C_C(G)$ onto $C(C) \setminus \{C\}$. The restrictions to $\Pi_C(G)$, $\Gamma_C(G)$, and $\mathcal{K}_C(G)$ map one-to-one onto $\Pi(C) \setminus \{C\}$, $\Gamma(C) \setminus \{C\}$, and $\mathcal{K}(C) \setminus \{C\}$ respectively, with inverse $B:C$. If $V \in \Gamma_C(G)$, then $V \triangleleft V^*$ if and only if $V \cap C \triangleleft V^* \cap C = (V \cap C)^*$, and in this case $V^*/V \cong (V \cap C)^*/V \cap C$.*

Proof. If $B \in C(C)$, then $C \cap (B:C) = B$ by the above. Hence $B:C \in C_C(G)$ if $B \neq C$. If $B \in \Pi(C)$, let $f,g \in G$ with $f \wedge g = e$. Now $(G(f) \vee (B:C)) \cap C$ and $(G(g) \vee (B:C)) \cap C$ both belong to $C(G)$ and contain $(B:C) \cap C = B \in \Pi(C)$. By Proposition 1.1.5, without loss of generality $(G(f) \vee (B:C)) \cap C \subseteq (G(g) \vee (B:C)) \cap C$. But $(G(f) \vee (B:C)) \cap C = (G(f) \vee (B:C)) \cap C \cap (G(g) \vee (B:C)) \cap C = ((G(f) \cap G(g)) \vee (B:C)) \cap C = (B:C) \cap C = B$ using Lemma 1.1.6 and Theorem 1.1.3. Thus $B:C \in \Pi(G)$. Clearly, if $A \in \Pi_C(G)$, then $(A \cap C):C \supseteq A$. If $g \in (A \cap C:C)^+$, let $c \in C^+ \setminus A$. Now $g \wedge c \in ((A \cap C):C) \cap C = A \cap C \subseteq A$ and so $c(g \wedge c)^{-1} \notin A$. But A is prime and $g(g \wedge c)^{-1} \wedge c(g \wedge c)^{-1} = e$; therefore $g \in A$. Consequently, $A \cap C:C = A$ so the maps are inverse to each other. The rest of the theorem follows from the second isomorphism theorem.

Note that if $G = \mathbf{Z} \boxplus \mathbf{Z}$ and $C = \{0\} \boxplus \mathbf{Z}$, then $\{(0,0)\} \cap C = (\mathbf{Z} \boxplus \{0\}) \cap C$ $(\{(0,0)\}:C = \mathbf{Z} \boxplus \{0\} \neq \{(0,0)\})$ so the map is not one-to-one if we consider all of $C_C(G)$.

As in group theory, direct summands play an important role. If G is an ℓ-group, then $C \in C(G)$ is called a *cardinal summand* of G if $G = C \boxplus D$ for some $D \in C(G)$. It is easy to see

LEMMA 1.1.12. *If* $G = \sum\{A_\lambda \mid \lambda \in \Lambda\} = \sum\{B_\mu \mid \mu \in M\}$, *then* $G = \sum\{A_\lambda \cap B_\mu \mid \lambda \in \Lambda, \mu \in M\}$

As in vector spaces, summands arise from orthogonality considerations. For $g \in G$, let $g^\perp = \{f \in G \mid |f| \wedge |g| = e\}$, the *polar* of g. By Lemma 0.1.8(a),(h), $g^\perp \in C(G)$ for all $g \in G$. If $X \subseteq G$, $X^\perp = \cap\{x^\perp \mid x \in X\}$, the polar of X; so $X^\perp \in C(G)$. Moreover, since $G(g)^\perp = g^\perp$, $X^\perp = C^\perp$ where C is the convex ℓ-subgroup of G generated by X. It is easy to see, in fact, that $C^\perp = \{e\} : C$ for each $C \in C(G)$ and if C is a cardinal summand of G, then $G = C \boxplus C^\perp$ and $C^{\perp\perp} = C$. Thus every cardinal summand is a polar. If every polar is a cardinal summand, then G is said to be *strongly projectable*. Let $P(G)$ denote the set of polars of G. Then it is straightforward to verify

THEOREM 1.1.13. *For any* ℓ-*group* G, $P(G)$ *is a complete Boolean algebra under inclusion, where the join is* $\mathbb{W}\{P_\lambda \mid \lambda \in \Lambda\} = (\cap\{P^\perp \mid \lambda \in \Lambda\})^\perp = (\cup\{P_\lambda \mid \lambda \in \Lambda\})^{\perp\perp}$.

By Zorn's Lemma, each prime subgroup contains a minimal prime subgroup. For each $g \in G \setminus \{e\}$, let P_g be a minimal prime subgroup of G contained in some value of g. Since $g \notin P_g$, $\cap\{P_g \mid g \in G \setminus \{e\}\} = \{e\}$. Hence the intersection of all minimal prime subgroups of G is $\{e\}$. Minimal prime subgroups are closely related to polars.

LEMMA 1.1.14. *If* $P \in C(G)$, *the following are equivalent:*
    (a) *P is a minimal prime subgroup of G;*
    (b) $P = \cup\{g^\perp \mid g \notin P\}$;
    (c) $P \in \Pi(G)$ *and* $h^\perp \not\subseteq P$ *whenever* $h \in P$.

Proof. (a) $\Rightarrow$ (b). If $f, g \in G \setminus P$, then $f \wedge g \neq e$ since P is prime. Hence there is an ultrafilter F on G^+ containing $G^+ \setminus P$. Let $Q = \cup\{g^\perp \mid g \in F\}$. Observe that if $g \in F$, then $g^\perp \subseteq G \setminus F \subseteq P$; thus $Q \subseteq P$. We now show $Q \in C(G)$. If $a, b \in Q$, then $a \in f^\perp$ and $b \in g^\perp$ for some $f, g \in F$. Now $e \leq |a \vee b| \wedge f \wedge g \leq |a| |b| \wedge f \wedge g$ and $e \leq |ab| \wedge f \wedge g \leq |a| |b| |a| \wedge f \wedge g$ by Lemma 0.1.8(g),(h). Moreover, $|a| |b| |a| \wedge f \wedge g \leq (|a| \wedge f)(|b| \wedge g)(|a| \wedge f) = e$. Hence $ab, a \vee b \in Q$ so $Q \in C(G)$ being clearly convex. If $c \notin Q$ and $c \wedge d = e$, then as $c \wedge g > e$ for all $g \in F$, $c \in F$ (F is an ultrafilter.) Thus $d \in c^\perp \subseteq Q$, proving that Q is prime. By the minimality of P, $Q = P$ and $F = G^+ \setminus P$. (If $c \in P$, then $c \in d^\perp$ for some $d \in F$; hence $c \notin F$.)

    (b) $\Rightarrow$ (c). Clear.

(c) $\Rightarrow$ (a). Let $Q \in \Pi(G)$ be minimal contained in P. By (a) $\Rightarrow$ (b) applied to Q, if $g \in P^+ \backslash Q$, then $g^\perp \subseteq Q \subseteq P$ contradicting (c). Thus $P = Q$.

COROLLARY 1.1.15. *If* $\{e\} \neq P \in C(G)$, *then* $P^\perp$ *is a minimal prime subgroup of* G *if and only if* P *is an o-group.*

Proof. If $P^\perp$ is a minimal prime subgroup of G, let $f,g \in P$ with $f \wedge g = e$ and $f \neq e$. As $P \cap P^\perp = \{e\}$, $g \notin P^\perp$ if $g \neq e$. By Lemma 1.1.14, $g \in f^\perp \subseteq P^\perp$; so $g = e$. Thus P is an o-group.

Conversely, if P is an o-group, let $f \wedge g = e$. If $f,g \notin P^\perp$, there are $c_1, c_2 \in P$ with $f \wedge c_1, g \wedge c_2 > e$. But $(f \wedge c_1) \wedge (g \wedge c_2) = e$, so $f \wedge c_1 = e$ or $g \wedge c_2 = e$ since P is an o-group. This contradiction shows that $P^\perp$ is a prime subgroup. If $h \in P^\perp$, then $h^\perp \supseteq P$; so $h^\perp \not\subseteq P^\perp$. By Lemma 1.1.14, $P^\perp$ is a minimal prime subgroup of G.

1.2. Structure Theorems

In this section we use some of the ideas of the previous section to obtain further information about normal-valued, Abelian and Archimedean ℓ-groups and subdirect products of o-groups.

An ℓ-group that is a subdirect product of o-groups is called a *representable* ℓ-group. This property is determined by the polars:

THEOREM 1.2.1 [Lorenzen, 1939]. *For any* ℓ-group G, *the following are equivalent:*

    (1) G *is representable;*
    (2) *For all* $a,b \in G$, $(a \wedge b)^2 = a^2 \wedge b^2$;
    (3) *For all* $a,b \in G$, $a^+ \wedge b^{-1} a^- b = e$;
    (4) *For all* $g \in G$, $g^\perp \lhd G$;
    (5) *Each minimal prime subgroup of* G *is normal.*

Proof. (1) $\Rightarrow$ (2). The identity holds in all o-groups and hence in all representable ℓ-groups.

(2) $\Rightarrow$ (3). By (2), $(ba)^2 \wedge b^2 = (ba \wedge b)^2 \leq ba \cdot b$. Hence $ba \wedge a^{-1} b \leq b$. Thus $(a \wedge b^{-1} a^{-1} b) \vee e = e$. Consequently, $(a \vee e) \wedge b^{-1}(a^{-1} \vee e)b = e$ for all $a,b \in G$.

(3) $\Rightarrow$ (4). If $f \wedge |g| = e$, then by Lemma 0.1.8(a),

$$f = f|g| \cdot |g|^{-1} = (f \vee |g|)|g|^{-1} = f|g|^{-1} \vee e;$$

similarly, $|g| = |g| f^{-1} \vee e$. But, by (3),

$$(\,|g\,|f^{-1})^+ \wedge h^{-1}(\,|g\,|f^{-1})^- \cdot h = e \text{ for all } h \in G;$$

i.e., $|g| \wedge h^{-1}fh = e$ for all $h \in G$. Thus $g^{\perp} \lhd G$.

(4) $\Rightarrow$ (5). By Lemma 1.1.14.

(5) $\Rightarrow$ (1). G/P is an o-group for each minimal prime P. Since the intersection of all minimal prime subgroups is $\{e\}$, G is a subdirect product of o-groups.

COROLLARY 1.2.2. *(i) Abelian ℓ-groups are representable.*

(ii) Representable ℓ-groups are normal valued.

Proof. (i) Immediate from the theorem.

(ii) Let G be representable, $V \in \Gamma(G)$ and $g \in V^*$. Now $V \supseteq P$ for some minimal prime prime P in G and $g^{-1}Pg = P$ by Theorem 1.2.1. By Proposition 1.1.5, $V \subseteq g^{-1}Vg \subset g^{-1}V^*g = V^*$ or $g^{-1}Vg \subseteq V \subset V^* = g^{-1}V^*g$. Since V^* covers V, $g^{-1}Vg = V$; i.e., $V \lhd V^*$.

We conclude this section by applying Theorem 1.2.1 to the study of Abelian and Archimedean ℓ-groups. In any ℓ-group, if $g^n \le f$ for all $n \in \mathbf{Z}$, we write $g \ll f$. An ℓ-group G is said to be *Archimedean* if for all $f,g \in G$, $g \ll f$ implies $g = e$. Equivalently, G is Archimedean if $g^n \le f$ for all $n \in \mathbf{Z}^+$ implies $g \le e$.

We begin with a characterization of Archimedean o-groups.

THEOREM 1.2.3 [Hölder,1901]. *Let H be an o-group. The following are equivalent:*

      (a) H is a subgroup of the real numbers $\mathbf{R}$;
      (b) H has no convex subgroups;
      (c) H is Archimedean.

Proof. (a) $\Rightarrow$ (b) $\Rightarrow$ (c) is obvious. (If $g \ll f$, then $G(g) \subset G(f)$; hence $G(g) = \{e\}$, so $g = e$.)

(c) $\Rightarrow$ (a). Let $e < h \in H$. If h is the least element of $H^+ \setminus \{e\}$, then $H = \{h^n : n \in \mathbf{Z}\} \cong \mathbf{Z}$. Therefore, we may assume that $H^+ \setminus \{e\}$ has no least element. Then for every $h > e$, there is $g \in H$ such that $e < g < h$. Either $g \le hg^{-1}$ or $hg^{-1} \le g$; so $e < g^2 \le h$ or $e < (hg^{-1})^2 \le h$. Thus for each $h \in H^+ \setminus \{e\}$, there is $f \in H^+ \setminus \{e\}$ with $f^2 \le h$.

We next prove that H is Abelian. Let $a,b \in H^+$ with $[a,b] \ne e$; without loss of generality, $h = [a,b] > e$. Let $e < f \in H$ with $f^2 \le h$. By the Archimedean property, there are $m,n \in \mathbf{Z}^+$ with $f^m \le a < f^{m+1}$ and $f^n \le b < f^{n+1}$. Hence $h =$

$a^{-1}b^{-1}ab < f^{-m}f^{-n}f^{m+1}f^{n+1} = f^2$, a contradiction. Thus $[a,b] = e$ for all $a,b \in H^+$, and so for all $a,b \in H$.

Fix $h_0 \in H$ with $h_0 > e$ and let $\varphi : H \to \mathbf{R}$ be defined by: $h\varphi = \sup\{\frac{m}{n} : h_0^m \leq h^n$ and $n > 0\}$. A routine verification establishes that φ is an embedding of the Abelian o-group H into the additive o-group $\mathbf{R}$.

If G is a normal-valued $\boldsymbol{\ell}$-group and $V \in \Gamma(G)$, then V^*/V is a o-group by Corollary 1.1.7. It has no proper convex subgroups (otherwise pulling back will give a convex $\boldsymbol{\ell}$-subgroup of G strictly between V and its cover V^*) and so is isomorphic to a subgroup of $\mathbf{R}$ by Theorem 1.2.3. Sewing together these copies of $\mathbf{R}$ is covered in the next chapter (Section 2.3). However, if G is Abelian (whence clearly normal valued), it is possible to obtain the following representation theorem:

THEOREM 1.2.4 [Conrad, Harvey, and Holland, 1963]. *If G is an Abelian $\boldsymbol{\ell}$-group, then G can be embedded in $V(\Gamma(G),\mathbf{R})$, where V is defined in Example 6 of Chapter 0.*

In the special case that G is an Abelian o-group, $\Gamma(G)$ is a totally ordered set. In this case, $V(\Gamma(G),\mathbf{R})$ is called a *Hahn group* in honour of:

THEOREM 1.2.4 (o-group case) [Hahn, 1907]. *If G is an Abelian o-group then G can be embedded in a Hahn group $V(\Gamma(G),\mathbf{R})$ for $\Gamma(G)$ a chain.*

LEMMA 1.2.5. *Let G be an $\boldsymbol{\ell}$-group. Then the following statements are equivalent:*

    (i)   *G is normal valued;*
    (ii) *If V is a regular subgroup of G and $f,g \in V^*$, then $[f,g] \in V$;*
    (iii) *For all $f,g \in G$, $[f,g] \ll |f| \vee |g|$;*
    (iv) *For all $f,g \in G^+$, $fg \leq g^2f^2$;*
    (v) *For all convex $\boldsymbol{\ell}$-subgroups A,B of G, $AB = BA$.*

Proof. (i) $\Rightarrow$ (ii). We just observed that if $V \lhd V^*$, then V^*/V is isomorphic to a subgroup of $\mathbf{R}$ and hence is Abelian. Thus if $f,g \in V^*$, $[f,g] \in V$.

(ii) $\Rightarrow$ (iii). Let $f,g \in G$ and $c = |[f,g]|^n$, for some positive integer n. Let M be any value of c. Then M is a value of $[f,g]$ and therefore either $f \notin M$ or $g \notin M$ so that $|f| \vee |g| \notin M$. Hence $M \subseteq N$, for some value N of $|f| \vee |g|$. Then $f,g \in N^*$ and N is normal in N^*. Since N^* covers N,

N^*/N is abelian so that $[f,g] \in N$ and also $c \in N$. Hence $NcP < N(|f| \vee |g|)$. Therefore $Mc < M(|f| \vee |g|)$ and (iii) follows from Lemma 1.1.8.

(iii) $\Rightarrow$ (iv). For all $f,g \in G^+$, $[g,f^{-1}] \leq |g| \vee |f^1| = |g| \vee |f| \leq gf$ from which (iv) follows.

(iv) $\Rightarrow$ (v). Let $f = ab$ where $a \in A$, $b \in B$. Then $|a|^{-1}|b|^{-1} \leq f \leq |a||b|$ so that, from (iv),

$$|b|^{-2}|a|^{-2} \leq f \leq |b|^2|a|^2.$$

Hence $e \leq |b|^2 f|a|^2 \leq |b|^4|a|^4$. By the Riesz decomposition (Lemma 0.1.10), $|b|^2f|a|^2 = b'a'$ for some $e \leq b' \leq |b|^4$, $e \leq a' \leq |a|^4$ so that $a' \in A$, $b' \in B$. Thus $f = |b|^{-2}b'a'|a|^{-2} \in BA$ and, by symmetry, (v) holds.

(v) $\Rightarrow$ (i). Let M be a value of some element in G and let $f \in (M^*)^+\backslash M$. To show that M is normal in M^* it suffices to show that $M^f = M$. Suppose that $M^f \neq M$. Then M and M^f are distinct values of f and so must be incomparable. Hence there must exist $a \in M$ such that $g = a^f \notin M$. Clearly $g \in M^*$. By (v), $MG(g) = G(g)M$ so that $MG(g)$ is an ℓ-subgroup and, by the Riesz decomposition a convex ℓ-subgroup. Since $MG(g)$ contains both M and g, $MG(g) = M^*$. By Theorem 1.2.3, $f \leq bg^n$, for some $b \in M$, $n \in Z$. Thus $f \leq bf^{-1}a^n f$, so that $f^1 \geq b^{-1}(a^{-1})^n$ and therefore $f \leq a^n b$. Hence $f \in M$, a contradiction. Therefore $M = M^f$ and M is normal in M^*, as required.

Note that $[f,g]^2 \leq |f| \vee |g|$ does not hold in all ℓ-groups: Let $f,g \in A(\mathbf{R})$ be given by $\alpha f = \alpha$-1 for all $\alpha \in \mathbf{R}$ and

$$\alpha g = \begin{cases} \dfrac{\alpha}{3} & \text{if } \alpha \geq 0, \\ \alpha & \text{if } \alpha \leq 0. \end{cases}$$

Then $0(|f| \vee |g|) = 1 < \frac{4}{3} = 0[f,g]^2$, so $[f,g]^2 \not\leq |f| \vee |g|$. Other characterisations for "normal valued" can be found in Chapter 2 Section 3 and Chapter 10 Section 4.

We can now fulfill our promise and use Lorenzen's Theorem to prove the following generalization of Hölder's Theorem:

THEOREM 1.2.6. *Every Archimedean ℓ-group is Abelian.*

Proof. Let G be an Archimedean ℓ-group. We first prove that G is representable by showing that $g^\perp \lhd G$ for each $g \in G$. If $f \wedge |g| = e$ and $h \in G^+$, let $a = h^{-1}fh \wedge |g| \geq e$. Now $e \leq a \wedge hah^{-1} \leq |g| \wedge f = e$, so $e = a^n \wedge ha^nh^{-1} \geq eva^nh^{-1} \geq e$ for all $n \in \mathbf{Z}^+$. Thus $a \ll h$ whence $a = e$. Therefore $g^\perp$ is normalized by every element of G^+ and hence by every element of G.

By Corollary 1.2.2, G is normal valued. Now let $f, g \in G$. By Lemma 1.2.5, $[f, g] \ll |f| \vee |g|$. Since G is Archimedean, $[f, g] = e$.

3. Free Abelian ℓ-groups

Finally, we consider free Abelian ℓ-groups and obtain two representations for them. In this section we write the group additively.

Let X be a set of symbols; say $X = \{x_i : i \in I\}$. An abelian ℓ-group H containing X is a *free Abelian ℓ-group on* X if every map from X into an Abelian ℓ-group L can be uniquely extended to an ℓ-homomorphism of H into L. Clearly, if it exists, the free Abelian ℓ-group on X is unique to within isomorphism; we will denote it by $F_{\mathcal{A}}(X)$. Clearly too, if $|X| = |Y|$, then the free Abelian ℓ-groups on X and Y are isomorphic, and we call $|X|$ the *rank* of the free Abelian ℓ-group on X. If $X = \varnothing$, then $\{0\}$ is the free Abelian ℓ-group of rank 0. The free Abelian ℓ-group of rank 1 is $\mathbf{Z} \boxplus \mathbf{Z}$, a free generator being $(1, -1)$. We will now confine our attention to obtaining free Abelian ℓ-groups of rank at least 2.

The first construction of $F_{\mathcal{A}}(X)$ was due to Birkhoff [1967, page 325] who observed that the ℓ-subgroup of $C(\mathbf{R}^X, \mathbf{R})$ (the ℓ-group of all continuous maps from $\mathbf{R}^X$ into $\mathbf{R}$ under the pointwise ordering), generated by the projections π_x ($x \in X$) is free on X. This gives a beautiful geometric description of $F_{\mathcal{A}}(X)$ and demonstrates immediately that it is Archimedean. Moreover, for each $w \in F_{\mathcal{A}}(X)$, let $Z(w) = \{r \in \mathbf{R}^X : w(r) = 0\}$, the *zero set of* w. If w has the special form $\sum_{j=1}^{n} m_j x_{i_j}$ with $m_j \in \mathbf{Z}$ and $x_{i_j} \in X$, then $Z(w \vee 0) = \{r \in \mathbf{R}^X : \sum_{j=1}^{n} m_j r(x_{i_j}) \leq 0\}$, a hyperspace containing the origin. Since each element of $F_{\mathcal{A}}(X)$ is a finite meet of finite joins of such elements w, we see that the zero set of any element of $F_{\mathcal{A}}(X)$ is a closed polyhedral cone with vertex the origin, i.e., a finite union of finite intersections of hyperspaces through 0. Conversely, given any closed polyhedral cone with vertex the origin given by integer points, there is an element of $F_{\mathcal{A}}(X)$ with zero set the given cone. This important duality is pursued in [Beynon, 1977] and [Glass and Madden, 1984].

Further, $n|w_1| \leq |w_2|$ for some positive integer n if and only if $Z(w_2) \subseteq Z(w_1)$, as is easily established. Hence if $I(w)$ denotes the ideal of $F_{\mathcal{A}}(X)$ generated by w, it follows that $F_{\mathcal{A}}(X)/I(w)$ is isomorphic (as an ℓ-group) to the ℓ-group $\{u \cap Z(w) : u \in F_{\mathcal{A}}(X)\}$ of elements of $F_{\mathcal{A}}(X)$ restricted in domain to the zero set of w. From this we can easily obtain two important consequences:

(a) $F_{\mathcal{A}}(X) \cong F_{\mathcal{A}}(Y)$ if and only if $|X| = |Y|$. Indeed, $F_{\mathcal{A}}(X)$ can be embedded (as an ℓ-group) in $F_{\mathcal{A}}(Y)$ if and only if $|X| \leq |Y|$. Caution: Since $F_{\mathcal{A}}(X)$ is isomorphic as a group to the free Abelian group on $\aleph_0$ generators if X is finite and $|X| \geq 2$, $F_{\mathcal{A}}(X)$ and $F_{\mathcal{A}}(Y)$ can be isomorphic as *groups* without X and Y being of the same cardinality.

(b) $F_{\mathcal{A}}(X)/I(w_1) \cong F_{\mathcal{A}}(X)/I(w_2)$ if and only if there is a piecewise linear homeomorphism between $Z(w_1)$ and $Z(w_2)$ where the linear pieces have integer coefficients. This correspondence will be used in Chapter 13.

The second construction of $F_{\mathcal{A}}(X)$ is due to Weinberg [1963, 1965] and allows generalizations. Let G be the free Abelian *group* on X. Then for any order $\leq$ on G such that $(G, \leq)$ is an o-group, the identity map of G equipped with the trivial order onto $(G, \leq)$ is an order-preserving group homomorphism. Let H be the cardinal product, $H = \Pi\{(G, \leq) : (G, \leq)$ is an o-group$\}$. Then $F_{\mathcal{A}}(X)$ is the ℓ-subgroup of H generated by the image of G under the diagonal map φ where $(g\varphi)_{\leq} = g$ for each total order $\leq$ of G. More generally, if G is any torsion-free Abelian po-group, let T be the positive cone of a total order on G and $T \supseteq G^+$. Since the intersection of all such T can be shown to be G^+, we get an order-preserving group embedding of G into the cardinal product $H = \Pi\{(G, T) : (G, T)$ is an o-group and $T \supseteq G^+\}$. The ℓ-subgroup $F_{\mathcal{A}}(G)$ of H generated by the image of G is the free Abelian ℓ-group over G; i.e., if L is any Abelian ℓ-group, any order-preserving group homomorphism of G into L can be uniquely lifted to an ℓ-group homomorphism of $F_{\mathcal{A}}(G)$ into L. If $G^+ = \{0\}$, that is if G is trivially ordered, then $F_{\mathcal{A}}(G)$ is Archimedean. This is obvious by the first construction if G is the free Abelian group on a set X; indeed, in this case, $F_{\mathcal{A}}(X) = F_{\mathcal{A}}(G)$ is a subdirect sum of copies of $\mathbf{Z}$. More generally, the group G can be embedded in its divisible closure $\bar{G}$ and $F_{\mathcal{A}}(G)$ can be embedded as an ℓ-group in $F_{\mathcal{A}}(\bar{G})$. Hence it is enough to assume that G is divisible (and so a rational vector space) and show that $F_{\mathcal{A}}(G)$ is a subdirect sum of copies of $\mathbf{Q}$. The key technical result required is that if $g_1, g_2, ..., g_n \in G$ are positively independent ($\Sigma m_i g_i = 0$ with each $m_i \in \mathbf{Q}^+$ implies each $m_i = 0$), there exists a group homomorphism $\theta : G \to \mathbf{Q}$ such that $g_i \theta > 1$. This is proved by induction on the dimension of the vector subspace generated by $\{g_1, ..., g_n\}$. Now if $x \in F_{\mathcal{A}}(G)^+$ and $x \neq 0$, then as $x = \bigvee_i \bigwedge_j (g_{ij}\varphi)$ for some finite set $\{g_{ij}\} \subseteq G$, $\{g_{i_0 j} : j \in J\}$ is a positively independent subset of G for some $i_0 \in I$. It follows that there is a group homomorphism $\theta : G \to \mathbf{Q}$ such that $g_{i_0 j}\theta > 1$ for all $j \in J$ and θ lifts to an ℓ-group homomorphism of $F_{\mathcal{A}}(G)$ into $\mathbf{Q}$ in which $x\theta > 1$. Therefore $F_{\mathcal{A}}(G)$ is a subdirect sum of copies of subgroups of $\mathbf{Q}$ and, consequently, is Archimedean.

W. Charles Holland

CHAPTER 2

LATTICE-ORDERED PERMUTATION GROUPS

1. Introduction

Although most of the elementary theorems about o-groups are as easy to prove in the general case as in the commutative case, there are no natural examples of non-commutative o-groups. Of course, it is easy to construct examples of non-commutative o-groups, but all of these are artificial and without much interest outside of the context of ordered groups. The situation is quite different for general lattice-ordered groups, however, because there is a class of non-commutative examples of great intrinsic interest, independent of the fact that they are lattice-ordered. These are the groups $A(\Omega)$ of order-preserving permutations of totally ordered sets Ω, endowed with the pointwise order. This means that for $f, g \in A(\Omega)$, we declare that $f \leq g$ iff for all $\alpha \in \Omega$, $\alpha f \leq \alpha g$. It is easily checked that this makes $A(\Omega)$ a lattice-ordered group in which $\alpha(f \wedge g) = \alpha f \wedge \alpha g$. Only rarely is $A(\Omega)$ commutative.

Interest in ordered permutation groups goes back at least to H. Kneser [1924] who proved that if $\mathbf{R}$ is the real line, then any two positive members of $A(\mathbf{R})$ which fix no points are conjugate, and to J. Schreier and S. Ulam [1935] who proved the conjugacy lemma (Lemma 2.5.2) for $A(\mathbf{R})$. Everett and Ulam [1945] investigated properties of the group $A((0,1))$ of automorphisms of the real unit interval. In his book on lattice theory , G. Birkhoff [1960] asked which lattice-ordered groups can be realized as $A(\Omega)$ for some Ω (problem 95 in the 1960 edition, problem 111 in the 1967 edition).

23

A. M. W. Glass and W. C. Holland (eds.), Lattice-Ordered Groups, 23–40.

The question remains without a satisfying answer. F. Šik [1958] made an extensive investigation of general $A(\Omega)$ and its subgroups. However, interest in $A(\Omega)$ as an ℓ-group was greatly increased when it was discovered [Holland, 1963a] that *every* ℓ-group can be embedded in some $A(\Omega)$ (Theorem 2.2.5). This provided a new intuitive standpoint from which to view ℓ-groups, making many of their basic properties obvious and facilitating the investigation of others. It is especially useful in looking for counterexamples, where one's search is supported by the knowledge that if there *is* a counterexample it must be in some $A(\Omega)$. There are theorems, as well, whose proof involves the representation in $A(\Omega)$ in an essential way (for example, Theorem 2.2.10).

2. Some Notation and Examples

Throughout this chapter Ω will denote a totally ordered set and $A(\Omega)$ the lattice-ordered group of all order-preserving permutations (automorphisms) of Ω with the pointwise order. We use multiplicative notation and write permutations to the right of their arguments. The identity of $A(\Omega)$ is denoted by e. A very extensive discussion and an annotated bibliography concerning the topics of this chapter are in the book by A. M. W. Glass [1981b], to which the interested reader is referred. Here we intend only to survey the subject, concentrating on transitive ℓ-subgroups of $A(\Omega)$, and abbreviating or omitting many proofs.

We begin with some examples.

EXAMPLE 2.2.1. Let $\mathbf{Z}$ denote the naturally totally ordered set, or group, of all integers. Then each automorphism of the ordered set $\mathbf{Z}$ is simply a translation and $A(\mathbf{Z})$ is isomorphic (as an ℓ-group) to the ordered group of integers, that is, to $\mathbf{Z}$.

EXAMPLE 2.2.2. Let Ω denote two copies of $\mathbf{Z}$, one entirely above the other. Then $A(\Omega)$ is the cardinal direct product of two copies of $\mathbf{Z}$.

EXAMPLE 2.2.3. Let Ω denote the lexicographically ordered product $\mathbf{Z} \times \mathbf{Z}$. Then Ω is a disjoint union of infinitely many convex sets, each isomorphic to $\mathbf{Z}$, and these convex sets are preserved as blocks by the elements of $A(\Omega)$; $A(\Omega)$ is then the wreath product of $\mathbf{Z}$ with $\mathbf{Z}$.

EXAMPLE 2.2.4. Let $\mathbf{R}$ denote the naturally ordered set of real numbers. Then $A(\mathbf{R})$ is not commutative and not totally ordered. In fact, $A(\mathbf{R})$ is not even representable, for if $f \in A(\mathbf{R})$ moves each point of the interval $(0,1)$

up and fixes all other points, and $g \in A(\mathbf{R})$ moves each point of $\mathbf{R}$ up one unit, then $g^{-1}fg$ moves each point of the interval $(1,2)$ up and fixes all other points, so that $f \wedge g^{-1}fg = e$. The situation is easily pictured, noting that the elements of $A(\mathbf{R})$ are just strictly monotone increasing onto functions.

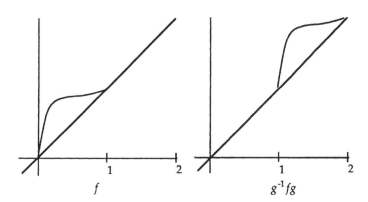

Note that every totally ordered group G permutes itself in an order-preserving manner (in the right regular representation, say) so that G is an ℓ-subgroup of $A(G)$. Of course, every ℓ-group also permutes itself in the same way, but then G is only a lattice, not a totally ordered set, and only rarely is $A(\Omega)$ an ℓ-group when Ω is not totally ordered. A study of the special cases when $A(\Omega)$ is an ℓ-group for certain partially ordered Ω was done by M. A. Bardwell [1980].

THEOREM 2.2.5 [Holland,1963a]. *Every ℓ-group is an ℓ-subgroup of $A(\Omega)$ for some totally ordered set Ω.*

Proof. Let G be an ℓ-group. For each $e \neq g \in G$, let $K(g)$ be some value of g. Then the set $G/K(g)$ of right cosets is totally ordered (Proposition 1.1.5). Let Ω be the union (which we may assume disjoint),

$$\Omega = \bigcup_{e \neq g \in G} (G/K(g)).$$

Order Ω in any way so that each subset $G/K(g)$ is convex. Then G acts on each of the sets $G/K(g)$, and hence also on Ω, by right translation; that is, for $f,g,x \in G, f:K(g)x \mapsto K(g)xf$. The resulting map, for each f, is a member of $A(\Omega)$, and this correspondence represents G (faithfully) as an ℓ-subgroup of $A(\Omega)$.

We can now note that the following facts about ℓ-groups follow trivially, since each is easy to verify in $A(\Omega)$ (these are also proved in Chapter 0).

COROLLARY 2.2.6(Corollary 0.1.2). *Every ℓ-group is torsion free.*

COROLLARY 2.2.7(Lemma 0.1.8(f)). *In an ℓ-group, if $g^n \geq e$ then $g \geq e$.*

COROLLARY 2.2.8(Lemma 0.1.8(b)). *In an ℓ-group, $|g| \geq e$.*

COROLLARY 2.2.9(Lemma 0.1.8(a)). *In an ℓ-group, if $f \wedge g = e$ then $fg = gf$.*

All of the above corollaries are easy to prove without reference to the theorem (though somewhat less intuitively), but the next theorem would be difficult to achieve otherwise. An ℓ-group G is said to be *laterally complete* if for every set P of pairwise disjoint elements in G there exists a least upper bound of P in G.

THEOREM 2.2.10 [Holland, 1963a]. *Every ℓ-group can be embedded in a laterally complete ℓ-group.*

Proof. Immediate from Theorem 2.2.5 since $A(\Omega)$ is clearly laterally complete.

A group G is *divisible* if for every $g \in G$ and every positive integer n, there is $x \in G$ such that $x^n = g$. In the same spirit as the previous theorem, we also have

THEOREM 2.2.11 [Holland, 1963a]. *Every ℓ-group can be embedded in a divisible ℓ-group.*

Proof. From Theorem 2.2.5, it suffices to prove the theorem for $G = A(\Omega)$. By (2.3.2) of the next section, Ω can be embedded in a 2-homogeneous ordered set Λ (definition in section 3) in such a way that $A(\Omega)$ embeds into $A(\Lambda)$, and by (2.5.1), $A(\Lambda)$ is divisible.

The following can also be shown using a permutation argument, though the proof is considerably more difficult.

THEOREM 2.2.12 [Pierce, 1972a]. *Every ℓ-group can be embedded in an ℓ-group in which any two strictly positive elements are conjugate.*

3. Transitivity and Primitivity

An examination of the proof of Theorem 2.2.5 may leave the reader feeling that the representation of G in $A(\Omega)$ is somewhat arbitrary since there is a lot of freedom in the ordering of Ω, and hence that the connection between G and Ω is a bit tenuous. In general, we cannot expect Ω to be completely determined by G, even when G is a very large ℓ-subgroup of $A(\Omega)$. For example, if Ω^+ denotes Ω with one new point adjoined at the upper end, it should be clear that $A(\Omega) = A(\Omega^+)$ since the new point is necessarily fixed by each member of $A(\Omega^+)$. A convenient and natural assumption tying G and Ω more closely together is that of transitivity. We use the notation (G,Ω) to indicate that the ℓ-group G is (isomorphic to) an ℓ-subgroup of $A(\Omega)$. In this case, we say that (G,Ω) *is an ℓ-permutation group.* To avoid trivial cases we will assume that Ω is infinite. Then (G,Ω) is *transitive* if for all $a,b \in \Omega$, there exists $g \in G$ such that $ag = b$. More generally, (G,Ω) is *n-transitive* if for all $a_1 < a_2 < ... < a_n$, $b_1 < b_2 < ... b_n$ in Ω, there exists $g \in G$ such that $a_i g = b_i$ for each i. For example, $A(\mathbf{R})$ is *n*-transitive (on $\mathbf{R}$) for all n, while $A(\mathbf{Z})$ is transitive (on $\mathbf{Z}$) but not 2-transitive. Conveniently, there is no distinction between 2-transitive and *n*-transitive for $n \geq 2$:

THEOREM 2.3.1. *For* $m \geq n \geq 2$, *an ℓ-permutation group* (G,Ω) *is m-transitive if and only if* (G,Ω) *is n-transitive* .

Proof. If (G,Ω) is *m*-transitive, the assumption that Ω is infinite easily implies that (G,Ω) is also *n*-transitive. Conversely, let (G,Ω) be *n*-transitive, and hence by the first statement, 2-transitive. Let $a_1 < a_2 < ... <$

a_m, $b_1 < b_2 < ... < b_m$ be given. By induction, there exists $f \in G$ such that $a_i f = b_i$ for $i = 1, 2, ..., m-1$. We may assume $a_m f \geq b_m$ (the other case being similar). There exists $g \in G$ such that $a_1 g = b_{m-1}$ and $a_m g = b_m$. Then for each $i = 1, 2, ..., m-1$, $a_i \, g \geq a_1 g = b_{m-1} \geq b_i$, and so for all $i = 1, 2, ..., m$, $a_i (g \wedge f) = b_i$.

Caution: Simon Thomas [a] has shown that for each n, there is a subgroup of $A(Q)$ that is n-transitive but not $n+1$-transitive. So the assumption in Theorem 2.3.1 that G be an ℓ-subgroup of $A(\Omega)$ is crucial.

If $A(\Omega)$ is transitive (on Ω) we say that Ω is *homogeneous*, and if $A(\Omega)$ is 2-transitive, that Ω is *2-homogeneous*.

LEMMA 2.3.2 [Holland, 1963a]. *Every ordered set Ω can be embedded in a 2-homogeneous ordered set Λ in such a way that $A(\Omega)$ embeds in $A(\Lambda)$.*

Proof. First observe that Ω embeds in an obvious way in the antilexicographically ordered group

$$\Gamma = \sum_{\alpha \in \Omega} \mathbf{R}_\alpha$$

where each $\mathbf{R}_\alpha$ is the ordered group of real numbers; and $A(\Omega)$ embeds in $A(\Gamma)$ via "shifts". In a similar way, the ordered group Γ embeds in the antilexicographically ordered field of formal power series (sequences of inversely well ordered support)

$$\Lambda \subseteq \prod_{\gamma \in \Gamma} \mathbf{R}_\gamma.$$

Again, the automorphisms extend so that $A(\Gamma)$ is embedded in $A(\Lambda)$. A consideration of the linear maps $x \mapsto ax+b$ shows that Λ is 2-homogeneous.

For a given ℓ-group G, it is not difficult to find the transitive representations (G, Ω).

THEOREM 2.3.3 [Holland, 1965]. *If C is a prime convex ℓ-subgroup of an ℓ-group G such that $\bigcap_{g \in G} g^{-1} C g = \{e\}$, then $(G, G/C)$ is a transitive ℓ-permutation group. Every transitive representation of G arises in this way.*

Proof. For the first statement, we use the method of Theorem 2.2.5 to see that G acts transitively as an ℓ-permutation group on G/C. The representation is faithful (one-to-one) because the intersection of the conjugates of C is trivial. Conversely, if (G,Ω) is transitive, let $\alpha \in \Omega$, and let $C = G_\alpha = \{g \in G \mid \alpha g = \alpha\}$ be the *stabilizer* at α. Then C is a prime convex ℓ-subgroup. For any $g \in G$,

$$g^{-1}Cg = \{f \in G \mid \alpha gf = \alpha g\} = G_{\alpha g}$$

so the intersection of all conjugates of C fixes each point of Ω (by transitivity) and hence is $\{e\}$. The correspondence $Cx \leftrightarrow \alpha x$ is an isomorphism between the ordered sets G/C and Ω, and the action of G on each set is respected by this isomorphism.

If (G,Ω) is an ℓ-permutation group, an equivalence relation $\sim$ on Ω is called a *convex G-congruence* if each $\sim$-class is convex and for all $\alpha,\beta \in \Omega$, $g \in G$, $\alpha \sim \beta$ iff $\alpha g \sim \beta g$. Each $\sim$-class B is then a *G-block*, that is, B is convex and for all $g \in G$, if $Bg \cap B \neq \emptyset$ then $Bg = B$. Conversely, If B is a block then all translates Bg ($g \in G$) constitute the classes of a convex G-congruence. In this case $\Omega/\sim$ is a totally ordered set and G induces a natural action $(G/L,\Omega/\sim)$ where $L = \{g \in G \mid \alpha \sim \alpha g$ for all $\alpha\}$ is the *lazy subgroup*. The ℓ-permutation group $(G/L,\Omega/\sim)$ is, in a certain sense, a homomorphic simplification of (G,Ω). We are interested in those (G,Ω) for which no non-trivial simplification is possible. A transitive (G,Ω) is called *primitive* if there is no non-trivial convex G-congruence on Ω; equivalently, there is no non-trivial G-block. Any 2-transitive ℓ-permutation group is primitive, as is $A(\mathbf{Z})$. On the other hand, if $\Omega = \mathbf{R} \times \mathbf{R}$ is the antilexicographically ordered product then $(A(\Omega),\Omega)$ has a proper convex congruence where $(x,y) \sim (a,b)$ iff $y = b$; $(G/L,\Omega)$ is essentially $(A(\mathbf{R}),\mathbf{R})$. Let $\bar{\Omega}$ denote the Dedekind completion (without end points) of the totally ordered set Ω. Each $g \in A(\Omega)$ has a unique extension to an element of $A(\bar{\Omega})$ which we will also denote by g. Under this extension, $A(\Omega)$ is an ℓ-subgroup of $A(\bar{\Omega})$. If Δ is an equivalence class of a proper convex G-congruence on Ω and $\bar{\delta}$ is the least upper bound of Δ in $\bar{\Omega}$, then $\bar{\delta}G = \{\bar{\delta}g \mid g \in G\}$ must consist entirely of upper end-points of classes of the same congruence, and so $\bar{\delta}G$ is not dense in $\bar{\Omega}$. The following theorem shows that this property can be used to characterize primitive (G,Ω).

THEOREM 2.3.4 [Holland, 1965]. *If (G,Ω) is transitive, the following are equivalent.*

   1. (G,Ω) *is primitive.*
   2. *For each $\bar{\alpha} \in \bar{\Omega}$, $G_{\bar{\alpha}}$ is a maximal convex ℓ-subgroup of G.*
   3. *For each $\alpha \in \Omega$, G_{α} is a maximal convex ℓ-subgroup of G.*
   4. *For each $\bar{\delta} \in \bar{\Omega}$, $\bar{\delta}G$ is a dense subset of $\bar{\Omega}$.*

From a slightly different point of view, if Δ is a class of a convex G-congruence and $\alpha \in \Delta$, then $G_{(\Delta)} = \{g \in G \mid \Delta g = \Delta\}$ is a convex ℓ-subgroup of G and $G_{\alpha} \subseteq G_{(\Delta)}$. Every convex ℓ-subgroup containing G_{α} arises in this way. Clearly G_{α} is prime, so the convex ℓ-subgroups containing G_{α} form a tower under inclusion (see Proposition 1.1.5). Moreover, $G_{(\Delta)} \subseteq G_{(\Delta')}$ iff $\Delta \subseteq \Delta'$. Hence

THEOREM 2.3.5. *If (G,Ω) is transitive, the convex G-congruences on Ω form a tower under inclusion.*

Suppose $\sim$ and $\approx$ are convex G-congruences on Ω for some transitive (G,Ω), and that $\sim$ *covers* $\approx$, that is, $\sim$ is the smallest member of the tower of convex G-congruences that properly contain $\approx$. We say that $(\approx,\sim)$ is a *covering pair*. Then if Δ is a $\sim$-class, $G_{(\Delta)}$ acts transitively on Δ. Letting

$$G'_{\Delta} = \{g \in G \mid \delta g \approx \delta \text{ for all } \delta \in \Delta\},$$

we see that G'_{Δ} is an ℓ-ideal of $G_{(\Delta)}$ and that $(G_{(\Delta)}/G'_{\Delta}, \Delta)$ is an ℓ-permutation group in a natural way, which we will refer to as the *component* of (G,Ω) at the pair $(\approx,\sim)$ (by transitivity, it is independent of the choice of $\approx$-class.) For example, if $\Omega = \mathbf{R} \times \mathbf{Z}$ antilexicographically ordered, and $G = A(\Omega)$ then there are exactly three convex G-congruences (including the trivial ones), $\equiv$, $\approx$, $\sim$, where $(x,y) \approx (z,w)$ iff $y = w$. The $(\equiv,\approx)$-component is $(A(\mathbf{R}),\mathbf{R})$ and the $(\approx,\sim)$-component is $(A(\mathbf{Z}),\mathbf{Z})$. The original ℓ-group of permutations (G,Ω) can, in a certain sense, be recaptured from these two components. The ordered set Ω is the antilexicographically ordered product of the set parts of the components, while G is the *wreath product* of $A(\mathbf{R})$ and $A(\mathbf{Z})$. That is, the action, for $g \in G$, $(r,n) \in \Omega$, is $(r,n)g = (rg_n, n\bar{g})$ where $\bar{g} \in A(\mathbf{Z})$ and $g_n \in A(\mathbf{R})$ (independently for each $n \in \mathbf{Z}$.) The algebraic structure of G is seen in the rules $\overline{gh} = \overline{g}\,\overline{h}$ and $(gh)_n = g_n h_{n\overline{g}}$, while the order on G is determined as $e \leq g$ iff for each $n \in \mathbf{Z}$, either $n < n\overline{g}$ or both $n = n\overline{g}$ and $e \leq g_n$. We write $(G,\Omega) = (A(\mathbf{R}),\mathbf{R})\mathrm{Wr}\,(A(\mathbf{Z}),\mathbf{Z})$.

For the general case of a transitive (G,Ω) the convex congruences form a (complete) tower and we let $\mathcal{K}$ be the set of covering pairs of congruences. If $k = (\approx,\sim) \in \mathcal{K}$, then as in the example there is an induced component (G_k,Ω_k) which is a primitive transitive ℓ-permutation group. If the set $\mathcal{K}$ is totally ordered in the natural way then $(\mathcal{K}, \{(G_k,\Omega_k) \mid k \in \mathcal{K}\})$ is the *skeleton* of (G,Ω), $\mathcal{K}$ is the *spine*, and each (G_k,Ω_k) is a *rib*. The paper [Holland and McCleary, 1969] investigates the question of how to recapture (G,Ω) from its skeleton. For a given totally ordered set $\mathcal{K}$ and transitive ℓ-permutation groups (G_k,Ω_k), $k \in \mathcal{K}$, the *wreath product* (W,Ω) is constructed as follows. Fix any point $0 \in \prod_{k \in \mathcal{K}} \Omega_k$ and let

$$\Omega = \{\alpha \in \textstyle\prod_{k \in \mathcal{K}} \Omega_k \mid \{k \in \mathcal{K} \mid \alpha_k \neq 0_k\} \text{ is inversely well ordered}\}.$$

That is, Ω consists of the elements of $\prod \Omega_k$ of inversely well ordered support. Order Ω antilexicographically. For each $k \in \mathcal{K}$ define equivalence relations $=_k$, $=^k$ on Ω by: $\alpha =_k \beta$ iff $\alpha_d = \beta_d$ for all $d \geq k$; and $\alpha =^k \beta$ iff $\alpha_d = \beta_d$ for all $d > k$. Let $U \subseteq A(\Omega)$ consist of all those automorphisms of Ω which preserve each of the relations $=_k$, $=^k$, $(k \in \mathcal{K})$. Then (U,Ω) is an ℓ-permutation group and each $=_k$ and $=^k$ is a convex U-congruence on Ω. Thus (U,Ω) induces components (not necessarily primitive) (U_k,Ω_k), $k \in \mathcal{K}$ where Ω_k is the same as the original. If we restrict U and consider only those elements which induce a member of G_k for each k, we obtain an ℓ-subgroup W of U and (W,Ω) is an ℓ-permutation group called the *wreath product* $(W,\Omega) = \mathrm{Wr}_{k \in \mathcal{K}}(G_k,\Omega_k)$. If each of the given (G_k,Ω_k) is primitive, then the skeleton of (W,Ω) is $(\mathcal{K},\{(G_k,\Omega_k)\})$.

THEOREM 2.3.6 [Holland and McCleary, 1969]. *If (G,Ω) is a transitive ℓ-permutation group with skeleton $(\mathcal{K},\{(G_k,\,\Omega_k) \mid k \in \mathcal{K}\})$, then (G,Ω) can be embedded in $\mathrm{Wr}_{k \in \mathcal{K}}(G_k,\Omega_k)$ in a natural way.*

A version of this theorem for unordered permutation groups can be found in [Holland, 1969]. In that case, the spine $\mathcal{K}$ is only partially ordered.

The wreath product construction can be used to characterize normal-valued ℓ-groups in the following theorem.

THEOREM 2.3.7 [J. A. Read, 1975]. *An ℓ-group G is normal-valued if and only if G can be embedded in* $\mathrm{Wr}_{k \in \mathcal{K}}(G_k, \Omega_k)$ *where each* (G_k, Ω_k) *is an ℓ-subgroup of the real numbers permuting itself regularly.*

Much of the above material can be generalized to the intransitive case. The interested reader should consult [McCleary, 1976] or [Glass, 1981b].

4. Classification of primitive groups

In this section we will give a classification of primitive ℓ-permutation groups due mostly to S. H. McCleary [1972a, 1976]. Although McCleary's work is more general, we restrict ourselves here to the transitive case.

If G is any ℓ-subgroup of the real numbers then the translation action (G,G) is primitive (from Theorem 2.3.4) because for each $\alpha \in G$, $G_\alpha = \{e\}$ is a maximal convex ℓ-subgroup of G. Any transitive (G,Ω) with $G_\alpha = \{e\}$ is said to be *regular*. Another type of primitive group, as noted earlier, is any 2-transitive (G,Ω). Furthermore, if $\Omega = \bar{\Omega}$ is Dedekind complete, then any transitive (G,Ω) is primitive by Theorem 2.3.4, part 4. An example of this type which is neither regular nor 2-transitive is the following. Let

$$P = \{g \in A(\mathbf{R}) \mid \alpha g + 1 = (\alpha + 1)g \text{ for all } \alpha \in \mathbf{R}\}.$$

That is, P is the centralizer of the element $z \in A(\mathbf{R})$, where $\alpha z = \alpha + 1$. P is said to be *periodic* with *period* z. Clearly, P is not 2-transitive. To see that P is also not regular, let g' be any member of $A(\mathbf{R})$ which fixes 0 and 1 but does not fix each point of the interval [0,1] (possible because $A(\mathbf{R})$ is 3-transitive.) Define $g \in A(\mathbf{R})$ by periodically extending the action of g' on [0,1] to $\mathbf{R}$ as follows. For $\alpha \in [n,n+1)$, let $\alpha g = \alpha z^{-n} g' z^n$. Then $e \neq g \in P$.

It is a remarkable fact (see Theorem 2.4.1) that every primitive transitive ℓ-permutation group which is neither regular nor 2-transitive is periodic in the sense of the preceding example. (A precise definition follows.) Since the proof is quite typical of arguments in the subject and has considerable depth, we give it in detail. The key to the distinction between the three types of primitive ℓ-permutation groups lies in the behavior of the stabilizer subgroups G_α. If (G,Ω) is regular then $G_\alpha = \{e\}$. If (G,Ω) is 2-transitive, then G_α acts transitively (in fact, 2-transitively) on the ray (α, ∞). Indeed, this last property is equivalent to 2-transitivity, as can easily be verified. The following lemma gives an even weaker-seeming equivalent condition.

LEMMA 2.3.8. *If (G,Ω) is transitive, then (G,Ω) is 2-transitive iff for each $\alpha < \beta < \gamma$, there exists $g \in G_\alpha$ such that $\gamma \leq \beta g$.*

Proof. The condition is obviously implied by 2-transitivity. Conversely, there exists $f \in G_\alpha$ such that $\beta f = \gamma$, and then $(f \vee e) \wedge g \in G_\alpha$ and $\beta((f \vee e) \wedge g) = \gamma$. Thus, G_α acts transitively on (α, ∞).

A transitive ℓ-permutation group (G,Ω) is *periodic* with *period* z if $z \in A(\bar{\Omega})$, $zg = gz$ for all $g \in G$ (extended to $\bar{\Omega}$), any other member of $A(\bar{\Omega})$ which commutes with each $g \in G$ is a power of z, and for each $\alpha \in \Omega$, $\{\alpha z^n \mid n \in \mathbf{Z}\}$ has no upper or lower bound in $\bar{\Omega}$. In this case, $G_\alpha \subseteq G_{\alpha z}$ and as we shall see, if (G,Ω) is also primitive, $G_\alpha = G_{\alpha z}$ and G_α acts 2-transitively on $(\alpha, \alpha z)$. Thus, αz is the "next" fixed point of G_α.

For the remainder of this section, let (G,Ω) be a primitive ℓ-permutation group which is neither regular nor 2-transitive. Suppose for some (hence each) $\alpha \in \Omega$, $\beta \leq \alpha$ implies $G_\alpha \subseteq G_\beta$. Then also the conditions $\beta \leq \alpha$ and $\alpha f = \alpha g$ imply $\beta f = \beta g$. Comparing f and $f \vee e$, we see that if $\alpha \leq \alpha f$ for some α, and $\beta \leq \alpha$ then $\beta \leq \beta f$. A dual observation results from comparing f and $f \wedge e$, so we see that for each $f \in G$, either $e \leq f$ or $f \leq e$, whence G is totally ordered. By Theorem 2.3.4, all stabilizers are maximal convex ℓ-subgroups of G, and hence they are all equal, there being only one. Thus G is regular, contrary to assumption.

(*)   Therefore, for some $\beta < \alpha$, G_α is not contained in G_β. Let Δ be the convexification of βG_α in $\bar{\Omega}$ and let $\bar{\alpha}$ be the least upper bound (in $\bar{\Omega}$) of Δ. Since Δ is non-trivial, the density of αG (Theorem 2.3.4) implies there exists $g \in G$ such that $\alpha g \in \Delta$. Then $\bar{\alpha} g^{-1} > \alpha$.

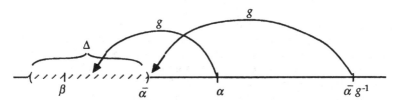

We now claim that G_α has a first non-trivial orbit above α, indeed, that $\bar{\alpha}g^{-1}$ is in the convexification of that orbit. To show this, let $\alpha < \gamma < \bar{\alpha}g^{-1}$. We will find a member of G_α which maps $\bar{\alpha}g^{-1}$ below γ as follows. Since (G,Ω) is primitive, $\bar{\alpha}G$ is dense in $\bar{\Omega}$ (Theorem 2.3.4), so there exists $h \in G$ with $\alpha < \bar{\alpha}h < \gamma$. Then $\alpha h^{-1} < \bar{\alpha}$ (a worst case is illustrated in the next picture), and so there exists $f \in G_\alpha$ such that $\alpha g f > \alpha h^{-1}$. Clearly, $G_\alpha \subseteq G_{\bar{\alpha}}$, and since G_α is a maximal prime convex ℓ-subgroup of G (Theorem 2.3.4 again), $G_\alpha = G_{\bar{\alpha}}$. Therefore $f \in G_{\bar{\alpha}}$.

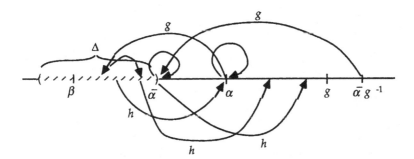

It follows that $\alpha h^{-1} < \alpha g f < \bar{\alpha}$, and hence that $\alpha < \alpha g f h < \bar{\alpha}h < \gamma$. We then see that $g f h \wedge e \in G_\alpha$ while $(\bar{\alpha}g^{-1})(g f h \wedge e) < \gamma$.

We are now able to define a mapping $z : \Omega \longrightarrow \bar{\Omega}$ by letting αz be the least upper bound of the first non-trivial orbit of G_α above α (alternatively, αz is the first fixed point of G_α above α.) An easy computation will convince the reader that for all $g \in G$, $\alpha z g = \alpha g z$.

We next claim that z is one-to-one and preserves order. Suppose that $\beta_1 < \beta_2$ are elements of Ω. There exists $e < f \in G$ such that $\beta_1 f = \beta_2$. Then $\beta_2 z = \beta_1 f z = \beta_1 z f \geq \beta_1 z$. But if $\beta_1 z f = \beta_1 z$ then $f \in G_{\beta_1 z} = G_{\beta_1}$

(by Theorem 2.3.4 and the fact that $G_{\beta_1} \subseteq G_{\beta_1 z}$), contradicting $\beta_1 < \beta$ $_1 f$. Hence $\beta_1 z < \beta_2 z$. Finally, we show that Ωz is dense in $\bar{\Omega}$. Fix $\alpha \in \Omega$. If $\bar{\beta} < \bar{\gamma}$ are any two members of $\bar{\Omega}$, then since $\alpha z G$ is dense in $\bar{\Omega}$ (Theorem 2.3.4) there exists $g \in G$ such that $\bar{\beta} < \alpha z g < \bar{\gamma}$. But $\alpha z g = \alpha g z \in \Omega z$. It follows now that z has a unique extension to $z \in A(\bar{\Omega})$, and that $zg = gz$ for all $g \in G$.

Next we show that any orbit $\{\alpha z^n\}$ has no bounds in Ω. Let Λ be the convexification of the orbit $\{\alpha z^n\}$ in Ω. If $\alpha g \in \Lambda$ for some g, then $\alpha z^n \leq \alpha g < \alpha z^{n+1}$ for some n. Hence for any m, $\alpha z^{n+m} \leq \alpha g z^m = \alpha z^m g \leq \alpha z^{n+m+1}$ so that $\alpha z^m g \in \bar{\Lambda}$ and consequently $\Lambda g = \Lambda$. It follows that the different translates Λf $(f \in G)$ form the classes of a convex G-congruence on Ω. By primitivity, $\Lambda = \Omega$, and the orbit has no bounds.

We finish the proof of the theorem by showing that any $w \in A(\bar{\Omega})$ which commutes with each $g \in G$ must be a power of z. Note that since $G_\alpha = G_{\alpha z}$, the latter part of the foregoing argument (from (*) on) can be applied with αz in place of α to show that the only fixed points of G_α are precisely the points $\{\alpha z^n\}$. If $w \in A(\bar{\Omega})$ commutes with each element of G, it is clear that αw must be a fixed point of G_α, and so $\alpha w = \alpha z^n$ for some n. Then $wz^{-n} \in G_\alpha$, and commutes with each element of G. If $\zeta \in \Omega$, there exists $g \in G$ with $\alpha g = \zeta$, and so

$$\zeta w z^{-n} = \alpha g w z^{-n} = \alpha w z^{-n} g = \alpha g = \zeta,$$

from which follows $w = z^n$, and the proof is complete. Thus:

THEOREM 2.4.1 [McCleary, 1972a]. *If (G,Ω) is a transitive primitive ℓ-permutation group, then (G,Ω) is either regular, 2-transitive, or periodic.*

5. The Full ℓ-Group $A(\Omega)$

An element $f \in A(\Omega)$ is called a *positive convex cycle* if for some $\alpha \in \Omega$, $\alpha < \alpha f$ and if $\beta \neq \beta f$ then for some n, $\alpha f^n \leq \beta < \alpha f^{n+1}$. In other words, the *support* of f (points moved) consists of a single interval, and the images of any point of that interval under the powers of f are coterminal in the interval. A *negative convex cycle* is defined similarly. We sometimes say that a convex cycle is a *bump*. Each positive member of $A(\Omega)$ is uniquely the join of a

disjoint set of bumps, and the support of each of these bumps is called a
supporting interval of f.

LEMMA 2.5.1 [Holland,1963a].   *If $A(\Omega)$ is 2-transitive, then $A(\Omega)$ is
divisible.*

Proof. Let $g \in A(\Omega)$ and a positive integer n be given. It will suffice to
assume that g is one positive bump. We choose any point α in the supporting interval of g, so that the points $...,\alpha g^i,...$ are coterminal in that interval.
We then choose points $\beta_1,\beta_2,...,\beta_n$ with $\alpha < \beta_1 < \beta_2 < ... < \beta_n = \alpha g$. Since
$A(\Omega)$ is 2-transitive, it is n-transitive also by Lemma 2.3.1. Hence we can find
$x \in A(\Omega)$ such that $\alpha x = \beta_1$, and $\beta_i x = \beta_{i+1}$. Taking x as defined only on the
interval $[\alpha,\beta_{n-1}]$, we can extend its action to $[\beta_{n-1},\beta_n]$ by letting $x = x^{n-1}g$. We
continue in this fashion both to the right and to the left, so that on the
entire supporting interval, $x^n = g$.

Any conjugate of a bump is a bump of the same type. Moreover, the
bump structure of every member of $A(\Omega)$ is preserved by conjugation, in the
following sense. For each $f \in A(\Omega)$ let $B(f)$ be the set of all pairs (I,σ) such
that I is a fixed point of f and $\sigma = 0$, or I is a supporting interval of f and
$\sigma = +$ if f is positive on I and $\sigma = -$ if f is negative on I. The symbol σ is
the *parity* of I. The set $B(f)$ has a natural total order where
$(I_1,\sigma_1) < (I_2,\sigma_2)$ if each point of I_1 is less than each point of I_2. If f and g
are conjugate, it is easy to verify that $B(f)$ and $B(g)$ are isomorphic by a map
that preserves both parity and order. The following conjugacy lemma, special cases of which were proved by Kneser [1924] and Schreier and Ulam
[1935], shows that the converse is true in many $A(\Omega)$.

LEMMA 2.5.2. *If $A(\Omega)$ is 2-transitive, $f,g \in A(\Omega)$, and $B(f)$ is isomorphic to
$B(g)$, then f and g are conjugate.*

Proof. By patching together on the different bumps, the lemma follows from the special case when each of f and g is a single bump of the
same parity with both or neither bounded on the left, and both or neither
bounded on the right. Suppose that $\alpha < \alpha f$ and that $\beta < \beta g$. By double
transitivity, the intervals $[\alpha,\alpha f]$ and $[\beta,\beta g]$ are isomorphic by some function
ϕ. We extend to the support of f as follows. If $\gamma \neq \gamma f$ then there is a unique
n such that $\alpha f^n \leq \gamma < \alpha f^{n+1}$. We define $\gamma \phi = \gamma f^{-n} \phi g^n$. Finally, we let ϕ fix
each point of Ω not in the support of f. Then $\phi \in A(\Omega)$ and $g = \phi^{-1} f \phi$.

In the proof of the previous lemma, we see that we may assume that ϕ has support bounded on the left or right if both f and g do.

Let $R(\Omega)$ be the set of all $f \in A(\Omega)$ such that the support of f is bounded on the left (f "lives on the right".) $L(\Omega)$ is defined dually, and $B(\Omega) = R(\Omega) \cap L(\Omega)$, the members of bounded support. Parts of the following theorem were proved by G. Higman [1954], J. T. Lloyd [1964], and in [Holland, 1963a].

THEOREM 2.5.3. *Each of $R(\Omega), L(\Omega)$, and $B(\Omega)$ is an ℓ-ideal of $A(\Omega)$. If $A(\Omega)$ is 2-transitive, then every normal subgroup of $A(\Omega)$ is an ℓ-ideal and $B(\Omega)$ is a simple group, and so has no proper ℓ-ideals. If Ω has a countable unbounded (above and below) subset, then $R(\Omega), L(\Omega)$, and $B(\Omega)$ are the only proper normal subgroups of $A(\Omega)$, and $B(\Omega)$ is the only proper normal subgroup of either $L(\Omega)$ or $R(\Omega)$.*

In the case that every countable subset of Ω has an upper bound, the situation is much more complex. This has been investigated by Ball [1974], Ball and Droste [1985], Droste [1985], and Droste and Shelah [1985]. The ℓ-simple ℓ-groups B (R) and $L(R)/B(R)$ are not ℓ-isomorphic since every positive element of $B(R)$ is disjoint from one of its conjugates, while this is not true of $L(R)/B(R)$. It is not known whether $L(R)/B(R)$ and $R(R)/B(R)$ are ℓ-isomorphic.

It is interesting to consider the problem of how close the connection between Ω and $A(\Omega)$ is. Obviously, Ω completely determines $A(\Omega)$ as an ℓ-group, while the converse is certainly not true without transitivity. (See the remarks at the beginning of section 3.) A more interesting example is the relationship between $A(R \backslash Q)$ and $A(Q)$. Each of these acts transitively, even 2-transitively on its set, and yet they are isomorphic as ℓ-groups, for every member of $A(Q)$ extends uniquely to a member of $A(R)$ which then restricts to a member of $A(R \backslash Q)$ and conversely, and the correspondence is clearly an ℓ-isomorphism. Although Q and $R \backslash Q$ have different cardinality, they are closely related as complementary subsets of R. In the following theorem, it is shown that this is the worst that can happen:

THEOREM 2.5.4 [Holland, 1965]. *If $(A(\Omega), \Omega)$ is transitive and $A(\Omega) \approx A(Q)$ then either $\Omega \approx Q$ or $\Omega \approx R \backslash Q$.*

For the case when $\Omega = \mathbf{R}$, an even stronger result is true:

THEOREM 2.5.5 [Holland, 1965]. *If $(A(\Omega),\Omega)$ is transitive and $A(\Omega) \approx A(\mathbf{R})$ then $\Omega \approx \mathbf{R}$.*

The conclusions of the preceding two theorems remain true under even weaker hypotheses.

THEOREM 2.5.4' [Gurevich and Holland, 1981]. *There is a sentence δ in the elementary language of groups such that δ holds in $A(\mathbf{Q})$ (and in $A(\mathbf{R} \backslash \mathbf{Q})$) and if $(A(\Omega),\Omega)$ is transitive and δ holds in $A(\Omega)$ then either $\Omega \approx \mathbf{Q}$ or $\Omega \approx \mathbf{R} \backslash \mathbf{Q}$.*

THEOREM 2.5.5' [Gurevich and Holland, 1981]. *There is a sentence τ in the elementary language of groups such that τ holds in $A(\mathbf{R})$ and if $(A(\Omega),\Omega)$ is transitive and τ holds in $A(\Omega)$ then $\Omega \approx \mathbf{R}$.*

For the details of the proofs, the reader is referred to the paper [Gurevich and Holland, 1981]. Here, we give just a few observations which may make these rather surprising conclusions more believable. Addressing ourselves to Theorem 2.5.5', the essential task is to show that the facts that Ω is Dedekind complete and has a countable dense subset can be captured by a sentence in the elementary language of groups. Because of an observation of McCleary ([1978]; see also Jambu-Giraudet [1983]) we may assume the language also contains the lattice symbols. First, we can describe the positive convex cycles of bounded support. They are just those positive elements of $A(\Omega)$ which are not the join of disjoint elements, and which are disjoint from one of their conjugates. For $g,h \in A(\Omega)$, the support of g lies to the left of the support of h (we say g *is to the left of* h) precisely when every conjugate of h by a positive element is disjoint from g. An element has bounded support when it is to the left of one of its conjugates. The supports of elements g,h of bounded support have the same least upper bound in the Dedekind completion $\bar{\Omega}$ of Ω when for all f, g is to the left of f iff h is to the left of f. Then Ω is Dedekind complete iff for every two elements g,h of bounded support, the supports of g and some conjugate of h have the same least upper bound.

To see that Ω must have a countable dense subset, note first that if $f,g \in A(\mathbf{R})$ are, respectively, translation by 1 and by $\sqrt{2}$, then $fg = gf$ and any element that commutes with both g and f must be a translation. In [Gurevich and Holland, 1981] it is shown that if f and g are elements of $A(\Omega)$ which commute with each other and such that every element which commutes with both must be a convex cycle which is disjoint from only the identity, then the orbit of any point of Ω under the subgroup generated by $\{g,h\}$ is dense in Ω (and, of course, countable.)

The proof of Theorem 2.5.4' is similar but somewhat more subtle. Many similar and more general results can be found in [Jambu-Giraudet, 1983], [Glass, Gurevich, Holland, and Jambu-Giraudet, 1981], and [Glass, 1981a].

The general question of which ℓ-groups arise as $A(\Omega)$ has not been answered, but several special cases are known. The earliest of these is due to Ohkuma. We recall that (G,Ω) is *regular* if for every $\alpha,\beta \in \Omega$ there is a unique $g \in G$ such that $\alpha g = \beta$.

THEOREM 2.5.6 [Ohkuma, 1954]. *If $(A(\Omega),\Omega)$ is regular then $A(\Omega)$ is ℓ-isomorphic to a subgroup of the real numbers and $A(\Omega) \approx \Omega$ as ordered sets.*

Proof: If $f \in A(\Omega)$, then $f \vee e$ must either fix all points or none. Hence, either $f \leq e$ or $f \geq e$. Thus, $A(\Omega)$ is totally ordered. If α is any point of Ω, then the correspondence $g \longleftrightarrow \alpha g$ gives an isomorphism of $A(\Omega)$ with Ω. Under this identification, the action of $A(\Omega)$ is just translation. If C is a proper convex subgroup of $A(\Omega)$ and $e \neq g \in C$, then defining $g^* = g$ on C and $g^* = e$ otherwise contradicts the regularity. Hence there are no proper convex subgroups, which implies that $A(\Omega)$ is archimedean, and by Hölder's theorem (Theorem 1.2.3), $A(\Omega)$ is a subgroup of the real numbers.

The subsets (or subgroups) of the reals which arise as in Theorem 2.5.6 are called *Ohkuma sets* (or *Ohkuma groups*). They are also known as *rigidly homogeneous chains*. The most obvious example is any set isomorphic to the integers **Z**. The major result of Ohkuma [1954] is that there are other examples. In [Glass, Gurevich, Holland, and Shelah, 1981] many examples of Ohkuma groups with special properties are constructed.

It is easy to see that any commutative transitive permutation group must be regular. Hence, Theorem 2.5.6 may be restated as: *$A(\Omega)$ is commutative and transitive iff Ω is an Ohkuma set.* If the assumption of transitivity is dropped, we may ask which groups $A(\Omega)$ are commutative. The answer was supplied by Chang and Ehrenfeucht:

THEOREM 2.5.7 [Chang and Ehrenfeucht, 1962]. *If $A(\Omega)$ is commutative then $A(\Omega)$ is isomorphic as a group to a complete direct product of subgroups of the real numbers; conversely, any such product is isomorphic to some $A(\Omega)$.*

If $A(\Omega)$ is commutative, one direction of Theorem 2.5.7 follows easily from Ohkuma's theorem by considering its action on the individual orbits. The converse is proved by an ingenious "tagging" to produce separate orbits for each direct factor. Theorem 2.5.7 has been generalized to give a complete description of those $A(\Omega)$ which are solvable in [Holland, 1985a].

W. Charles Holland
Bowling Green State University
Bowling Green, Ohio 43403
U. S. A.

Volker Weispfenning

CHAPTER 3

MODEL THEORY OF ABELIAN ℓ-GROUPS

So blickt man klar, wie selten nur,
Ins innere Walten der Natur.

--W. Busch, Maler Klecksel

All groups in this chapter will be commutative, written additively. So an ℓ-*group* is an Abelian lattice-ordered group and an *o-group* is an Abelian linearly ordered group.

The model theory of ℓ-groups is distinguished from the purely algebraic theory of these groups by the study of formal statements of a restricted nature about ℓ-groups. The classical and still most prominent kind of such statements is given by the formulas of first-order logic. They provide a formal framework for the concept of an *elementary algebraic statement* about ℓ-groups - as opposed to, e.g., infinitary statements, higher-order statements (about subsets, ideals, homomorphisms, etc.), or statements involving generalized quantifiers (e.g., "there exist infinitely many"). Throughout this chapter we will stick largely to this framework. In order to present substantial results in some depth, we will further concentrate on three major, interrelated model theoretic topics:

(1) elementary equivalence,

(2) decidability,

41

A. M. W. Glass and W. C. Holland (eds.), Lattice-Ordered Groups, 41–79.
© 1989 by Kluwer Academic Publishers.

(3) algebraically and existentially closed ℓ-groups.

In regard to (1): We refer to formulas without parameters (free variables) as *sentences*. Given a set Φ of formulas, one may classify ℓ-groups according to their properties expressible in Φ: G and H are *Φ-equivalent* if the same sentences in Φ are valid in G and in H (notation: $G \equiv H (\Phi)$). For the set Φ = SE of *all* sentences, this yields the concept of *elementary equivalence* between G and H, denoted by $G \equiv H$. Isomorphic ℓ-groups are elementarily equivalent, but not vice versa. In fact, a fundamental result of first-order model theory, the *Löwenheim-Skolem-Tarski theorem* (see Keisler, 1977; section 4]), guarantees that any non-trivial ℓ-group G has elementarily equivalent ℓ-groups H of any infinite cardinality. So elementary equivalence and Φ-equivalence for smaller and smaller classes Φ of sentences provide a whole hierarchy of coarser and coarser classifications of ℓ-groups. While the isomorphism problem for ℓ-groups seems to be of insurmountable difficulty (see [Glass and Madden, 1984]), the classification of ℓ-groups up to Φ-equivalence may well be feasible for many sets Φ of sentences. The general problem of classifying *all* ℓ-groups up to elementary equivalence is still wide open; the existing evidence (see [Gurevich, 1967a] and [Burris, 1985]) suggests that this is a very hard problem. On the other hand, this problem has been solved for substantial classes of ℓ-groups:

The classification of Archimedean o-groups began with Presburger's fundamental paper [1929], and was completed by Robinson and Zakon in [Robinson and Zakon, 1960] and [Zakon, 1961]. After intermediate results by Kargapolov [1963], Gurevich [1965] extended this classification in a giant step to the class of all o-groups. A very accessible account of this classification was presented (among other things) by Schmitt in [1982] and [1984].

For the set of universal elementary sentences, a classification of all ℓ-groups up to universal equivalence was obtained in [Khisamiev, 1966] and [Khisamiev and Kokorin, 1966], based on the earlier result of Gurevich and Kokorin [1963] that all non-trivial o-groups are universally equivalent (see also [Brignole and Ribeiro, 1965] and [Glass, a]).

The classification of non-linearly ordered ℓ-groups according to elementary equivalence is much less advanced. At present, it concerns only *projectable ℓ-groups*, i.e., ℓ-groups G in which the lattice $Pp(G)$ of principal polars is relatively complemented. But even for this class, the classification is far from being complete; it is applicable only to projectable ℓ-groups, whose o-group factors in a representation have "simple" elementary properties such as divisible o-groups and Z-groups ([Weispfenning 1981], [Glass and Pierce, 1980a], [Point, 1983]).

In regard to (2): Decidability is concerned with the question, to what extent the elementary properties of a class $\mathcal{K}$ of ℓ-groups can be determined algorithmically. For a given set Φ of formulas, let $\mathrm{Th}_{\Phi}(\mathcal{K})$, the *$\Phi$-theory of* $\mathcal{K}$ denote the set of all sentences in Φ that hold in all ℓ-groups $G \in \mathcal{K}$; *in particular, for* $\Phi = SE$, $\mathrm{Th}(\mathcal{K}) = \mathrm{Th}_{\Phi}(\mathcal{K})$ denotes the *(elementary) theory of* $\mathcal{K}$. $\mathrm{Th}_{\Phi}(\mathcal{K})$ is *decidable* if there is an algorithm which, on input of a sentence $\varphi \in \Phi$, decides whether or not φ is valid in all $G \in \mathcal{K}$. Thus, in principle, a decidable theory is trivial. Fortunately (for mathematicians), decidable classes of algebraic structures are rare, and ℓ-groups in general are no exception to this rule: By [Gurevich, 1967a] (see also [Burris, 1985]), the theory of the class $\mathcal{A}r$ of all Archimedean ℓ-groups is hereditarily undecidable. This means that for any class $\mathcal{K}$ of ℓ-groups extending $\mathcal{A}r$ - in particular for the class of all ℓ-groups - $\mathrm{Th}(\mathcal{K})$ is undecidable. Nevertheless, there are large and interesting classes of ℓ-groups whose theory is decidable: By [Presburger, 1929], [Robinson and Zakon, 1960], and [Zakon, 1961], the theories of all dense or discrete Archimedean o-groups, as well as many related theories, are decidable. By [Gurevich, 1965], the theory of the class of all o-groups is decidable; this holds even for the monadic second-order theory of this class, where elementary properties of the chain of convex subgroups are expressible (see [Gurevich, 1977 and 1980]). For non-o-groups, a number of theories of classes of projectable ℓ-groups such as the class of divisible projectable ℓ-groups, or the class of projectable Z-groups, are decidable by [Weispfenning , 1981], [Glass and Pierce, 1980a] (compare also [Burris and Werner, 1979] and [Point, 1983]).

If we restrict our attention to universal sentences only, the situation changes radically: The universal theory of the class of all ℓ-groups, the universal theory of the class of all o-groups, and the universal theory of every single ℓ-group are decidable ([Gurevich and Kokorin, 1963], [Khisamiev, 1966], [Khisamiev and Kokorin, 1966]).

Decidability as such does not have the same algebraic significance as the classification according to Φ-equivalence. As long as the efficiency or feasibility of a decision algorithm is not specified, it remains largely a philosophical issue. A first coarse measure of efficiency is the asymptotic time or space complexity of the algorithm, when performed on some abstract machine model such as a Turing machine. Most of the known decision algorithms for theories of ℓ-groups are *primitive recursive*: Let the functions $f_n : \mathbf{N} \to \mathbf{N}$ be defined recursively by $f_1(x) = 2^x$, $f_{n+1}(x) = f_n^{(x)}(1)$; then an algorithm is *primitive recursive (elementary recursive)* if for almost all $m \in \mathbf{N}$, its computing time for an input of length $\le m$ is bounded by $f_n(m)$ (by $f_1^{(n)}(m)$) for some $n \in \mathbf{N}$. Any non-elementary recursive decision problem can safely be considered non-feasible; this applies in particular to the decision problem for the theory of all o-groups (by [Rabin, 1977; Theorem

19]). On the other hand, the theory of divisible o-groups, as well as the theory of many other Archimedean o-groups turns out to be elementary recursive ([Ferrante and Rackoff, 1975, 1979], [Weispfenning, 1985]); so these decision algorithms may potentially be useful in algebraic practice. The decision problem for the decidable classes of projectable ℓ-groups mentioned above is not known to be elementary recursive.

In regard to (3): The concepts of algebraically and existentially closed ℓ-groups are defined in analogy with algebraically closed fields and algebraically closed abstract (non-commutative) groups (cf. [Macintyre, 1977]): Let G be an ℓ-group in a class $\mathcal{K}$ of ℓ-groups; then G is *algebraically (existentially) closed in $\mathcal{K}$* if every finite system of equations $w_i(x_1,...,x_n) = 0$ (and inequations $w_i'(x_1,...,x_n) \neq 0$) with parameters from G, that is solvable in some extension H of G in the class $\mathcal{K}$, is also solvable in G itself. (For non-trivial fields and abstract groups these concepts coincide.) Sometimes, one wants to restrict the extension-ℓ-groups H admitted in this definition. This is achieved by distinguishing in G extra elements, operations, or relations; the extension groups H then have to respect these distinguished constants, operations, or relations. If, e.g., we distinguish all congruences $\equiv_n \bmod(nG)$ on G, then an extension H of G respecting these relations has to be pure.

The main goal is to characterize algebraically and existentially closed ℓ-groups in more tangible algebraic terms. For the class OG of non-trivial o-groups the solution is straightforward: $G \in$ OG is algebraically closed (a.c.) iff G is existentially closed (e.c.) iff G is divisible. For the class $\mathcal{K}$ of discrete o-groups with distinguished smallest positive element 1, $G \in \mathcal{K}$ is a.c. iff G is e.c. iff G is a Z-group, i.e., elementarily equivalent to the o-group Z of integers [Glass and Pierce, 1980a].

For the class $\mathcal{K}$ of arbitrary ℓ-groups the characterization is much more involved: $G \in \mathcal{K}$ is a.c. iff G is divisible and satisfies some elementary lattice-theoretic conditions; in order to be e.c., G must satisfy in addition some elementary lattice-theoretic conditions and also a non-elementary condition involving the smallest convex ℓ-subgroups $G(g)$ containing the element $g \in G$. In particular, e.c. is a much stronger condition that a.c. (compare [Glass and Pierce, 1980a,b], [Point, 1983], [Saracino and Wood, 1983 1984], [Lavaca, 1980]).

For projectable ℓ-groups with distinguished projector operation and some related classes, a.c. and e.c. groups can again be characterized nicely by elementary axioms (compare [Glass and Pierce, 1980a], [Point, 1983], [Weispfenning, 1976, 1978]).

The plan of this chapter is as follows:

Section 1   introduces some basic model theoretic concepts and methods and illustrates their use by means of the easy example of divisible o-groups.

Section 2   treats the universal properties of o-groups and ℓ-groups in general on the basis of the results of section 1.

Section 3   studies o-groups elementarily equivalent to the o-group Z of integers.

Section 4   studies the elementary properties of dense Archimedean o-groups.

Section 5   outlines the elementary classification and decidability for o-groups in general and o-groups with distinguished chain of convex subgroups.

Section 6   studies a.c. and e.c. ℓ-groups and the elementary properties of projectable ℓ-groups.

Section 7   provides some suggestions for further research.

I am very much indebted to the organizers and participants of this workshop. They created an immensely lively, stimulating, and fruitful atmosphere for mathematical discussions. Above all, I feel grateful to Andrew Glass who introduced me into this distinguished circle of ℓ-group adepts.

3.1. Divisible o-Groups

We begin by making the model theoretic concepts mentioned in the introduction more precise. (Abelian) *group words* are - as usual - formal expressions of the form 0 or $k_1 x_1 + ... + k_n x_n$, where k_i are integers and x_i are variables. Addition and subtraction are handled as in the free Abelian group generated by $x_1,...,x_n$. ℓ-group words are formal expressions built up from 0 and variables by means of the ℓ-group operation symbols $+,-,\wedge,\vee$, and appropriate use of brackets. (In this chapter, the logical operation symbols which might otherwise be confused with $\wedge$ and $\vee$ will be denoted by $\bigwedge$ and $\bigvee$ respectively.) We do not insist on ℓ-group words being presented in a normal form such as $\bigvee_i \bigwedge_j w_{ij}$ or $\bigwedge_i \bigvee_j w_{ij}$, where w_{ij} are group words (see Corollary 0.1.7).

Atomic formulas (in the language of ℓ-groups) are equations $w = w'$ and inequalities $w \leq w'$ between ℓ-group words. *Quantifier-free formulas* are finite expressions obtained from atomic formulas by means of the Boolean operation symbols $\wedge$ (and), $\vee$ (or), $\neg$ (not), with appropriate use of brackets $(\ ,\ )$. In arbitrary *formulas* we admit in addition *existential and universal quantification* of the form $\exists x(\varphi)$ and $\forall x(\varphi)$, where x is a variable and φ is a formula; φ is then called the *scope* of the *quantifier* $\exists x$ or $\forall x$. Any occurrence of x in this scope is said to be *bound*. A variable x is *free* in a formula φ if it has at least one non-bound occurrence in φ. By renaming bound variables we may always assume that no free variable of φ has a bound occurrence in φ. We write $\varphi = \varphi(x_1,...,x_n)$ to indicate that all free variables of φ are among $x_1,...,x_n$. When formulas are interpreted in an ℓ-group G by giving all logical symbols their intended meaning, free variables behave as parameters: We say $\varphi(a_1,...,a_n)$ *holds in G at* $a_1,...,a_n \in G$ (notation $G \vDash \varphi(a_1,...,a_n)$) if the interpretation of the formula $\varphi(x_1,...,x_n)$ becomes a true statement about G when the elements a_i of G are substituted for the variables x_i. A formula without free variables is a *sentence*. The interpretation of a sentence φ in G does not depend on any parameters; we write $G \vDash \varphi$ for "φ is true in G". We let FO, SE, QF, QFSE, AT, ATSE be the sets of formulas, sentences, quantifier-free formulas, quantifier-free sentences, atomic formulas, atomic sentences, respectively. For $\Phi \subseteq$ FO, two ℓ-groups G, H are *Φ-equivalent over a common ℓ-subgroup K* iff for all $\varphi(x_1,...,x_n) \in \Phi$, $a_1,...,a_n \in K$, $G \vDash \varphi(a)$ iff $H \vDash \varphi(a)$. We write $G \underset{K}{\equiv} H (\Phi)$. Φ-equivalence without parameters is defined analogously using sentences $\varphi \in \Phi$. Notice that for any extension H of G, $G \underset{G}{\equiv} H$ (QF).

Among all formulas, we single out those that contain no occurrence of the lattice operations $\vee, \wedge$; they form the set oFO of *o-group formulas*. oSE, oQF, etc., are defined similarly using the sets SE, QF, etc., above. It should be clear that any formula is equivalent in the class of all o-groups to some o-group formula. For use in section 2, we need to elaborate this fact a little more: We associate with any word w a finite set $A(w)$ of group words as follows (induction on the length $\ell(w)$ of w) : $A(0) = \{0\}$, $A(x) = \{x\}$ for a variable x, $A(w \vee w') = A(w \wedge w') = A(w) \cup A(w')$, $A(w \pm w') = \{v \pm v' \mid v \in A(w), v' \in A(w')\}$. For a formula φ, we let $A(\varphi) = \bigcup \{A(w) \mid w \text{ occurs in } \varphi\}$.

LEMMA 3.1.1. *For every (quantifier free) formula* $\varphi(x_1,...,x_n)$ *one can construct a (quantifier free) o-group formula* $\varphi^o(x_1,...,x_n)$ *such that* φ *and* φ^o *are equivalent in all o-groups, and all words v occurring in* φ^o *are in* $A(\varphi)$.

Proof. We construct φ° by induction on the number k of occurrences of the lattice operations $\wedge, \vee$ in φ. If $k = 0$, $\varphi^\circ = \varphi$. Otherwise, pick the first occurrence $w_1 \diamondsuit w_2$ of $\wedge$ or $\vee$ in a word or subword of φ. Let φ_1 (φ_2) result from φ by replacing $w_1 \diamondsuit w_2$ by w_1 (w_2), and put

$$\varphi^\circ = (w_1 \underset{\leqq}{\lessgtr} w_2 \wedge \varphi_1^\circ) \vee (w_1 \underset{\geqq}{\gtrless} w_2 \wedge \varphi_2).$$

We begin our model theoretic study of ℓ-groups with the simple and instructive case of divisible o-groups. Here the three fundamental problems presented in the introduction have a simple, clear-cut answer. Let D_0 (D) be the class of all (non-trivial) divisible o-groups.

THEOREM 3.1.2. *Any non-trivial divisible o-group is elementarily equivalent to the group* $\mathbf{Q}$ *of rationals. So* $\{0\}$ *and* $\mathbf{Q}$ *are representatives of the elementary equivalence classes* $\{\{0\}\}$ *and* D *of* D_0.

THEOREM 3.1.3. Th(D) *is decidable. In fact, there is a decision procedure for* Th(D) *that on input of a sentence* φ *(* $\varphi \in$ oSE*) runs in space bounded by* $2^{2^{O(\ell(\varphi))}}$

(by $2^{O(\ell(\varphi))}$).

THEOREM 3.1.4. D *is the class of existentially closed o-groups and* D_0 *is the class of algebraically closed o-groups.*

We prove these theorems by the method of *quantifier elimination* (q.e.). If $\mathcal{K}$ is a class of ℓ-groups, then $\mathcal{K}$ admits quantifier elimination if every formula $\varphi(x_1,...,x_n)$ is equivalent in $\mathcal{K}$ to a quantifier-free formula $\varphi'(x_1,...,x_n)$. An algorithm that computes φ' from φ is called a *quantifier elimination procedure*. The complexity of such a procedure can be measured e.g. in terms of the space and time it requires for input φ of length $\ell(\varphi)$, when implemented on a Turing machine. A simple argument shows that the running time $T(\ell)$ can always be bounded by $2^{cS(\ell)}$, where c is a constant and $S(\ell)$ is the space required for input of length ℓ (see [Garey and Johnson, 1979]). The same applies of course for decision procedures. The immediate model theoretic applications of quantifier elimination are gathered in the following trivial lemma.

LEMMA 3.1.5. (1) *Suppose* $\mathcal{K}$ *admits quantifier elimination. Then for all* $G, H \in \mathcal{K}$ *with common ℓ-subgroup* K, $G \underset{K}{\equiv} H$; *in particular, any extension between*

ℓ-groups in $\mathcal{K}$ is elementary. Moreover, two ℓ-groups G and H in $\mathcal{K}$ are elementarily equivalent iff they are QFSE-equivalent.

(2) Let $\mathcal{K}' \subseteq \mathcal{K}$ and suppose $\mathcal{K}'$ admits q.e. and every $G \in \mathcal{K}$ has an extension $H \in \mathcal{K}'$. Then every $H \in \mathcal{K}'$ is existentially closed in $\mathcal{K}$.

(3) Suppose $\mathcal{K}$ admits a q.e. procedure running in space bounded by $exp_k(O(\ell(\varphi)))$, $k \geq 1$, on input φ. ($exp_k(n)$ is the iterated exponential function, given by $exp_1(n) = 2^n$, $exp_{k+1}(n) = 2^{exp_k(n)}$.) Assume further that the validity of a sentence $\varphi' \in$ QFSE in $\mathcal{K}$ can be decided in space polynomial in $\ell(\varphi')$. Then the validity of a sentence φ in $\mathcal{K}$ can be decided in space bounded by $exp_k(O(\ell(\varphi)))$.

We only prove (2) in order to recollect the precise meaning of "existentially closed" : Let $H \in \mathcal{K}'$, $\psi(x_1,...,x_m,y_1,...,y_n) \in$ QF, $\varphi(y_1,...,y_n) = \exists x_1...\exists x_m \psi$, let $b_1,...,b_n \in H$, $H \subseteq G \in \mathcal{K}$ and assume $G \models \varphi(b)$. Find $H' \in \mathcal{K}'$ extending G. Then $H' \models \varphi(b)$. So if $\varphi'(y) \in$ QF is equivalent to $\varphi(y)$ in $\mathcal{K}'$, then $\varphi'(b)$ holds in H' and hence in H; so $\varphi(b)$ holds in H.

There is one less immediate, but important application of quantifier elimination that uses model theoretic diagrams and the Compactness Theorem (3.4.1 below) of first-order logic (see [Keisler, 1977]):

LEMMA 3.1.5. (4) Suppose $\mathcal{K}$ is axiomatized by a set of sentences and admits quantifier elimination; let Sub($\mathcal{K}$) be the class of all ℓ-subgroups of ℓ-groups in $\mathcal{K}$. Then Sub($\mathcal{K}$) has the amalgamation property, i.e., in Sub($\mathcal{K}$), any diagram of embeddings can be completed as shown (compare [Pierce, 1972a]):

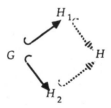

For the class D we are going to show:

LEMMA 3.1.6. D admits a quantifier elimination procedure that on input of a formula φ runs in space and time bounded by $exp_2(O(\ell(\varphi)))$.

Proof. It suffices to show the following claim:

There is a procedure that on input of a formula $\varphi(y_1,...,y_n) \in$ oFO produces a quantifier-free formula $\varphi'(y_1,...,y_n)$ in time bounded by $exp_2(O(\ell(\varphi)))$, such that $y_n > 0 \Rightarrow (\varphi(y) \Leftrightarrow \varphi'(y))$ holds in D.

From this claim we construct a quantifier elimination procedure for o-group formulas by induction on the number n of variables in the input formula $\varphi(y_1,...,y_n)$: If $n = 0$, then φ is a sentence, and so φ is equivalent in D to $\exists z(z > 0 \wedge \varphi)$, and so by the claim to $\exists z(z > 0 \wedge \varphi'(z))$. We may assume that z occurs in φ' only in inequalities of the form $nz \geq 0$ and $nz \leq 0$ with $n > 0$. Let φ^* be the sentence resulting from φ' by replacing these inequalities by $0 = 0$ and $0 \neq 0$, respectively. Then $\exists z(z > 0 \wedge \varphi'(z))$ is equivalent in D to $\exists z(z > 0 \wedge \varphi^*)$, and so to $\exists z(z > 0) \wedge \varphi^*$, and so to φ^*.

If $n > 0$, let $\varphi_1(y_1,...,y_n)$ be the formula $\varphi(y_1,...,y_{n-1},-y_n)$ and let $\varphi_2(y_1,...,y_{n-1})$ be the formula $\varphi(y_1,...,y_{n-1},0)$. Then by the induction assumption, a quantifier-free formula φ_2^* equivalent to φ_2 in D has already been constructed. So $\varphi(y_1,...,y_n)$ is equivalent in D to

$$(\varphi'(y) \wedge y_n > 0) \vee (\varphi_1'(y) \wedge -y_n > 0) \vee \varphi_2^*.$$

In this way we can produce a quantifier-free equivalent $\varphi^*(y)$ of and o-group formula $\varphi(y)$ within the time indicated. For an ℓ-group formula $\varphi(y)$, we first pass to the associated o-group formula $\varphi^o(y)$ (see Lemma 3.1.1) and then to $\varphi^{o*}(y)$. Since $\ell(\varphi^o) \leq \exp(O(\ell(\varphi)))$, the time required in this case increases by one exponential.

The claim is proved by induction on the number of quantifiers in φ. So it suffices to treat the case where $\varphi(y)$ is of the form $\exists x\psi(x,y)$, ψ quantifier free. We assume that ψ has atomic formulas of the form

(*) $\quad n_i x \underset{\geq}{\overset{\leq}{}} w_i(y)$ $\quad(i \in I)$, where n_i are non-negative integers.

Let $n = \mathrm{lcm}(n_i : i \in I, n_i \neq 0)$. By multiplying the inequalities (*) by $2n/n_i$, we may assume that all atomic formulas in ψ containing x are of the form

(**) $\quad 2nx \underset{\geq}{\overset{\leq}{}} w_i'(y)$ $\quad(i \in J)$ with $J \subseteq I$, $w_i' = (2n/n_i)w_i$,

and hence that $\psi(x,y)$ is of the form $\psi_1(2nx,y)$. In the class D, $\exists x\psi_1(2nx,y)$ is equivalent to $\exists x'\psi_1(2x',y)$.

Let now $G \in D$, $a,b_1,...,b_n \in G$, $b_n > 0$, and assume $G \models \psi_1(2a,b)$. Replace the subscripts $i \in J$ by natural numbers such that the elements $w_i(b)$ are arranged in increasing order, $w_1(b) \leq ... \leq w_k(b)$. Then the following cases can occur:

(1) $a = w_h(b)$,

(2) $w_h(b) < a < w_{h+1}(b)$ for some h,

(3) $a < w_1(b)$,

(4) $a > w_k(b)$.

In the first case, $G \models \psi_1(2w_h(b),b)$, in the second $G \models \psi_1(w_h+w_{h+1}(b),b)$, in the third $G \models \psi_1(2w_1(b)-b_n, b)$, and in the last case $G \models \psi_1(2w_k(b)+b_n, b)$. So we have shown that in D, $\quad y_n > 0$ implies the equivalence of $\exists x \psi(x,y)$, $\exists x' \psi_1(2x',y)$ and

$$\bigvee_{i,j \in J} \psi_1(w_i + w_j, y) \vee \bigvee_{i \in J} \psi_1(2w_i \pm y_n, y).$$

Using this construction of φ', it is now not too difficult to verify that the procedure runs in double exponential time.

The combination of 3.1.6 and 3.1.5 yields now an easy proof of Theorems 3.1.2 - 3.1.4: For 3.1.2 it suffices to remark that any two o-groups are QFSE-equivalent since quantifier-free sentences are just Boolean combinations of atomic formulas of the form $0 = 0$, $0 \leq 0$. For 3.1.4, it remains to show that every algebraically closed o-group is divisible, and that the trivial o-group is algebraically closed, which is obvious from the definitions. Theorem 3.1.3 with a space bound on exponential higher than stated follows readily from 3.1.6 and 3.1.5 (3). The better space bound results from the special nature of our quantifier elimination procedure, in which quantifiers ranging over the whole group are replaced by quantifiers ranging over an explicitly described finite set of test points depending on the parameters of the formula. So in order to decide $\varphi \in$ oSE, we replace all quantifiers in φ by quantifiers ranging over the set of all rational multiples $(m/n)1$ of a new constant $1 > 0$, where the binary length of m,n is bounded by $\exp(c\, \ell(\psi))$ for a suitable positive c. Then all the resulting "instances" of φ can be decided in the space indicated (compare [Ferrante and Rackoff, 1975], [Weispfenning, 1988]).

REMARK 3.1.7. It can be shown that *any* decision procedure for Th(D) requires a running time bounded below by $\exp(O(\ell(\varphi)))$ for infinitely many o-group sentences φ (see [Ferrante and Rackoff, 1979]), and that *any* quantifier elimination procedure for D requires a space bounded below by $\exp_2(O(\ell(\varphi)))$ for infinitely many o-group formulas φ (see [Weispfenning, 1988]). So the upper bounds stated are not completely unrealistic.

3.2. Universal Properties of ℓ-groups

The results and techniques of section 3.1 put us in a position to classify all ℓ-groups with respect to their universal properties, and to determine these properties algorithmically.

A formula φ is *universal (existential)* if it is of the form

$$\forall x_1...\forall x_n \psi \quad (\exists x_1...\exists x_n \psi),$$

where ψ is quantifier-free. We denote the set of all universal formulas, universal sentences, existential formulas, existential sentences by U, USE, E, ESE, respectively; the corresponding sets in oFO get a prefix "o". OG_0 (OG) is the class of all (non-trivial) o-groups, LG_0 (LG) is the class of all (non-trivial) ℓ-groups.

THEOREM 3.2.1 ([Gurevich and Kokorin, 1963]; compare also [Brignole and Ribeiro, 1965]). *Any two non-trivial o-groups are E-equivalent and hence U-equivalent.*

Proof. Notice that in Lemma 3.1.1, $\varphi \in$ E implies $\varphi^\circ \in$ oE. So it suffices to consider sentences $\varphi = \exists x_1...\exists x_n \psi(x)$ with $\psi \in$ oQF. Let $0 \neq G \in$ OG. If $\mathbf{Z} \models \varphi$, then $G \models \varphi$, since $\mathbf{Z}$ embeds into G. Conversely, if $G \models \varphi$, let $\bar{G}$ be the divisible hull of G. Then $\bar{G} \models \varphi$, and so by 3.1.2, $\mathbf{Q} \models \varphi$, say

$$\mathbf{Q} \models \psi(a_1/b,...,a_n/b), \quad \text{where } a_i, b \in \mathbf{Z},\ b > 0.$$

Then $\mathbf{Z}(1/b) \models \varphi$ and $\mathbf{Z}(1/b) \cong \mathbf{Z}$; so $\mathbf{Z} \models \varphi$. Therefore $\mathbf{Z}$ and G satisfy the same sentences in oE and hence the same sentences in E.

Let W be the set of all universal sentences of the form

$$\forall x_1...\forall x_n(w_1(x) = 0 \wedge ... \wedge w_m(x) = 0 \Rightarrow w(x) = 0).$$

These sentences express the *uniform word problem* for ℓ-groups. Their negations can be put into the form

$$\exists x_1...\exists x_n(w_1(x) = 0 \wedge ... \wedge w_m(x) = 0 \wedge w(x) \neq 0).$$

Let us call these sentences *special existential*, and let SpE denote the set of special existential sentences.

Then we have the following variant of Theorem 3.2.1:

THEOREM 3.2.2. *Any two non-trivial ℓ-groups are SpE-equivalent and hence W-equivalent.*

Proof. Let $\varphi = \exists x_1...\exists x_n \psi(x)$, $\psi \in$ QF, be special existential, and let $G \in$ LG. If $\mathbf{Z} \models \varphi$, then $G \models \varphi$, since $\mathbf{Z}$ embeds into G. Conversely, assume $G \models \varphi$, say $G \models \psi(a_1,...,a_n)$ for some $a_i \in G$. Let $G \subseteq \prod_{i \in I} G_i$, where G_i are o-groups (see Theorem 1.2.1). Then there exists $j \in I$ such that $G_j \models \psi(a(j))$. Hence $G_j \models \varphi$, and so by 3.2.1, $\mathbf{Z} \models \varphi$.

COROLLARY 3.2.3. *In order to test the validity of a sentence $\varphi \in W$ in LG (i.e., an instance of the uniform word problem for LG) it suffices to test the validity of φ in Z.*

Notice that $\mathrm{Th}_E(LG) = \mathrm{Th}_E(Z)$, since Z embeds into any non-trivial ℓ-group. This does not mean, however, that all $G \in LG$ have the same existential properties: Let δ_n be the sentence

$$\exists x_1 ... \exists x_n (\bigwedge_{1 \le i \le n} x_i > 0 \wedge \bigwedge_{1 \le i \ne j \le n} x_i \wedge x_j = 0).$$

Then $G \vDash \delta_n$ iff G has dimension $\ge n$, and $G \vDash \delta_n$ for all $0 < n \in N$ iff G has infinite dimension (see [Bigard, Keimel, and Wolfenstein, 1977; Section 7.4]). This shows that any two E-equivalent ℓ-groups have the same dimension (dimensions being taken in $N \cup \{\infty\}$). We are going to show that the converse holds as well. So $\dim(G) \in N \cup \{\infty\}$ is an invariant for the universal properties of an ℓ-group.

THEOREM 3.2.4 ([Khisamiev and Kokorin, 1966], [Khisamiev, 1966]). *Two ℓ-groups are E-equivalent (or equivalently, U-equivalent) iff they have the same dimension.*

This will be a consequence of the following lemma. Call $\psi \in QF$ *normal* if it is obtained from equations and inequations by means of $\wedge$ and $\vee$ only; $s(\psi)$ is then the maximum of 1 and the number of inequations in ψ.

LEMMA 3.2.5. *Let $\varphi = \exists x_1 ... \exists x_n \psi(x) \in E$ with $\psi \in QF$ normal. Let $G \in LG$, $G \subseteq \prod_{i \in I} G_i$, $G_i \in OG$, and assume $G \vDash \varphi$. Put $t = \min(|I|, s(\psi))$. Then $Z^t \vDash \varphi$.*

Proof. Pick $a_1, ..., a_n \in G$ with $G \vDash \psi(a)$. Let S (S') be the set of all equations (inequations) $\sigma(x)$ in ψ such that $G \vDash \sigma(a)$. For every $\sigma \in S'$ pick $i_\sigma \in I$ such that $G_{i_\sigma} \vDash \sigma(a(i_\sigma))$, and put $J = \{i_\sigma \mid \sigma \in S'\}$. Then

$$\exists x (\bigwedge \{\rho(x) \mid \rho \in S\} \wedge \bigwedge \{\sigma(x) \mid \sigma \in S', i_\sigma = j\})$$

holds in G_j for each $j \in J$, and hence by 3.2.1 in Z. So the formula

$$\exists x (\bigwedge \{\rho(x) \mid \rho \in S\} \wedge \bigwedge \{\sigma(x) \mid \sigma \in S'\})$$

holds in $Z^{|J|}$, and hence in Z^t, since $t \ge |J|$. Consequently $Z^t \vDash \varphi$.

Proof of 3.2.4. Let $\varphi = \exists x \psi \in ESE$, $\psi \in QF$, and $G \in LG$. If $\dim G = d < \infty$, we show $Z^d \vDash \varphi$ iff $G \vDash \varphi$: Assume without loss of generality that ψ is normal. Let $a_1, ..., a_d$ be pairwise disjoint positive elements if G. If $Z^d \vDash \varphi$, then

$$Z^d \ni (k_1, ..., k_d) \mapsto k_1 a_1 + ... + k_d a_d \in G$$

is an embedding of $\mathbf{Z}^d$ into G, and so $G \models \varphi$. Conversely, assume $G \models \varphi$. Since G has dimension d, G can be embedded into a direct product of d o-groups G_i $(i \in I)$ (see [Bigard, Keimel, and Wolfenstein, 1977; Section 7.4]), and so $\prod_{i \in I} G_i \models \varphi$.

If $\dim G = \infty$, we show $G \models \varphi$ iff $\mathbf{Z}^s \models \varphi$, where $s = s(\psi)$. The argument is as above, except that I may be arbitrary, and so $\mathbf{Z}^d$ has to be replaced by $\mathbf{Z}^s$.

Next we study the algorithmic aspects of the universal theory of ℓ-groups. The main tool is the following result on the complexity of integer linear programming.

THEOREM 3.2.6 [von zur Gathen and Sieveking, 1978]. *Let $Ax \geq B$ $(A,B$ matrices of size $m \times n$ and $m \times 1$, respectively) be a system of linear diophantine inequalities. Let c be the maximum of the absolute values of all entries of A and B, and suppose the system has a solution $a \in \mathbf{Z}^n$. Then it has a solution $b \in \mathbf{Z}^n$ with $|b_i| \leq n^{n/2}(n+1)c^n$.*

REMARK. [von zur Gathen and Sieveking, 1978] gives a more specific result, from which 3.2.6 is obtained by means of *Hadamard's inequality*: $|\det(c_{ij})| \leq n^{n/2}(\max|c_{ij}|)^n$ for a square matrix (c_{ij}).

If $\varphi(x_1,...,x_n) \in$ FO, we let rank(φ) be the maximum of all absolute values of coefficients $k_i \in \mathbf{Z}$ such that $k_1x_1+...+k_nx_n$ is a group word occurring in the associated o-group formula φ^o. The proof of 3.1.1 shows that rank$(\varphi) \leq \ell(\varphi)$.

LEMMA 3.2.7. *Let $\mu(x_1,...,x_n) \in$ QF, normal of rank $\leq r$, where $r \geq 1$. If $\mathbf{Z} \models \exists x \mu(x)$, then there exist $a_1,...,a_n \in \mathbf{Z}$ with $|a_i| \leq n^{n/2}(n+1)r^n$ such that $\mathbf{Z} \models \mu(a)$.*

Proof. Let $\mu^o(x)$ be the o-group formula associated with $\mu(x)$ in 3.1.1. Introducing the new constant 1 into the language, we may rewrite μ^o equivalently in $\mathbf{Z}$ in such a way that the resulting formula μ' contains no negation, and has atomic subformulas of the form $k_1x_1+...+k_nx_n \geq k$, where $k_i, k \in \mathbf{Z}$, $|k_i| \leq r$, $k \in \{0,+1,-1\}$. Pick $b_1,...,b_n \in \mathbf{Z}$ such that $\mu(b)$, and hence $\mu^o(b)$ and $\mu'(b)$ hold in $\mathbf{Z}$. Let S be the set of all atomic subformulas $\sigma(x)$ of μ' such that $\mathbf{Z} \models \sigma(b)$. Then S is a system of linear diophantine inequalities with coefficients bounded in absolute value by r. So by 3.2.6, there exist $a_1,...,a_n \in \mathbf{Z}$ with $|a_i| \leq n^{n/2}(n+1)r^n$ such that $\sigma(a)$ holds in $\mathbf{Z}$ for all $\sigma \in S$. Consequently, $\mu'(a)$, $\mu^o(a)$, and $\mu(a)$ hold in $\mathbf{Z}$.

THEOREM 3.2.8. *Let* $\varphi = \forall x_1...\forall x_n \psi(x_1,...,x_n) \in$ *USE*, $\psi \in$ *QF*, *normal of rank* $\leq r$, *where* $r \geq 1$. *Then* φ *holds in all* ℓ-*groups (of dimension* $\leq d < \infty$) *iff for* $t = s(\psi)$ $(t = \min(d,s(\psi))$, $\mathbf{Z}^t \models \psi(a)$ *for all* $a_1,...,a_n \in \mathbf{Z}^t$ *with* $|a_i| \leq n^{n/2}(n+1)r^n$.

Proof. In view of 3.2.4 and 3.2.5, it suffices to show that if $\mathbf{Z}^t \models \neg\varphi$, then there exist $a_1,...a_n \in \mathbf{Z}^t$ with $|a_i| \leq n^{n/2}(n+1)r^n$ such that $\mathbf{Z}^t \models \neg\psi(a)$. Suppose $\mathbf{Z}^t \models \neg\psi(b)$, let S (S') be the set of all equations (inequations) $\sigma(x)$ in $\neg\psi$ such that $\mathbf{Z}^t \models \sigma(b)$, and let $\psi_i(x)$ $(1 \leq i \leq t)$ be the formula

$$\bigwedge\{\sigma(x) \mid \sigma \in S\} \wedge \bigwedge\{\sigma(x) \mid \sigma \in S', \mathbf{Z} \models \sigma(b(i))\}.$$

Then $\mathbf{Z} \models_t \psi_i(b(i))$ and $\mathrm{rank}(\psi_i) \leq r$ for all $1 \leq i \leq t$. So by 3.2.7, there exist $a_1,...,a_n \in \mathbf{Z}^t$ with $|a_i| \leq n^{n/2}(n+1)r^n$ such that $\mathbf{Z} \models_t \psi_i(a(i))$ for all $1 \leq i \leq t$. Consequently, $\mathbf{Z}^t \models \sigma(a)$ for all $\sigma \in S \cup S'$, and so $\mathbf{Z}^t \models \neg\psi(a)$.

COROLLARY 3.2.9. *The universal theory of the class of all* ℓ-*groups (of specified dimension* $d \in \mathbf{N} \cup \{\infty\}$) *is decidable; in fact it is in the complexity class co-NP (see [Garey and Johnson, 1979]).*

Proof. It suffices to test the validity of a universal sentence φ as in 3.2.8 for all $a_1,...,a_n \in \mathbf{Z}^t$ with $|a_i| \leq n^{n/2}(n+1)r^n$, where t,r may be taken as $\ell(\psi)$. So the binary length of the vectors a_i is bounded by a polynomial in $\ell(\varphi)$.

REMARK 3.2.10. It can be shown that all the theories considered in 3.2.9, in fact even the uniform word problem for ℓ-groups W are also co-NP-hard, e.e., belong to the hardest problems in this complexity class (see [Weispfenning, 1986b], [Garey and Johnson, 1979]).

3.3. Z-Groups

In this section we study the elementary properties of the o-group $\mathbf{Z}$ of integers and characterize the o-groups elementarily equivalent to $\mathbf{Z}$.

Among the elementary properties of the o-group $G = \mathbf{Z}$, the following are fundamental:

3.3.1. (i) *G is a discrete group.*

(ii) *For all* $1 < n \in \mathbf{N}$, G/nG *is cyclic of order* n.

So it is natural to study the class ZG of all o-groups G sharing these properties with $\mathbf{Z}$. Such an o-group G is called a Z-*group.* For any discrete o-group G, we let 1 denote the smallest positive element of G. The following characterization of Z-groups is useful and easy to verify (compare [Conrad, 1962]):

LEMMA 3.3.2. *Let G be a discrete o-group. Then G is a Z-group iff $G/\mathbf{Z}\cdot 1$ is divisible.*

It shows in particular, that all lexicographic products of a divisible o-group with $\mathbf{Z}$ are Z-groups. It should be clear, how to write down an infinite set of sentences in oSE axiomatizing ZG. Since $\mathbf{Z} \in$ ZG, any ℓ-group elementarily equivalent to $\mathbf{Z}$ must be a Z-group. As the reader may have suspected by now, the converse is also true:

THEOREM 3.3.3. *Let G be an ℓ-group. Then $G \equiv \mathbf{Z}$ iff G is a Z-group.*

The proof uses again the technique of quantifier elimination. This time, however, quantifier elimination is impossible within the set FO or oFO of formulas considered so far. Indeed, q.e. would imply by 3.1.5(1) that $2\mathbf{Z} \subset \mathbf{Z}$ is an elementary extension, which is nonsense. To exclude extensions of this kind, we add the constant 1 to our language. But then 3.1.5(1) shows that even in the extended language, q.e. is still impossible: Let $G = \mathbf{Q} \times \mathbf{Z}$, $K = \mathbf{Z} \times \mathbf{Z}$. Then $K \subset G$ and the map $(a,b) \mapsto (a,a+b)$ is an embedding of K into G. On the other hand, the element $(1,2)$ is divisible by 2 in G, but $f(1,2) = (1,3)$ is not. To exclude this phenomenon, we add congruences modulo all natural numbers $n > 1$ as new relation symbols to our language. The set ZFO of Z-*formulas* is then defined as follows: Z-*words* are of the form $k1 + k_1 x_1 + ... + k_n x_n$, atomic Z-*formulas* are of the form $w = w'$, $w \le w'$, $w \equiv w'(n)$ for $1 < n \in \mathbf{N}$, where w, w' are Z-words. Arbitrary Z-formulas are obtained from atomic Z-formulas as before by Boolean operations and quantification. ZSE, ZQF, ... are defined similarly as before.

THEOREM 3.3.4. *ZG admits a quantifier elimination procedure in ZFO which on input φ runs in time and space bounded by $\exp_3(O(\ell(\varphi)))$.*

Proof. As in the proof of 3.1.6, it suffices to produce a quantifier-free equivalent of a formula φ of the form $\exists x \psi(x,y)$, $\psi \in$ ZQF. We may assume that the atomic subformulas of ψ do not contain the variable x, or are of the form

(*) $n_i x\ \rho_i\ w_i(y)$, where w_i are Z-words, ρ_i is one of the relations $\le$, $\ge$, $\equiv(m_i)$, and n_i are positive integers.

Multiplying these relations (including the moduli) by suitable integers as in the proof of 3.1.6, we may assume that all n_i are equal, say $n_i = n$. Then we may write $\psi(x,y)$ as $\psi_1(nx,y)$, and replace $\exists x\psi(x,y)$ equivalently in ZG by $\exists x'(\psi_1(x',y) \wedge x' \equiv 0(n))$. This reduces our task to dealing with formulas of the form $\exists x\mu(x,y)$, where $\mu\in$ ZQF and has atomic subformulas of the form

(**)　$x\ \rho_i\ w_i(y)$　$(i\in I)$, where ρ_i is on of the relations $\leq, \geq, \equiv (m_i)$.

Suppose $G \models \mu(a,b)$ for some Z-group G and $a,b_1,...,b_n\in G$. Replace the subscripts $i\in I$ by natural numbers such that $w_i(b)$ are arranged in increasing order, say $w_1(b),...,w_k(b)$. Then the following cases can occur:

(1)　$a = w_h(b)$,

(2)　$w_h(b) < a < w_{h+1}(b)$　for some h,

(3)　$a < w_1(b)$,

(4)　$w_k(b) < a$.

Let m be the least common multiple of all moduli m_i occurring in μ. In the first case, $G \models \mu(w_h(b),b)$. In the last case, 3.3.2 guarantees that $a \equiv w_k(b)+j1(m)$ for some $1 \leq j \leq m$; so $G \models \mu(w_k(b)+j1,b)$. The third case is treated symmetrically. In the second case, we find again by 3.3.2 that $G \models \mu(w_h(b)+j1,b)$　for some $1 \leq j \leq m$.

Thus we have shown that the following equivalence holds in ZG:

$$\exists x\mu(x,y) \Leftrightarrow \bigvee_{i\in I} \bigvee_{\substack{j\in Z \\ |j|\leq m}} \mu(w_i(y)+j1,y).$$

This completes the description of the quantifier elimination procedure for ZG. A straightforward but somewhat tedious calculation yields the desired upper bound on the running time of the procedure (see [Ferrante and Rackoff, 1975, 1979] and [Weispfenning, a]).

The conclusions to be drawn from this result via Lemma 3.1.5 are by now routine (compare the proof of Theorem 3.1.3). Besides the proof of 3.3.3, we obtain:

THEOREM 3.3.5. *The elementary theory of Z-groups is decidable. In fact there is a decision procedure running in space bounded by* $\exp_2(O(\ell(\varphi)))$ *and in time bounded by* $\exp_3(O(\ell(\varphi)))$ *on input* $\varphi\in$ ZSE.

For Theorems 3.3.3 and 3.3.5, it is immaterial whether we consider sentences in ZSE or in oSE. For sentences in SE, 3.3.3 remains valid by Lemma 3.1.1; in 3.3.5 the bounds have to be increased by one exponential (compare the proof of 3.1.3). For the study of algebraic and existential closedness, the choice of the language is, however, crucial. Let us denote by 1FO the set of Z-formulas φ that contain no congruences.

THEOREM 3.3.6. *ZG equals the class of all algebraically closed discrete o-groups and the class of all existentially closed discrete o-groups with respect to formulas in 1FO.*

For the proof we need the following lemma.

LEMMA 3.3.7 [Glass and Pierce, 1980a]. *For every discrete o-group G there is a Z-group H that extends G and has the same smallest positive element as G.*

Proof. Let $\bar{G}$ be the divisible hull of G, $g \in G$, and p prime. An elementary calculation (see [Glass and Pierce, 1980a]) shows that for some $0 \leq i < p$, the subgroup H_i of $\bar{G}$ generated by $G \cup \{(g\text{-}i1)/p\}$ has smallest positive element $1 \in G$. So by a chain construction, there exists an o-group $G \subseteq H \subseteq \bar{G}$ such that H has smallest positive element 1 and $H/\mathbf{Z}1$ is p-divisible for all primes p. So $H/\mathbf{Z}1$ is divisible, and hence by 3.3.2, H is a Z-group.

Proof of 3.3.6. We show that any $\varphi \in$ 1FO is equivalent in ZG to a universal formula $\varphi^* \in$ 1FO: Let $\varphi' \in$ ZQF be equivalent to φ in ZG, and assume that in φ' all negations immediately precede atomic formulas. Replace any negated (unnegated) congruence in φ' as follows:

$\neg w \equiv w'\,(m)$ by $\forall z(mz + w \neq w')$,

$w \equiv w'\,(m)$ by $\forall z(\bigwedge_{0 < i < m} mz + i1 + w \neq w')$.

Then the resulting formula φ^* is universal and equivalent to φ in ZG. From this fact together with Lemma 3.3.7, we may conclude as in the proof of 3.1.5(2) that every Z-group is existentially closed (with respect to 1FO) in the class of discrete o-groups.

Conversely, assume G is an algebraically closed discrete o-group, and let H be a Z-group extending G with smallest positive element $1 \in G$. Let $g \in G$, $1 < n \in \mathbf{N}$ be arbitrary. Then for $\varphi(y) = \bigvee_{0 \leq i < n} \exists x(nx + i1 = y)$, $H \models \varphi(g)$, and so $G \models \varphi(g)$. By 3.3.2, this shows that G is a Z-group.

3.4. Dense Regular Groups

This section is devoted to the elementary properties of o-subgroups of the reals. Recall Theorem 1.2.3 that an o-group G is isomorphic to an o-subgroup of the reals iff G is *Archimedean*, i.e., satisfies the infinitary sentence

(α) $\forall x \forall y (0 < x \wedge 0 < y \Rightarrow \bigvee_{n \in N} nx > y)$.

Thus the Archimedean property completely characterized the class $S(R)$ of o-subgroups of the reals from an algebraic viewpoint.

On the other hand, this property is not elementary as it stands. We are going to show that it is in fact not even equivalent to a set of elementary sentences; so the class of Archimedean o-groups cannot be axiomatized by a set of sentences. This gives us an appropriate occasion to introduce another fundamental result of elementary model theory: the compactness theorem.

THEOREM 3.4.1. Compactness Theorem. *Let Φ be a set of elementary sentences (involving possibly extra constants besides 0). If every finite subset Φ' of Φ holds in some ℓ-group $G = G(\Phi')$, then Φ holds in some ℓ-group H.*

The validity of this theorem is of course not restricted to ℓ-groups. An 'algebraic' proof can be obtained e.g. by employing ultraproducts of ℓ-groups (see [Eklof, 1977]).

Let us now assume for a contradiction, that (α) is equivalent to a set Φ of elementary sentences in the class of ℓ-groups. Let ψ be the sentence $\forall x(x > 0 \vee x < 0)$, and let c,d be new constants. Then

$\Phi \cup \{\psi\} \cup \{0 < c,\ 0 < d,\ nc \le d \mid n \in N\}$

is unsatisfiable in any ℓ-group G. So by 3.4.1, there exists a positive integer N such that

$\Phi \cup \{\psi\} \cup \{0 < c,\ 0 < d,\ nc \le d \mid n \le N\}$

is not satisfiable in any ℓ-group G, which is nonsense, since it holds in the o-group of reals with $c = 1$, $d = N1$.

So we have to find an elementary replacement for the Archimedean property. It results from the following easy and well-known fact.

PROPOSITION 3.4.2. *Let G be an o-subgroup of the reals.*

(i) If G is discrete, then G is isomorphic to **Z.**

(ii) If G is dense (in itself), then G is dense in **R,** *i.e., G intersects any non-empty open interval in* **R.**

As far as discrete subgroups of the reals are concerned, our task is thus completed by section 3.3: An o-group G is elementarily equivalent to a discrete subgroups of the reals iff G is a Z-group. Let now G be a dense subgroup of the reals. Then for any positive integer n, nG is also a dense subgroup of the reals and so by 3.4.2, nG is dense in **R** and hence dense in G. This suggests the following definition due to Robinson and Zakon [1960]: An o-group is *dense n-regular* for some positive integer n, if $G \models \forall x \forall y \exists z (x < y \Rightarrow x < nz < y)$, or equivalently, $G \models \rho_n$, where ρ_n is the sentence $\forall x \forall y \exists z (0 < x \Rightarrow 0 < z < y \wedge z \equiv x \ (n))$. G is *dense regular* if it is dense n-regular for all positive integers n. So any dense Archimedean o-group is dense regular and $\{\rho_n \mid 0 < n \in \mathbf{N}\} \cup \{\psi\}$ is a set of axioms for dense regular o-groups. By our discussion above, not every dense regular o-group is Archimedean. Explicit counterexamples are e.g. lexicographic products of the form $G = D \overset{\leftrightarrow}{\times} H$, where D is a non-trivial divisible o-group and H is a dense Archimedean o-group. This follows from the following obvious characterization of dense regular o-groups.

PROPOSITION 3.4.3. *Let G be a dense o-group. Then G is dense regular iff for all non-trivial convex subgroups G' of G, G/G' is divisible.*

Recall from 3.3.2 that the corresponding property is true for Z-groups as well. This suggests the following definition: An o-group G is *regular* if for all non-trivial convex subgroups G' of G, G/G' is divisible. Then a regular group is either dense regular or a Z-group. Our goal is now to show that regularity is the correct elementary substitute for the Archimedean property.

THEOREM 3.4.4. *Let G be an o-group. Then G is elementarily equivalent to an o-subgroup of the reals iff G is regular.*

COROLLARY 3.4.5. *Let Σ be the elementary theory of S(**R**), i.e., the set of all elementary sentences true in all o-subgroups of the reals, and let G be an ℓ-group. Then Σ holds in G iff G is regular.*

To prove the theorem, it suffices by the remarks above to show that any dense regular group G is elementarily equivalent to an o-subgroup G' of the reals. In order to construct such a group G', we will have to determine elementary invariants for dense regular groups:

For any prime number p, G/pG is an elementary p-group, and hence a vector space over the prime field Z/pZ; we let $\dim(p,G)$ be the dimension of this vector space. Then the cardinal numbers $\dim(p,G)$ determine the structure of the groups G/nG for any positive integer n:

LEMMA 3.4.6. *Let G be a torsion-free group and let m,n be positive integers and p a prime.*

(i) *If $(n,m) = 1$, then G/nmG is isomorphic to $G/nG \times G/mG$.*

(ii) *If C is a set of elements of G that are pairwise incongruent modulo pG such that $\{c+pG \mid c \in C\}$ is a basis of G/pG, then G/p^mG is a free Z/p^mZ-module and has $\{c+p^mG \mid c \in C\}$ as a basis.*

Proof. (i) The homomorphism $f : G \to G/nG \times G/mG$, $f(g) = (g+nG, g+mG)$ has kernel G/nmG and is surjective by the Chinese remainder theorem.

(ii) Induction on m shows:

(1) For $c_1,...,c_n \in C$ and integers $k_1,...,k_n$, $k_1c_1+...+k_nc_n \equiv 0$ (p^m) implies $p^m \mid k_j$ for $1 \leq j \leq n$.

(2) For any g in G there exist $k_1,...,k_n$ in Z and $c_1,...,c_n$ in C such that $g \equiv k_1c_1+...+k_nc_n$ (p^m).

A routine application of the compactness theorem shows that the cardinalities $\dim(p,G)$ are not elementary invariants unless they are finite. Hence we consider instead the modified numbers $\beta(p,G) = \min(\dim(p,G),\omega)$ that do not distinguish between infinite cardinalities. The $\beta(p,G)$ are known as the *Szmielew invariants* of G; they were introduced in [Szmielew, 1955] together with other invariants in order to classify arbitrary (Abelian) groups up to elementary equivalence. Since the property $\beta(p,G) \geq n$ can be expressed by an elementary sentence, these numbers are indeed elementary invariants. Theorem 3.4.4 is now an immediate consequence of the following facts:

THEOREM 3.4.7 [Zakon, 1961]. *Let Prim be the set of primes and let $B : Prim \to N \cup \{\omega\}$ be an arbitrary map. Then there exists an o-subgroup G of the reals with $\beta(p,G) = B(p)$ for all primes p.*

THEOREM 3.4.8 [Robinson and Zakon, 1960]. *Let G,H be non-trivial dense regular o-groups. Then G and H are elementarily equivalent iff $\beta(p,G) = \beta(p,H)$ for all primes p.*

Proof of 3.4.7. For any prime p, let $Z(p) = \{a/b \in \mathbf{Q} \mid (b,p) = 1\}$ be the group of rational p-adic integers. Let $\{r(p,m) \mid p$ prime, m a positive integer$\}$ be a system of real numbers linearly independent over $\mathbf{Q}$. Let $G(p) = Z(p) \cdot r(p,1) + \ldots Z(p) \cdot r(p,B(p))$ (i.e., $G(p) = \{0\}$ for $B(p) = 0$), and let G be the sum over all primes of the o-groups $G(p)$. Then G satisfies the theorem.

To prove 3.4.8, we establish a partial quantifier elimination procedure for the class DR of dense regular o-groups. As in the case of Z-groups, we admit congruences as atomic formulas. Accordingly, we define *c-formulas* as formulas obtained from equations $w = w'$, inequalities $w \leq w'$ and congruences $w \equiv w' \ (m)$ between group words by means of the Boolean operations and quantification. We call a c-formula φ *restricted* if no variable bound in φ by a quantifier occurs in an equation of inequality of φ. So a restricted sentence is equivalent in any o-group to a sentence containing no equations and inequalities.

PROPOSITION 3.4.9. *There is a procedure assigning to any c-formula φ a restricted c-formula φ' such that φ and φ' are equivalent in DR.*

Proof. φ' is defined by induction of the number of quantifiers $\exists x$, $\forall x$ in φ such that the variable x occurs in some equation or inequality of φ. So it suffices to treat the case that φ is of the form $\exists x \psi(x,y_1,\ldots,y_r)$, where ψ is restricted. As in the proof of 3.1.6, we may assume that x occurs in ψ only in the form $nx \leq w_i$, $nx \geq w_i$, $nx \equiv w_k \ (m_k)$ for $i \in I$, $k \in K$, where n is a fixed positive integer. Then $\exists x \psi$ can be written as $\exists x \psi^\sim(nx,y_1,\ldots,y_r)$, which is equivalent to $\exists x(\psi^\sim(x,y_1,\ldots,y_r) \wedge x \equiv 0 \ (n))$. So we may assume from now on that x occurs in ψ only in the form $x \leq w_i$, $x \geq w_i$, $x \equiv w_k \ (m_k)$ for $i \in I, k \in K$. The idea now is to simulate the truth values of the inequalities $x \leq w_i$, $x \geq w_i$ for all possible values of x by inequalities not involving x, and thus to eliminate the occurrence of x in inequalities. Accordingly, we let ψ' be the restricted formula

$$\psi^+ \vee \psi^- \vee \bigvee_{h,k \in I} \exists x \psi_{h,k}$$

where ψ^+ results from ψ by replacing $x \leq w_i$ $(x \geq w_i)$ by $0 \neq 0$ $(0 = 0)$, ψ^- results from ψ by replacing $x \leq w_i$ $(x \geq w_i)$ by $0 = 0$ $(0 \neq 0)$, and $\psi_{h,k}$ results from ψ by replacing $x \leq w_i$ $(x \geq w_i)$ by $w_h + w_k \leq 2w_i$ $(w_h + w_k = 2w_i)$. The proof of the equivalence between φ and φ' proceeds now as for 3.1.6; the dense regularity is used crucially in order to conclude that $x < y \wedge \exists z \rho$ is equivalent in DR to $\exists z(x < z < y \wedge \rho)$, whenever $\exists z \rho$ is restricted. This completes the proof of 3.4.9.

Proof of 3.4.8. If G and H are elementarily equivalent, then $\beta(p,G) = \beta(p,H)$, since these numbers are elementary invariants. Conversely, suppose G and H have the same Szmielew invariants. By the Löwenheim-Skolem theorem (see [Keisler, 1977]), there exist countable o-groups G' and H' elementarily equivalent to G and H, respectively. Then $\dim(p,G') = \beta(p,G') = \beta(p,H') = \dim(p,H')$, and so by 3.4.6, G'/mG' and H'/mH' are isomorphic for any positive integer m. Let now φ be a c-sentence that holds in G and let φ' be the restricted c-sentence associated with φ by 3.4.9. Then φ' holds in G and hence in G'. Let now m be the lcm of all moduli m_k occurring in φ'. Then the truth value of φ' depends only on the structure of G/mG, and so φ' holds in H'. Hence φ holds in H.

In contrast to the situation for divisible o-groups and Z-groups, the algorithmic properties of a dense regular group G depend on the map $p \mapsto \beta(p,G)$. As may be expected, the theory of G, Th(G), is decidable, roughly speaking, if the Szmielew invariants of G are computable. Given positive integers p and n with p prime and $p \leq n$, let $\gamma(p,n,G) = \min(3n,\beta(p,G))$. Then the following holds:

THEOREM 3.4.10 [Weispfenning, 1985]. *(i)* *If Th(G) is decidable, then the map $(p,n) \mapsto \gamma(p,n,G)$ is computable.*

(ii) *If the map $(p,n) \mapsto \gamma(p,n,G)$ is computable in time $t(p,n)$, then Th(G) is decidable in time $\exp_2 O(n) \cdot \max\{t(p,n) \mid p \leq n\}$.*

Proof. (i) It suffices to decide for every prime $p \leq n$ and $k \leq 3n$, whether $\beta(p,G) \geq k$.

(ii) (Sketch) Let φ' be the restricted sentence associated with a c-sentence φ in 3.4.9. Then φ' can be computed from φ in time double exponential in the length of φ; moreover, the number s of congruences is at most doubled in this process. We may assume that φ' contains no equation or inequality and has length $\leq n$. Put $k = 3s$. Then by the Chinese remainder theorem there exist $c_1,...,c_k$ in G such that for every prime $p \leq n$, $c_1,...,c_h$ are independent modulo pG for $h \leq \beta(p,G)$ and $c_{h+1},...,c_k \equiv 0$ (p). Rewrite φ' such that all congruences in φ' have prime power moduli and introduce $c_1,...,c_k$ as new constants into c-formulas. Then by 3.4.6, φ', and hence φ, can be decided in Th(G) by replacing quantifiers $\exists x$ ($\forall x$) in φ by disjunctions (conjunctions) over all linear combinations $r_1 c_1 + ... + r_k c_k$, where r_i are nonnegative integers of binary length $\leq n$. A more detailed analysis of this procedure yields the desired time bound.

Essentially the same procedure yields the following corollary.

COROLLARY 3.4.11. *The theory Th(DR) of all non-trivial dense regular o-groups is decidable in double exponential time.*

3.5. General o-Groups and their Convex Subgroups

In this section we present a very rough sketch of the elementary classification of o-groups in general, a decision procedure for the class of o-groups, and the extension of these results to o-groups with their distinguished chain of convex subgroups. Almost all the results are due to Y. Gurevich [1965,1977] and to P. Schmitt [1982,1984]. The amount of material and its technical difficulty admits only a vague outline of the results and ideas involved.

In the case of regular groups, we have obtained elementary invariants and a decision procedure by means of effective quantifier elimination for Z-formulas and c-formulas. The reason for the success was essentially due to the fact that in a regular o-group G, the congruence classes $g+nG$ ($g \in G$, $n \in \mathbf{N}$) are very evenly distributed in G with respect to the ordering $<$ of G. This changes radically for non-regular groups: Consider e.g. the lexicographic sum $G = \mathbf{Z}(2) \overrightarrow{\times} \mathbf{Z}(3)$, where $\mathbf{Z}(p)$ is the o-group of rational p-adic integers. Then $(1,1)+3G$ is dense in G, whereas any two elements of $(1,1)+2G$ are "far apart". So the basic problem is to find (preferably elementarily defined) concepts in the class OG of o-groups, that describe the distribution of congruence classes $g+nG$ with respect to the ordering $<$ as closely as possible by elementary means.

In a first step, let us consider o-groups G that are composed of a fixed finite number of regular groups; in other words, we assume that G has a distinguished chain $\{0\} = S_0 \subset S_1 \subset \dots \subset S_K = G$ of convex subgroups such that S_{k+1}/S_k is a regular o-group for $0 \leq k < K$. Let Σ_K be the class of all these structures $(G, S_1, \dots, S_{k-1})$. One may expect that Σ_K is decidable and that the sequence of elementary invariants for the regular o-groups S_{k+1}/S_k ($0 \leq k < K$) form elementary invariants for Σ_K. Both expectations turn out to be true. A proof has been carried out in [Weispfenning, 1981] using quantifier elimination: The class of formulas is extended by unary predicates S_k for the distinguished convex subgroups, by the relations $x \overset{k}{\underset{m}{\equiv}} y \Leftrightarrow \exists u \exists z (x - y - u = mz$ $\wedge\ S_k(u))$ representing the congruences $\underset{m}{\equiv}$ in G / S_k, and by constants representing the smallest positive element of S_{k+1}/S_k, if this group happens to be discrete. Then the quantifier elimination problem can be reduced to the corresponding problem for the factors S_{k+1}/S_k, where it is solved by regularity.

This approach fails for infinite towers of regular groups. In fact, the full lexicographic product $G = \mathbf{Z}^N$ and the lexicographic sum $\mathbf{Z}^{(N)}$ are not elementarily equivalent (see [Schmitt, 1984; example p. 397] and [Schmitt, 1982; Theorem 4.2). On a more basic level, this approach is inappropriate for the elementary classification of o-groups, since the convex subgroups S_k are in general not elementary definable in G.

So we are thrown back to the basic question, how to describe the distribution of congruence classes $g+nG$ in an o-group by elementary means. Clearly, convex subgroups of G should provide a good tool; we should, however, take care to use only such convex subgroups that can be described by elementary formulas (depending possibly of finitely many parameters from G). Thus obvious candidates like $G(g)$ (the smallest convex subgroup of G containing the element g) for $g \in G$ have to be discarded. (The compactness theorem 3.4.1 can be used to show that these groups are not elementarily definable in terms of the parameter g.) Our experience from Archimedean o-groups in sections 3.3 and 3.4 has taught us to replace Archimedean o-groups like the "Archimedean jump" $G(g)/V(g)$ by regular o-groups, or even more cautiously (compare [Schmitt, 1984; page 392]) by n-regular o-groups for fixed $n \in \mathbf{N}$. Accordingly, we replace $V(g)$ by $A_n(g)$, the smallest convex subgroup of G such that $G(g)/A_n(g)$ is n-regular. This definition still refers to the group $G(g)$, which is not elementarily definable; nevertheless $A_n(g)$ is definable by the following formula:

$$x \in A_n(g) \text{ iff } \forall z \forall u (0 \le z < ng \ \wedge \ 0 < z+nu \Rightarrow |nx| < z+nu).$$

The reason is simply that any congruence class $h+ng$ with $h \in G(g)$ contains an element h' with $0 \le h' < ng$.

Another kind of elementarily definable convex subgroup comes up in connection with the question, how close a fixed congruence class $g+nG$ approaches 0: Put $F_n(g) = \{h \in G \mid \forall z (z \underset{n}{\equiv} g \Rightarrow |nh| < |z|\}.$

Then $F_n(n)$ is empty or a convex subgroup of G defined elementarily in terms of g. The groups $F_n(g)$ $(g \in G)$ are called the n-fundaments of G. The n-regularity of G can now be expressed by the condition that $F_n(g) \subseteq \{0\}$ for all

$g \in G$. The n-fundaments of G readily introduce other elementarily defined subgroups of G:

$$E_n(g) = \{h \in G \mid F_n(h) \subseteq F_n(g)\},$$
$$\overset{*}{E_n}(g) = \{h \in G \mid F_n(h) \subset F_n(g)\},$$
$$\overset{*}{F_n}(g) = E_n(g)/\overset{*}{E_n}(g).$$

$\overset{*}{F_n}(g)$ is of exponent n since $F_n(ng) = \varnothing$. The elementary properties of (abstract) groups H of bounded exponent are classified according to W. Szmielew [1955] by the elementary invariants $\alpha(p,k)(H) = \dim_{Z_p}(p^{k-1}H[p]/p^kH[p])$ (without any distinction between infinite dimensions of different cardinalities).

For fixed $n \geq 2$, the convex subgroups considered so far form a chain (linearly ordered set) $Sp_n(G) = \{A_n(g) \mid g \in G\} \cup \{F_n(g) \mid g \in G\}$. The additional elementary information attached to these convex subgroups can be coded by countably many unary predicates (colors) on $Sp_n(G)$: For $C \in Sp_n(G)$,

A(C)   iff $C = A_n(g)$ for some $g \in G$,

F(C)   iff $C = F_n(g)$ for some $g \in G$,

D(C)   iff G/C is discrete,

$\alpha(p,k,m)(C)$ iff for some $g \in G$, $C = F_n(g)$ and $\alpha(p,k)(\overset{*}{F_n}(g)) \geq m$.

The chain $Sp_n(G)$ furnished with these "colors" forms a "colored chain", called the *n-spine* of G. Since n-spines are defined from G in an elementary way, the following is obvious:

PROPOSITION 3.5.1. *For any two o-groups G,H, $G \equiv H$ implies $Sp_n(G) \equiv Sp_n(H)$ for all $n \geq 2$.*

Gurevich's deep theorem states that the converse is also true:

THEOREM 3.5.2 [Gurevich, 1965]. *If $Sp_n(G) \equiv Sp_n(H)$ for all $n \geq 2$, then $G \equiv H$.*

We sketch the method of proof employed by P. Schmitt [1982]: The idea is to eliminate quantifiers ranging over group elements in favor of quantifiers ranging over n-spines. Two major problems arise in this approach and have to be overcome with considerable effort and ingenuity:

(1) In order to eliminate quantifiers ranging over group elements one by one, one has to prove a much stronger result ("the great transfer theorem"; see [Schmitt, 1984; Theorem 1.7]) translating formulas with free

variables $x_1,...,x_k$ ranging over group elements into formulas containing only "simple relations" between the x_i's and elementary relations in some n-spine between certain convex subgroups (F_n's and A_n's) computed from $x_1,...,x_k$.

(2) In this translation, elementary properties of n-spines for different natural numbers n may come up; they have to be reduced to elementary properties of a single m-spine, where m is a common multiple of the n's in question.

How useful are n-spines $\{Sp_n(G) \mid 2 \leq n\}$ as elementary invariants for an o-group G? Without any doubt, these colored chains are simpler than the o-groups they classify. Nevertheless, their structure is still highly complex. In fact, an intrinsic characterization of all colored chains $(S, <, ...)$ such that $(S, <, ...)$ is isomorphic to some $Sp_n(G)$ for given $n \geq 2$ is still lacking (see [Schmitt, 1982; Chapter 5, page 77]). On the other hand, these elementary invariants lead to the second major result on the elementary properties of o-groups in general:

THEOREM 3.5.3 [Gurevich, 1965]. *The elementary theory of all o-groups is primitive recursively decidable.*

Proof sketch. A logical compactness argument (see 3.4.1) applied to Theorem 3.5.2 shows that any o-group sentence φ is equivalent in the class OG of all o-groups to an elementary sentence φ^* about some n-spine for a certain $n = n(\varphi)$. A more detailed analysis following [Gurevich, 1977] reveals that φ^* and $n(\varphi)$ can be obtained from φ by a primitive recursive procedure. So in order to decide the validity of φ in OG_0 , it suffices to decide the validity of φ^* in the class $\Sigma_n = \{Sp_n(G) \mid G \in OG_0\}$. This seems to lead straight back to the unsolved problem mentioned above. The way out is as follows: Whereas the class Σ_n may not be axiomatizable, its elementary theory, $Th(\Sigma_n)$, turns out to be primitive recursively axiomatizable (see Schmitt, 1982; Chapter 5]). Moreover, if we "forget" all but finitely many colors on $Sp_n(G)$, the elementary properties of $\{Sp_n \mid G \in OG_0\}$ can be axiomatized by a single axiom, say α. Now let Σ be the class of all colored chains with the fixed number of colors occurring in φ^*. Applying a device of Gurevich's (see [Gurevich, 1977; Section 15]), the theory of Σ can be reduced in a primitive recursive way to the theory of ordinary (uncolored) chains. The latter was shown to be primitive recursively decidable in [Ehrenfeucht, 1959] and [Lauchli and Leonard, 1966]. As a consequence, the validity of φ in OG_0 can now be decided by the validity of the sentence $\alpha \Rightarrow \varphi^*$ in Σ.

We have seen so far that *certain* convex subgroups of an o-group G play a decisive role in the elementary classification of o-groups. In the rest of this section, we indicate how this technique can be extended to include elementary properties of *all* convex subgroups of G.

The set $C(G)$ of all convex subgroups of G forms a chain under the inclusion relation $\subseteq$. Accordingly, elementary properties of G together with $C(G)$ are expressed formally as follows: We introduce a new sort of variable, denoted by capital letters $X,Y,Z,...,$ intended to range over $C(G)$. In addition to atomic o-group formulas we admit as new atomic formulas equations $X = Y$ and inequalities $X < Y$ as well as element-relations $a \in X$, where a is a group word. Formulas in the extended sense are obtained from these atomic formulas by means of the logical operators $\wedge$, $\vee$, $\neg$, and quantification $\exists X, \forall X$ over convex subgroups of G. We refer to these formulas as *second-order o-group formulas* or *o^*-formulas* for short.

Similarly as for the n-spines $\mathrm{Sp}(G)$, the chain $C(G)$ is now furnished with countably many "colors": For $X \in C(G)$,

$\quad$ $D(X)$ $\qquad$ iff G/X is discrete,

$\quad$ $\alpha(p,k,m)(X)$ $\quad$ iff $\alpha(p,k)(F^*_{p^k}(X)) \geq m$.

Here the $\alpha(p,k)$ are the Szmielew invariants defined above, and for $n \in \mathbf{N}$, $F^*_n(X)$ is a quotient group obtained similarly to $F^*(g)$ above: Let $E_n(X) = \{h \in G \mid F_n(h) \subseteq X\}$, $E_n(X) = \{h \in G \mid F_n(h) \subset X\}$; then $F^*_n(X) = E_n(X)/E^*_n(X)$. The technique of replacing quantifiers ranging over group elements by quantifiers ranging over convex subgroups sketched above in the proof of Theorem 3.5.2 can be extended to o^*-formulas (see [Gurevich, 1977; Theorem 7.1]). As a result, the elementary properties of an o-group G together with $C(G)$ are completely determined by the elementary properties of the colored chain $(C(G), <, ...)$:

THEOREM [Gurevich, 1977; theorem 7.2]. *There is a primitive recursive procedure assigning to any o^*-sentence φ a sentence φ^* in the language of colored chains such that for any o-group G, $(G,C(G)) \vDash \varphi$ iff $C(G) \vDash \varphi^*$.*

We illustrate the method of proof for the special case of dense regular o-groups; similar arguments are valid for Z-groups. In both cases, no colors are required; so in the resulting sentences φ^*, $<$ and $=$ are the only relations.

Let a *c^*-formula* be an o^*-formula in which congruences are admitted as atomic relations. By induction on the number of quantifiers over group elements (first-order quantifiers), it suffices to eliminate the quantifier $\exists x$ in c^*-formulas of the form $\exists x \varphi$, where φ contains no first-order quantifier; moreover, we may assume that in φ all quantifiers over convex subgroups (second-order quantifiers) have been moved to the front, and that x occurs in φ only in the form $nx \; r_i \; a_i$ or $nx + b_j \in X_j$, where each r_i is one of the relations $=, <, >, \underset{m_i}{\equiv}$, and a_i, b_j are group words not containing the variable x, and $n \in N$ is fixed. Adjoining a large list of case distinctions about the order relation between the a_i, b_j and X_j, and adjoining furthermore a congruence $\underset{n}{\equiv}$, we can reduce $\exists x \varphi$ equivalently in DR to a formula $\exists x \varphi'$ as above, where x occurs only in the form $x \; r_i \; a_i'$ and $x \in Y_j$ with r_i as above. We assume now that negations occur in φ' only in the form $b \notin Y$. Construing X as $V(x)$ and X' as $G(x)$, we find that $\exists x \varphi'$ is equivalent in DR to $\exists X \exists X'(X < X' \wedge \exists x(x \notin X \wedge x \in X' \wedge \varphi''))$, where φ'' results from φ' by replacing $x \in Y_j$ by $X' \leq Y_j$ and $x \notin Y_j$ by $Y_j \leq X'$. So x occurs in φ'' only in the form $x \; r_i \; a_i'$. Adjoining enough case distinctions about the element relation between a_i', X, and X', the quantifier $\exists x$ can now be eliminated essentially as in section 3.4. (This is the only step where dense regularity is used.)

Concerning the theory of all colored chains occurring as $C(G)$ for some o-group G, facts similar to those described for n-spines hold (see [Gurevich, 1977; Part 3]) and lead to a corresponding decidability result:

THEOREM 3.5.5 [Gurevich, 1977]. *The theory of all pairs $(G, C(G))$, where G is an arbitrary o-group, is primitive recursively decidable. In fact this remains true when, in addition, quantification over finite subsets of $C(G)$ is admitted in formulas.*

3.6. Existentially Closed ℓ-Groups and ℓ-Groups with Projector

Encouraged by the easy success in the case of o-groups, we begin our model theoretic study of ℓ-groups with algebraically and existentially closed ℓ-groups. Since any ℓ-group is representable as an ℓ-subgroup of a direct product of divisible o-groups, any ℓ-group can be extended to a divisible ℓ-group. So we may conclude as for o-groups:

LEMMA 3.6.1 [Glass and Pierce, 1980a]. *Any algebraically closed ℓ-group is divisible.*

Conversely, the proof of 3.1.4 shows that divisibility is also sufficient for an ℓ-group G to be algebraically closed *qua* (torsion-free) group. So the question remains, which properties characterize an algebraically and existentially closed ℓ-group G *qua* lattice. Since G can be extended to a direct product of o-groups, it is not unreasonable to expect that G has a "sufficient" supply of pairwise disjoint (orthogonal) elements. In order to phrase variants of this property concisely, we recall the following definitions (see Section 1.1): Let $a, b \in G$. Then a and b are *orthogonal* ($a \perp b$) if $a \wedge b = 0$. For $a >$ 0, $G(a) = \{g \in G \mid \exists n \in N$ with $-na < g < na\}$ denotes the smallest convex ℓ-subgroup of G containing a. We say a is *basic* if the interval $[0,a] = \{g \in G \mid 0 \leq g \leq a\}$ is linearly ordered; this is equivalent to saying that $[0,a]$ does not contain two positive orthogonal elements. The element a is a *weak unit* if there is no positive b in G orthogonal to a. If $0 \leq a$ and $b \perp c$, then we say a *splits over* b,c if there exist b', c' in G such that $b' \perp c'$, $b' + c' = a$ (equivalently, $b' \vee c' = a$), $b' \perp c$, and $b \perp c'$. G has the *splitting property* (SP) if for all $0 \leq a$ and $b \perp c$ in G, a splits over b,c. G has the *disjunction property* (DP) if for all positive a, b in G with $a \notin G(b)$ there exists $c \in G$ such that $0 < c \leq a$ and $c \perp b$. (This is equivalent to the Wallman disjunction property in the sense of [Birkhoff, 1967] for the lattice of principal convex ℓ-subgroups of G.)

LEMMA 3.6.2. *Let G be an ℓ-group.*

(1) If G is algebraically closed, then G has the splitting property.

(2) If G is existentially closed, then G contains no basic elements and no weak unit, and G has the disjunction property.

Proof. Notice that any direct product H of o-groups extending G has the splitting property and the disjunction property. This proves (1) and the second half of (2). If $0 < a \in G$, we may assume (by duplicating some factor of H) that a is not basic in H. So a is not basic in G. By extending H to $H' = H \boxplus Q$ we see that a cannot be a weak unit in G.

It turns out that these algebraic properties suffice to characterize algebraically and existentially closed ℓ-groups:

THEOREM 3.6.3. *Let G be an ℓ-group.*

(1) G is algebraically closed iff G is divisible and has the splitting property.

(2) G is existentially closed iff G is non-trivial, divisible, has no basic elements and no weak unit, and has the splitting property and the disjunction property.

Notice that all the properties mentioned in 3.6.3 except the disjunction property are elementary, i.e., can be expressed by sets of sentences. In particular, the class AC of algebraically closed ℓ-groups is *elementary*, i.e., is axiomatized by a set of sentences. For the class EC of existentially closed ℓ-groups this in not the case:

PROPOSITION 3.6.4. *EC cannot be axiomatized by a set of sentences.*

Proof. Assume for a contradiction that Φ axiomatizes EC. Let c,d be new constants and let Ψ be the following set of sentences:

$$\Phi \cup \{d > 0, \ c > nd \mid n \in \mathbf{N}\} \cup \{\neg \exists x(0 < x \le c \wedge x \bot d\}.$$

Then by 3.6.2(2) there is no ℓ-group satisfying Ψ. So by the Compactness Theorem 3.4.1, there exists a positive integer m such that the set

$$\Psi' = \Phi \cup \{d > 0, \ c > md, \ \neg \exists x(0 < x \le c \wedge x \bot d\}$$

holds in no ℓ-group G. On the other hand, Ψ' holds in any $G \in$ EC when d is interpreted as an arbitrary positive element and $c = (m+1)d$, a contradiction.

The essential fact underlying this proof is that in EC the infinitary property $x \in G(y)$ is equivalent to the elementary formula

$$\alpha(x,y): \qquad \neg \exists v(0 < v \le |x| \wedge v \bot |y|).$$

As a consequence (compare [Glass and Pierce, 1980a]), the Archimedean property can be expressed in EC by the elementary sentence

$$\forall x \forall y(x > 0 \Rightarrow \exists z(\alpha(z,x) \wedge z \ge y)).$$

Theorem 3.6.3 provides a number of "concrete" examples of algebraically and existentially closed ℓ-groups:

3.6.5 [Glass and Pierce, 1980a]. Any divisible o-group (including the trivial one) and any direct sum and direct product of divisible o-groups is algebraically closed.

3.6.6 [Glass and Pierce, 1980a; Saracino and Wood, 1983]. Let A be a non-trivial o-group furnished with the discrete topology, let X be a Hausdorff space with a basis of compact clopen sets, and let $\bar{C}(X,A)$ be the ℓ-group of all continuous functions $f : X \to A$ with compact support. Then $\bar{C}(X,A)$ is algebraically closed. Moreover, if in addition X is perfect and not compact, then $\bar{C}(X,A)$ is existentially closed. It was shown in [Saracino and Wood, 1983] that among these examples there is a unique ℓ-group $G =$

$\bar{C}(X,Q)$ prime with respect to EC (i.e., $G \in$ EC and G is embeddable into any $H \in$ EC).

A rich source of examples are hyperarchimedean ℓ-groups: Recall that G is *hyperarchimedean* if every homomorphic image of G is Archimedean. By [Bigard, Keimel and Wolfenstein, 1977; Theorem 14.1.2], any hyper-archimedean ℓ-group is isomorphic to an ℓ-subgroup G of a direct power $\mathbf{R}^I$ of the o-group of reals with the following property:

3.6.7. For all $0 \le a,b$ in G there exists a positive integer n such that $[\![a \le nb]\!] \supseteq [\![b \ne 0]\!]$. (Here and in the following, the bracket notation is a convenient way of expressing local properties in ℓ-subgroups G of a direct product $\prod_{i \in I} G_i$ of o-groups G_i : Whenever $\varphi(x_1,...,x_n)$ is a formula and $a_1,...,a_n \in G$, then $[\![\varphi(a_1,...,a_n)]\!] = \{i \in I \mid G_i \models \varphi(a_1(i),...,a_n(i))\}$; in particular $[\![b \ne 0]\!] = \text{support}(b)$.)

Using 3.6.7, it is not difficult to verify:

LEMMA 3.6.8. *Any hyperarchimedean ℓ-group has the splitting property and the disjunction property.*

As a consequence, 3.6.3 yields:

THEOREM 3.6.9. *Let G be a hyperarchimedean ℓ-group.*

(1) $G \in$ AC iff G is divisible.

(2) [Saracino and Wood, 1983] $G \in$ EC iff G is nontrivial, divisible, without basic elements and weak units.

Based on 3.6.6, it was shown in [Glass and Pierce, 1980a] that there are $2^{\aleph_0}$ pairwise nonisomorphic countable hyperarchimedean ℓ-groups. A classification of existentially closed ℓ-groups by elementary equivalence is, however, not in sight. All we know at present is that there are infinitely many pairwise inequivalent existentially closed ℓ-groups. The crux of the argument is to find two inequivalent ℓ-groups $G, H \in$ EC. We are going to distinguish between G and H by an elementary property strengthening the splitting property:

Let $a,b,c \in G$. Then c is the *projector* of a and b (notation: $c = \mathrm{pr}(a,b)$), if $|c| \perp |b|$ and for all d in G, $|d| \perp |b|$ implies $|d| \perp |a-c|$. G is *projectable* if every pair of elements a,b of G has a (necessarily unique) projector (cf Section 1.1). So any projectable ℓ-group has the splitting property. Again the following is easy to verify:

LEMMA 3.6.10. *Any hyperarchimedean ℓ-group is projectable.*

So it suffices to exhibit a nonprojectable ℓ-group $G \in$ EC: Let X be a noncompact, perfect Hausdorff space with a basis of compact clopen sets. Let $A = \mathbf{Q} \times \mathbf{Q}$ be the lexicographic product of two copies of the o-group of rationals and consider $H = A^{(X \times X)}$. If $x \in X$, and Y, Z are compact clopen subsets of X, then $\{x\} \times X$ is a *line* and $Y \times Z$ is a *rectangle*. Let B (C) be the set of characteristic functions of lines (of rectangles), whose unique non-zero value is in $\{0\} \times \mathbf{Q}$ (in $\mathbf{Q} \times \{0\}$), and let G be the ℓ-subgroup of H generated by $B \cup C$. Then 3.6.3(2) shows that G is existentially closed. On the other hand, if $l \neq \varnothing$ is a line and $r \neq \varnothing$ is a rectangle such that $l \subseteq r$ and a (b) is the characteristic function of r (l) with non-zero value $(1,0)$ $((0,1))$, then $\mathrm{pr}(a,b)$ does not exist in G.

The proof of Theorem 3.6.3 is too long too be included here in detail. We only outline a proof of the first half and give some hints for the second half. Two tools are required: *patching* and *instantiation*.

Let G be an ℓ-subgroup of the direct product $H = \prod_{i \in I} H_i$ of o-groups H_i; let $c, a_1, \ldots, a_n$, $0 \leq b_1, \ldots, b_n$ be in G. Then we say c *patches* a *over* b if

$$\bigcup_{1 \leq k \leq n} (\llbracket c = a_k \rrbracket \cap \llbracket b_k = 0 \rrbracket) \supseteq \llbracket \bigwedge_{1 \leq k \leq n} b_k = 0 \rrbracket.$$

G has the *patchwork property* (PP) if for all $a_1, \ldots, a_n$, $0 \leq b_1, \ldots, b_n$ in G there exists c in G that patches a over b.

LEMMA 3.6.11. *Suppose $G \subseteq \prod_{i \in I} H_i$. If G has the splitting property, then G has the patchwork property.*

Proof. By induction on n. We sketch the case $n = 2$. First one generalizes SP to the case where a is arbitrary, by applying SP to a^+ and a^- separately. Let now a_1, a_2, $0 \leq b_1, b_2$ be in G. Put $b_i' = b_i - (b_1 \wedge b_2)$; then $b_1' \perp b_2'$. Choose c_i, d_i splitting a_i over b_1', b_2', and put $c = c_1 + d_2$.

INSTANTIATION LEMMA 3.6.12. *Let $\varphi(x_1,...,x_n,y_1,...,y_m)$ be a quantifier-free o-group formula (without negations). Then one can construct finitely many conjunctions $\psi_k(y)$, $k \in K$, of atomic o-group formulas and their negations (of atomic o-group formulas only) and corresponding rational linear combinations of $y_1,...,y_m$, $t_{1k}(y),...,t_{nk}(y)$ such that the following holds in all divisible o-groups:*

$$(y_1 \neq 0 \wedge \varphi(x,y)) \Rightarrow \bigvee_{k \in K}(\varphi(t_{1k}(y),...,y_{nk}(y),y) \wedge \psi_k(y)).$$

Proof. By induction on n; the essential case is $n = 1$, where $t_{hk}(y)$ and $\psi_k(y)$ are obtained by an analysis of the proof of 3.1.6.

Proof of 3.6.3(1). Let $G \subseteq H = \prod_{i \in I} H_i$ with $H_i \in D$. Let $\varphi(x_1,...,x_n,y_1,...,y_m)$ be a conjunction of atomic formulas, and let $a_1,...,a_n \in H$, $b_1,...,b_m \in G$ such that $\varphi(a,b)$ holds in H. We show that there exist $c_1,...,c_n \in G$ such that $\varphi(c,b)$ holds in H and hence in G:

Let $\varphi^\circ(x,y)$ be the quantifier-free o-group formula without negations associated with $\varphi(x,y)$ by 3.1.1. Then $H \models \varphi(a,b)$ iff $H_i \models \varphi(a(i),b(i))$ for all $i \in I$ iff $H_i \models \varphi^\circ(a(i),b(i))$ for all $i \in I$. Applying the instantiation lemma to $\varphi^\circ(x,y)$, we find rational linear expressions $t_{hk}(y)$ $(1 \leq h \leq n)$ and corresponding conjunctions of atomic o-group formulas $\psi_k(y)$ $(k \in K)$ such that

$$\bigvee_{k \in K}(\varphi^\circ(t_{1k}(b)(i),...,t_{nk}(b)(i),b(i)) \wedge \psi_k(b(i)))$$

holds in each of the divisible o-groups H_i. Each $\psi_k(y)$ determines in a straightforward way an ℓ-group word $w_k(y)$ such that $H_i \models \psi_k(b)$ iff $w_k(b)(i) = 0$. Put $a_{hk} = t_{hk}(b)$, $d_k = w_k(b)$ for $1 \leq h \leq n$, and $k \in K$. Applying the patchwork property to a_{hk}, d_k $(k \in K)$ for fixed h, we find $c_h \in G$ with

$$\bigcup_{k \in K}(\llbracket c_h = a_{hk} \rrbracket \cap \llbracket d_k = 0 \rrbracket) \supseteq \llbracket \bigwedge_{k \in K} d_k = 0 \rrbracket) = I.$$

Consequently,

$$\bigvee_{k \in K}(\varphi^\circ(c_1(i),...,c_n(i),b(i)) \wedge \psi_k(b(i)))$$

holds in each H_i. Therefore $\varphi^\circ(c(i),b(i))$ and $\varphi(c(i),b(i))$ hold in each H_i, and so $\varphi(c,b)$ holds in G.

The proof of 3.6.3(2) is similar. The factors H_i, where negated atomic formulas hold, require additional attention: Here one uses DP, SP, and the absence of basic elements in G repeatedly to 'separate' these factors by suitable elements of G. For the factors H_i, where all parameters b vanish, one applies the absence of a weak unit in G together with lemma 3.2.5.

A more detailed study of the class EC can be conducted in terms of finite and infinite forcing in model theory. For these matters, we refer the reader to [Glass and Pierce, 1980a], [Saracino and Wood, 1983], [Point, 1983].

We direct our attention instead to the class Pr of *projectable ℓ-groups,* that admits a satisfactory model theoretic analysis.

To begin with, we remark that projectability is a natural, purely lattice theoretical condition on an ℓ-group G (see [Grätzer, 1971; part 3] for all facts concerning pseudocomplemented distributive lattices.):

LEMMA 3.6.13. *Let G be an ℓ-group.*

(1) *For $0 \leq a,b \in G$, $\mathrm{pr}(a,b)$ is the pseudocomplement of $a \wedge b$ in the interval $[0,a]$ and $\mathrm{pr}(a,b) \vee \mathrm{pr}(a,\mathrm{pr}(a,b)) = a$.*

(2) *The following are equivalent:*
   (i) *G is projectable;*
   (ii) *Every interval $[0,a]$ in G is pseudocomplemented;*
   (iii) *Every interval $[0,a]$ in G is a Stone algebra.*

Proof. Easy.

Let now PU (DPU) be the class of all (divisible) projectable ℓ-groups with a distinguished weak unit u. Then by 3.6.13, for $G \in$ PU, $[0,u]$ is a Stone algebra with $\mathrm{pr}(u,a)$ as the pseudocomplement a^* of a for $a \in [0,u]$. Consequently $B(G) = \{a^* \mid a \in [0,u]\}$, the *skeleton* of $[0,u]$, is a Boolean subalgebra of G. When G is an ℓ-subgroup of a direct product of o-groups G_i, then the elements of $B(G)$ take on only the values 0 and $u(i)$. We have now the following criterion for elementary equivalence in DPU:

THEOREM 3.6.14. *Let $G,H \in$ DPU. Then G and H are elementarily equivalent iff $B(G)$ and $B(H)$ are elementarily equivalent qua Boolean algebras.*

This result provides satisfactory elementary invariants for the class DPU, since Boolean algebras are classified up to elementary equivalence by the *Tarski invariants* (see [Chang and Keisler, 1973; section 5.5]): Let Inv be the set of all pairs (m,n) in the set $\{(\infty,0)\} \cup (\mathbf{N} \times (\mathbf{Z} \cup \{\infty\}))$. Then there is a 1 - 1 correspondence between elementary equivalence classes of non-trivial Boolean algebras and Inv.

Moreover, every non-trivial Boolean algebra B is realized as some $B(G)$ for $G \in$ DPU: We may assume that $B \in \{0,1\}^I$ for some index set I. Regarding $\{0,1\}$ as a subset of the rationals $\mathbf{Q}$, B becomes a sublattice of the ℓ-group $H = \mathbf{Q}^I$. Let G be generated by B in H; then $B(G) = B$. So we may conclude:

COROLLARY 3.6.15. Inv *is a complete set of elementary invariants for the class* DPU.

The validity of theorem 3.6.14 is established via an effective quantifier elimination procedure relative to $B(G)$: We extend the language of ℓ-group formulas as follows (cf. [Weispfenning, 1975]): First, we admit scalar multiplication with fixed rational numbers, u as a new constant, and pr as a new binary operation symbol. Call the resulting terms and formulas L-terms and L-formulas. Second, we adjoin to L the language B of Boolean algebras with new symbols $0,1,\wedge,\vee,^*$ and new variables $\xi,\eta,\ldots$. Third, we connect the two languages by a new unary operation symbol V for $V(a) = pr(u,a)$, regarded as mapping L-terms onto B-terms. Call the resulting language L*. We refer to quantified L-variables (B-variables) as L-quantifiers (B-quantifiers). Then the following holds:

THEOREM 3.6.16. *There is an effective procedure assigning to any L*-formula φ an L-quantifier-free L*-formula φ' such that φ and φ' are equivalent in* DPU.

For an L*-sentence φ, φ' is an L-quantifier-free L*-sentence; so its validity in some $G \in$ DPU depends only on $B(G)$. This proves theorem 3.6.14.

Proof of 3.6.16. By induction on the number of L-quantifiers in φ, it suffices to consider the case where φ is of the form $\exists x \psi(x,y,\eta)$, where ψ is L-quantifier-free. In a preliminary step we can associate with any quantifier-free L-formula $\rho(x,y)$ a B-term $v\rho(x,y)$ such that whenever $G \in$ DPU is represented as a subdirect product of o-groups G_i , $i \in I$, and a,b are in G, then $[\![\rho(a,b)]\!] = [\![v\rho(a,b) = 1]\!]$. (For atomic $\rho = (t = 0)$, $v\rho$ is e.g. $V(t)^*$.) By introducing new B-variables $\xi_1,\ldots,\xi_n$, $\exists x \psi(x,y,\eta)$ can now be reduced to the form $\exists \xi(\exists x \psi'(x,y,\xi) \wedge \mu(\xi,\eta))$, where ψ' is of the form $\bigwedge_{h \in H} v\psi_h(x,y) \geq \xi_h$, and μ implies that the ξ_i are pairwise orthogonal. Hence it suffices to treat $\exists x \psi' \ldots$. As in 3.1.1, we can associate with each of the quantifier-free L-formulas ψ_h a quantifier-free o-group formula ψ_h° with additional constant u, such that ψ_h and ψ_h° are equivalent in the class D. The proof of 3.1.6 provides finitely many rational linear combinations $t_{hk}(y,u)$ ($k \in K_h$) such that $\exists x \psi_h^\circ(x,y)$ is equivalent in all divisible o-groups with distinguished positive element u to $\bigvee_{k \in K_\eta} \psi_h^\circ(t_{hk}(y),y)$. Using the fact that every $G \in$ DPU has the splitting property and hence the patchwork property (when represented), we may conclude that $\exists x \psi'$ is equivalent in DPU to

$$(\exists \zeta_{hk})_{h \in H, k \in K_\eta} \bigwedge_{h \in H} (\bigvee_{k \in K_h} \zeta_{hk} \geq \xi_h \wedge \bigwedge_{k \in K_h} v \psi_h(t_{hk}(y), y) \geq \zeta_{hk})).$$

This eliminates the quantifier $\exists x$ and completes the proof.

The model theoretic analysis of Boolean algebras that leads to the Tarski invariants (see [Chang and Keisler, 1973; Section 5.5]) shows also that the theory of every single Boolean algebra as well as the theory of all Boolean algebras is decidable. Combining this fact with 3.6.16, we get:

COROLLARY 3.6.17. *The theory of every single $G \in$ DPU as well as the theory of the whole class DPU is decidable.*

This is in contrast to a result of Gurevich (see [Gurevich, 1967a] and [Burris, 1985]) saying that the theory of all ℓ-groups is hereditarily undecidable.

Results similar to these results on the class DPU hold for two variants of this class.

The first variant is the class DP of non-trivial divisible projectable ℓ-groups without weak unit. Here we have to find a suitable replacement for the Boolean algebra $B(G)$: If G is an ℓ-group and $0 \leq a \in G$, then recall from Chapter 5 that $a^{\perp\perp}$ denotes the *principal polar* of a, and that the the set of all principal polars in G forms a distributive lattice $Pp(G)$. The map $a \mapsto a^{\perp\perp}$ is a lattice homomorphism of G^+ onto $Pp(G)$. If $G \in$ PU, then this map restricts to a Boolean isomorphism on $B(G)$. If $G \in$ DP, then $Pp(G)$ is a relatively complemented distributive lattice with smallest element $0 = 0^{\perp\perp}$ (a generalized Boolean algebra, gBa), and for fixed $0 < b \in G$, the map $a \mapsto a^{\perp\perp}$ restricts to a lattice embedding on $\{pr(b,a) \mid a \in G\}$. To obtain quantifier elimination relative to $Pp(G)$ in a language L* similar to that used in 3.6.16 (without u and with * replaced by an operation symbol for relative complementation in $Pp(G)$), it suffices to replace the role of the weak unit u in the proof of 3.6.16 by a 'large' element of $Pp(G)$ introduced via an existential quantifier $\exists \sigma$ (compare also [Point, 1983; Section 2.3]). Again there is a modified set of elementary invariants (the *Ershov invariants*, see [Ershov, 1964]) for the class of all gBa's and every gBa is realized as $Pp(G)$ for some $G \in$ DP. This yields:

THEOREM 3.6.18. *Let $G,H \in DP$. Then $G \equiv H$ iff $Pp(G) \equiv Pp(H)$. The theory of every single $G \in DP$ as well as the theory of the whole class DP is decidable.*

In the last variant of DPU the role of divisible o-groups as factors in the representation is replaced with Z-groups (compare [Glass and Pierce, 1980a]): A positive element s in an ℓ-group G is *singular* if for all $0 \le a \le s$ in G, a is orthogonal to s-a. Let PZ, the class of projectable Z-ℓ-groups, $\mathcal{K}$ be the class of all $G \in PU$ in which u is singular and for every prime p, every element $a \in G$ is congruent mod pG to a sum of at most p singular elements. If $G \in ZP$ is represented as a subdirect product of o-groups G_i, then each G_i is a Z-group (as studied in section 3.4) with smallest positive element $1 = u(i)$, and $B(G)$ coincides with the interval $[0,1]$.

In order to get a counterpart of 3.6.16, we drop rational scalar multiplication in L and add instead congruences $\equiv (n)$ for every positive integer n as new binary relation symbols. Notice that if $G \in PZ$ is represented as a subdirect product of Z-groups G_i, $i \in I$, and $a \in G$, then $a \equiv 0$ (n) iff $[\![a \equiv 0\ (n)]\!] = I$. This is a consequence of the patchwork property in G. So all one has to do in the proof of 3.6.16 is to replace the rational linear combinations $t_{hk}(y)$ by suitable Z-words obtained from the proof of 3.3.4. Furthermore, any Boolean algebra can be realized as $B(G)$ for some $G \in PZ$. This yields the following result (compare [Glass and Pierce, 1980a; Section 2]).

THEOREM 3.6.19. *Let $G,H \in PZ$. Then $G \equiv H$ iff $B(G) \equiv B(H)$. The set* Inv *of Boolean invariants is also a compete set of elementary invariants for PZ. The theory of any single $G \in PZ$ as well as the theory of the whole class PZ is decidable.*

We close this section by relating the classes DPU, DP, PZ to existential closedness. Let DPU*, DP*, PZ* be the class of all $G \in DPU$, DP, PZ, respectively, that have no basic elements. Let PDt (the class of projectable discrete ℓ-groups) be the class of all $G \in PU$ in which u is singular. Regard the structures in Pr, PZ, PU as ℓ-groups with pr as distinguished operation, with u as distinguished element, and with both distinguished, respectively. Then the following can be shown from the results above without much effort using [Schmid, 1979].

THEOREM 3.6.20 (compare [Glass and Pierce, 1980a], [Burris and Werner, 1979], [Weispfenning, 1978]). *DPU*, DP*, PZ* are the classes of existentially closed structures in PU, Pr and PDt, respectively.*

3.7. Concluding Remarks

As indicated in the introduction, we have deliberately restricted the model theoretic scope of this chapter to a few central topics. More advanced or specialized model theoretic issues considered in the literature on ℓ-groups include:

(1) More general notions of formulas, including e.g. generalized quantifiers ([Baudisch, Seese, Tuschik, and Weese, 1980], [Gurevich, 1977], [Lenski, 1984 and 1988], [Schmitt, 1982; chapter 8]).

(2) Model theoretic stability issues ([Schmitt, 1982; chapter 7]).

(3) Model theoretic forcing, finitely and infinitely generic ℓ-groups ([Point, 1983], [Glass and Pierce, 1980a], [Saracino and Wood, 1983, 1984]).

(4) Structure theorems for ℓ-groups admitting quantifier elimination in specified languages ([Point, 1983], [Weispfenning, 1984,1986c]).

(5) Relative existential closedness of one o-group in another ([Weispfenning, 1986a]).

Finally, we mention some problems that are suggested by the results presented in this chapter:

(1) The proof of the undecidability of the theory of all ℓ-groups in [Burris, 1985] suggests that the main obstacle to decidability lies in the too general lattice structure of ℓ-groups. For projectable ℓ-groups this obstacle is removed: G *qua* lattice is a relative Stone algebra. So there may be some hope to obtain elementary invariants and decidability for the class of projectable ℓ-groups. Even more optimism may suggest an extension of Gurevich's monumental result [1977] to the class of projectable ℓ-groups G with distinguished complete Heyting algebra $C(G)$ of convex ℓ-subgroups. On a more modest scale, we conjecture that satisfactory elementary invariants and decidability can be obtained for the theory of hyperarchimedean ℓ-groups.

(2) We conjecture that the theories DPU, DP, PZ of projectable groups shown to be decidable in section 3.6 have in fact an elementary recursive decision problem. What are tight bounds for the asymptotic complexity of these theories?

(3) Is there an extension of the results of section 3.2 to the set of $\forall\exists$-sentences, i.e., of sentences of the form $\forall x_1...\forall x_n \exists y_1...\exists y_m \varphi$, where φ is quantifier free?

(4) Find a characterization of relative existential closedness of one ℓ-group G in another H, generalizing [Weispfenning, 1986a].

Volker Weispfenning
University of Heidelberg

Current address:
University of Passau
8390 Passau
F. R. G.

Joe L. Mott

CHAPTER 4

GROUPS OF DIVISIBILITY:
A UNIFYING CONCEPT FOR INTEGRAL DOMAINS AND
PARTIALLY ORDERED GROUPS

4.1. Introduction

The theory of divisibility and factorization in an integral domain is essentially the study of a partially ordered abelian group. This is because if D is an integral domain with identity, K a field containing D, and K^* the set of nonzero elements of K, then the divisibility relation (a divides b for $a, b \in K^*$ if and only if $b/a \in D$) determines a partial order on K^* modulo the units of D. To be sure, the divisibility relation is reflexive and transitive on K^* and so determines a preordering compatible with the group structure of K^* (if a divides b, then ac divides bc for all $c \in K^*$).

But, while the divisibility relation is not antisymmetric on K^*, it is nearly so. For if a divides b and b divides a, then $b = ad_1$, and $a = bd_2$, where $d_1, d_2 \in D$. Then $a = ad_1d_2$, $1 = d_1d_2$, and $d_1, d_2 \in U(D)$, the set of units of D. Thus, the divisibility relation can be made antisymmetric if we identify elements of K^* that differ only by a unit factor of D. In short: the partial order $\leq$ defined on $K^*/U(D)$ by $aU(D) \leq bU(D)$ if and only if a divides b makes $K^*/U(D)$ into a partially ordered abelian group with positive cone $[K^*/U(D)]^+ = D^*/U(D)$.

80

A. M. W. Glass and W. C. Holland (eds.), Lattice-Ordered Groups, 80–104.
© 1989 by Kluwer Academic Publishers.

If K is the quotient field of D, we designate this group as the *group of divisibility* of D and denote it by $G(D)$. But in the case where K need not be the quotient field of D, the partially ordered group is called the *semi-value group* of D in K.

Next, observe that the group of divisibility of D is order isomorphic to the multiplicative group $P_K(D)$ of principal nonzero fractional ideals of K partially ordered by <u>reverse</u> containment. To understand this statement, recall that if $a \in K^*$, then the set $aD = \{ad \mid d \in D\}$ is a principal (nonzero) fractional ideal of K. (If $a \in D$, then aD is called an *integral* principal ideal of D.) Then $aU(D) \leq bU(D)$ if and only if $bD \subseteq aD$. Moreover the map $\alpha : G(D) \to P_K(D)$ defined by $\alpha(aU(D)) = aD$ is an order isomorphism.

Let $v : K^* \to K^*/U(D)$ be the canonical group homomorphism such that $v(a) = aU(D)$ for any $a \in K^*$. Then properties of elements of K^* and principal fractional ideals of K can be translated readily to properties of $G(D)$ via the map v. For example, an element $d \in K^*$ is a common divisor (multiple) of $a,b \in K^*$ if and only if $v(d)$ is a lower (upper) bound of $v(a)$ and $v(b)$ in $G(D)$. In addition, the greatest common divisor (least common multiple) of a,b corresponds to the greatest lower bound (least upper bound) of $v(a)$ and $v(b)$ in $G(D)$. Therefore, each pair of nonzero elements of D have a greatest common divisor in D if and only if each pair of elements in the positive cone of $G(D)$ have a greatest lower bound or, in other words, if and only if $G(D)$ is lattice ordered. Likewise, the (integral) principal ideals of D are linearly ordered under containment if and only if $G(D)$ is totally ordered. In the former case, D is called a *GCD-domain* and in the latter, a *valuation ring*. If $G(D)$ is order isomorphic to the infinite cyclic group $\mathbf{Z}$ of integers, then D is said to be a *discrete valuation ring*.

4.2. Unique Factorization

Of course, the nicest kind of divisibility is illustrated by the so-called *unique factorization domains*, those domains D that satisfy the *fundamental theorem of arithmetic*. To clarify this statement, let us recall the distinction between irreducible and prime elements of an integral domain D. Any nonunit of D^* that cannot be written as a product of two nonunits of D is said to be *irreducible* or *nonfactorable*. In other words, $q \in D^*$ is irreducible if $q = bc$, where $b,c \in D$, implies either b or c is a unit of D. Clearly, an element $q \in D^*$ is irreducible if and only if $v(q) = qU(D)$ is a minimal positive element (atom) of the group of divisibility $G(D)$. Thus, it is appropriate to designate irreducible elements as *atoms* of D, and then to say that D is *atomic* if each nonzero nonunit of D is a finite product of atoms.

On the other hand, a nonzero nonunit $p \in D$ is called a *prime* element if when p divides a product of elements of D, then p divides one of the factors. If p is a prime element of D, then, of course, pD is a prime ideal of D in the sense that the residue class ring D/pD is an integral domain.

A prime element is always irreducible, but conversely an atom need not be prime. (In $Z[\sqrt{-5}]$, each factor of $2 \cdot 3 = (1+\sqrt{-5})(1-\sqrt{-5})$ is irreducible but not prime.)

Nevertheless, in some circumstances the two concepts of atom and prime are equivalent. In particular, in a GCD-domain they are the same, and this fact follows from a well-known result in lattice-ordered groups that I refer to as:

EUCLID'S LEMMA. *Let G be a lattice-ordered group under multiplication, where e denotes the identity element of G. Then a positive element $p \in G \setminus \{e\}$ is an atom of G if and only if $p \leq a \cdot b$, where a and b are positive, implies $p \leq a$ or $p \leq b$.*

For a proof see Fuchs [1963, page 17].

But there are weaker conditions under which atoms are prime. For example, suppose that D is an integral domain such that the group of divisibility is a *Riesz group*, that is, $G(D)$ is a directed group which satisfies the following *interpolation property*: if $a_1, a_2, b_1, b_2 \in G(D)$ where $a_1 \leq b_1, a_1 \leq b_2$, $a_2 \leq b_1, a_2 \leq b_2$, then there exists some $c \in G(D)$ such that $a_1 \leq c \leq b_1, a_2 \leq c \leq b_2$. In particular, if $G(D)$ is a Riesz group and $e \leq h \leq g_1 g_2$ where $e \leq g_1$ and $e \leq g_2$, then $h = h_1 h_2$ where $e \leq h_1 \leq g_1$ and $e \leq h_2 \leq g_2$ [Fuchs, 1965]. These facts show that if p is an atom of D, and if p divides ab, where $a, b \in D$, then $e \leq v(p) \leq v(a)v(b)$ in $G(D)$, and since $G(D)$ is a Riesz group, $v(p) = v(x)v(y)$ where $e \leq v(x) \leq v(a)$ and $e \leq v(y) \leq v(b)$. But since $v(p)$ is a minimal positive element of $G(D)$, $v(x) \leq v(p)$ and $v(y) \leq v(p)$ imply that $v(p) = v(x)$ or $v(p) = v(y)$. Of course, this means that p divides either a or b and, therefore, that p is a prime element of D.

DEFINITION. A *unique factorization domain (UFD)* is an integral domain D that satisfies

(A) D is atomic, and

(U) any two factorizations of a nonzero nonunit of D into products of atoms differ only in the order of factorization and by unit factors.

Thus if $c = a_1a_2...a_r = b_1b_2...b_s$, where a_i, b_j are atoms, then $r = s$ and after suitable renumbering of the b's, $b_i = a_iu_i$ where u_i is a unit of D for each i.

The best known examples of unique factorization domains are:

(i) The ring Z of integers or, more generally, any Euclidean domain;

(ii) The ring of polynomials in any number of variables over a field;

(iii) Principal ideal domains, that is, domains for which each ideal is principal.

Proving that an integral domain is a UFD is rarely trivial, so it is useful to have as many characterizations as possible. We list a few in the following theorem. Conditions (1)-(5) are well known; (6) is new. Cohn [1968, page 255] observed (6) under the additional assumption of integral closure. The additional assumption is unnecessary.

THEOREM 4.2.1. *For an integral domain D, the following are equivalent:*

(1) D is a UFD;

(2) Each nonzero nonunit of D is a finite product of prime elements;

(3) D is atomic and atoms are prime;

(4) D satisfies the ascending chain condition on principal ideals and atoms are prime;

(5) D is a GCD-domain that satisfies the ascending chain condition on principal ideals;

(6) The group of divisibility of D is a Riesz group and D is atomic.

Proof. To show (1) implies (2) we need only observe that in a UFD, atoms are primes. Let q be an atom and suppose q divides ab. Thus, $ab = qc$ where $c \in D$. Write a, b, c as products of atoms,

$$(a_1a_2...a_n)(b_1b_2...b_m) = q(c_1c_2...c_d).$$

Use (U) in the definition of UFD to conclude $q = a_iu_i$ or $q = b_iu_i$ for some i where u_i is a unit. Then we conclude q divides a or q divides b.

Clearly (2) implies (3). Moreover, that (3) implies (1) follows by induction on the number of irreducible factors.

A UFD is a GCD-domain by the familiar result that $d = \prod p_i^{\min(a_i, b_i)}$ is the greatest common divisor of $a = \prod p_i^{a_i}$ and $b = \prod p_i^{b_i}$, where p_i is irreducible and a_i and b_i are nonnegative integers for each i. Furthermore (A) and (U) clearly imply that a principal ideal of a UFD has only finitely many distinct principal ideal divisors. Thus, (1) implies (5); (5) implies (4) by Euclid's Lemma.

To observe that (4) implies (3) we first note that the ascending chain condition on principal ideals implies the maximal condition on principal ideals. Thus, if some nonzero nonunit element of D is not a finite product of atoms, let aD be a proper principal ideal maximal with respect to the property that a is not a product of atoms. Then since a is not an atom, $a = bc$, where b and c are nonunits of D. Thus $aD \subset bD$ and $aD \subset cD$. But the maximality of aD implies b and c are products of atoms, and hence that $a = bc$ is too. This contradiction proves that if (4) holds then D is atomic.

Clearly (5) implies (6) since a lattice-ordered group is a Riesz group. Moreover, (6) implies (3) since atoms are prime if $G(D)$ is a Riesz group. This completes the proof of Theorem 4.2.1.

EXAMPLE. Let $D = k+xK[x]$ where K is a proper field extension of the field k. Then D is an integral domain in which x is irreducible but not prime since x divides $(x/\lambda)(\lambda x)$, for any $\lambda \in K \backslash k$, but x does not divide either factor. Nevertheless, D satisfies the ascending chain condition on principal ideals since if $(0) \subset f(x)D \subset g(x)D$, then the degree of $g(x)$ is less than the degree of $f(x)$. Thus, D satisfies (A) but not (U) in the definition of unique factorization domains. In fact, if K is an infinite extension of k, then D is not Notherian since the ideal $xK[x]$ is not finitely generated.

In my opinion since polynomials and their properties should be familiar, the above example should be more accessible to students than the examples , $\mathbf{Z}[\sqrt{-5}]$ or $\mathbf{Z}[\sqrt{10}]$, usually given to illustrate that factorizations of products of irreducibles need not be unique.

THE IMPORTANCE OF UNIQUE FACTORIZATION.

Unique factorization is important for several reasons.

(i)   The solution of the Diophantine equation $ax+by = d$, where $a,b,d,x,$ and y are integers, depends on the Euclidean algorithm and hence on unique factorization.

(ii) Unique factorization also plays a role in the solution of certain quadratic diophantine equations. Probably the first step toward a general theory was taken by Fermat in 1640 when he proved the following result:

For the binary quadratic form $Q(x,y) = x^2+y^2$ and for any positive integer m, all relatively prime solutions (x,y) to $Q(x,y) = m$ can be obtained by successive applications of two results called the genus and compostion theorems.

GENUS THEOREM. For any prime integer p, $Q(x,y) = p$ can be solved for integers x and y if and only if $p \equiv 1 \bmod 4$ or $p = 2$. The representation is unique except for obvious changes in sign and rearrangements of x and y.

COMPOSITION THEOREM. $Q(x,y)Q(x',y') = Q(xx'-yy',x'y+xy')$.

In the intervening years between Fermat and Gauss, other mathematicians including Euler, Lagrange, and Legendre discovered for a variety of binary quadratic forms results analogous to Fermat's. Gauss [1801] achieved a complete generalization of Fermat's genus and composition theorems for an arbitrary quadratic form $Q(x,y) = ax^2+bxy+cy^2$ where a,b, and c are relatively prime integers and $d = b^2-4ac$ is the nonzero discriminant of $Q(x,y)$. Gauss showed how binary quadratic forms with a given discriminant can be divided into equivalence classes so that two forms belong to the same class if and only if there exists an integral unimodular substitution relating them, and how these classes can be partitioned into genera so that two forms are in the same genus if and only if they are rationally equivalent. But the number of quadratic forms required in Gauss' genus and composition theorems is greater than 1 if the ring D of algebraic integers in the field $Q(\sqrt{d})$ fails to be a UFD. In retrospect, we see now that the reason behind the simplicity of Fermat's genus and composition theorems is that $Z[i] = \{a+bi \mid a,b \in Z\}$ is a Euclidean domain and, hence, a UFD.

(iii) Unique factorization also entered into Kummer's attempt to prove Fermat's Last Theorem. Since primitive Pythagorean triples, relatively prime integers x,y, and z such that $x^2+y^2 = z^2$, can be completely characterized because $Z[i]$ is a UFD, Kummer attempted to generalize these ideas to the case where n is a prime integer and then to factor $x^n+y^n = z^n$ as $(x+y)(x+\xi y)...(x+\xi^{n-1}y)$ in $Z[\xi] = \{a_0+a_1\xi+...+a_{n-1}\xi^{n-1} \mid a_i \in Z\}$, where ξ is a primitive n-th root of unity. Now $Z[\xi]$ satisfies the ascending chain condition on principal ideals because, in fact, it is Noetherian. Therefore, $Z[\xi]$ is atomic, but, unfortunately, it need not be a UFD. In fact, if ξ is a primitive twenty-third root of unity, then $Z[\xi]$ is not a UFD. Nevertheless, for those values of n for which $Z[\xi]$ is a UFD, Kummer was able to prove that $x^n+y^n = z^n$ has no integral solutions.

Kummer went on to derive other more general criteria. He introduced the concept of a regular prime and proved that Fermat's Last Theorem holds for all regular primes and for certain classes of irregular primes. In particular, he dealt with the three irregular primes (37, 59, and 67) less than 100 and concluded, therefore, that Fermat's Last Theorem was valid for all odd primes less than 100.

In an attempt to restore unique factorization and, consequently, to extend the list of values for which Fermat's Last Theorem holds, Dedekind [1871] gave a general definition of ideals and proved that in the ring of algebraic integers in a number field (a finite extension of the rational numbers), each ideal is a unique product of prime ideals. (Such domains have since been called *Dedekind domains*.) In Dedekind domains the emphasis is shifted from divisibility of elements to that of ideals and from prime elements to prime ideals.

Undoubtedly Dedekind knew that failure of unique factorization in a Dedekind domain corresponds to the absence of a lattice order on $G(D)$. In fact, the crucial step that Dedekind took was to embed $G(D)$ into the lattice-ordered group of all ideals of the quotient field K of D.

(iv) Many other developments in algebraic number theory were motivated by the desire to restore unique factorization by some extension process. Actually, Dedekind's ideal theory was at least the third major attempt to cope with non-unique factorization. The first was Gauss' composition of quadratic forms theorem, proved in 1801; the second was Kummer's 1847 explanation that unique factorization for algebraic integers can be achieved by "ideal numbers". It seems to me that Kummer was close to the discovery of the concept of valuation on the field $Q(\xi)$ and to the observation that the decomposition of an element $x \in Z[\xi]$ is attained in the principal ideal domain $B = \bigcap_{i=1}^{n} V_i$, where $\{V_i\}$ is the collection of valuation overrings of $Z[\xi]$ that contain x. For a systematic exposition of Kummer's ideal numbers (now called *divisors*), see Chapter 3 of [Borevich and Shafarevich, 1966].

A fourth attempt at restoring unique factorization can be found in *class field theory*, the study of abelian extensions of an arbitrary number field. Here the main idea is centered around the observation that, in the ring D of algebraic integers in a number field K, each ideal A, though not necessarily principal in D, becomes principal in $\bar{D}$, the ring of integers in a finite extension $\bar{K}$ of K (in particular, in the maximal unramified abelian extension of K called the Hilbert class field of K). This led to the conjecture, the so-called "embedding problem" for K, that D can be embedded in the ring of integers D' in some finite extension K' of K where D' is a UFD. The Hilbert class field K_1 of K is such that the Galois group of K_1 over K is isomorphic to the class group of D, the group of all fractional ideals of K modulo the principal fractional ideals. (The domain D is a PID if and only if the class group of D is trivial so the class group is an attempt at measuring how far away D is from unique factorization.) But there may be non-principal ideals in the ring D_1 of integers in K_1 so let K_2 be the Hilbert class field of K_1. Continuing, we obtain a tower of fields

$$K \subseteq K_1 \subseteq K_2 \subseteq \dots \subseteq K_n \subseteq \dots$$

where each field is the Hilbert class field of its predecessor. Let K_∞ be the union of these fields.   Around 1926, Furtwängler posed the "class field tower problem", when he asked: Is K_∞ finite over K? In fact, K_∞ is finite over K if and only if the ring of integers in K_∞ is a UFD. Thus, the embedding problem for K is equivalent to the class field tower problem. Golod and Safarevich [1964] showed that K_∞ could have infinite degree over K.

Around fifty years after Kummer's and Dedekind's pioneering work, Arnold [1929], Prüfer [1932], Clifford [1938], Lorenzen [1939], and others started a fifth line of investigation, the study of *ideal systems*. Since then, these studies have been continued by Jaffard [1960], Aubert [1979], and Močkor [1983]. I appeal to some properties of the system of v-ideals in section 4.3 of this paper.

(v)   Unique factorization has its place in algebraic geometry, too. Zariski [1947] discovered that there are two distinct concepts of a *simple* (or *non-singular*) *point* on an r-dimensional algebraic variety V in affine n-space. Traditionally, a point P of V had been defined to be simple if and only if rank $J_P = n\text{-}r$, where J_P is the Jacobian matrix $[\delta f_i / \delta x_j]$ at P, where $\{f_1(x_1,x_2,\dots,x_n), f_2(x_1,x_2,\dots,x_n),\dots,f_t(x_1,x_2,\dots,x_n)\}$ is a basis for the defining prime ideal of V. Zariski defined P to be simple if and only if the corresponding local ring (R,M), where M is the unique maximal ideal of R, is a regular local ring. Then he observed that the two definitions are equivalent if the ground field k has characteristic 0 or is a perfect field of characteristic $p \neq 0$. Now it is common practise to call P a *regular* point if P has a corresponding regular local ring, and to call P a *smooth* point if the Jacobian criterion is satisfied at P.

But the regular local rings (R,M) that occur in algebraic geometry as local rings corresponding to regular points are special in that they contain the ground field k. Moreover, R^* (the completion of R with respect to the topology determined by the powers of the maximal ideal M) is isomorphic to the ring of formal power series in r indeterminants over the residue field R/M. Thus, R^* is a UFD and Zariski observed that R is a UFD as well. For several years it was unknown whether this observation could be extended, but Auslander and Buchsbaum [1959] proved that an arbitrary regular local ring is a UFD. The line of investigation started by Zariski reached its zenith in Hironaka's [1964] spectacular proof of the resolution of singularities theorem for algebraic varieties over algebraically closed fields of characteristic 0.

Now after this brief review of the importance of unique factorization, let us return to our general discussion and observe that unique factorization domains can be characterized by the group of divisibility.

THEOREM 4.2.2. *An integral domain D is a UFD if and only if the group of divisibility $G(D)$ is isomorphic to a cardinal sum of copies of the infinite cyclic group* **Z**.

This follows immediately from the correspondence $xD = \prod_\lambda p_\lambda^{n_\lambda} D \leftrightarrow (n_\lambda)$, where $\{p_\lambda\}$ is the set of prime elements of D and $n_\lambda \in \mathbf{Z}$ for each λ.

Theorem 4.2.2, part (5) of Theorem 4.2.1, and the observation that the ascending chain condition of principal ideals of D is equivalent to the descending chain condition of positive elements in the group of divisibility $G(D)$ gives a proof of the following theorem of Ward [1940] and Birkhoff [1942].

THEOREM 4.2.3 (Ward and Birkhoff). *An abelian lattice-ordered group G is isomorphic to a cardinal sum of copies of* **Z** *if and only if G satisfies the descending chain condition on positive elements.*

Finally, let me mention some generalizations of Theorem 4.2.3 to more general partially ordered groups and therefore, via groups of divisibility, to generalizations of unique factorization domains. Mott and Zafrullah [a and b] and Zafrullah [a] have investigated partially ordered groups in which each element is uniquely decomposed as a finite sum of elements whose definition generalizes the notion of prime elements.

Although Ward and Birkhoff proved their theorem in 1940-42, I rather suspect that Dedekind anticipated it in the 1870's because of his idea to restore unique factorization to the ring of algebraic integers.

4.3. Quotient Rings and Groups of Divisibility

A subset S of a ring R is said to be a *multiplicative system* if S is a subsemigroup of the multiplicative semigroup of R, that is, if S is closed under multiplication and contains the multiplicative identity 1 of R. If, moreover, any factor of an element of S again lies in S, then S is said to be *saturated*. Now if D is a subring of a field K, and S is a multiplicative system of D, then the set $D_S = \{a/s \mid a \in D, s \in S\}$ is a subring of K, called the *quotient ring of D with respect to S*. Moreover, $D_S = D_T$, where

$$T = \{x \in D \mid x \text{ is a unit in } D_S\} = U(D_S) \cap D$$

is a saturated multiplicative system in D. If S is saturated, $U(D_S) = \{s/t \mid s, t \in S\}$ and $S = U(D_S) \cap D$.

To describe the group of divisibility of a quotient ring we review the definitions of *directed* po-group. An abelian po-group G is said to be *directed* if it satisfies any of the equivalent properties (1)-(4) in the following theorem of Clifford [1940].

THEOREM (Clifford). *For an abelian po-group G under addition, the following are equivalent.*
> *(1) Each pair $a, b \in G$ has an upper bound in G;*
> *(2) For each $a \in G$, $G = \{y - z \mid y \geq a \text{ and } z \geq a\}$;*
> *(3) For each $a \in G$, $a = y - z$, where $y \geq 0$ and $z \geq 0$;*
> *(4) For each $a \in G$, there is an upper bound for a and 0.*

What is more, there is a natural relationship between directed subgroups of the group of divisibility of an integral domain D and quotient rings of D. This relationship is described in the following theorem proved originally in [Mott, 1974b].

THEOREM 4.3.1. *Suppose D is an integral domain with quotient field K, $G(D)$ is its group of divisibility, and v is the canonical homomorphism $v : K^* \to G(D)$. There is a one-to-one correspondence τ between saturated multiplicative systems in D and convex directed subgroups of $G(D)$ where if S is a saturated multiplicative system in D, then the group $G_S = \{v(s_1)/v(s_2) \mid s_i \in S\} = \tau(S)$ is the corresponding* convex directed subgroup in $G(D)$. Moreover, $G(D)/G_S$ is the group of divisibility of D_S.

Proof: If S is a saturated multiplicative system in D, then clearly $v(S)$ is a subsemigroup of the positive cone $D^*/U(D) = v(D^*)$ of $G(D)$. Further, if $e \leq g \leq v(s)$, where $e = v(1)$ is the identity of $G(D)$ and $s \in S$, then $g = v(x)$ where x divides s. But since S is saturated, $x \in S$. Thus, $g \in v(S)$ and $v(S)$ is a convex subset of $v(D^*)$ and the subgroup $G_S = \{g_1 g_2^{-1} \mid g_i \in v(S)\}$ is a convex directed subgroup of $G(D)$.

Conversely, if H is any convex directed subgroup of $G(D)$, let $S = v^{-1}(H^+)$. Then S is clearly a saturated multiplicative system in D. Moreover, the group of divisibility of D_S,

$$K^*/U(D_S) \cong (K^*/U(D))/(U(D_S)/U(D))$$

is isomorphic to $G(D)/G_S$.

Since a UFD is a GCD-domain with the ascending chain condition on principal ideals we need conditions that imply that D is a GCD-domain (or, in other words, that the group of divisibility is lattice ordered). Since a po-group G is lattice ordered if and only if each pair of positive elements has a least upper bound it is immediate that an integral domain D is a GCD-domain if and only if each pair of elements of D have a least common multiple. But then we conclude the well-known fact that:

(1) *An integral domain D is a GCD-domain if and only if the intersection of two principal ideals of D is a principal ideal.*

The following is a result I proved in 1970; I do not believe that it has ever appeared in the literature. First, let me review some notation. If D is an integral domain with quotient field K, and A and B are subsets of K, then $A:B = \{x \in D \mid xB \subseteq A\}$, while $[A:B]_K = \{x \in K \mid xB \subseteq A\}$. Obviously, $A:B = D \cap [A:B]_K$.

THEOREM 4.3.2. *An integral domain D is a GCD-domain if and only if $aD:bD$ is principal for all a and $b \in D^*$.*

Proof. Let $d = \gcd(a,b)$. Then $a = a'd$ and $b = b'd$ where $a',b' \in D$. Moreover, $\gcd(a',b') = 1$. Let us prove that $aD:bD = a'D$. Clearly $a'D \subseteq aD:bD$ since $a'b = a'(db') = ab' \in aD$. Moreover, if $x \in aD:bD$, then $xb = ra$ for some $r \in D$. But then $xb' = ra' \in (a'D) \cap (b'D)$. Since $\gcd(a',b') = 1$, $(a'D) \cap (b'D) = a'b'D$ and $xb' = sa'b'$, where $s \in D$. Therefore $x = sa' \in a'D$ and we conclude $aD:bD \subseteq a'D$.

Conversely, suppose $xD:yD$ is principal for each $x,y \in D^*$. Let $a,b \in D^*$. Consider $A = (aD) \cap (bD)$. We show $A = cD$ where $c = a'b$ and $aD:bD = a'D$. First of all, $c \in A$, so we need only show $A \subseteq cD$. But if $x \in A$, then $x = ra = sb$ where $r,s \in D$. This implies $s \in aD:bD = a'D$ and that $s = a's'$, where $s' \in D$. But then $x = a's'b \in cD$.

In general, any prime p in D either becomes a unit in a quotient ring D_S or it remains prime, depending on whether or not p divides an element of S. Contrary to a statement in [Cohn, 1973, page 6], atoms of D_S need not come from atoms of D. For example, the element x is not an atom in $D = \mathbf{Z}_{(p)} + x\mathbf{Q}[x]$, where p is a prime integer, but x becomes an atom in $D_S = \mathbf{Q}[x]$, where $S = \{p^i\}_{i=0}^{\infty}$.

The following well-known theorem then is immediate from Theorems 4.2.1 and 4.3.2.

THEOREM 4.3.3. *If D is a UFD (respectively, a GCD-domain) and if S is a multiplicative system in D, then D_S is a UFD (respectively, a GCD-domain).*

For certain special multiplicative systems, the converse of Theorem 4.3.3 holds. Nagata observed that fact for UFD's [1957], and I and Schexnayder [1976] and Močkor [1983, page 102] observed analogous results for GCD-domains.

An ideal P of a ring R is said to be a *prime (maximal) ideal* of R if the residue class ring R/P is an integral domain (field). Moreover, if P is a prime ideal of R, the the complement of P in R is a multiplicative system of R. If D is an integral domain and if M is a maximal ideal of D, let D_M denote the quotient ring of D with respect to the multiplicative system determined by the complement of M in D.

Frequently properties of an integral domain are inherited by each of the quotient rings D_M, where M is a maximal ideal of D. In short: global properties often imply local properties. Usually the converse situation does not apply, but on occasion it does. That is, properties that hold in each D_M, where M is a maximal ideal of D, can sometimes be verified in the domain D itself. When that is the case, we say that we have obtained a *local characterization* of the property.

While it is impossible to obtain a local characterization of GCD-domains, we can come close. To see how close, we need the concept of *v-ideal* (sometimes called *divisorial ideal*).

If A is a subset of K, the quotient field of D, then $A_v = (A^{-1})^{-1}$ is the v-ideal of A, where $A^{-1} = \{x \in K \mid xA \subseteq D\} = [D:A]_K$. Moreover, A_v is the intersection of all principal fractional ideals of K that contain A. Thus, if d is the greatest common divisor of a and b, then $(a,b)_v = dD$. Two different sets can determine the same v-ideal; for example, A_v is determined by any set B, where $A \subseteq B \subseteq A_v$. An ideal B is called a v-ideal of finite type if $B = A_v$ for some finite set A, and B is said to be invertible if $BB^{-1} = D$. An invertible ideal is always a v-ideal, and B^{-1} is always a v-ideal for any ideal B.

We have the following characterization of GCD-domains in terms of v-ideals.

(2) An integral domain D is a GCD-domain if and only if each v-ideal of finite type is principal.

In fact, if D is a GCD-domain, the v-ideal of a finite set $A = \{a_1, a_2, ..., a_n\}$ is principal generated by any greatest common divisor d of the a_i's. Moreover, in this case, $A^{-1} = \frac{1}{d}D$. But, in general, a v-ideal of finite type need not be finitely generated if D is not a GCD-domain.

THEOREM 4.3.4. An integral domain D is a GCD-domain if and only if D satisfies the following conditions:

(1) D_M is a GCD-domain for each maximal ideal M of D;

(2) For each v-ideal A of finite type, A^{-1} is a v-ideal of finite type;

(3) Invertible ideals are principal.

Proof. If D is a GCD-domain, then D_M is a GCD-domain since the intersection of two principal ideals of D_M is the extension of the intersection of two principal ideals of D. Since D is a GCD-domain, this intersection must be principal. Thus (1) holds. Moreover, as we have mentioned (2) is obvious when D is a GCD-domain. For a proof of (3), see [Kaplansky, 1970, page 42, Exercise 15].

To prove the converse, we need two lemmas.

LEMMA 1. If A is a v-ideal of D of finite type and if S is a multiplicative system in D, then $(AD_S)^{-1} = A^{-1}D_S$.

Proof. Let B be a finitely generated ideal such that $A = B_v$. Then $A^{-1} = B^{-1}$. In general $[C:B]_K = [CD_S:BD_S]_K$ for any ideal C since B is finitely generated. Thus, $A^{-1}D_S = B^{-1}D_S = (BD_S)^{-1}$. Moreover, $(BD_S)^{-1} = ((BD_S)_v)^{-1} = ((B_vD_S)_v)^{-1}$ by Lemma 4 of [Zafrullah, 1978]. But clearly $((B_vD_S)_v)^{-1} = (B_vD_S)^{-1} = (AD_S)^{-1}$.

LEMMA 2. *If A is a v-ideal of D of finite type such that A^{-1} is also of finite type, then AD_S is a v-ideal of D_S of finite type.*

Proof. By Lemma 1, $(AD_S)^{-1} = A^{-1}D_S$ and therefore $(AD_S)_v = (A^{-1}D_S)^{-1} = A_v D_S = AD_S$ since $A = A_v$. Thus AD_S is a v-ideal of D_S. Now let B be a finitely generated ideal such that $A = B_v$. Then $(BD_S)_v = (B_v D_S)_v = (AD_S)_v = AD_S$ so that AD_S is a v-ideal of finite type.

Now let us show that (1)-(3) of Theorem 4.3.4 imply that D is a GCD-domain. Let A be a v-ideal of finite type. We show A is principal by showing A is invertible. Let M be any maximal ideal of D. Since A^{-1} is also a v-ideal of finite type by (2), Lemma 2 implies that AD_M is a v-ideal of D_M of finite type and since D_M is a GCD-domain, AD_M is principal. Therefore, by [Bouvier and Zafrullah,a], A is invertible. Then by (3), A is principal, and the proof of Theorem 4.3.4 is complete.

Remarks. (1) Lemmas 1 and 2 are due to M. Zafrullah. Originally I proved Theorem 4.3.4 if D satisfied the additional property 4: For each v-ideal A of finite type, A is finitely generated. But Zafrullah showed me that with the aid of his two lemmas (4) was superfluous.

(2) As a rule, v-ideals do not extend to v-ideals of quotient rings. To understand the following example we need the concept of Krull dimension. The *Krull dimension* of an integral domain D is d if d is the maximal integer k where $(0) = P_0 \subset P_1 \subset ... \subset P_k$ is a chain of proper prime ideals of D. Now let D be a valuation ring of Krull dimension two where the maximal ideal M of D is principal. Then each ideal of D is a v-ideal; but if the nonmaximal proper prime P of D is such that PD_P is not principal, then PD_P is not a v-ideal of D_P. (Robert Gilmer and William Heinzer pointed out this example to me.)

Of course, Lemma 2 shows that some v-ideals do always extend to v-ideals in a quotient ring (for example, principal ideals and invertible ideals extend respectively to principal and invertible ideals of a quotient ring), but the general question about extensions of v-ideals of finite type is still unsettled.

An integral domain D is called a *Mori domain* if D satisfies the ascending chain condition on v-ideals. In particular, in a Mori domain every v-ideal is a v-ideal of finite type. Moreover, a UFD is both a GCD-domain and a Mori domain. Besides that, any *Krull domain*, an integral domain D that is an intersection of discrete valuation rings V_α where each nonzero nonunit of D is a nonunit in only finitely many V_α's, is a Mori domain. Therefore, we can obtain the following corollary to Theorem 4.3.4.

COROLLARY 4.3.5. *An integral domain D is a UFD if and only if D satisfies the following properties:*

(1) D_M *is a UFD for each maximal ideal M of D;*

(2) *D is a Mori domain;*

(3) *Invertible ideals are principal.*

Proof. The proof is obvious once we observe that a Mori domain satisfies condition (2) of Theorem 4.3.4.

4.4. Semi-valuations

Groups of divisibility can be studied via certain special homomorphisms called semi-valuations. To understand this concept, let us recall a property of the canonical homomorphism $v: K^* \to K^*/U(D)$, where D is a subring of the field K. Note that for $a,b,c \in K^*$, if c divides a and b, then c divides $a+b$. In other words, if $v(c) \leq v(a)$ and $v(c) \leq v(b)$, then $v(c) \leq v(a+b)$. But this means that $v(a+b)$ is an upper bound of the set of lower bounds of $\{v(a),v(b)\}$ and we write $v(a+b) \in \mathcal{UL}\{v(a),v(b)\}$. In summary: addition in K does place some limitations on the partial order of $K^*/U(D)$.

DEFINITION. A *semi-valuation* of a field K is a homomorphism $w: K^* \to G$ into a partially ordered group G such that

(i) $w(a+b) \in \mathcal{UL}\{w(a),w(b)\}$ for all $a,b \in K^*$,

and   (ii) $w(-1) = e$, the identity of G.

The subgroup $w(K^*)$ is called the *semi-value group* of w. Two semi-valuations w_1, w_2 of K having respective semi-value groups G_1 and G_2 are called *equivalent* if there exists an order preserving isomorphism $\phi: G_1 \to G_2$ such that $\phi w_1 = w_2$ or, in other words, so that the diagram

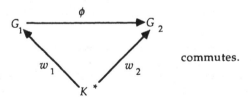

commutes.

If D is a subring of K, then the canonical homomorphism $v: K^* \to K^*/U(D)$ is a semi-valuation of K. And conversely, if w is any semi-valuation of K, then $R_w = \{x \in K^* \mid w(x) \geq e\} \cup \{0\}$ is a subring of K, called the *semi-valuation ring* of w. Moreover, the semi-valuation w is equivalent to the natural semi-valuation $n_w : K^* \rightarrow K^*/U(R_w)$.

Therefore, there is a one-to-one correspondence between equivalence classes of semi-valuations of K and subrings of K. Besides that, the field K is the quotient field of R_w if and only if each element of the semi-value group $w(K^*) = G$ is the difference (if G is considered a group under addition) of two positive elements of G. By applying the theorem of Clifford [1940] mentioned in section 3, we see that $w(K^*)$ is order isomorphic to the group of divisibility of R_w if and only if $w(K^*)$ is a directed po-group.

In short, the study of groups of divisibility reduces to the study of semi-valuations with directed semi-value groups.

What is more, we have the following simple theorem about semi-value groups. If G is a partially group and if $a,b \in G$, we write $a \parallel b$ to mean a and b are not comparable under the ordering in G.

THEOREM 4.4.1. *Suppose that v is a semi-valuation on a field K with semi-value group G. Suppose moreover that $x,y \in K^*$ are such that $v(x) \parallel v(y)$. If $x+y \in K^*$, then*

$$v(x+y) \in \mathcal{UL}\{v(x),v(y)\} \setminus (\mathcal{U}\{v(x)\} \cup \mathcal{U}\{v(y)\}).$$

Proof. We know $v(x+y) \in \mathcal{UL}\{v(x),v(y)\}$ since v is a semi-valuation. But if $v(x+y) \in (\mathcal{U}\{v(x)\} \cup \mathcal{U}\{v(y)\})$, then $v(x) \leq v(x+y)$ or $v(y) \leq v(x+y)$. Of course, this means that for the principal fractional ideals of K either $xR_v \supseteq (x+y)R_v$ or $yR_v \supseteq (x+y)R_v$. But then we conclude $xR_v \supseteq yR_v$ or $yR_v \supseteq xR_v$ or, equivalently, $v(x) \leq v(y)$ or $v(y) \leq v(x)$. This last conclusion is contrary to the hypothesis.

Clearly any directed group that violates the above "box condition" of Theorem 4.4.1 cannot be a group of divisibility.

EXAMPLE 4.4.2. Jaffard [1956] originally observed that the subgroup of the cardinal sum of two copies of $\mathbf{Z}$

$$J = \{(a,b) \mid a \text{ and } b \text{ are integers such that } a+b \text{ is even}\}$$

cannot be a group of divisibility. Clearly J is directed but $v(x) = (2,2)$ and $v(y) = (3,1)$ violate the condition of Theorem 4.4.1 since

$$\mathcal{UL}\{(2,2),(3,1)\} = \mathcal{U}\{(2,2)\} \cup \mathcal{U}\{(3,1)\}.$$

Condition (i) of the definition of semi-valuation seems to play a crucial role. Let us incorporate that property in the following definition of V-homomorphism (whose definition and properties are due originally to Ohm [1969]).

DEFINITION. If B and C are partially ordered groups and h is a homomorphism of B into C, then h is a *V-homomorphism* if for any $b_0, b_1, ..., b_n \in B$, $b_0 \in \mathcal{U}_{LB}\{b_1, ..., b_n\}$ implies that $h(b_0) \in \mathcal{U}_{LC}\{h(b_1), ..., h(b_n)\}$. A *V-isomorphism* is an isomorphism such that h and h^{-1} are V-homomorphisms. A *V-embedding* of B in C is a V-homomorphism which is one-to-one. A subgroup B of C is a *V-subgroup* if the identity map is a V-homomorphism.

Properties of V-homomorphisms:

(1) A V-homomorphism is an o-homomorphism. This is because if $x \in \mathcal{U}_{LB}\{e_B\}$, then $h(x) \in \mathcal{U}_{LC}\{h(e_B)\}$ or $x \geq e_B$ implies $h(x) \geq e_C$, where e_B, e_C are the identities of B and C respectively.

(2) The composition of two V-homomorphisms is a V-homomorphism.

(3) If B and C are lattice ordered, then a homomorphism h is a V-homomorphism if and only if h is an ℓ-homomorphism.

For my purposes, the most important properties of V-homomorphisms are the following:

(4) If $v : K^* \rightarrow B$ is a semi-valuation of K with semi-value group B and $h : B \rightarrow C$ is a V-homomorphism, then the composition hv is a semi-valuation.

(5) Let B be a po-group and let h be an embedding of B into the cardinal product $C = \prod C_\lambda$ of directed po-groups C_λ. Let $p_\lambda : C \rightarrow C_\lambda$ be the projection of C onto C_λ, and $h_\lambda = p_\lambda h$. Then, of course $h = \prod h_\lambda$. But h is a V-homomorphism if and only if each h_λ is a V-homomorphism. Moreover, if $v : K^* \rightarrow B$ is a homomorphism onto the po-group B and if $h : B \rightarrow C = \prod C_\lambda$ is a V-embedding of B into C, then v is a semi-valuation of K if and only if each $v_\lambda = h_\lambda v$ is a semi-valuation of K. In this case, the subring R_v of K can be realized as $R_v = \cap R_{v_\lambda}$.

(6) If B is a group of divisibility of a domain and $h : B \rightarrow C$ is a V-homomorphism onto C such that $h(B^+) = C^+$ (h is an o-epimorphism), then C is a group of divisibility. This is because if $v : K^* \rightarrow B$ is the semi-valuation onto B, then, by (4), hv is a semi-valuation onto C. Since B is directed and $h(B^+) = C^+$, C is directed by Clifford's result.

We can apply the contrapositive of (6) to conclude that B cannot be a group of divisibility if we know the image of B under some V-homomorphism is not a group of divisibility.

THEOREM 4.4.3. *Suppose $h : A \to B$ is an o-epimorphism of abelian partially ordered groups (that is, $h(A) = B$ and $h(A^+) = B^+$). If the kernel H of h is directed, then h is a V-homomorphism.*

Proof. Suppose $a_0 \in \mathcal{UL}\{a_1,...,a_n\}$ and suppose $h(b) \leq h(a_j)$ for each j. We show $h(b) \leq h(a_0)$. Since $h(a_j)-h(b) \geq 0$ for each j, there is an element $k_j \in H$ such that $a_j-b+k_j \in A^+$. Since H is directed we can write $k_j = k'_j-k''_j$, where $k'_j,k''_j \in H^+$. But if $k = k'_1+...+k''_n$, then $a_j-b+k \in A^+$. This means that $b-k \leq a_j$ for each j and by definition of a_0, $a_0 \geq b-k$. Hence $h(a_0) \geq h(b-k) = h(b)$.

The following theorem announced in [Spikes, 1971, page 60] will be useful. I include a proof supplied by A. M. W. Glass.

THEOREM 4.4.4. *Suppose that A and B are abelian po-groups (under addition) and $h : A \to B$ is an o-homomorphism of A onto B. Suppose that B is totally ordered. Then h is a V-homomorphism if $B^+ \backslash h(A^+)$ contains a maximal element.*

Proof. Let $a_0 \in \mathcal{UL}\{a_1,...,a_n\}$, where $a_i \in A$. Since B is totally ordered, we may assume $h(a_1) \leq h(a_2) \leq ... \leq h(a_n)$. Hence, it is enough to show that if $a_0 \in \mathcal{UL}\{a_1\} = \mathcal{U}(a_1)$, then $h(a_0) \in \mathcal{U}(h(a_1))$. So assume $a_1 \leq a_0$ but $h(a_0) < h(a_1)$. Let d' be a maximal element of $B^+ \backslash h(A^+)$, and choose $d \in A$ such that $h(d) = d'$. Since $h(a_1-a_0+d) > h(d) = d'$, $h(a_1-a_0+d) = h(b)$ for some $b \in A^+$. But then $b = a_1-a_0+d+k$ for some element $k \in H$, the kernel of h. Since $a_1 \leq a_0$, $a_0-a_1+b \in A^+$; i.e., $d+k \in A^+$. But $h(d+k) = d'$ so $d' \in h(A^+)$, a contradiction.

COROLLARY 4.4.5. *Suppose that $h : B \to C = \prod C_\lambda$ is an o-embedding of the po-group B into the cardinal product C of totally ordered groups C_λ where for each λ, the kernel H_λ of $h_\lambda = p_\lambda h$ is a directed o-subgroup of B, and where $p_\lambda : C \to C_\lambda$ is the projection map onto C_λ. Let $T_\lambda = (C_\lambda)^+$. Suppose that for each λ, either $T_\lambda \backslash h_\lambda(B^+)$ contains a maximal element or $T_\lambda = h_\lambda(B^+)$. Then h is a V-homomorphism.*

Proof. By property (5) of V-homomorphisms, we need only observe that each h_λ is a V-homomorphism. But this follows from Theorems 4.4.3 and 4.4.4.

Let us recall two theorems from [Glastad and Mott, 1982] and review a definition. A *Bézout domain* is an integral domain such that finitely generated ideals are principal. In particular, a Bézout domain is a GCD-domain.

THEOREM 4.4.6. *Suppose that D is an integral domain with quotient field K such that the group of divisibility G(D) has finite torsion-free rank n. Then there are at most n incomparable valuation rings V_i such that $D \subseteq V_i \subseteq K$. Moreover each of the valuation rings V_i has Krull dimension at most n. The integral closure*

$$\bar{D} = \{t \in K \mid t \text{ satisfies a monic polynomial in } D[x]\}$$

is a Bézout domain with finitely many prime ideals.

COROLLARY 4.4.7. *If G(D) is a torsion-free abelian group of rank one, the D is a one-dimensional valuation ring and G(D) is totally ordered.*

The following unpublished result of R. Pendleton is immediate.

COROLLARY 4.4.8 (Pendleton). *If the infinite cyclic group Z is a group of divisibility, then Z must be totally ordered and hence Z^+ is either the set of nonnegative integers or the set of nonpositive integers.*

THEOREM 4.4.9 *Suppose that $h : B \to C = \prod C_\lambda$ is a V-embedding of a directed po-group B into the cardinal product C of the totally ordered groups C_λ where $C_\lambda \cong Z$ for each λ. Suppose, further, that for each λ, $p_\lambda h$ maps B onto C_λ, where $p_\lambda : C \to C_\lambda$ is the projection map. If for some λ, $(C_\lambda)^+ \setminus p_\lambda h(B^+)$ is nonempty, then B is not a group of divisibility.*

Proof. If so, then let $v : K^* \to B$ be a semi-valuation onto B. By (5), hv and $p_\lambda hv$ must be semi-valuations for each λ, since h is a V-homomorphism. Hence, for each λ, $p_\lambda h(B)$ must be the group of divisibility for some domain $R_\lambda \subseteq K$. But then $p_\lambda h(B)$ must be a totally ordered subgroup of Z for each λ. Therefore, by Corollary 4.4.8, $p_\lambda h(B^+) = Z^+$ or Z^- and $(C_\lambda)^+ = p_\lambda h(B^+)$ for each λ. This contradicts the hypothesis that there is some λ_0 such that $(C_{\lambda_0})^+ \setminus p_{\lambda_0} h(B^+)$ is nonempty.

EXAMPLE 4.4.10. Let $B = \{(a,b) \mid a,b \in Z\}$ and let

$$B^+ = \{(0,0)\} \cup \{(x,y) \mid x \geq 1, y \geq 2\}.$$

We assert: B cannot be a group of divisibility of an integral domain. All we need do to prove this assertion is to prove directly that B is a V-subgroup of $C = Z_1 \oplus_c Z_2$, the cardinal product of two copies of the integers. For let $h : B \to C$ be the identity map and let $h_i = p_i h$ where the p_i's are the projection maps. If $a_1, a_2, ..., a_n$ are distinct elements of B, where $a_i = (x_i, y_i)$, let $L = \{(x,y) \mid x \leq \min\{x_i\} - 1$ and $y \leq \min\{y_i\} - 2\}$. Then $L_B(a_1, a_2, ..., a_n) = L_B(a_1) \cap ... \cap L_B(a_n)$ can be described in two ways, according as to whether or not some a_j is a lower bound of $\{a_1, a_2, ..., a_n\}$: $L_B(a_1, a_2, ..., a_n) = L$ or $L \cup \{a_j\}$. In the latter case, note that $x_j = \min\{x_i\}$ and $y_j = \min\{y_i\}$. But in either case, $\mathcal{U}L_B(a_1, a_2, ..., a_n) = \{(x,y) \mid \min\{x_i\} \leq x$ and $\min\{y_i\} \leq y\} = \mathcal{U}L_C(a_1, a_2, ..., a_n)$.

Then h is a V-embedding and since $(Z_2)^+ \backslash h_2(B^+) = \{1\}$ and $(Z_1)^+ = h_1(B^+)$, Theorem 4.4.9 then shows that B cannot be a group of divisibility.

EXAMPLE 4.4.11. In [Mott and Schexnayder, 1976], a composite construction was used to show that the subgroup B of the cardinal sum $Z_1 \oplus_c Z_2$ of two copies of Z is a group of divisibility where $B^+ = \{(0,0)\} \cup \{(a,b) \mid a \geq 1, b \geq 1\}$. Curiously enough, (B, B^+) is almost lattice ordered is the sense that the addition of the set $\{(0,k),(k,0) \mid k$ a positive integer$\}$ to B^+ makes B equal to $Z_1 \oplus_c Z_2$ and hence lattice ordered. But (B, B^+) is not a Riesz group because if $a_1 = (0,0)$, $a_2 = (0,1)$, $b_1 = (1,2)$, and $b_2 = (2,2)$, there is no $c \in B$ such that $a_j \leq c \leq b_i$ for $i = 1, 2$, $j = 1, 2$.

Nevertheless, the same composite construction makes (A, A^+) a group of divisibility where A is the subgroup of the cardinal sum of two copies of the group of real numbers and $A^+ = \{(0,0)\} \cup \{(a,b) \mid a > 0$ and $b > 0\}$. Now, then, (A, A^+) is a Riesz group, and in fact, an anti-lattice (see [Fuchs, 1965] for the definition).

THEOREM 4.4.12. *Suppose* $h : B \to C = \prod_{i=1}^{n} C_i$ *is a V-embedding of the directed po-group* B *into the cardinal product* C *of finitely many totally ordered groups* C_i *such that* $p_i h$ *is an o-epimorphism for each* i. *If* B *is a group of divisibility, then* B *must be lattice ordered.*

Proof. If B is a group of divisibility of a domain D, then the semivaluation $v : K^* \to B$ onto B is such that for each i, $v_i = p_i h v$ is also a semivaluation. Moreover, $D = \bigcap_{i=1}^{n} R_{v_i}$ is the intersection of valuation rings R_{v_i}

(since the semi-value group of v_i is C_i). But by a theorem of Nagata [1962], D must be a Bézout domain and hence its group of divisibility is lattice ordered.

4.5. Some Open Questions

QUESTION 1. *What partially ordered groups* (G,G^+) *can occur as groups of divisibility?*

Of course, a necessary condition is that G is directed, but this condition is not sufficient as shown by several examples in this paper and other examples in the literature [Anderson, David, and Ohm, 1981], [Cohen and Kaplansky, 1946], [Jaffard, 1956], [Martinez, 1973a], [Ohm, 1969].

On the contrary, some partial orders are sufficient for the existence of a domain D so that G is isomorphic to $G(D)$. In fact, Krull [1931] showed that any totally ordered group is the group of divisibility of an integral domain (necessarily a valuation ring). His result is a special case of the following more general result of Jaffard, Kaplansky, and Ohm. See [Mott, 1972] for the history of the development of the theorem; Nakayama [1942,1946] may have known a version of the theorem as well.

THEOREM 4.5.1. *If* G *is an abelian lattice-ordered group, then there exists a Bézout domain* D *such that* $G(D)$ *is order isomorphic to* G.

Sketch of proof. Let k be an arbitrary field and let $k[G]$ be the group algebra over k. Then $k[G]$ is an integral domain with identity. Let $K = k(G)$ be the quotient field of $k[G]$. Define $w : k[G] \to G$ by

$$w(\sum_{i=1}^{n} a_i x^{g_i}) = \inf\{g_1,...,g_n\}.$$

Extend w to K by $w(f/g) = w(f) - w(g)$. Then w is a semi-valuation on K. Moreover, if D is the semi-valuation ring of w, then the group of divisibility of D is order isomorphic to G.

This theorem has been used quite successfully in the following general context. When possible, phrase a given problem for an integral domain D in terms of the group of divisibility $G(D)$, solve the problem in $G(D)$ and (hopefully) pull back the solution to D. On the other hand, Theorem 4.5.1 has been used to construct counterexamples to questions about domains D; for examples, see [Lorenzen, 1939], [Sheldon, 1971, 1973, 1974, a], [Heinzer, 1969], and [Lewis, 1973]. The method works with sufficient frequency to merit any commutative ring theorist learning to use it.

QUESTION 2. *What partially ordered groups, besides ℓ-groups, are groups of divisibility by virtue of their order properties alone?*

For example, under what conditions can a pseudo-lattice-ordered group be a group of divisibility? The abelian group G is said to be a *pseudo-lattice-ordered group* (abbreviated pseudo ℓ-group) if and only if for each $g \in G$, $g = a - b$, where a and b are pseudo-disjoint. For the basic results concerning pseudo ℓ-groups, see [Conrad, 1966a] and [Conrad and Teller, 1970]. In [Conrad, 1966a], it is shown that in a pseudo ℓ-group G, two elements $a, b \in G^+$ are pseudo-disjoint if and only if

(*) $c \leq a, b$ implies $nc \leq a, b$ for each positive integer n.

Obviously, any ℓ-group is a pseudo ℓ-group. Moreover, a po-group G is a pseudo ℓ-group if and only if G is a Riesz group such that each $g \in G$ is the difference of two elements a and b in G^+ that satisfy (*) ([Jakubik, 1970] and [Conrad and Teller, 1970]).

In [1973a], Martinez proved the following theorem.

THEOREM 4.5.2 (Martinez). *Let Λ be a partially ordered set and let $V = V(\Lambda, R_\lambda)$ be the additive group of real-valued functions whose support satisfies the ascending chain condition. If V is a group of divisibility, then Λ is a root system and V is a lattice-ordered group.*

Every pseudo ℓ-group can be embedded in some $V(\Lambda, R_\lambda)$, and for any partially ordered set Λ the group $V(\Lambda, R_\lambda)$ is always a pseudo ℓ-group. Thus, some pseudo ℓ-groups are not groups of divisibility. Moreover, Martinez's result might lead one to conjecture that any pseudo ℓ-group that is a group of divisibility must be lattice ordered. This conjecture would be further substantiated by the following simple observation:

An integral domain D is a UFD if and only if D satisfies

(1) D is atomic, and

(2) the group of divisibility is a pseudo ℓ-group.

This observation follows from Theorem 4.2.1 (6) since a pseudo ℓ-group is a Riesz group.

But be that as it may, the conjecture is false without additional assumptions. Let me explain why.

Say a domain D is a *P-domain* if $G(D)$ is a pseudo ℓ-group. If $\{D_\lambda\}$ is a directed system of P-domains, then $\cup D_\lambda = D$ is a P-domain, as I now show.

Let K and K_λ denote the quotient fields of D and D_λ respectively. We need only show that $G(D)$ is a Riesz group that satisfies (*). If $v(a_i) \le v(b_j)$ in $G(D)$ for $i = 1,2$, $j = 1,2$ where $a_i, b_j \in K$, there is a λ such that $a_i, b_j \in K_\lambda$. Since D_λ is a P-domain, there is a $c \in K_\lambda$ such that $b_j D_\lambda \subseteq c D_\lambda \subseteq a_i D_\lambda$. Then clearly $c = a_i d_i$ and $b_j = c d_j'$ where $d_i, d_j' \in D_\lambda$. But then $b_j D \subseteq cD \subseteq a_i D$. Likewise if $x \in K$, $x \in K_{\lambda_0}$ for some λ_0 so that $x = d_1/d_2$ where $d_i \in D_{\lambda_0}$. But D_{λ_0} being a P-domain implies that $x = a/b$ where $v(a)$ and $v(b)$ are pseudo positive in $G(D_{\lambda_0})$. If $yD \supseteq aD + bD$ where $y \in K$ then $y \in K_{\lambda_1}$; let λ_2 be such that $K_{\lambda_2} \supseteq K_{\lambda_1}$, $K_{\lambda_2} \supseteq K_{\lambda_0}$, $D_{\lambda_2} \supseteq D_{\lambda_1}$, and $D_{\lambda_2} \supseteq D_{\lambda_0}$. Then in K_{λ_2}, $y^n D_{\lambda_2} \supseteq aD_{\lambda_2} + bD_{\lambda_2}$ so that $y^n a_n = a$ and $y^n b_n = b$ where $a_n, b_n \in D_{\lambda_2} \subseteq D$. Thus, $y^n D \supseteq aD + bD$ for each positive integer n. Therefore $G(D)$ satisfies (*).

EXAMPLE: Cohn gives the following example in [1968]. Let D be the semigroup algebra $F[\Gamma]$ where F is a field and

$$\bigcup \Gamma_n = \Gamma = \{(s,t) \mid s,t \text{ are positive rational numbers}\} \cup \{(0,0)\}$$

and where

$$\Gamma_n = \{(u,v) \mid u,v \text{ are nonnegative rational numbers where } v/n \le u \le vn\}.$$

Then Cohn observes that $D = \bigcup D_n$ where $D_n = F[\Gamma_n]$ is a polynomial ring in two variables. Thus, each D_n is a UFD. In particular, each $G(D_n)$ is lattice ordered so that each D_n is a P-domain. But D is not a GCD-domain as $(1,2)$ and $(2,1)$ have no greatest common divisor in D. Thus, D is a P-domain but not a GCD-domain. In other words, $G(D)$ is a pseudo ℓ-group that is not lattice ordered.

Let me refine question 2 to be more specific. In particular, I am interested in questions like:

(2a) *Which abelian anti-lattice groups are groups of divisibility?*

(2b) *What are necessary and sufficient conditions on an integral domain D so that $G(D)$ is an anti-lattice?*

(2c) *What tight Riesz groups can be groups of divisibility?* See [Loy and Miller, 1972] for the definition. The second example in Example 4.4.11 is a tight Riesz group.

The following example due to A. M. W. Glass shows that an abelian anti-lattice tight Riesz group need not be a group of divisibility. Let $G =$

$Q \times (\mathbf{Z}/2\mathbf{Z})$ where $G^+ = \{(q,x) \mid q > 0\} \cup \{(0,0)\}$. Note that $(q,x) \parallel (q',y)$ if and only if $q = q'$ and $x \neq y$. Thus, if $x \neq y$, $\mathcal{L}\{(q,x),(q,y)\} = \{(q',z) \mid q' < q\}$ and then $\mathcal{U}\mathcal{L}\{(q,x),(q,y)\} = \{(q',z) \mid q' \geq q\}$. Moreover, $\mathcal{U}\{(q,x)\} = \{(q',z) \mid q' > q\} \cup \{(q,x)\}$ and $\mathcal{U}\{(q,y)\} = \{(q',z) \mid q' > q\} \cup \{(q,y)\}$. Hence $\mathcal{U}\{(q,x)\} \cup \mathcal{U}\{(q,y)\} = \mathcal{U}\mathcal{L}\{(q,x),(q,y)\}$. But then Theorem 4.4.1 implies that G is not a semi-value group and hence is not a group of divisibility. It is easy to verify that G is a directed antilattice that is a tight Riesz group.

Relative to (2b) I can say that if $G(D)$ is an anti-lattice, then D must contain only one maximal ideal M. To see this we must only observe that the set of nonunits of D is an ideal. If $x, y \in D$ are nonunits, we show $x \cdot y$ is a nonunit by contradiction. If $x \cdot y$ is a unit, then $(x-y)D = D$ and this means $\inf\{v(x),v(y)\} = e$ in $G(D)$. Since $G(D)$ is an anti-lattice, either $v(x) = e$ or $v(y) = e$. But then we conclude that x or y is a unit in D. In fact, M is the union of v-ideals of finite type, that is, M is a *t-ideal* [Mott and Zafrullah, b].

QUESTION 3. *If A and C are groups of divisibility, is the cardinal sum of A and C necessarily a group of divisibility?*

In some special situations the answer is yes. For example, Schexnayder and I showed [1976] that the group of divisibility $G(D[x])$ of the polynomial ring over a domain D is the cardinal sum of $G(D)$ and $G(K[x])$ if and only if D is a GCD-domain. A similar question is investigated by Ohm [1969] and by Anderson and Ohm [1981].

To explain their results let us recall some definitions. A short exact sequence of po-groups $\{e\} \to A \xrightarrow{\alpha} B \xrightarrow{\beta} C \to \{e\}$ is called *order exact* if $\alpha(A^+) = \alpha(A) \cap B^+$ and $\beta(B^+) = C^+$. In particular, α and β are then order homomorphisms. Moreover, the exact sequence is called *lexicographically exact* if $B^+ = \{b \in B \mid \beta(b) > e$ or $b \in \alpha(A^+)\}$. A lexicographically exact sequence is always order exact.

Ohm and Anderson [1981] showed that if $\{e\} \to A \to B \to C \to \{e\}$ is a lexicographically exact sequence of po-groups where A is a semi-value group and C is totally ordered, then B is a semi-value group. But if C is a special kind of lattice-ordered group and $A \neq \{e\}$, then B is not a group of divisibility even if B is the lexicographic direct sum of A and C. Thus, if A and C are groups of divisibility it need not be the case that the lexicographic sum $A \oplus C$ is a group of divisibility.

QUESTION 4. *What are the most general conditions on $G(D)$ such that atoms of D are primes.* See Corollary 6.5 in [Mott and Zafrullah, 1981].

QUESTION 5. *What are the most general conditions on an integral domain D and a multiplicative system S of D such that v-ideals of finite type in D extend to v-ideals of finite type in D_S?*

QUESTION 6. In [1982] Glastad and I showed that if $D = k+xK[[x]]$ is the ring of formal power series over a field K with constant terms in a subfield k, then $G(D)$ is the direct sum of K^*/k^* and $\mathbf{Z}$. Thus, we ask: *What is the general structure of K^*/k^*?*

Brandis [1965] showed that K^*/k^* cannot be finitely generated except in the case that K is a finite field. Note that if k is the field Z_2 of two elements, then question 6 becomes:

(6a) *What is the general structure of K^* where K is a field?*

Of course, if K is finite then it is well known that K^* is cyclic.

In [1971, page 298] Samuel makes the following assertion: "The fact that, for an infinite field K, the multiplicative group K^* and the additive group K^+ are finitely generated, is true and very easy to prove." The statement is false as it stands because the multiplicative group of $Z_2(x)$, where x is transcendental, is a free group of countable rank. Undoubtedly, Samuel meant to say that K^* and K^+ are *not* finitely generated, for I can prove that if K^* is finitely generated, then K is finite.

(6b) *Is it possible to describe all fields K for which K^* is torsion free?*

It is easy to see that K^* is torsion free if and only if the characteristic of K is 2 and the prime subfield Z_2 is algebraically closed in K. More can be said here; I intend to enlarge on these observations in a future paper.

Finally, I ask:

(6c) *Is it possible to describe all torsion-free abelian groups that are the multiplicative groups of fields?*

Joe L. Mott
Florida State University
Tallahassee, Florida, 32306-3027
U. S. A.

Marlow Anderson, Paul Conrad, and Jorge Martinez

CHAPTER 5

THE LATTICE OF CONVEX ℓ-SUBGROUPS
OF A LATTICE-ORDERED GROUP

In this chapter we shall examine in some detail the lattice $C(G)$ of convex ℓ-subgroups of a lattice-ordered group G. Our emphasis is on determining how much information is available from a strictly lattice-theoretic consideration of $C(G)$. Many important classes of ℓ-groups are described in these terms, but it is impossible to so describe varieties of ℓ-groups; in section 5.4 we shall consider examples which reveal this limitation to our approach.

5.1. Classes of Convex ℓ-subgroups

In this first section we shall describe important classes of convex ℓ-subgroups for structure and representation theory using the language of the lattice $C(G)$; in many cases we shall need only to remind the reader of results from Chapter 1.

The *primes* $\Pi(G)$ are the finitely meet-irreducible elements of the lattice $C(G)$ ([Proposition 1.1.5]), while the *values* $\Gamma(G)$ are the meet-irreducible elements (see the remarks preceding Lemma 1.1.8.). Since $C(G)$ is a Brouwerian lattice ([Theorem 1.1.3]), it admits a pseudo-complementation operator $\perp$; the *polars* $\mathcal{P}(G)$ are those elements $A \in C(G)$ for

A. M. W. Glass and W. C. Holland (eds.), Lattice-Ordered Groups, 105–127.
© *1989 by Kluwer Academic Publishers.*

which $A^{\perp\perp} = A$, and this set is a complete Boolean algebra. Those polars A which are complemented (that is, $G = A \vee A^{\perp}$) are precisely the *cardinal summands* $S(G)$; this set is a subalgebra of $\mathcal{P}(G)$.

The *principal convex ℓ-subgroups* $C_p(G) = \{G(g) \mid g \in G\}$ are conveniently described by [Lemma 1.1.6]. The elements of this sublattice are also describable in lattice-theoretic terms. For, recall that an element x of a lattice L is *compact* if $x \leq \vee \{x_\alpha \mid \alpha \in A\}$ implies that $x \leq \vee \{x_\beta \mid \beta \in F\}$ for some finite subset F of A. Clearly, every principal convex ℓ-subgroup is compact in $C(G)$. On the other hand, if $C \in C(G)$, then $C = \cup \{G(c) \mid e \leq c \in C\}$. (Since $G(x) \vee G(y) = G(x \vee y)$, this family of convex ℓ-subgroups is directed.) Therefore, if C is compact,

$$C = \overset{n}{\underset{i=1}{\vee}} G(c_i) = G(c_1 \vee c_2 \vee ... \vee c_n).$$

Thus, $C_p(G)$ consists precisely of the compact elements of $C(G)$.

We've also shown that each $C \in C(G)$ is a (directed) join of compact elements of this lattice; this means that $C(G)$ is a *compactly generated* lattice (see [Pierce, 1968], page 50). Since $C(G)$ is also complete, this makes it an *algebraic* lattice (see [Birkhoff, 1967], page 187).

An element x of a lattice L is *join-irreducible* if whenever $x = \vee D$, with D directed upward, it follows that $x \in D$. A compact element is always join-irreducible, and the converse is true in an algebraic lattice (see [Birkhoff, 1967]).

We now examine some other important classes of convex ℓ-subgroups which are distinguishable in the lattice $C(G)$.

A convex ℓ-subgroup is *essential* if it contains all the values of some non-identity element; it is *special* if it is the only value of some non-identity element. It is evident that an essential subgroup is prime, and that a special subgroup is both regular and essential. Furthermore, an essential subgroup is the intersection of essential values. Note that $C \in C(G)$ is special if and only if there exists $G(g)$ such that $C \cap G(g)$ is the unique maximal convex ℓ-subgroup of $G(g)$. Also, C is essential if and only if there exists $G(g)$ such that $G(g) \not\subseteq A \in C(G)$ implies that $A \subseteq C$. Thus, the special and essential subgroups are distinguishable in the lattice $C(G)$.

For each $C \in C(G)$, let

$$I(C) = \cap \{S \in C(G) \mid S \not\subseteq C\}.$$

Notice that if $K \in C(G)$, then either $K \subseteq C$ or $I(C) \subseteq K$. We can now use $I(C)$ to determine whether an element of $C(G)$ is essential or special:

PROPOSITION 5.1.1. *For any* $C \in C(G)$,

$$I(C) = \{g \in G \mid \text{each value of } g \text{ is contained in } C\} \cup \{e\}.$$

Thus, $I(C) \neq \{e\}$ if and only if C is essential. Furthermore, C is special if and only if $I(C) \not\subseteq C$, and if this is the case then $I(C) = G(g)$, where g is special with value C.

Proof. Suppose $g \in I(C)^+$ and M_γ is a value of g. Since $I(C) \not\subseteq M_\gamma$, we have that $M_\gamma \subseteq C$. Conversely, suppose that each value of g is contained in C. Then if $S \in C(G)$ and $S \not\subseteq C$, $g \in S$. We conclude that $g \in I(C)$. It follows immediately that $I(C) \neq \{e\}$ if and only if C is essential.

If C is special and the only value of $s \in G^+$, then it is clear that $I(C) = G(s)$, and that $I(C) \cap C$ is the value of s in $G(s)$. Evidently, $I(C) \not\subseteq C$. Conversely, if $I(C) \not\subseteq C$, then $I(C)$ is compact. (For if $I(C) \subseteq \vee \{A_\lambda \mid \lambda \in \Lambda\}$, where each $A_\lambda \in C(G)$ and no A_λ contains $I(C)$, then each $A_\lambda \subseteq C$, whence it follows that $I(C) \subseteq C$. Consequently some $A_\mu \supseteq I(C)$, and in particular, $I(C)$ is compact.) But then $I(C) = G(k)$ for some $k > e$. By the first part of the proposition, C is the only value of k.

A subset Δ of $\Gamma(G)$ is called *plenary* if Δ is a filter of Γ and $\cap \{M_\delta \mid \delta \in \Delta\} = \{e\}$. Let

$$E(G) = \{\gamma \in \Gamma(G) \mid M_\gamma \text{ is essential}\}.$$

Now clearly $E(G)$ is a filter, and consequently is plenary if and only if $\cap \{M_\delta \mid \delta \in E(G)\} = \{e\}$. If $\gamma \in E(G)$ then M_γ contains all the values of some g and so M_γ must belong to every plenary subset Δ of $\Gamma(G)$. Moreover, if $\delta \in \Delta \backslash E(G)$ and we delete $\{\gamma \in \Delta \mid \gamma \leq \delta\}$ from Δ then the resulting subset is still plenary.

Thus, if Δ is a plenary subset of $\Gamma(G)$ then $\Delta \supseteq E(G)$ and if $\Gamma(G)$ has a minimal plenary subset then it must equal $E(G)$. In particular, $\Gamma(G)$ admits a plenary subset if and only if $\cap \{M_\delta \mid \delta \in E(G)\} = \{e\}$.

Note that in general $\cap \{M_\delta \mid \delta \in E(G)\}$ is an invariant of G and is in fact normal in G. We shall describe this convex ℓ-subgroup further in the next proposition.

We first need some definitions. For each $e \neq g \in G$, let

$$R_g = \vee \{M_\alpha \mid \alpha \text{ is a value of } g\}, \text{ and}$$

$$R(G) = \cap \{R_g \mid e \neq g \in G\}.$$

We call $R(G)$ the *radical* of G; see [Conrad, 1964].

PROPOSITION 5.1.2. $R(G) = \cap \{M_\gamma \mid \gamma \in E(G)\}$, and so $R(G)$ is an invariant of $C(G)$. $\Gamma(G)$ has a minimal plenary subset Δ if and only if $R(G) = \{e\}$ and if this is the case then $\Delta = E(G)$.

Proof. If $g \notin R(G)$ then $g \notin R_h$ for some h and so R_h is contained in a value M of g. Then g has an essential value. Thus if $x \in \cap M_\gamma$ then x has no essential values, and so $x \in R(G)$ and we have $\cap M_\gamma \subseteq R(G)$. If $\gamma \in E(G)$ then $M_\gamma \supseteq R_{h_\gamma}$ for some h_γ. Thus

$$\cap M_\gamma \supseteq \cap R_{h_\gamma} \supseteq \cap R_g = R(G).$$

The last proposition is essentially all contained in [Conrad, 1964]; the reader should also consult [Byrd and Lloyd, 1967]. Notice that

> in case G is a normal-valued ℓ-group, then $R(G)$ is the distributive radical $D(G)$ (see Theorem 1.1.10), and so in this case $R(G) = \{e\}$ if and only if G is completely distributive.

However, the distributive radical is not in general distinguishable in the lattice $C(G)$, as we shall see in the fourth section of this chapter.

The last class of convex ℓ-subgroups which we wish to describe is the collection of lex-subgroups. Given an ℓ- group G, let N be the set of all elements of G which are not order units; the subgroup $\langle N \rangle$ of G generated by N is called the *lex-kernel* of G. The following theorem shows that it is recognizable in the lattice $C(G)$.

THEOREM 5.1.3. $\langle N \rangle = \vee \{M \mid M \text{ is a minimal prime}\}$
$$= \vee \{a^{\perp\perp} \mid a^{\perp\perp} \neq G\}.$$

An ℓ-group G is a *lex-extension* of $C \in C(G)$ if 1) C is prime, and 2) $g > c$ for all $g \in G^+ \backslash C$ and $c \in C$. We shall henceforth abbreviate condition 2) by writing $g > C$. (Note that if $C \neq \{e\}$, then 2) implies 1).) If G is a lex-extension of C, we shall write $G = \text{lex}(C)$. The following theorem shows that the set of convex ℓ-subgroups C of G for which $G = \text{lex}(C)$ is distinguishable in the lattice $C(G)$.

THEOREM 5.1.4. *For* $\{e\} \neq C \in C(G)$ *the following are equivalent:*
1) $G = \text{lex}(C)$.
2) $C \supseteq \langle N \rangle$.
3) C *is comparable to every member of* $C(G)$.

The proofs of Theorems 5.1.3 and 5.1.4 are straightforward and appear in[Conrad, 1968]; see also [Lavis, 1963].

A convex ℓ-subgroup C of G is called a *lex-subgroup* of G if $C = \text{lex}(U)$ for some $U \subset C$. If, in addition, there does not exist a proper lex-extension of C in G, we say that C is a *maximal* lex-subgroup. Note that C is a lex-subgroup if and only if

$$C \supseteq \vee\{M \cap C \mid M \text{ is a minimal prime and } M \not\supseteq C\}.$$

Therefore the lex-subgroups (and the maximal lex-subgroups) are distinguishable in $C(G)$; denote by $M(G)$ the root system of maximal lex-subgroups of G. We list some basic properties of lex-subgroups. Proofs may be found in [Conrad, 1968].

5.1.5. *Suppose that A and B are incomparable lex-subgroups. Then $A \cap B = \{e\}$. Thus, the lex-subgroups of G form a root system.*

5.1.6. *Let A be a lex-subgroup. Then $A^{\perp\perp}$ is a lex-extension of A and a maximal lex-subgroup.*

5.1.7. (Clifford) *Suppose that A is a lex-subgroup and $e < g \in G$. Then $g \notin A \boxplus A^\perp$ if and only if $g > A$.*

5.1.8. *For $e < g \in G$ the following are equivalent:*
  1) $G(g)$ is a lex-subgroup;
  2) g is special (in G);
  3) g is special in $G(g)$.

If any of these hold, and C (respectively, D) is the value of g in G (respectively, $G(g)$), then
  i) D is the largest convex ℓ-subgroup of $G(g)$,
  ii) $D = C \cap G(g)$ is normal in $G(g)$, $G(g) = \text{lex}(D)$ and $G(g)/D$ is Archimedean,

and   *iii) $C = D \boxplus g^\perp = D \boxplus D^\perp$ is normal in its cover $C^* = G(g) \boxplus g^\perp$.*

Call a compact element of an algebraic lattice *special* if it is finitely join irreducible. In [Martinez, 1973b] it is shown that a compact element c is special if and only if it is *join irreducible*, that is, precisely when c covers a unique element. Thus in $C(G)$ a compact element A is special if and only if there exists $e < g$, special in G, for which $A = G(g)$.

5.1.9. *For $A \in C(G)$ the following are equivalent:*
  1) A is a lex-subgroup;
  2) $G(a) \subseteq A \subseteq a^{\perp\perp}$, for some special element $e < g \in G$.

COROLLARY. *The following are equivalent for $A \in C(G)$:*
  1) *A is a maximal lex-subgroup;*
  2) *$A = a^{\perp\perp}$ for some special element $e < a \in G$;*
  3) *A is a lex-subgroup and a polar.*

We can now establish a root system isomorphism between the special elements of $\Gamma(G)$ and the special elements of $C_p(G)$.

PROPOSITION 5.1.10. *The map $M \mapsto I(M)$ is an order isomorphism from the set of special elements of $\Gamma(G)$ onto the root system of special elements $C(G)$. The inverse map is the one which assigns to $G(c)$ (with c special) the unique value of c.*

Proof. If $c > e$ is special let $Sp(G(c))$ be the value of c. If $G(c) = G(d)$ then $c^\perp = d^\perp$, and if N is the maximal convex ℓ-subgroup of $G(c)$, $Nc^\perp = Nd^\perp$ is the value of cd; the mapping Sp is therefore well-defined. It now follows from the proof of Proposition 5.1.1 that $Sp(I(M)) = M$ and $I(Sp(G(c))) = G(c)$.

5.2. The Boolean Algebra $\mathcal{P}(G)$ of Polars of G and its Subalgebra $\mathcal{S}(G)$ of Cardinal Summands

In this section we shall emphasize classes of ℓ-groups which can be described primarily in terms of the Boolean algebra of polars. The first result in this direction is an important link between polars and primes. The equivalence of 3), 4), 6), and 7) is valid in any Brouwerian lattice; see [Martinez, 1973b; Lemma 2.1]. The theorem in this form is due to Conrad [1968].

THEOREM 5.2.1. *For $\{e\} \neq A \in C(G)$ the following are equivalent:*
  1) *A is an o-group;*
  2) *If $e < a \in A$ then $a^\perp = A^\perp$;*
  3) *$A^\perp$ is a prime subgroup;*
  4) *$A^\perp$ is a minimal prime subgroup;*
  5) *$A^{\perp\perp}$ is a maximal convex o-subgroup (that is, $A^{\perp\perp}$ is maximal in $C(G)$ with respect to being an o-group);*
  6) *$A^{\perp\perp}$ is a minimal polar;*
  7) *$A^\perp$ is a maximal polar;*
  8) *Each $e < a \in A$ is special.*

Recall that an element $e < s \in G$ is *basic* if $\{x \in G \mid e \leq x \leq s\}$ is totally ordered. The following theorem occurs in [Conrad, 1968]:

THEOREM 5.2.2. *For $e < s \in G$ the following are equivalent:*
  1) *s is basic;*
  2) *$G(s)$ is an o-group;*
  3) *$s^{\perp\perp}$ is a minimal polar;*
  4) *$s^{\perp\perp}$ is the largest convex o-subgroup containing s.*

COROLLARY [Lloyd, 1965]. *If $s > e$, then s is basic iff $s^{\perp}$ is the unique minimal prime not containing s.*

A subset S of G is a *basis* if S is a maximal set of pairwise disjoint elements of G all of which are basic (the maximality being with respect to disjointness only!). If S is a basis let $B = \vee \{s^{\perp\perp} \mid s \in S\}$; this is the *basis subgroup* of G. It is easy to see that B is independent of the choice of basis; B is in fact the join of all the maximal convex o-subgroups of G. Note that since the $s^{\perp\perp}$ are pairwise disjoint, they generate their cardinal sum.

Hölder's Theorem (Theorem 1.2.3) asserts that a nontrivial ℓ-group G is o-isomorphic to an additive subgroup of the reals $\mathbf{R}$ if and only if $|C(G)| = 2$; these are archimedean o-groups. Thus the atoms in $C(G)$ are the nonzero archimedean convex o-subgroups of G. Now suppose that each nonzero element in $C(G)$ exceeds an atom. Let $\{G(x_\alpha) \mid \alpha \in A\}$ be the set of atoms of $C(G)$. Then $\mathcal{P}(G)$ is *atomic* (meaning that each polar is the join (relative to the operations in $\mathcal{P}(G)$) of the atoms it contains.) Moreover, $\{x_\alpha^{\perp\perp} \mid \alpha \in A\}$ is the set of atoms in $\mathcal{P}(G)$, and

$$\vee G(x_\alpha) = \boxplus \, G(x_\alpha) \subseteq \boxplus \, x_\alpha^{\perp\perp} = \text{the basis subgroup of } G.$$

THEOREM 5.2.3. *For an ℓ-group G the following are equivalent:*
  1) *G has a basis;*
  2) *Each $e < g \in G$ exceeds a basic element;*
  3) *The intersection of all prime polars of G is zero;*
  4) *$\mathcal{P}(G)$ is atomic;*
  5) *$\Gamma(G)$ has a minimal plenary subset Δ satisfying: for each $\alpha \in \Delta$ there exists $\beta \in \Delta$ such that $\alpha \geq \beta$ and $\{\delta \in \Delta \mid \delta \leq \beta\}$ is a chain.*

THEOREM 5.2.4. *G has a finite basis if and only if each pairwise disjoint set in G is finite.*

Proofs of these theorems may be found in [Conrad, 1970a]. Part of Theorem 5.2.3 occurs in [Conrad, 1960], as does Theorem 5.2.4.

We shall now examine more closely the set $S(G)$ of cardinal summands; we've seen earlier (Chapter 1, page 16) that it is a subset of $P(G)$; we next show that it is in fact a sublattice of $C(G)$.

For this purpose, suppose that

$$G = \boxplus A_\lambda = \boxplus B_\mu$$

are two cardinal sum decompositions of G. Since $C(G)$ is distributive (Theorem 1.1.2), $G = \boxplus (A_\lambda \cap B_\mu)$; that is, the two decompositions have a common refinement. In particular, if $A, B \in S(G)$, then

$$G = (A \cap B) \boxplus (A \cap B^\perp) \boxplus (A^\perp \cap B) \boxplus (A^\perp \cap B^\perp),$$

which implies that $A \cap B \in S(G)$. Since $A^\perp \cap B^\perp = (A \vee B)^\perp$ we also conclude that $A \vee B \in S(G)$.

We can express this in the language of lattice theory by saying that $S(G)$ is the *center* of the lattice $C(G)$ (see [Birkhoff, 1967; page 67]); from this one can deduce that $S(G)$ is a sublattice of $C(G)$.

A natural class of ℓ-groups to consider is those for which every polar is a cardinal summand; such ℓ-groups are called *strongly projectable* (see Chapter 1, page 16).

THEOREM 5.2.5. *For an ℓ-group G the following are equivalent:*
  1) *$P(G)$ is a sublattice of $C(G)$;*
  2) *$P(G) = S(G)$;*
  3) *$A^\perp \vee B^\perp = (A \cap B)^\perp$ for all $A, B \in C(G)$;*
  4) *Any two minimal prime ideals in the lattice of ideals of $C(G)$ are coprime.*

The equivalence of 1), 2), and 3) holds in any Brouwerian lattice (see [Birkhoff, 1967; page 130]). The equivalence of 2) and 4) for Brouwerian lattices was proved by Grätzer and Schmidt [1957]. Brouwerian lattices with the properties of Theorem 5.2.5 are called *Stone lattices*.

If G is the countable cardinal product of copies of the reals, then clearly $P(G) = S(G)$. However, $S(G)$ is not a complete sublattice of $C(G)$ (since the cardinal sum is not a summand). Thus, the center of a complete lattice need not be a complete sublattice. This answers a question posed by S. Holland ([Birkhoff, 1967; page 131]). In the next few pages we will show that

$\mathcal{P}(G)$ is a complete sublattice of $C(G)$ if and only if G is a cardinal sum of o-groups.

The following examples show that $S(G)$ need not be a complete subalgebra of $\mathcal{P}(G)$.

EXAMPLE 1. The eventually constant sequences of integers form an ℓ-group for which the algebra of cardinal summands is not a complete sublattice of the algebra of polars.

EXAMPLE 2. Let G be the group of all pairs $((x_n),y)$ where (x_n) is an eventually constant sequence of integers, and $y \in \mathbf{Z}$. Addition is coordinatewise; $0 \leq ((x_n),y)$ if each $x_n \geq 0$, and if (x_n) is finitely nonzero then $y \geq 0$. Set

$$A_m = \{((x_n),y) \mid x_1 = x_2 = \dots = x_m = 0\}.$$

Then each A_m is a cardinal summand but

$$\bigcap_{m=1}^{\infty} A_m = \{((0),y) \mid y \in \mathbf{Z}\},$$

which does not belong to $S(G)$.

Now suppose that $\{A_\lambda \mid \lambda \in \Lambda\}$ is a subset of $S(G)$. For $g \in G$ we have

$$g = g_\lambda + \bar{g}_\lambda \in A_\lambda \boxplus A_\lambda^\perp.$$

The map τ taking $g \mapsto (\dots, g_\lambda, \dots)$ is an ℓ-homomorphism taking G into ΠA_λ with kernel $\cap A_\lambda^\perp$. If the A_λ's are pairwise disjoint, then $\vee A_\lambda = \boxplus A_\lambda$, and

$$\boxplus A_\lambda \subseteq G\tau \subseteq \Pi A_\lambda.$$

Note that the atoms of $S(G)$ are the cardinally indecomposable summands, and that $A \cap B = \{e\}$ if A and B are distinct atoms. This gives 1) $\Rightarrow$ 2) in:

THEOREM 5.2.6. *For an ℓ-group, the following are equivalent:*

    1) $S(G)$ *is atomic and* $\cap\{A_\lambda^\perp \mid A_\lambda \text{ is an atom in } S(G)\} = \{e\}$;

    2) *There is an ℓ-isomorphism τ of G into ΠA_λ such that $\boxplus A_\lambda \subseteq G\tau$, where the A_λ's are cardinally indecomposable ℓ-groups.*

Proof. 2) $\Rightarrow$ 1). Without loss of generality

$$\boxplus A_\lambda \subseteq G \subseteq \Pi A_\lambda.$$

Clearly $G = A_\lambda \boxplus A_\lambda^\perp$, and each A_λ is an atom in $S(G)$. If $\{e\} \neq C \in S(G)$ then $C \cap A_\lambda \neq \{e\}$ for some λ, which means that $C \supseteq A_\lambda^\perp$.

COROLLARY 1. $S(G)$ is atomic and a complete sublattice of $C(G)$ if and only if $G = \boxplus A_\lambda$, where the A_λ's are cardinally indecomposable. If so, then $\{A_\lambda \mid \lambda \in \Lambda\}$ is the set of all atoms of $S(G)$.

Proof. Suppose $S(G)$ is atomic and a complete sublattice of $C(G)$. Let $\{A_\lambda \mid \lambda \in \Lambda\}$ be the set of atoms in $S(G)$. Then $\boxplus A_\lambda = \vee A_\lambda \in S(G)$, and so $G = \boxplus A_\lambda$.

Conversely, if $G = \boxplus A_\lambda$, and each A_λ is cardinally indecomposable, then by Theorem 5.2.6 $S(G)$ is atomic. If $C \in S(G)$ then $C = \boxplus (C \cap A_\lambda)$, and since each A_λ is cardinally indecomposable, for each λ either $C \supseteq A_\lambda$ or $C \cap A_\lambda = \{e\}$. Hence C is the join of some of the A_λ's, and this is enough to show that $S(G)$ is a complete sublattice of $C(G)$.

Note that Theorem 4.3 of [Byrd, Conrad, and Lloyd, 1971] asserts that if an ℓ-group G is characteristically simple and A is an atom of $S(G)$, then $G = \boxplus A_\lambda$, where each $A_\lambda \cong A$. However, isomorphy of atoms is not recoverable in $C(G)$.

COROLLARY 2. $S(G)$ is atomic and contains only a finite number of atoms if and only if $G = A_1 \boxplus A_2 \boxplus \ldots \boxplus A_n$, where the A_i's are cardinally indecomposable.

THEOREM 5.2.7. For an ℓ-group G the following are equivalent:
1) $P(G)$ is atomic and each atom in $P(G)$ is a cardinal summand of G;
2) There is an ℓ-isomorphism τ of G so that

$$\boxplus B_\lambda \subseteq G\tau \subseteq \Pi B_\lambda ,$$

where each B_λ is an o-group.

Proof. 1) $\Rightarrow$ 2). By Theorem 5.2.3, G has a basis. So by hypothesis $B = \boxplus B_\lambda$ where B is the basis subgroup of G. If $g \in G$ then for each $\lambda \in \Lambda$, $g = g_\lambda \bar{g}_\lambda \in B_\lambda \boxplus B_\lambda^\perp$. By Theorem 5.2.5, the map τ taking $g \mapsto (\ldots, g_\lambda, \ldots)$ is the desired ℓ-isomorphism.

2) $\Rightarrow$ 1). It is clear that G has a basis and that $\boxplus B_\lambda$ is the basis subgroup. Moreover, for each $\lambda \in \Lambda$, $G\tau = B_\lambda \boxplus B_\lambda^\perp$. Theorem 5.2.3 implies that $P(G)$ is atomic.

We finally arrive at the desired characterization of ℓ-groups where $P(G)$ is a complete sublattice of $C(G)$:

THEOREM 5.2.8. $P(G)$ is a complete sublattice of $C(G)$ if and only if G is a cardinal sum of o-groups.

Proof. Suppose first that $P(G)$ is a complete sublattice of $C(G)$. If $a_1, a_2, \ldots$ is a maximal pairwise disjoint set in $G(g)$, then

$$g^\perp = \cap a_i^\perp \quad \text{and} \quad g^{\perp\perp} = \boxplus a_i^{\perp\perp},$$

forcing the set of a_i's to be finite. By Theorem 5.2.4, $G(g)$ has a basis, and so G has a basis $\{b_\lambda \mid \lambda \in \Lambda\}$. Finally, $G = \boxplus b_\lambda^{\perp\perp}$, since $P(G)$ is a complete sublattice of $C(G)$.

Conversely, if G is a cardinal sum of o-groups, then it is clear that $S(G) = P(G)$. By Corollary 1 to Theorem 5.2.6, $P(G)$ is a complete sublattice of $C(G)$.

COROLLARY. For an ℓ-group G, the following are equivalent:
  1) G is a cardinal sum of Archimedean o-groups;
  2) $S(G) = C(G)$;
  3) $P(G) = C(G)$.

Proof. 2) and 3) are equivalent by the Corollary to Theorem 5.2.5.

1) $\Rightarrow$ 2). If $C \in C(G)$, then

$$C = C \cap G = C \cap \boxplus A_\lambda = \boxplus (C \cap A_\lambda),$$

and since each A_λ is an Archimedean o-group, $C \cap A_\lambda$ is A_λ or $\{e\}$, and so $C \in S(G)$.

3) $\Rightarrow$ 1). By Theorem 5.2.8, $G = \boxplus A_\lambda$, where each A_λ is an o-group. Since each convex ℓ-subgroup of A must be a polar in G, each A_λ is Archimedean.

The equivalence of 1) and 2) is a theorem of Bigard's [1969].

Now let $Pp(G) = \{g^{\perp\perp} \mid g \in G\}$ be the set of all *principal polars* of G, and let $CoPp(G) = \{g^\perp \mid g \in G\}$. As the polars (and double polars) of the compact elements $G(g)$ of $C(G)$, these sets are identifiable in the lattice $C(G)$.

THEOREM 5.2.9. The sets $Pp(G)$ and $CoPp(G)$ are sublattices of $P(G)$. Moreover, for $a, b \in G^+$

$$1) \quad a^{\perp\perp} \cap b^{\perp\perp} = (a^\perp \vee b^\perp)^\perp = (a \wedge b)^{\perp\perp}, \qquad \text{and}$$

$$2) \quad a^{\perp\perp} \mathbb{w} b^{\perp\perp} = (a^\perp \wedge b^\perp)^\perp = (a^{\perp\perp} \vee b^{\perp\perp})^{\perp\perp} = (a \vee b)^{\perp\perp},$$

where $\mathbb{w}$ is the polar join.

This result is valid for any two principal polars in any algebraic Brouwerian lattice (see [Birkhoff, 1967].) Note also that the map $a^{\perp\perp} \mapsto a^{\perp}$ is clearly a lattice anti-isomorphism between $Pp(G)$ and $CoPp(G)$.

Now consider $A \in \mathcal{P}(G)$ and let $\{a_\lambda \mid \lambda \in \Lambda\}$ be a maximal pairwise disjoint subset of A. Then $A = \mathsf{w} a_\lambda^{\perp\perp}$, and so if $Pp(G)$ is a complete sublattice of $\mathcal{P}(G)$, then $Pp(G) = \mathcal{P}(G)$.

PROPOSITION 5.2.10. *For an $\boldsymbol{\ell}$-group G, the following are equivalent:*
1) *$Pp(G)$ is a subalgebra of $\mathcal{P}(G)$;*
2) *$CoPp(G)$ is a subalgebra of $\mathcal{P}(G)$;*
3) *$Pp(G)$ is a Boolean algebra;*
4) *$Pp(G) = CoPp(G)$;*
5) *If $e < g \in G$, then $g \wedge h = e$ and $(g \vee h)^\perp = e$ for some $e < h \in G$.*

PROPOSITION 5.2.11. *$Pp(G)$ is a complete sublattice of $C(G)$ if and only if G is a finite cardinal sum of o-groups.*

Proof. If $\{a_\lambda \mid \lambda \in \Lambda\}$ is a disjoint subset of G then $\mathsf{w} a_\lambda^{\perp\perp} = \boxplus a_\lambda^{\perp\perp} = a^{\perp\perp}$, for some $a \in G$ provided that $Pp(G)$ is a complete sublattice of $\mathcal{P}(G)$. But this forces Λ to be finite. Thus G has a finite basis, and we are done because G is the basis group.

COROLLARY. *$Pp(G) = C(G)$ if and only if G is a finite cardinal sum of Archimedean $\boldsymbol{\ell}$-groups.*

Proof. If $Pp(G) = C(G)$ then G is a finite cardinal sum of o-groups. But since each convex subgroup of these o-groups is a polar, the o-groups must be Archimedean.

We shall now inquire into when $Pp(G)$ and $CoPp(G)$ are sublattices of $C(G)$. This question is answered by the following theorem, due to Bigard; a proof appears in [Bigard, Keimel, and Wolfenstein, 1977; page 142].

THEOREM 5.2.12. *$CoPp(G)$ is a sublattice of $C(G)$ if and only if each proper prime subgroup of G exceeds a unique minimal prime.*

This is the so-called "stranded primes" property. Note that since $(a \vee b)^\perp = a^\perp \wedge b^\perp$ for all $a, b \in G^+$, $CoPp(G)$ is a sublattice of $C(G)$ if and only if $(a \wedge b)^\perp = a^\perp \vee b^\perp$. This condition has been called *semiprojectable*.

To answer the analogous question for $Pp(G)$, we must introduce another class of convex ℓ-subgroups. We call $C \in C(G)$ a *z-subgroup* if $g \in C$ implies that $g^{\perp\perp} \subseteq C$. Such subgroups were first studied by Bigard [1968]; they are evidently identifiable in the lattice $C(G)$. Clearly each polar is a z-subgroup, and it follows from Lemma 1.1.14 that each minimal prime is a z-subgroup. Let $Z(G)$ denote the set of all z-subgroups of G. Bigard [1968] showed that $Z(G)$ is a complete Brouwerian lattice, but $Z(G)$ need not be a sublattice of $C(G)$. For example, let $G = C(\mathbf{R})$, the group of real-valued continuous functions on $\mathbf{R}$. Then

$$P = \{f \in G \mid f(x) = 0 \text{ for all } x \geq 0\}$$

is a polar (and hence a z-subgroup), with

$$P^\perp = \{f \in G \mid f(x) = 0 \text{ for all } x \leq 0\}.$$

But $P \vee P^\perp = \{f \in G \mid f(0) = 0\}$ is not a z-subgroup, because $f(x) = x$ belongs to $P \vee P^\perp$, whereas $f^{\perp\perp} = G$.

We can now express Theorem 5.2.8 in the language of z-subgroups:

PROPOSITION 5.2.13. *For an ℓ-group G, the following are equivalent:*
1) *G is a cardinal sum of o-groups;*
2) *$Z(G) = S(G)$;*
3) *G is semiprojectable and finite valued.*

Proof. 2) $\Rightarrow$ 1). Each minimal prime is a summand. Thus if $\{N_\lambda \mid \lambda \in \Lambda\}$ is the family of minimal primes, then each $N_\lambda^\perp$ is an o-group (Corollary 1.1.15), and $G = \boxplus \, N_\lambda^\perp$ because $\boxplus \, N_\lambda^\perp \in Z(G) = S(G)$.

The remainder of the proof is reasonably clear (or see [Bigard, Keimel, and Wolfenstein, 1977; page 144].)

We can now determine when $Pp(G)$ is a sublattice of $C(G)$:

THEOREM 5.2.14. *For an ℓ-group G, the following are equivalent:*
1) *$Pp(G)$ is a sublattice of $C(G)$;*
2) *Each convex ℓ-subgroup C of G contains a unique maximum z-subgroup $Z(C)$;*
3) *Each proper prime P of G contains a unique maximum z-subgroup $Z(P)$ (which is necessarily prime.)*

Proof. 1) $\Rightarrow$ 2). Let $Z(C) = \{g \in G \mid g^{\perp\perp} \subseteq C\}$. It is evident that this is a z-subgroup, and thus it is clearly the largest z-subgroup contained in C.

2) $\Rightarrow$ 3). $Z(P)$ is prime because each minimal prime is a z-subgroup.

3) $\Rightarrow$ 1). Suppose that there exists

$$k \in (g^{\perp\perp} w h^{\perp\perp}) \setminus (g^{\perp\perp} \vee h^{\perp\perp}).$$

If $M \supseteq (g^{\perp\perp} \vee h^{\perp\perp})$ and M is a value of k, then $g,h \in Z(M)$, which implies that

$$g^{\perp\perp} w h^{\perp\perp} = (g \vee h)^{\perp\perp} \subseteq M,$$

a contradiction (note: $Z(M) \in C(G)$!).

Notice that $C(G) = Z(G)$ if and only if $Cp(G) = Pp(G)$ (see [Bigard, Keimel, and Wolfenstein, 1977; page 59].)

Recall that an $\boldsymbol{\ell}$-group is *projectable* if $Pp(G) \subseteq S(G)$. Observe that for a projectable $\boldsymbol{\ell}$-group, $Pp(G)$ is a sublattice of $C(G)$, because $S(G)$ is. A similar statement holds for $CoPp(G)$. However, even if both $Pp(G)$ and $CoPp(G)$ are sublattices of $C(G)$, G need not be projectable. For consider once again Example 2. The principal polars here are summands except for $Q = \{((0),y) \mid y \in \mathbf{Z}\}$; note that $Q^{\perp} = \{((x_n),0) \mid (x_n)$ is finitely nonzero$\}$, and so $QQ^{\perp} \neq G$. On the other hand, it is evident that $Pp(G)$ and $CoPp(G)$ are sublattices of $C(G)$.

The next result characterizes projectability in terms of z-subgroups. It is due to Bigard, and is proved in [Bigard, Keimel, and Wolfenstein, 1977; Theorem 7.5.4].

THEOREM 5.2.15. *G is projectable if and only if each proper prime exceeds a unique prime z-subgroup.*

COROLLARY 1. *Each proper prime subgroup of a projectable $\boldsymbol{\ell}$-group G exceeds a unique minimal prime.*

Corollary 1 is proved directly by Conrad and McAlister [1969; Proposition 5.3]; the following corollary is Proposition 5.4 of that paper:

COROLLARY 2. *Suppose that $Pp(G)$ is a subalgebra of $\mathcal{P}(G)$. Then the following are equivalent:*

1) *G is projectable;*
2) *Each proper prime subgroup of G exceeds a unique minimal prime.*

Finally, note that $S(G) \subseteq Pp(G)$ if and only if G has an order unit. For $G \in S(G) \subseteq Pp(G)$ implies that $G = g^{\perp\perp}$ for some $g \in G$. Conversely, if $G = g^{\perp\perp}$ and $G = A \boxplus B$, then $g = ab$, with $a \in A$ and $b \in B$. It is easy to verify that $A = a^{\perp\perp}$. Thus, if G is strongly projectable, then $\mathcal{P}(G) = Pp(G)$ if and only if G has an order unit.

5.3. Finite-valued ℓ-groups and Lex-Sums

In this section we shall examine how conditions on the root systems $\Gamma(G)$ of values and $\mathcal{M}(G)$ of maximal lex-subgroups lead to important classes of ℓ-groups describable in the lattice $C(G)$.

We shall begin by investigating when $C(G)$ is uniquely determined by the root system $\Gamma(G)$ (recall that $\Gamma(G) \subseteq C(G)$.) For this purpose we need some terminology from lattice theory. Suppose L is a lattice and S is the subset of meet-irreducible elements of L. We say that S *generates* L if each element of L is the meet of a filter of S. If, in addition, for each pair S_1 and S_2 of filters of S, $\wedge S_1 = \wedge S_2$ implies that $S_1 = S_2$, we say that L is *freely generated* by S.

The next theorem is Theorem 2.1 in [Conrad, 1965].

THEOREM 5.3.1. *Suppose S generates L. The following are equivalent:*

1) *L is freely generated by S;*
2) *L satisfies the general distributive law:*
$$\bigwedge_{\delta \in \Delta} \bigvee_{a \in A_\delta} u_{\delta,a} = \bigvee_{\tau \in F} \bigwedge_{\delta \in \Delta} u_{\delta,\tau(\delta)}$$
and its join-meet dual, where for each $\delta \in \Delta$, A_δ is a set and F is the set of all maps τ of Δ into the union of the A_δ's such that $\tau(\delta) \in A_\delta$ for each $\delta \in \Delta$;
3) *$b \vee (\wedge_\lambda a_\lambda) = \wedge_\lambda (b \vee a_\lambda)$ for all $a_\lambda, b \in S$.*

We say that an ℓ-group G is *finite valued* if each non-identity element of G has only a finite number of values. In $C(G)$, Theorem 5.3.1 translates as follows:

THEOREM 5.3.2. *For an ℓ-group G, the following are equivalent:*
1) *$\Gamma(G)$ freely generates $C(G)$;*
2) *$C(G)$ satisfies the general distributive laws;*

3) $B\vee(\wedge_\lambda A_\lambda) = \wedge_\lambda(B\vee A_\lambda)$, where B and each A_λ are regular;

4) Each regular subgroup is special;

5) G is finite valued;

6) Each non-identity element of G can be written uniquely as a product of pairwise disjoint special elements;

7) $C(G)$ is a topological lattice under order convergence;

8) Each element in $C_p(G)$ covers only a finite number of elements in $C(G)$.

The equivalence of 1) through 6) is Theorem 3.9 in [Conrad, 1965], and Birkhoff [1967; page 249] proved that for a completely distributive lattice, 7) is equivalent to 3) and its dual. Finally, $e \neq g \in G$ is finite valued if and only if $G(g)$ has only a finite number of maximal convex ℓ-subgroups; thus, 5) and 8) are equivalent.

In summary then, whether G is finite valued is determined by the lattice $C(G)$, and in this case the root system $\Gamma(G)$ determines $C(G)$ completely.

For each root system Δ there is an abelian finite-valued ℓ-group G for which $\Gamma(G) \cong \Delta$; namely, set

$$\Sigma(\Delta,R) = \{f:\Delta\to R \mid supp(f) \text{ is finite}\}.$$

However, an abelian finite-valued ℓ-group may have the same lattice of convex ℓ-subgroups as a non-abelian ℓ-group, and thus the variety of abelian ℓ-groups is not describable by $C(G)$; we shall see later that no variety is so describable.

A particularly interesting class of finite-valued ℓ-groups is obtained by making the following definition:

An ℓ-group G is a *lex-sum* of the o-groups $\{A_\lambda \mid \lambda \in \Lambda\}$ if for some ordinal σ there exists a chain

$$A^0 \subset A^1 \subset ... \subset A^\alpha \subset ...$$

in $C(G)$ such that

1) $G = \bigcup_{\alpha<\sigma} A^\alpha$, $A^\alpha = \boxplus \{A^\alpha_\lambda \mid \lambda \in \Lambda_\alpha\}$, where each A^α_λ admits no proper lex-extension,

2) $\Lambda_0 = \Lambda$, $A^0_\lambda = A_\lambda$ (for $\lambda \in \Lambda$),

3) $A^{\alpha+1}_\lambda = A^\alpha_\mu$ for some $\mu \in \Lambda_{\alpha+1}$, and

4) if α is a limit ordinal then A^α_λ is a maximal lex-extension of the join of a chain of $A^\beta_{\lambda_\beta}$, one λ_β per $\beta < \alpha$.

If each cardinal sum referred to in 3) is finite we say that the lex-sum is *restricted*. Note that $A^0 = \boxplus\, A^0_\lambda$ is the basis subgroup, and that each A^α is normal in G.

As usual we will use DCC (ACC) as an abbreviation for "descending (ascending) chain condition."

The following theorem of Conrad shows that we can completely characterize ℓ-groups having the structure described above, using only lattice-theoretic hypotheses on $C(G)$:

THEOREM 5.3.3 [Conrad, 1968]. *An ℓ-groups G is a lex-sum of o-groups if and only if G is finite valued and $\mathcal{M}(G)$ satisfies the DCC.*

We will not prove this theorem here; see [Conrad, 1970a] for a proof.

We should point out two other characterizations of the ℓ-groups described in Theorem 5.3.3. A *filet chain* is a set of strictly positive elements $\{a_1, a_2, b_2, a_3, b_3, ...\}$ such that

$$a_n \wedge b_n = e \text{ for } n \geq 2, \text{ and}$$

$$a_n \geq a_{n+1} \vee b_{n+1} \text{ for all } n.$$

Notice that $a_1^{\perp\perp} \supset a_2^{\perp\perp} \supset ...$ since $a_n > a_{n+1}$ and $b_{n+1} \in a_n^{\perp\perp} \setminus a_{n+1}^{\perp\perp}$.

For a proof of the following theorem see [Conrad, 1970a].

THEOREM 5.3.3a. *For an ℓ-group G, the following are equivalent:*
1) *G is a lex-sum of o-groups;*
2) *G is finite valued and $\mathcal{M}(G)$ satisfies the DCC;*
3) *Each filet chain is finite;*
4) *$Pp(G)$ satisfies the DCC.*

An ℓ-group G satisfies *property F* if no positive element of G exceeds an infinite set of pairwise disjoint elements (such ℓ-groups are also called *orthofinite*.) This is quite evidently equivalent to requiring that $G(a)$ have a finite basis for each $e < a \in G$, and is consequently a property recognizable in the lattice $C(G)$. An *ω-lex-sum* of o-groups is a lex-sum of o-groups for which $A^\omega = G$. We now have:

COROLLARY 1 [Conrad, 1968]. *An ℓ-group G has property F if and only if G is a restricted ω-lex-sum of o-groups.*

Proof. The sufficiency is obvious. As for the necessity, property F clearly implies 3) in Theorem 5.3.3a. Hence G is a lex-sum of o-groups. In view of property F it must be restricted, and from this it is not hard to see that $G = A^\omega$.

COROLLARY 2 [Conrad, 1968]. *G has a finite basis if and only if G is a lex-sum of finitely many o-groups.*

We give a longer list of conditions which are equivalent to the presence of a finite basis.

THEOREM 5.3.4. *For an ℓ-group G, the following are equivalent:*
1) *G has a finite basis;*
2) *G is a lex-sum of finitely many o-groups;*
3) *Each disjoint subset of G is finite;*
4) *$\Gamma(G)$ contains only a finite number of roots;*
5) *G has only a finite number of minimal primes;*
6) *Each proper convex ℓ-subgroup of G has a finite basis;*
7) *$\mathcal{P}(G)$ is finite;*
8) *$Pp(G)$ is finite;*
9) *G is finite valued and $\mathcal{M}(G)$ is finite;*
10) *Each $C \in C(G)$ can be written $C = C_1 \boxplus ... \boxplus C_n$, with each C_i indecomposable (special);*
11) *$Pp(G)$ satisfies the ACC;*
12) *Each minimal prime belongs to $Pp(G)$;*
13) *G is a (finitary) lex-extension of a finite cardinal sum of o-groups.*

Most of these equivalences are proved in [Conrad, 1960] or [Conrad, 1970a]. For example (as illustrations): 8) clearly implies 11). And if G has an infinite set of pairwise disjoint elements $a_1, a_2, ...$ then $a^{\perp\perp} \subset (a_1 a_2)^{\perp\perp} \subset ... $. Hence 12) implies 3). The equivalence of 1) and 12) is a corollary to Proposition 5.3.5.

In addition, note the following equivalences (where n is a fixed integer):
a) *G has a basis of n elements;*
b) *$\Gamma(G)$ has exactly n roots;*
c) *Each $g \in G$ has less than or equal to n values, and at least one element has exactly n values;*

d) G contains n pairwise disjoint elements, but not $n+1$.

For a proof, see [Conrad, 1970a].

If L is an algebraic lattice, then L satisfies the ACC if and only if each $x \in L$ is compact (see [Birkhoff, 1967; pages 181-184].)

COROLLARY 1. *For an ℓ-group G, the following are equivalent:*
1) $C(G) = C_p(G)$;
2) $C(G)$ satisfies the ACC;
3) $C_p(G)$ is a complete sublattice of $C(G)$;
4) $C_p(G)$ satisfies the ACC;
5) $\Gamma(G)$ has only finitely many roots, and satisfies the ACC;
6) G is a lex-sum of finitely many o-groups, and each o-group used in the construction satisfies the ACC on convex subgroups.

For the proof of 5) $\Rightarrow$ 1), see [Birkhoff, 1967].

Notice that if G is the ℓ-group of sequences of integers with finite range, then $C_p(G) \neq C(G)$, while $C_p(G)$ is a complete lattice (but *not* a complete *sublattice* of $C(G)$).

COROLLARY 2. *For an ℓ-group G, the following are equivalent:*
1) $C(G)$ is finite;
2) $\Gamma(G)$ is finite;
3) $C(G)$ satisfies the ACC and the DCC;
4) G has a finite basis, and each o-group used in the construction of the lex-sum has a finite number of convex subgroups.

Note that Birkhoff [1967; page 183] shows that an algebraic lattice L with the DCC is isomorphic to the family of trivially ordered subsets of $\Gamma(L)$, the set of meet-irreducible elements of L.

COROLLARY 3. *For an ℓ-group G, the following are equivalent:*
1) $C(G)$ satisfies the DCC;
2) $\Gamma(G)$ satisfies the DCC and has only a finite number of roots;
3) G has a finite basis, and each o-group used in the construction of the lex-sum has the DCC on convex subgroups.

PROPOSITION 5.3.5. *For a prime subgroup M of G the following are equivalent:*

1) $M \in Pp(G)$;
2) *M is a minimal prime with an order unit (for M).*

Proof. 1) $\Rightarrow$ 2) is clear.

2) $\Rightarrow$ 1). Let u be an order unit in M. From Theorem 5.1.6 (since $G^+\backslash M$ is an ultrafilter) there is an element $a \in G^+\backslash M$ such that $u \wedge a = e$. If a is not basic it bounds two disjoint positive elements x,y; since M is prime, one of these, say x, belongs to M. But then $u \wedge x = e$, which contradicts the fact that u is a unit for M. Thus a is basic. Now $a^\perp \subseteq M$ since $a \notin M$ and consequently $a^\perp = M$. This makes M a polar with unit u; hence $u^{\perp\perp} = M$.

COROLLARY 1. *If $e < g \in G$ and $G(g)$ is a minimal prime, then $g^{\perp\perp} = G(g)$ and $g^\perp = a^{\perp\perp}$ for a suitable basic element $e \leq a \in G$.*

Proof. Since $G(g)$ has an order unit, we may apply 5.3.5 and Theorem 5.2.1 part 5.

COROLLARY 2 (Theorem 5.3.4, 1 iff 12). *G has a finite basis if and only if each minimal prime is a principal polar.*

Proof. If G has a basis $\{a_1,...,a_n\}$ then $a_i^\perp = (a_1...a_{i-1}a_{i+1}...a_n)^{\perp\perp}$, and clearly the $a_i^\perp$ are the only primes.

Conversely, if $e < g \in G$ and M is a minimal prime with $g \notin M$, then find $e \leq a,b \in G$ (with a basic) so that $M = b^{\perp\perp} = a^\perp$. Then $g \geq g \wedge a > e$ and $g \wedge a$ is basic; thus, G has a basis.

If $\{a_1,a_2,...\}$ is an infinite set of pairwise disjoint basic elements, we may choose $b_n > e$ (for each n) such that $b_n^{\perp\perp} = a_n^\perp$ ($a_n^\perp$ is a minimal prime), and $b_m \wedge b_n > e$ for each pair m,n. The set of finite meets $b_{n_1} \wedge ... \wedge b_{n_k}$ forms a filter. Let T be an ultrafilter containing it and set $P = \bigcup \{t^\perp \mid t \in T\}$. Then P is a minimal prime and $P \supseteq b_n^\perp = a_n^{\perp\perp}$ for all n. Hence, if G has an infinite basis there is a minimal prime that contains the basis group. This is a contradiction, since such a prime is not in $Pp(G)$.

Note that in the ℓ-group of eventually constant sequences of integers, each minimal prime *except* the basis group is a principal polar.

COROLLARY 3. *An ℓ-group G has property F if and only if each $C \in C_p(G)$ has a finite basis.*

Recall that $u \in G$ is a *strong order unit* if $G = G(u)$. The final proposition and corollary of this section are easy to prove. A proof of 5.3.6 may be found on page 3.32 of [Conrad, 1970a].

PROPOSITION 5.3.6. *Suppose G has a finite basis $\{a_1,...,a_n\}$ and $A_i = a_i^{\perp\perp}$. If $C \in \mathcal{P}(G)$, then*

$$C = (A_{i_1} \boxplus ... \boxplus A_{i_k})^{\perp\perp}$$

for suitable indices $1,...,k$. Also,

$$C = (\boxplus_{i \neq i_t} A_i)^{\perp} = \bigcap_{i \neq i_t} A_i^{\perp}.$$

If $e \neq C \in \mathcal{M}(G)$ then it is an indecomposable polar. Also, $Pp(G) = \mathcal{P}(G)$, and $\mathcal{M}(G)$ is precisely the set of indecomposable polars of G.

COROLLARY. *For an ℓ-group G, the following are equivalent:*

1) *Each minimal prime belongs to $C_p(G)$;*
2) *G has a finite basis, and if a is special then either $a^{\perp\perp} \in C_p(G)$ or $a^{\perp\perp} = G$;*
3) *G has a finite basis, and each o-group used in the construction of the lex-sum, except possibly the top, has a strong order unit.*

5.4. Kenoyer's Example

We now show that the lattice $C(G)$ does not determine whether or not the ℓ-group G is archimedean, completely distributive, or normal valued. Indeed, $C(G)$ does not determine whether or not G belongs to any proper variety of ℓ-groups. Also, closed prime subgroups are not recognizable in $C(G)$. By Theorem 1.1.10 and ([McCleary,1972b] or [Glass, 1981b, Theorem 8.3.2]), it suffices to exhibit ℓ-groups G and C with $C(G)$ and $C(C)$ isomorphic lattices such that

(1) G is Archimedean and G is the only closed prime subgroup of G, and   (2) C is completely distributive but not normal valued.

Indeed, the C we give has the property that every non-identity element has a normal value even though C is not normal valued. The following example is due to Kenoyer [1984].

We write $C(\mathbf{R})$ for $C(\mathbf{R},\mathbf{R})$, the ℓ-group of continuous functions from $\mathbf{R}$ into $\mathbf{R}$ under pointwise addition and ordering (Chapter 0, Example 7).

Let $G = \{g \in C(\mathbf{R}) \mid \exists a_0 < \ldots < a_n$ such that $g \mid [a_j, a_{j+1}]$ is linear for $0 \leq j < n$ and $g(x) = 0$ if $x \notin (a_0, a_n)\}$, an Archimedean ℓ-group. We "turn the picture through $45°$ " to obtain the ℓ-group C of Chehata defined in Example 8 of Chapter 0. Using the compactness of the closures of supports in G and C (If $g \in G$, $\mathrm{supp}(g) = \{x \in \mathbf{R} \mid g(x) \neq 0\}$; if $g \in C$, $\mathrm{supp}(g) = \{x \in \mathbf{R} \mid xg \neq x\}$.) it is not difficult to show that the maximal convex ℓ-subgroups of G are $G_a = \{g \in G \mid g(a) = 0\}$ $(a \in \mathbf{R})$ and those of C are $C_a = \{g \in C \mid ag = a\}$ $(a \in \mathbf{R})$ – see below. Let

$$G_{a,r} = \{g \in G \mid (\exists \varepsilon > 0)(g(b) = 0 \; \forall b \in [a, a+\varepsilon])\} \text{ and}$$

$$G_{a,l} = \{g \in G \mid (\exists \varepsilon > 0)(g(b) = 0 \; \forall b \in [a-\varepsilon, a])\}.$$

Then $G_{a,r}$, $G_{a,l} \subseteq G_a$ and any prime subgroup of G properly contained in G_a is one of these two convex ℓ-subgroups. The proof again uses compactness and we give it below. Analogous results hold for C setting up an isomorphism between $\Pi(G)$ and $\Pi(C)$, and hence $C(G)$ and $C(C)$. If $f \notin C_a$, then C_a is a value of f and as $f^{-1}C_a f = C_{af} \neq C_a$, C is not normal valued although G is, being abelian. Note that $\Pi(C)$ has all its roots finite and every non-identity element of C has a value normal in its cover (if $a = \mathrm{sup}(\mathrm{supp}(g))$, then $C_{a,r}$ is a value of g, and $C_{a,r}$ is normal in its cover C_a.)

We now prove the results about G, the proofs for C being analogous. First note that if $g \in G$, then the closure of $\mathrm{supp}(g) = \mathrm{cl}(\mathrm{supp}(g))$ is a compact set. Hence $\{g(x) \mid x \in \mathbf{R}\}$ is bounded for each $g \in G$. Suppose that $L \in C(G)$ and $L \not\subseteq G_a$ for all $a \in \mathbf{R}$. Let $g > 0$ and $m = \mathrm{sup}\{g(x) \mid x \in \mathbf{R}\}$. By hypothesis, for each $a \in \mathrm{cl}(\mathrm{supp}(g)) = K$, there is $f_a \in L^+ \backslash G_a$. Now $f_a(a) > 0$ and as f_a is continuous, $U_a = \{x \in \mathbf{R} \mid f_a(x) > \frac{1}{2}f_a(a)\}$ is an open neighborhood of a. $\{U_a \mid a \in K\}$ is an open cover of K and so there is a finite subcover $\{U_{a_j} \mid j = 1,\ldots,q\}$. Let $n = \mathrm{min}\{\frac{1}{2}f_{a_j}(a_j) \mid 1 \leq j \leq q\}$ and $0 < p \in \mathbf{Z}$ such that $pn > m$. Now $f = \vee_j f_{a_j} \in L^+$ and $0 \leq g \leq pf$ so $g \in L$. Thus $L = G$. Since $G_a \subseteq G_b$ only if $a = b$, each proper subgroup of G is contained in G_a for a unique $a \in \mathbf{R}$.

Now if $P \in \Pi(G)$ and $P \subset G_a$, we claim $P \subseteq G_{a,r}$ or $P \subseteq G_{a,l}$. If this were false, there would be $h_1 \in P^+ \backslash G_{a,r}$ and $h_2 \in P^+ \backslash G_{a,l}$. Let $g \in G_a^+$ and $m = \mathrm{sup}\{g(x) \mid x \in \mathbf{R}\}$. By hypothesis, there are $\varepsilon_j > 0$ and $m_j > 0$ such that $h_1(x) = m_1(x-a)$ if $x \in [a, a+\varepsilon_1]$ and $h_2(x) = -m_2(x-a)$ if $x \in [a-\varepsilon_2, a]$. Choose ε_1 and ε_2 sufficiently small so that g has slope n_1, say, throughout $[a, a+\varepsilon_1]$ and slope $-n_2$, say, throughout $[a-\varepsilon_2, a]$. Let $0 < n \in \mathbf{Z}$ be such that $m_j n \geq n_j$ $(j = 1,2)$. Then

$0 \le g(x) \le n(h_1 \vee h_2)(x)$ for all $x \in [a\text{-}\varepsilon_2, a+\varepsilon_1]$. Let $K_0 = \mathrm{cl}(\mathrm{supp}(g)) \backslash (a\text{-}\varepsilon_2, a+\varepsilon_1)$, a compact set containing a. Since P is prime, $P \not\subseteq G_b$ for all $b \in K_0 \backslash \{a\}$ and, as in the previous paragraph, we can find $f \in P^+$ with $f(x) \ge g(x)$ for all $x \in K_0$. Then $0 \le g \le f \vee n(h_1 \vee h_2)$ so $g \in P$. Hence $P = G_a$, the desired contradiction.

To complete the proof that $\Pi(G) = \{G_a, G_{a,r}, G_{a,l} \mid a \in \mathbf{R}\}$, it is enough to show that if $P \in \Pi(G)$ and $P \subseteq G_{a,r}$, then $P = G_{a,r}$ (the proof for $G_{a,l}$ being the same). For reductio ad absurdum, assume $P \subset G_{a,r}$ and let $g \in G_{a,r}^+ \backslash P$. Let $\varepsilon > 0$ with $g(x) = 0$ for all $x \in [a, a+\varepsilon]$.

Let

$$f(x) = \begin{cases} x\text{-}a & \text{if } x \in [a, a+\varepsilon/2] \\ a+\varepsilon\text{-}x & \text{if } x \in [a+\varepsilon/2, a+\varepsilon] \\ 0 & \text{if } x \notin [a, a+\varepsilon] \end{cases}$$

Then $f \in G$ and $f \wedge g = 0$. Since P is prime and $g \notin P$, it follows that $f \in P \subseteq G_{a,r}$ a contradiction.

Finally, since G is archimedean, any closed convex ℓ-subgroup of G is a polar [Bigard, Keimel and Wolfenstein, 1977; Theorem 11.1.10]. Clearly no G_a, $G_{a,l}$, or $G_{a,r}$ is a polar, so G is the only closed prime subgroup of G. On the other hand, since C acts doubly transitively on $\mathbf{R}$, C_a is closed for every $a \in \mathbf{R}$ [Glass, 1981b; Theorems 8D & 8C]. Since $\cap \{C_a \mid a \in \mathbf{R}\} = \{e\}$, C is completely distributive (Theorem 1.1.10).

Marlow E. Anderson
The Colorado College
Colorado Springs, Colorado 80903
U. S. A

Paul F. Conrad
University of Kansas
Lawrence, Kansas 66044
U. S. A

Jorge Martinez
University of Florida
Gainesville, Florida 32611
U. S. A

Jorge Martinez

CHAPTER 6

TORSION THEORY OF ℓ-GROUPS

6.1. Introduction

This will undoubtedly turn out as a very biased survey of the subject of torsion classes of ℓ-groups, not accidentally tilted in the direction of questions
having to do with extensions of one ℓ-group by another. It is by no means
intended to be a complete treatment. Rather, we will incorporate here the
central ideas together with a generous list of examples. The chapter will
close with some very recent contributions. Almost all proofs will be omitted; nearly all the principal results appear in [Martinez, 1975a; 1980], and
[Holland and Martinez, 1979]. The new ones take off on the concept of split
subgroups, appearing in [Martinez, a].

All classes C of ℓ-groups that we consider are assumed to be non-
empty and closed with respect to isomorphisms; i.e., if $A \in C$ and $A \cong B$ then
$B \in C$ as well.

A class of ℓ-groups T is called a *torsion class* if it is closed with respect
to

        (t1) taking ℓ-homomorphic images of ℓ-groups in the class,
        and
        (t2) for any ℓ-group G, forming arbitrary joins of convex ℓ-
            subgroups of G which belong to the class.

128

A. M. W. Glass and W. C. Holland (eds.), Lattice-Ordered Groups, 128–141.
© 1989 by Kluwer Academic Publishers.

If T is a torsion class we use $T(G)$ to denote the join of all the convex ℓ-subgroups of G belonging to T. $T(G)$ is called a *torsion radical of* G; it is invariant under all the ℓ-automorphisms of G, and in particular it is an ℓ-ideal.

We then have the following elementary lemma.

LEMMA 6.1.1. *Suppose T is a torsion class. Then*

   (i) *if C is a convex ℓ-subgroup of G then $T(C) \subseteq T(G)$;*

   (ii) *if $\phi : G \to H$ is an onto ℓ-homomorphism then*
   $\phi(T(G)) \subseteq T(H)$;

   (iii) $T(T(G)) = T(G)$.

Conversely, suppose we associate to each ℓ-group G a convex ℓ-subgroup $\mathcal{D}(G)$ subject to (i), (ii), and (iii) above, and set $T = \{G \mid \mathcal{D}(G) = G\}$, then T is a torsion class, and for each ℓ-group G, $T(G) = \mathcal{D}(G)$.

Proof. $T(C)$ is a convex ℓ-subgroup of G and belongs to T; therefore $T(C) \subseteq T(G)$. That proves (i). As for (ii), $\phi(T(G)) \in T$ and is a convex ℓ-subgroup of H, which implies $\phi(T(G)) \subseteq T(H)$. (iii) is obvious.

Suppose the function $\mathcal{D}$ satisfies (i), (ii), and (iii), and $T = \{G \mid \mathcal{D}(G) = G\}$. If $\phi : G \to H$ is an onto ℓ-homomorphism and $G \in T$ then

$$H = \phi(G) = \phi(\mathcal{D}(G)) \subseteq \mathcal{D}(H),$$

and so $H \in T$. Next suppose $\{C_\lambda \mid \lambda \in \Lambda\}$ is a family of convex ℓ-subgroups of an ℓ-group G, $C = \vee C_\lambda$, and each $C_\lambda \in T$. Then

$$C = \vee C_\lambda = \vee \mathcal{D}(C_\lambda) \subseteq \mathcal{D}(C),$$

which implies $C = \mathcal{D}(C)$ and $C \in T$. Therefore, T is a torsion class.

Since $\mathcal{D}(\mathcal{D}(G)) = \mathcal{D}(G)$, we have that $\mathcal{D}(G) \in T$, which implies $\mathcal{D}(G) \subseteq T(G)$. But $T(G) \in T$ as well, and so $T(G) = \mathcal{D}(T(G)) \subseteq \mathcal{D}(G)$, proving $T(G) = \mathcal{D}(G)$.

(Note: This proof is fairly typical of the arguments in [Martinez, 1980].)

If a torsion class T is also closed with respect to

(t3) taking convex ℓ-subgroups,

we say that T is *hereditary*. T is hereditary if and only if $T(C) = C \cap T(G)$ for every convex ℓ-subgroup C of G.

Lemma 6.1.1 says that in dealing with torsion classes we may either consider the classes themselves or else their torsion radicals.

Now suppose $\mathcal{T}$ is a torsion class and $\hat{\mathcal{T}} = \{G \mid \mathcal{T}(G) = \{e\}\}$. Then $\hat{\mathcal{T}}$ is closed with respect to

(tf1) taking convex ℓ-subgroups, and
(tf2) subdirect products.

Any class of ℓ-groups satisfying (tf1) and (tf2) will be called a *torsion-free class*. If X is any torsion-free class we let $X(G)$ stand for the intersection of all the ℓ-ideals K of G for which $G/K \in X$. We then get the following companion to Lemma 6.1.1 (see [Martinez, b].)

LEMMA 6.1.2. *Suppose X is a torsion-free class. Then*
    (a) $X(A) \subseteq X(G)$ for every convex ℓ-subgroup A of G,
    (b) $X(G/X(G)) = \{e\}$, and
    (c) if K is an ℓ-ideal of G such that $X(G/K) = \{e\}$ then
        $K \supseteq X(G)$.
Conversely, suppose $\mathcal{J}$ is an assignment associating to each ℓ-group G an ℓ-ideal $\mathcal{J}(G)$ satisfying (a), (b), and (c), and $X = \{G \mid \mathcal{J}(G) = \{e\}\}$. Then X is a torsion-free class and $X(G) = \mathcal{J}(G)$ for each ℓ-group G.

And so, as for torsion classes, we have two ways to regard torsion-free classes: either by handling the classes themselves of else their "verbal" ideals, to borrow a term from varieties of groups.

Before giving examples let us mention one further association to round out the symmetry between torsion classes and torsion-free classes. If X is any torsion-free class let $\hat{X}$ denote the class $\{G \mid X(G) = G\}$; then it is easily proved that $\hat{X}$ is a torsion class. In this manner we have set up a pair of correspondences, $\mathcal{T} \rightarrow \hat{\mathcal{T}}$ assigning a torsion-free class to a torsion class, and $X \rightarrow \hat{X}$ doing the reverse. We shall examine these correspondences shortly. For any torsion class $\mathcal{T}$, $\hat{\mathcal{T}}$ is called the *opposite* torsion-free class, and for any torsion-free class X, $\hat{X}$ is the *opposite* torsion class. Let $\hat{X} = X^*$ and $\hat{\mathcal{T}} = \mathcal{T}^*$. We will see that, in general $X \neq X^*$ and $\mathcal{T} \neq \mathcal{T}^*$. Both $\hat{\mathcal{T}}$ and $\hat{X}$ satisfy an additional property:

(ex) if G is any ℓ-group and A is an ℓ-ideal so that both A and G/A are in the class, then G is as well.

That is, $\hat{T}$ and $\hat{X}$ are closed under forming extensions. A class satisfying (ex) will be called *complete*.

EXAMPLES:

First, we give a list of popular torsion classes; unless otherwise noted the class is hereditary.

$\mathcal{H}$, the class of *hyper-archimedean* ℓ-groups. An ℓ-group belongs to $\mathcal{H}$ iff every ℓ-homomorphic image is archimedean. For more detailed information about $\mathcal{H}$ and its radical see [Conrad, 1974b] and [Martinez, 1974a]. For example, $e < a \in \mathcal{H}(G)$ if and only if for each $e < g \in G$ there is a positive integer n so that $a \wedge g^n = a \wedge g^{(n+1)}$. (Lemma 2.1 in [Martinez, 1974a].)

$\mathcal{F}$, the class of *finite-valued* ℓ-groups. An ℓ-group belongs to $\mathcal{F}$ iff each value of every element is special (Section 5.3).

Sp, the class of *Specker* groups. An ℓ-group belongs to Sp iff it is generated, as a group, by its singular elements. (Recall that an element $s > e$ is *singular* if $s = tu$ with $t, u > e$ only when $t \wedge u = e$.)

$\mathcal{F}_0$, the class of ℓ-groups satisfying *property (F)*. An ℓ-group belongs to $\mathcal{F}_0$ iff no positive element exceeds an infinite pairwise disjoint set (Section 5.3). Note that $\mathcal{F}_0 \subseteq \mathcal{F}$.

$\mathcal{P}w$, the class of *pairwise splitting* ℓ-groups. An ℓ-group G belongs to $\mathcal{P}w$ iff for each pair $e < x, y \in G$, there exist $x_1, x_2 \in G$ such that $x = x_1 x_2$, $x_1 \wedge x_2 = e$, $x_1 \in G(y)$, and $x_2 \wedge y \ll x_2$. Actually, this class is not well known, but it deserves a better fate. The basic facts concerning $\mathcal{P}w$ are in [Martinez, 1977]. Lemma 3 of that article says that $G \in \mathcal{P}w$ if and only if G is normal valued and for each $e < x \in G$, $G(x)/N_x \in \mathcal{E}$, where N_x is the intersection of all maximal convex ℓ-subgroups of $G(x)$. Observe that an archimedean ℓ-group is pairwise splitting iff it is hyperarchimedean.

And speaking of normal-valued ℓ-groups, we observed

LEMMA 6.1.3 [Martinez, 1975a]. *The variety $\mathcal{N}$ of normal-valued ℓ-groups is a torsion class.*

Holland then generalized this.

THEOREM 6.1.4 [Holland, 1979]. *Every variety of ℓ-groups is a torsion class.*

Moreover, every variety of ℓ-groups is also a torsion-free class, and it is obvious that a torsion class which is also a torsion-free class is necessarily a variety.

The following torsion classes are complete:

$\mathcal{N}$, the variety of normal-valued ℓ-groups. By a theorem of Holland [1976] (see Theorem 10.4.6), $\mathcal{N}$ is the largest proper variety of ℓ-groups.

$\mathcal{D}c$, the class of ℓ-groups satisfying the descending chain condition on prime subgroups. $G \in \mathcal{D}c$ if and only if every prime subgroup is a value.

$\mathcal{D}_s$, the class of ℓ-groups which have no infinite doubling chains. A *doubling chain* is a sequence $c_1 > c_2 > ... > c_n > ... > e$ so that $c_n \geq (c_{n+1})^2$. (See [Martinez, 1975b]; in fact, $\mathcal{D}_s = Sp^*$.)

Consider also the following construction. First, a definition: if G is an ℓ-group and $A \subseteq B$ are convex ℓ-subgroups with A normal in B we call B/A a *subquotient* of G. The subquotient B/A *dominates* the subquotient B_1/A_1 if $A \subseteq A_1 \subseteq B_1 \subseteq B$. Now if $\mathcal{T}$ is any torsion class, let $\mathcal{T}^{\perp\perp}$ stand for the class of all ℓ-groups G in which every nontrivial subquotient dominates a nontrivial subquotient in $\mathcal{T}$. (See the definition of $\mathcal{T}^{\perp}$ in the later section on polar classes, page 135.) Then (see Theorem 2.5 of [Martinez, 1975a]) for each heriditary torsion class $\mathcal{T}$, $\mathcal{T}^{\perp\perp}$ is a complete hereditary torsion class.

Now some examples of torsion-free classes:

$\mathcal{A}r$, the class of archimedean ℓ-groups.

$\mathcal{R}\ell$, the class of subdirect products of copies of subgroups of the real numbers. We note that the torsion class $\hat{\mathcal{R}}\ell$, consisting of those ℓ-groups having no convex ℓ-subgroups which are both normal and maximal is not hereditary: in any totally ordered ℓ-group having no maximal convex ℓ-subgroups, each principal convex ℓ-subgroup has a normal maximal convex ℓ-subgroup.

$\mathcal{Z}$, the class of subdirect products of copies of the integers.

One final example for this list: if α is an infinite cardinal, let $\mathcal{V}_\alpha$ be the class of ℓ-groups in which each non-identity element exceeds one having at most α values. This torsion class is also not hereditary. $\hat{\mathcal{V}}_\alpha$ consists of all ℓ-groups in which every non-identity element has more than α values.

6.2 The Main Theorem

THEOREM 6.2.1 (Connection Theorem). *The functions $T \to \hat{T}$ and $X \to \hat{X}$ between torsion and torsion-free classes form a Galois connection. In addition, $T(G) \subseteq \hat{T}(G) = T^*(G)$, for each torsion class T and each ℓ-group G, while $X(G) \supseteq \hat{X}(G) = X^*(G)$, for each torsion-free class X and each ℓ-group G. A torsion class T (respectively, torsion-free class X) is complete if and only if $T = T^*$ (respectively, $X = X^*$). In fact, $T \to \hat{T}$ is a one-to-one correspondence between complete torsion classes and complete torsion-free classes. T^* is the smallest complete torsion class containing the torsion class T, and X^* is the largest complete torsion-free class contained in the torsion-free class X.*

Sketch of the proof: It should be evident that $T \to \hat{T}$ and $X \to \hat{X}$ are order inverting. If $G \in T$, it certainly has no non-trivial ℓ-homomorphic images in $\hat{T}$, which implies $G \in T^*$. Thus $T(G) \subseteq T^*(G)$ for every ℓ-group G. We have $G/T^*(G) \in \hat{T}$, and if K is an ℓ-ideal of G such that $G/K \in \hat{T}$ then $(T^*(G))/(K \cap T^*(G))$ is an ℓ-homomorphic image of $T^*(G)$ belonging to $\hat{T}$; the completeness of T^* implies $T^*(G) \subseteq K$. All of this implies $\hat{T}(G) = T^*(G)$.

Now suppose U is a complete torsion class containing T. Then $\hat{U} \subseteq \hat{T}$, which implies that $G/U(G) \in \hat{T}$. Therefore, if $G \in T^*$ then $G = U(G)$, and so $T^* \subseteq U$.

Note: this version combines the Connection Theorem with Theorem 3 in [Martinez, 1980]. See that paper for the complete proof.

Now suppose that T is any torsion class. Let $T^1 = T$, and if σ is any ordinal greater than 1, we define $T^\sigma(G)$ for any ℓ-group G as follows:

$$T^\sigma(G) = \begin{cases} \bigcup\{T^\rho(G) \mid \rho < \sigma\} \text{ if } \sigma \text{ is a limit ordinal} \\ \text{the preimage in } G \text{ of } T(G/T^{\sigma-1}(G)) \text{ otherwise.} \end{cases}$$

For each ordinal σ, T^σ is a torsion class (and hereditary if T is). Note that for each ℓ-group G, the chain $T^1(G) \subseteq T^2(G) \subseteq \ldots$ is necessarily eventually constant. So, for sufficiently large σ (depending on G), $T^\sigma(G) = T^{\sigma+1}(G)$. For such σ, let $\bar{T}(G) = T^\sigma(G)$. Then $\bar{T} = T^*$, from which it follows that if T is hereditary, so is T^*. From [Martinez, 1975a] we quote Theorem 1.7 (where the double polar is in the ℓ-group G in the sense of Section 1.1).

THEOREM 6.2.2. *For each hereditary torsion class* T, $T^*(G) \subseteq T(G)^{\perp\perp}$ *in any* ℓ-*group G.*

There is, of course, an analogous transfinite construction of X^* for a given torsion-free class. Define $X^1 = X$, and if σ is an ordinal number greater than 1, let

$$X^{\sigma} = \begin{cases} \cap\{X \, P(G) \mid \rho < \sigma\} \text{ if } \sigma \text{ is a limit ordinal} \\ \\ X(X^{\sigma-1}(G)) \text{ otherwise.} \end{cases}$$

Then $X^*(G) = X^{\sigma}(G)$ for sufficiently large σ, which, naturally, depends on G.

6.3 Hereditary Torsion Classes

For the remainder of this chapter (unless otherwise noted) all torsion classes will be assumed hereditary. We first examine the consequences for $X = \hat{T}$.

PROPOSITION 6.3.1. (See [Martinez, b], Theorem 1.6.) *The opposite torsion-free class* $X = \hat{T}$ *satisfies:*
  (h1) $X(A) = A \cap X(G)$ *for each convex* ℓ-*subgroup A of G.*
  (h2) *If A is a dense convex* ℓ-*subgroup of G (that is, $0 < g \in G \Rightarrow$*
       $\exists a \in A$ *such that $0 < a \le g$) and $A \in X$, then $G \in X$.*
  (h3) *If P is a normal polar in G and $G \in X$, then $G/P \in X$.*
X also satisfies (t2): it is closed under joins of convex ℓ-*subgroups.*

Let us consider these properties for an arbitrary torsion-free class X. First, there are "ℓ-ideal" versions of (h1) and (h2):

  (hℓ1) $X(A) = A \cap X(A)$ *for every* ℓ-ideal A of G.
  (hℓ2) If A is a dense ℓ-ideal of G and $A \in X$ then $G \in X$.

Clearly (h1) implies (hℓ1) and (h2) implies (hℓ2). It is proved in [Martinez, b] that (hℓ1) and (hℓ2) are equivalent. Proposition 6 of [Martinez, b] says that (hℓ2) implies (h3), and the lemma immediately preceding it shows that any torsion-free class satisfying (h2) must be complete. It's also quite easy to see that (h1) implies (t2).

On the other hand, there are the following counterexamples: Any variety which is not complete (i.e., any proper variety other than $\mathcal{N}$) satisfies (t2) but not (h2) or (hℓ2); any variety also satisfies (h3); $\mathcal{R}\ell$ satisfies (h3) but neither (h2) nor (t2); the class of all ℓ-groups G having having no finite-valued elements satisfies (t2) but not (h3).

Let us summarize with a diagram indicating the valid implications and the only equivalence we know.

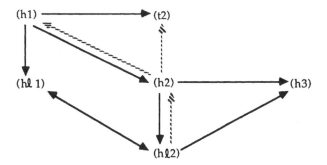

Notes: The broken lines indicate that we do not know whether the implications hold. To show that (hℓ1) does not imply (h1) take $\mathcal{J}$ to be the class of ℓ-groups in which every ℓ-ideal is a cardinal summand. Then both $\mathcal{J}$ and $\mathcal{J}^*$ satisfy (hℓ1) but neither one is hereditary. However, we are cheating a bit here: $\mathcal{J}$ and $\mathcal{J}^*$ are torsion classes, not torsion-free classes. But are we, after all?

PROPOSITION 6.3.2. *For a torsion-free class X the following are equivalent.*

    (1) *X is complete and $\hat{X}$ is closed with respect to taking ℓ-ideals.*

    (2) *X satisfies (hℓ1)*

    (3) *X satisfies (hℓ2).*

The proof of this proposition appears in [Martinez, b].

6.4 Polar Torsion Classes and Inaccessibility

Let $\mathcal{T}$ be a hereditary torsion class. $\mathcal{T}^\perp$ will denote the largest hereditary torsion class satisfying $\mathcal{T} \cap \mathcal{T}^\perp = \{\{e\}\}$. That such a class exists is a consequence of the following theorem from section 2 of [Martinez, 1975a].

THEOREM 6.4.1. *The class* **T** *of all hereditary torsion classes is a complete Brouwerian lattice(page 15). If* $\{T_\lambda \mid \lambda \in \Lambda\} \subseteq$ **T** *then for each ℓ-group G*

$$(\vee_\lambda T_\lambda)(G) = \Sigma_\lambda T_\lambda(G) \quad and$$

$$(\wedge_\lambda T_\lambda)(G) = \cap_\lambda T_\lambda(G).$$

$T^\perp$ is called the *polar* torsion class of T. In general, T is called a *polar torsion class* if $T = (T^\perp)^\perp$ $(= T^{\perp\perp}$, as is easy to verify). We then have the following results; again, refer to section 2 of [Martinez, 1975a].

THEOREM 6.4.2. *For any torsion class T,*
     $T^\perp = \{G \mid G$ *has no non-trivial subquotients in* $T\}$.
     $T^{\perp\perp} = \{G \mid$ *each non-trivial subquotient of G dominates a non-trivial subquotient in* $T\}$.
Moreover, $T^\perp \subseteq \hat{T}$ and $T^* \subseteq T^{\perp\perp}$; in particular, a polar class is complete.

A bit later we answer the question of when the equation $T^* = T^{\perp\perp}$ holds. For now, let us turn to the lattice **T**; we've discussed the supremum and infimum, and now let us introduce yet another binary operation. Suppose T and U are two not necessarily hereditary torsion classes; define $T \cdot U$ by: $G \in T \cdot U$ if $G/T(G) \in U$. Thus, if $T_1 = T \cdot U$, $T_1(G)$ is defined by the equation $T_1(G)/T(G) = U(G/T(G))$. $T \cdot U$ is a torsion class, and both T and U are subclasses of it. If T and U are hereditary so is $T \cdot U$. In this notation the ordinal sequence used to construct T^* can be viewed as follows: $T^{\sigma+1} = T^\sigma \cdot T$ for every ordinal σ.

This product is associative, in general far from commutative, and

PROPOSITION 6.4.3. (2.2 of [Martinez, 1975a]). *If T and $\{U_\lambda \mid \lambda \in \Lambda\}$ are torsion classes, then*

     (a) $\cap_\lambda T \cdot U_\lambda = T \cdot (\cap_\lambda U_\lambda)$ and $(\cap_\lambda U_\lambda) \cdot T \subseteq \cap_\lambda (U_\lambda \cdot T)$.

     (b) $\vee_\lambda (T \cdot U_\lambda) = T \cdot (\vee_\lambda U_\lambda)$ and $(\vee_\lambda U_\lambda) \cdot T \subseteq \vee_\lambda (U_\lambda \cdot T)$.
The containments in (a) and (b) can be strict.

Let us now turn to questions of inaccessibility in **T**, and look for torsion classes which (in various ways) can or cannot be obtained by combining two or more proper subclasses. There are at least three reasonably common types of inaccessible elements:

     (a) T is *indecomposable* if $T = T_1 \vee T_2$ with $T_1 \cap T_2 = \{\langle e \rangle\}$ implies that $T = T_1$ or $T = T_2$.

(b) $\mathcal{T}$ is *finitely join irreducible* if $\mathcal{T} = \mathcal{T}_1 \vee \mathcal{T}_2$ implies that $\mathcal{T} = \mathcal{T}_1$ or $\mathcal{T} = \mathcal{T}_2$.

(c) $\mathcal{T}$ is *join irreducible* if $\mathcal{T} = \vee_\lambda \mathcal{T}_\lambda$ implies that $\mathcal{T} = \mathcal{T}_\lambda$ for some λ.

Evidently, join irreducibility implies finite join irreducibility, which in turn implies indecomposability.

Our first result in this area comes from [Martinez, 1975a].

PROPOSITION 6.4.4. *If $\mathcal{T}$ is a torsion class which is also finitely meet irreducible, then $\mathcal{T}$ is indecomposable.*

The results on join irreducibility are from [Holland and Martinez, 1979]. A class C has property *LEX* if it contains a non-trivial o-group L so that for each $G \in C$ the lexicographic product $G \overleftarrow{\times} L \in C$. For example, $\mathcal{F}$, the class of all finite-valued ℓ-groups, and $\mathcal{F}_0$, the class of ℓ-groups satisfying property (F), are not complete, yet they do have property LEX.

Suppose G is an ℓ-group and $G_g = G$ for each $e < g \in G$. Form $\bar{G} = \Pi\{G_g \mid g > e\}$ and let δ be the element defined by $\delta_g = g$. Let Hz(G) be the ℓ-subgroup of $\bar{G}$ generated by the cardinal sum and δ. We say that a class C has the *horizontal closure property* if for each $G \in C$, Hz(G) $\in C$. Every variety of ℓ-groups has the horizontal closure property; the class $\mathcal{H}$ of hyperarchimedean ℓ-groups does as well.

THEOREM 6.4.5. (1,2, and 3 of [Holland and Martinez, 1979].) *Suppose $\mathcal{T}$ is a torsion class which is either complete, has the LEX property, or else the horizontal closure property. Then $\mathcal{T}$ is join irreducible.*

COROLLARY 6.4.6. $\mathcal{N}, \mathcal{H}, \mathcal{F}, \mathcal{F}_0, \mathcal{P}w, \mathcal{D}c$ *and all varieties of ℓ-groups are join irreducible.*

Next we pose the question: in forming $\mathcal{T}^*$, how long is the sequence of $\mathcal{T}^\sigma$'s? The answer is given by Theorem 6.4.8; a key step on the way is the preceding theorem and the next lemma.

LEMMA 6.4.7. *Suppose $\mathcal{T} = \mathcal{T}_1 \cdot \mathcal{T}_2$ and $\mathcal{T}$ is complete. Then either $\mathcal{T} = \mathcal{T}_1$ or $\mathcal{T} = \mathcal{T}_2$.*

THEOREM 6.4.8. *For any torsion class $\mathcal{T}$ in* T, *either $\mathcal{T} = \mathcal{T}^*$ or else $\mathcal{T}^\sigma \neq \mathcal{T}^*$ for every ordinal σ.*
(Note: the proofs of Lemma 6.4.7 and Theorem 6.4.8 appear in [Holland and Martinez, 1979].)

The following theorem, also from [Holland and Martinez, 1979], has even more dramatic consequences. We shall assume that the reader is acquainted with wreath products. In particular, we are interested here in wreath products over (possibly) large indexing sets: $\mathrm{Wr}_{\ell \in \mathcal{K}} G_\ell$, where each G_ℓ is an ℓ-group; the members of the wreath product may be thought of as matrices $(g_{\ell r})$ with entries $g_{\ell r} \in G_\ell$ and indexed over $\mathcal{K}$ and another set R about which we shall remain vague; see Section 2.3 for more details. One property of such a wreath product should be stressed; it is this: if $(g_{\ell r}) \in \mathrm{Wr}_{\ell \in \mathcal{K}} G_\ell$ and $g_{as} \neq e$ for some $a \in \mathcal{K}$ and $s \in R$, then for each $b < a$, the convex ℓ-subgroup generated by $(g_{\ell r})$ has a subquotient isomorphic to G_b. Using this property Holland and Martinez [1979] obtain :

THEOREM 6.4.9. *Suppose $\mathcal{U}$ is a torsion class closed under the following operation: for each family $\{G_n \mid n \in \mathbf{Z}\}$, $\mathrm{Wr}_n G_n \in \mathcal{U}$ if each $G_n \in \mathcal{U}$. If $\mathcal{U} = \mathcal{T}^*$ for some torsion class $\mathcal{T}$, it follows that $\mathcal{U} = \mathcal{T}$.*

COROLLARY 6.4.10. *If $\mathcal{U} = L$, the class of all ℓ-groups, $\mathcal{N}$ or any polar class then $\mathcal{U} = \mathcal{T}^*$ implies that $\mathcal{U} = \mathcal{T}$. In particular, $\mathcal{T}^* = \mathcal{T}^{\perp\perp}$ precisely when $\mathcal{T}$ is a polar class.*

The above results permit us to construct ascending chains of torsion classes of any (ordinal) length we please. For details see [Holland and Martinez, 1979].

Before moving on to results of a more recent vintage let us mention the paper of Jakubik [1982a], in which he discusses irreducibility criteria for singly generated torsion classes. (Compare also with [Holland and Martinez, 1979].) In that same article he also investigates covering properties. He shows, for example, that if a (hereditary) torsion class is singly generated it has infinitely many covers in **T**, and that each one is again singly generated. In addition, he shows (Proposition 4.9) that for each torsion class $\mathcal{T}$ there is a least torsion class $\mathcal{T}^c \supsetneq \mathcal{T}$ having no covers at all.

6.5 Locally Conditioned Torsion Classes

Recall that if G is an ℓ-group and $x \neq e$ in G, we use $G(x)$ for the convex ℓ-subgroup generated by x, and N_x for the intersection of all the maximal convex ℓ-subgroups of $G(x)$. We shall refer to the quotient $G(x)/N_x$ as the *local residue* of x. (Note: this is not exactly the same as the typical use of "local" in universal algebra.) N_x is invariant under every ℓ-automorphism of $G(x)$, and is therefore normal in $G(x)$. Observe that G is normal valued if and only if each local residue of G is archimedean if and only if each

$G(x)/N_x \in \mathcal{R}\ell$. In addition N_x is a maximally below subgroup (see [Ball, Conrad, and Darnel, 1986]) and is therefore closed.

Now let X stand for any class of archimedean ℓ-groups. Here, we will use additive notation. Loc(X) stands for the class of all ℓ-groups G for which each local residue belongs to X. For example, $\mathcal{P}w = $ Loc($\mathcal{H}$), from a result quoted earlier in the chapter, and $\mathcal{F} = $ Loc(O_R), where O_R is the class of all cardinal sums of subgroups of R. Also, $\mathcal{N} = $ Loc($\mathcal{A}r$) $= $ Loc($\mathcal{R}\ell$). In [Conrad and Martinez, a] Loc(Sp) is discussed in terms of discrete elements; $0 < d \in G$ is *discrete* if for each value V of d, $V+d$ is an atom of G/V. Then from Theorem 15 of [Conrad and Martinez, b], $G \in $ Loc(Sp) if and only if each $0 < g \in G$ can be written $g = \sum(n_i d_i + w_i)$ where the d_i are discrete and pairwise disjoint, $n_i \in N$ and $|w_i| << d_i$ for each $i = 1,2,...,t$.

A class of ℓ-groups is said to be a *radical class* [Jakubik, 1982a] if it satisfies (t2) and (t3) in the definition of torsion class. In the following proposition, $A+B = \{a+b \mid a \in A, b \in B\}$.

PROPOSITION 6.5.1. (2.5 of [Martinez, a].) Loc(X) *is closed under taking (a) convex ℓ-subgroups, (b) unions of chains of convex ℓ-subgroups belonging to* Loc(X), *and (c) images under split homomorphisms; (this to be explained presently). Further,* Loc(X) *is a radical class if X satisfies (d) $G = A+B$, $A,B \in X$, and A,B convex ℓ-subgroups of G imply $G \in X$.*

PROPOSITION 6.5.2. (2.9 of [Martinez, a].) *If X satisfies (d) above and for each $G \in X$ and each onto ℓ-homomorphism $\phi : G \rightarrow H \in \mathcal{A}r$ we have $H \in X$, then* Loc(X) *is a torsion class.*

However, let us pause for a moment to discuss the concept of a split subgroup. Suppose G is an ℓ-group and $A \subseteq G$ is an ℓ-subgroup. We say that A is a *split subgroup* if for each $0 < g \in G$ we can write $g = g_1 + g_2$ with $g_1 \wedge g_2 = 0$, $g_1 \in A$ and g_2 *lying above* A, which signifies that every value of g_2 contains A, or equivalently, $G(g_2) \cap A \subseteq N_{g_2}$.

Let us list some of the basic properties of split subgroups, with a few intriguing criteria which characterize this concept as well, citing the appropriate reference from [Martinez, a] as we go.

PROPOSITION 6.5.3 (Proposition 1.8). *Each split subgroup is necessarily convex and closed.*

PROPOSITION 6.5.4 (Proposition 1.4). *If A and B are split subgroups of G then so are $A \cap B$ and $A \vee B$.*

PROPOSITION 6.5.5 (Proposition 1.5). *If A is a split subgroup of G then $A+B = B+A$ for each convex ℓ-subgroup B of G.*

We give a proof of this curious result, so the reader may have a taste of the type of argument involved in most of these propositions. We show that $A+B \subseteq B+A$. So suppose $g = a+b$, with $0 \leq a \in A$ and $0 \leq b \in B$. Since A is split, write $g = y+z$, disjointly, with $y \in A$ and z lying above A. It's enough to prove that $z \in B$. Now as $z \leq g$, use the Riesz Interpolation Property to get $z = a_1+b_1$, with $0 \leq a_1 \leq a$ and $0 \leq b_1 \leq b$. Since z lies above A, z and b_1 are archimedean equivalent, which implies that $z \in B$.

PROPOSITION 6.5.6 (Proposition 1.1). *If either A is a cardinal summand of G or G is a lex-extension of A, then A is split in G.*

PROPOSITION 6.5.7 (Theorem 1.7). *G is finite valued if and only if each convex ℓ-subgroup of G is split.*

PROPOSITION 6.5.8 (Proposition 1.13). *Suppose G is normal valued. Then a convex ℓ-subgroup A of G is split if and only if for each $x \neq 0$ in G, $((G(x) \cap A)+N_x)/N_x$ is a cardinal summand of $G(x)/N_x$.*

PROPOSITION 6.5.9 (Theorem 1.12). *A convex ℓ-subgroup A of G is a split subgroup of G if and only if (i) $A+B = B+A$ for each convex ℓ-subgroup B of G, and (ii) for each family $\{B_\lambda \mid \lambda \in \Lambda\}$ of convex ℓ-subgroups of G,*

$$A+(\cap_\lambda B_\lambda) = \cap_\lambda(A+B_\lambda).$$

Incidentally, the last two criteria say, separately, that among normal-valued ℓ-groups a split subgroup is recognizable in the lattice of all convex ℓ-subgroups. We also point out that in reference to Proposition 6.5.4, the lattice of split subgroups of an ℓ-group is in general not a complete sublattice of the lattice of all convex ℓ-subgroups. Both infinite closures fail.

Now, a *split homomorphism* is simply an ℓ-homomorphism the kernel of which is split. In [Martinez, a] it is also shown that $\phi: G \to H$ is a split homomorphism if and only if for each $g \in G$ there is an $a \in G$ such that a lies above $\ker(\phi)$ and $\phi(g) = \phi(a)$. Thus, it follows that if ϕ is a split homomorphism and $\phi(a) \wedge \phi(b) = 0$ with a and b lying above the kernel, then $a \wedge b = 0$.

Any class of ℓ-groups $\mathcal{U}$ which arises as $\text{Loc}(X)$ for a suitable class X of archimedean ℓ-groups is said to be *locally conditioned*. The result which truly gives locally conditioned classes their special charm is the following (from [Martinez, a]):

THEOREM 6.5.10. *Suppose X is closed under forming finite cardinal sums. Then $\mathcal{U} = \text{Loc}(X)$ is split complete; that is, if A is an ℓ-ideal of G which is split in G and A and G/A belong to $\mathcal{U}$, then $G \in \mathcal{U}$.*

Here is yet another interesting locally conditioned class: let $\mathcal{A}r^b$ stand for the class of archimedean ℓ-groups with a basis. Then $G \in \text{Loc}(\mathcal{A}r^b)$ if and only if each $0 < g \in G$ has a special component; i.e., there is a special element $s \leq g$ so that $s \wedge (g-s) = 0$. We write $\pi Sv = \text{Loc}(\mathcal{A}r^b)$ and call the ℓ-groups in this class *pseudo-special-valued*. Every special-valued ℓ-group is indeed pseudo-special-valued, but $\pi Sv \neq Sv$, the class of special-valued ℓ-groups. Look at EC-Z, the group of pairs (s,n) where s is an eventually constant integer-valued sequence, and $n \in \mathbf{Z}$, with coordinatewise addition. Now set $(s,n) > 0$ if $s \geq 0$, and if s is eventually 0, then $n > 0$. Then EC-Z is an extension over a split ℓ-ideal of a special-valued ℓ-group by another. Thus Sv is not split complete. (Incidentally, both Sv and πSv are radical classes but neither one is a torsion class.) Indeed, Sv is not locally conditioned at all (see [Martinez, a], section 2.) Neither is Pr, the class of projectable (cf. page 118) ℓ-groups, nor $\mathcal{F}_o$, the torsion class of ℓ-groups satisfying property (F). And, more significantly perhaps, $\mathcal{D}c$, the torsion class of ℓ-groups which satisfy the descending chain condition on primes, is complete but not locally conditioned.

Let us close with some thoughts about extension theory of ℓ-groups, and how the study of radical classes might be helpful in the future. It seems clear to this author, especially after a considerable period of time during which many silly ideas led him essentially nowhere - a fate one, no doubt, deserves for indulging in them - that, in order to proceed intelligently vis-à-vis the topic of extensions, one needs to develop *particular theories* around specific kinds of kernels: for example, extensions of closed ℓ-ideals or - as we've seen in the final section - of split ℓ-ideals. Section 4 of [Martinez, a] explores this theme with yet other candidates. Most recently, with Conrad, we have explored the topic of extensions using kernels which satisfy vastly different requirements. Our results will appear in [Conrad and Martinez, c].

Finally, an open question. An ℓ-group is a *vector lattice* if it is also a real vector space in which positive scalar multiplication preserves order. Is the class of vector lattices a torsion class of ℓ-groups? The difficulty lies in forming suprema of convex ℓ-subgroups which carry a vector lattice structure.

J. Martinez
University of Florida
Gainesville, Florida 32611
U. S. A

Richard N. Ball

CHAPTER 7

COMPLETIONS OF ℓ-GROUPS

7.0. Introduction

Lattice-ordered groups admit a rich variety of completeness notions; some examples are conditional completeness (every bounded subset has a supremum), lateral completeness (every pairwise disjoint subset has a supremum), and projectability (each polar is a cardinal summand). For a completeness property C one then hopes for a completion result of the following form.

THEOREM PROTOTYPE. *Given any ℓ-group G there is an ℓ-group H unique up to isomorphism over G with respect to the following three properties:*

(a) G is an order dense ℓ-subgroup of H, i.e., for all $e < h \in H$ there is some $g \in G$ such that $e < g \leq h$;

(b) H has completeness property C;

(c) For all ℓ-subgroups K such that $G \leq K < H$, K lacks property C.

Two features of this paradigm distinguish this survey from the material in the next chapter. First, we only treat completions H in which G is order dense. This class includes almost all satisfactory known completions, but leaves out the recent archimedean results in [Ball and

142

A. M. W. Glass and W. C. Holland (eds.), Lattice-Ordered Groups, 142–174.
© 1989 by Kluwer Academic Publishers.

Hager, a,b] and [Madden and Vermeer, a]. And second, the methods we discuss apply generally to all ℓ-groups. We do not discuss the conditional completion, for instance, since a conditionally complete ℓ-group is necessarily archimedean, but instead treat the Dedekind-MacNeille completion which generalizes the conditional completion but which applies to all ℓ-groups.

This survey is organized in three sections, covering roughly the top, right and center columns and left edge, respectively, of the following diagram.

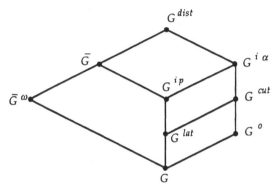

In this diagram each type of completeness implies the sorts of completeness below it, and the superior completion contains the inferior completion as an ℓ-subgroup. Therefore, if one shares the author's bias that bigger is better, then the distinguished completion G^{dist} is the best completion of all. More to the point, by constructing G^{dist} in Section 7.1 we solve what might be called the audacity problem: rather than requiring the reader to summon the audacity to imagine that G might have an extension with a particular completeness property C, we can instead simply check whether G^{dist} has property C. If so, then the required completion is usually

$$\bigcap \{K : G \leq K \leq G^{dist} \text{ and } K \text{ has property } C\}.$$

Besides conserving audacity, the latter approach avoids the frequently grisly details involved in individual *ad hoc* constructions of completions. In fact, we have avoided almost all grisly details by omitting the meat of the construction of G^{dist}, since its inclusion would have lengthened this chapter by half again. It is nevertheless true on balance that G^{dist} has taken a good deal of the pain out of the study of completions.

Section 7.2 is motivated by Cantor's construction of the real numbers by Cauchy sequences. The methods here are generalizations of uniform space techniques, and these methods are applied in particular to three natural convergence concepts. Order convergence gives rise to the order Cauchy completion G^o, shown to be the (ℓ-group) Dedekind-MacNeille completion of G. Polar convergence gives rise to the polar Cauchy completion G^{ip} (the i indicates that the Cauchy completion process may need to be iterated), which contains the lateral completion G^{lat}. α-convergence gives rise to the α-Cauchy completion $G^{i\alpha}$, and $G^{i\alpha}$ is *cut complete*: $G^{i\alpha}$ contains a supremum for any subset having a supremum in any ℓ-supergroup in which $G^{i\alpha}$ is order dense. The Theorem Prototype applied to cut completeness defines the cut completion G^{cut}.

Whereas the ideas of Section 7.1 are from distributive lattices and those of Section 7.2 from uniform spaces, Section 7.3 deals with concepts which cannot have scope much wider than ℓ-groups. The maximal piecewise approximated extension $\bar{G}$ and maximal finitely piecewise approximated extension $\bar{G}^\omega$ agree with the orthocompletion and projectable hull, respectively, in case G is representable. We prove, in fact, that neither completion could be constructed by the methods of Section 7.2. However, the completion ideas of Sections 7.2 and 7.3 jointly account for all of G^{dist} inasmuch as $(\bar{G}^\omega)^{cut} = G^{dist}$ whenever G is representable. This theorem (Theorem 7.3.11) is thus the finale of this survey.

7.1. The Distinguished Completion

In any (concrete) category C, an extension $G \leq H$ is *essential* provided that the only morphisms out of H with monic restriction to G are those monic on H. Clearly then, if C is a set of objects totally ordered by essential extension, and if $\cup C$ is an object of C, then $\cup C$ is an essential extension of any $C \in C$. One is led to ask whether for a given object G of C there is an essential extension H of G which admits no proper essential extension itself, a so-called *maximal essential extension*. For example, the maximal essential extension of an object in the category of archimedean ℓ-groups ([Conrad, 1971b]) contains all significant completions studied prior to [Madden and Vermeer, a] and [Ball and Hager, a,b]. However, as Conrad points out in [1971b], the category of lattice-ordered groups lacks maximal essential extensions because any ℓ-group is essentially extended by its lexicographic product with a totally ordered group. Nevertheless, we can achieve something like the same objectives by making use of the fact that maximal essential extensions exist in $\mathcal{D}$, the category of distributive lattices with lattice homomorphisms. Therefore, let us consider G to be only a distributive lattice for the time being, and in the next few results outline the important characteristics of the maximal essential extension B of G in $\mathcal{D}$. These characteristics are precisely that B is a complete Boolean algebra which essentially extends G, and B is unique with respect to these properties. Since the approach is purposely somewhat desultory, the reader who is either impatient or already knows the details should skip ahead to Theorem 7.1.13. A fuller account can be found in [Balbes, 1967] or [Balbes and Dwinger, 1974].

The existence of the maximal essential extension B of G is a consequence of its bounded cardinality (see the Zorn's Lemma argument that began this section), and this bound is a consequence of the plethora of lattice congruences on G, a point made by Corollary 7.1.3 and its proof. First we state a lemma whose proof is an easy lattice theory exercise.

LEMMA 7.1.1. *The finest lattice congruence $\approx$ on a distributive lattice G which identifies elements a and b is this: $x \approx y$ if and only if $(a \wedge b) \wedge (x \vee y) \le x \wedge y$ and $(a \vee b) \vee (x \wedge y) \ge x \vee y$.*

We shall say that elements g_1 and g_2 of G *distinguish* elements h_1 and h_2 of G provided that any lattice congruence on G which identifies h_1 with h_2 also identifies g_1 with g_2. This is evidently equivalent to $(h_1 \wedge h_2) \wedge (g_1 \vee g_2) \le g_1 \wedge g_2$ and $(h_1 \vee h_2) \vee (g_1 \wedge g) \ge g_1 \vee g_2$. The pivotal case of this relation is the one in which $h_1 < h_2$ and $g_1 < g_2$, and this can be pictured in the lattice $C(\mathbf{R})$ of continuous real-valued functions on $\mathbf{R}$ as follows.

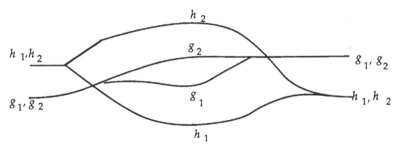

The reader who finds such sketches motivating can find more of them in the expository article [Ball, 1980a]; be warned, however, that the suggestions of proof found there are best ignored in favor of the treatment in [Ball, a], which is not only complete but is also more elegant. Evidently we have the following.

PROPOSITION 7.1.2. *H is an essential extension of G in the category $\mathcal{D}$ if and only if for every pair of distinct elements of H there is a pair of distinct elements of G distinguishing them.*

COROLLARY 7.1.3. *If H is an essential extension of G in $\mathcal{D}$ then $|H| \leq 2^{|G|}$.*

Proof. This corollary is true for all objects in $\mathcal{D}$ but of interest here only when G is infinite. In this case associate with each $h \in H$ the set $\{(a,b) \in G^2 : a \leq b \text{ and } h \vee a \geq b\}$. This association is one-to-one by Proposition 7.1.2.

PROPOSITON 7.1.4. *Every distributive lattice has at least one maximal essential extension in $\mathcal{D}$.*

Next we observe that any maximal essential extension of a distributive lattice must be a Boolean algebra.

PROPOSITION 7.1.5. *Any distributive lattice G has an essential extension B which is a Boolean algebra.*

Proof. Embed G in any Boolean algebra C and let $\approx$ be maximally coarse among those lattice congruences on C which do not identify distinct elements of G. Then B can be taken to be the Boolean algebra $C/\approx$.

In fact, any maximal essential extension of a distributive lattice must be a complete Boolean algebra. To see this most easily we must introduce the useful notion of join and meet density. Given subsets $X,G \subseteq H$ let

$$\mathcal{L}_G(X) = \{g \in G : g \leq x \text{ for all } x \in X\}, \text{ and}$$

$$\mathcal{U}_G(X) = \{g \in G : g \geq x \text{ for all } x \in X\},$$

the respective sets of lower and upper bounds of X in G. Then sublattice G is *join and meet dense* (jamd for short) in lattice H if and only if each $h \in H$ is the supremum (in H) of its lower bounds in G and simultaneously the infimum (in H) of its upper bounds in G, that is $h = \vee \mathcal{L}_G(h) = \wedge \mathcal{U}_G(h)$. The *Dedekind-MacNeille completion* of a lattice G we shall take to be $DM(G) = \{X \subseteq G : X = \mathcal{L}_G(\mathcal{U}_G(X))\}$ ordered by inclusion, with G embedded in $DM(G)$ by $g \to \mathcal{L}_G(g)$. Join and meet density is intimately related to the Dedekind-MacNeille completion ([MacNeille, 1937]) by the following proposition, whose proof is uncomplicated. The best general reference for this material is Chapter 12 of [Balbes and Dwinger, 1974].

PROPOSITION 7.1.6. *Sublattice G is jamd in lattice H if and only if there is a lattice injection $\theta: H \to DM(G)$ over G. θ is unique when it exists.*

It is important to understand that $DM(G)$ need not be distributive, even when G is.

LEMMA 7.1.7. *If G is a jamd sublattice of the distributive lattice H, then H is an essential extension of G.*

Proof. Consider $h_1 < h_2$ in H. Since G is jamd in H, there is some $g_2 \in \mathcal{L}_G(h_2) \backslash \mathcal{L}_G(h_1)$. Because $g_2 \wedge h_1 < g_2$ there must exist an element $g_1' \in \mathcal{U}_G(g_2 \wedge h_1) \backslash \mathcal{U}_G(g_2)$. Let $g_1 = g_1' \wedge g_2 < g_2$. Then $h_2 \vee g_1 \geq h_2 \geq g_2$ and $h_1 \wedge g_2 \leq g_1' \wedge g_2 = g_1$, meaning g_1 and g_2 distinguish h_1 from h_2. This proves the lemma by Proposition 7.1.2.

COROLLARY 7.1.8. *Any maximal essential extension B of a distributive lattice G is a complete Boolean algebra.*

Proof. This follows from Lemma 7.1.7, since the Dedekind-MacNeille completion $DM(B)$ of a Boolean algebra B is a complete Boolean algebra.

As we saw in the proof of Corollary 7.1.3, each element b of an essential extension of G is determined by the pairs $c \leq d$ in G for which $b \vee c \geq d$. Consequently it is useful in such a situation to define

$$[b] = \{(c,d) : c \leq d \text{ in } G \text{ such that } b \vee c \geq d\}.$$

In case B is a Boolean algebra we shall use $\vee_{[b]}(d \wedge c^\perp)$ to abbreviate $\vee\{d \wedge c^\perp : (c,d) \in [b]\}$.

PROPOSITION 7.1.9. *In a Boolean algebra B which is an essential extension of G, $\vee_{[b]}(d \wedge c^\perp) = b$ for all $b \in B$.*

Proof. A simple fact about Boolean algebras is that $b \vee c \geq d$ if and only if $b \geq d \wedge c^\perp$. Therefore $b \geq a = \vee_{[b]}(d \wedge c^\perp)$. The inequality cannot be strict, for if it were then we could find $k < f$ in G distinguishing a from b by Proposition 7.1.2. But $b \vee k \geq f$ means $(k,f) \in [b]$, so $a \geq f \wedge k^\perp$ and $a \vee k \geq f$. But since we also know $a \wedge f \leq k$, we get the contradiction $k = f$.

PROPOSITION 7.1.10. *The maximal essential extension B of a distributive lattice G is unique up to isomorphism over G.*

Proof. Suppose B and C are maximal essential extensions of G. Define $\theta: B \to C$ by declaring $\theta(b) = \vee_C\{d \wedge c^\perp : (c,d) \in [b]_B\}$ for each $b \in B$. This supremum exists in C because C is complete. θ, which can readily be seen to be a lattice homomorphism, is injective by Proposition 7.1.2 and surjective by Proposition 7.1.9.

It is a classical result of Sikorski [1948] that the complete Boolean algebras are the injectives in $\mathcal{D}$. Thus we can sum up the discussion to this point as follows. (See [Balbes and Dwinger, 1974] and [Balbes, 1967].)

THEOREM 7.1.11. *The following are equivalent for a distributive lattice B with sublattice G:*
  (a) B is the maximal essential extension of G;
  (b) B is an essential extension which is a complete Boolean algebra;
  (c) For any distributive lattice H, H is an essential extension of G if and only if there is an injection $\theta{:}H{\rightarrow}B$ over G;
  (d) B is an essential extension of G and an injective in $\mathcal{D}$.
The injection of part (c) is uniquely given by

$$\theta(h) = \vee_B \{d \wedge c^\perp : c \le d \text{ in } G \text{ and } h \vee c \ge d\}.$$

Let us awaken now from the slumber we imposed upon ourselves by ignoring the group structure of G. In particular, letters like G, H, and K now denote ℓ-groups, $G \le H$ now means that G is an ℓ-subgroup of H, and the term homomorphism now refers to ℓ-homomorphism. We shall say that an ℓ-group H is a *distinguished extension* of an ℓ-group G whenever $G \le H$ satisfies Proposition 7.1.2; that is, whenever H is an essential extension of G in $\mathcal{D}$. This relation has a couple of evocative characterizations in the richer context of lattice-ordered groups.

THEOREM 7.1.12. *The following are equivalent for ℓ-groups $G \le H$:*
  (a) H is a distinguished extension of G;
  (b) G is order dense in H and $e = \wedge\{ |hg^{-1}| : g \in G \}$ for all $e < h \in H$;
  (c) G is order dense in H, and for all $h_1 < h_2$ in H there is some $g \in G$ satisfying $(h_1 g^{-1} \wedge e)^\perp \cap (h_2 g^{-1} \vee e))^{\perp\perp} \ne \{e\}$.

We omit the moderately technical and lengthy proof of Theorem 7.1.12 (for which see [Ball, a]), and content ourselves with a couple of remarks on it. First, the naturalness of condition (b) argues in favor of the naturalness of the distinguished extension relation among ℓ-groups. Second, although condition (c) may not at first seem natural, it can be interpreted as asserting for each "needle's eye" formed by $h_1 < h_2$ in H the existence of a "thread" $g \in G$ passing through it. A sketch of this situation in $C(\mathbf{R})$ might look like this.

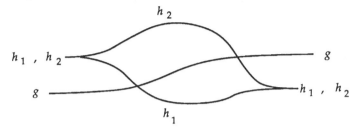

The central matter for the construction of G^{dist} is the extension of the group operation from G to a suitable sublattice of its maximal essential $\mathcal{D}$ extension B. For a start, it is clear what should be meant by the product of $b \in B$ with $g \in G$: after all, right (left) multiplication by g is a lattice isomorphism on G and so lifts to a unique isomorphism of B. Explicitly, for $b = \vee_{[b]}(d \wedge c^{\perp})$ we get $bg = \vee_{[b]}(dg \wedge (cg)^{\perp})$, and likewise for gb. Observe that $(c,d) \in [b]$ implies $(cg,dg) \in [bg]$, and conversely by multiplying on the right by g^{-1}.

Now if an element $b \in B$ is to lie in an ℓ-supergroup of G contained in B then certainly $bg > b$ and $gb > b$ for all $e < g \in G$. It is perhaps a little surprising then to be able to define so simply

$$G^{dist} = \{b \in B : bg > b \text{ and } gb > b \text{ for all } e < g \in G\}.$$

It is not as easy to see that the product of $b_1, b_2 \in G^{dist}$ must be

$$\vee\{d_1 d_2 \wedge (c_1 d_2 \vee d_1 c_2)^{\perp} : (c_i, d_i) \in [b_i]\}.$$

That is, $[b_1 b_2] \supset \{(c_1 d_2 \vee d_1 c_2, d_1 d_2) : (c_i, d_i) \in [b_i]\}$, and the two sets are equivalent in the sense that the Boolean differences of the pairs from each have the same supremum. The reader familiar with ℓ-permutation groups will find in [Ball, 1980a] a sketch which illustrates this definition in $A(\mathbf{R})$, the ℓ-group of order preserving permutations of $\mathbf{R}$. The following proposition shows that $[b_1 b_2]$ must contain at least the pairs indicated. Its proof uses the method of primes: $a \geq b$ in G if and only if $Pa \geq Pb$ for all prime subgroups P of G (Lemma 1.1.8).

PROPOSITION 7.1.13. *Suppose that h_i, c_i, and d_i (i = 1,2) are elements of an ℓ-group H such that $c_i \leq d_i$ and $h_i \vee c_i \geq d_i$. Then $h_1 h_2 \vee (c_1 d_2 \vee d_1 c_2) \geq d_1 d_2$.*

Proof. The conclusion could be false only if there were a prime subgroup P for which $P(h_1 h_2 \vee c_1 d_2 \vee d_1 c_2) < Pd_1 d_2$. In such a case $Pc_1 < Pd_1$ and $Pd_1 c_2 < Pd_1 d_2$. The first inequality with $P(h_1 \vee c_1) \geq Pd_1$ gives $Ph_1 \geq Pd_1$, and the second with $Pd_1(h_2 \vee c_2) \geq Pd_1d_2$ gives $Pd_1h_2 \geq Pd_1d_2$. Combining yields $Ph_1h_2 \geq Pd_1d_2$, contrary to supposition.

The uniqueness of the multiplication on G^{dist} is absolute in the following sense. For any sublattice H of B containing G to which the multiplication on G can be extended in such a way as to render H an ℓ-group, it must be true that H is contained in G^{dist}, and in fact contained as an ℓ-subgroup ([Ball, a]).

The proof that G^{dist} is a lattice-ordered group is long and technical; its inclusion would not only abuse the reader's patience but would unduly lengthen this article as well. In partial expiation of this lapse, and to convey some feel for the technicalities, we offer the following.

PROPOSITION 7.1.14. *The multiplication on G^{dist} is associative.*

Proof. Consider $(c_i, d_i) \in [b_i]$ for $i = 1,2,3$. Since b_1b_2 is the supremum of Boolean differences of pairs

$$(c_1d_2 \vee d_1c_2, d_1d_2),$$

$(b_1b_2)b_3$ is the supremum of differences of pairs

$$((c_1d_2 \vee d_1c_2)d_3 \vee (d_1d_2)c_3, (d_1d_2)d_3).$$

Since b_2b_3 is the supremum of differences of pairs

$$((c_2d_3 \vee d_2c_3), d_2d_3),$$

$b_1(b_2b_3)$ is the supremum of differences of pairs

$$(c_1(d_2d_3) \vee d_1(c_2d_3 \vee d_2c_3), d_1(d_2d_3)).$$

But both forms reduce to

$$(c_1d_2d_3 \vee d_1c_2d_3 \vee d_1d_2c_3, d_1d_2d_3).$$

THEOREM 7.1.15. *G^{dist} is an ℓ-group which is a distinguished extension of G. For any distinguished extension H of G, the canonical lattice injection of Theorem 7.1.11 maps H into G^{dist} and is in fact an ℓ-injection.*

COROLLARY 7.1.16. *$(G^{dist})^{dist} = G^{dist}$.*

Proof. Since the concatenation of distinguished extensions is itself a distinguished extension, there is a (unique) ℓ-injection $\theta: H \rightarrow G^{dist}$, where H is $(G^{dist})^{dist}$. But θ must be the identity map on the ℓ-subgroup $G^{dist} \leq H$ by uniqueness, and then the one-to-one quality of θ forces $G^{dist} = H$.

We shall term an ℓ-group G *distinguished* provided that $G = G^{dist}$. In the next two sections we develop most of the known completeness properties C along with the corresponding completion G^C (see the Theorem Prototype in the introduction). In each case it develops that G^C is a distinguished extension of G, with the result that we shall be proving at the same time that a distinguished ℓ-group enjoys all these completeness properties C.

7.2. Cauchy Completions

The guiding example of the sort of completion we consider here is Cantor's construction of the real numbers from the rational numbers by Cauchy sequences. The key idea is that the Cauchy Criterion gives a way of identifying those sequences of rationals which converge to a real number without knowing which real number. If $\mathcal{F}$ is the filter associated with a sequence $\{q_n\}$ of rationals, i.e., $\mathcal{F}$ has as base sets of the form $F_m = \{q_n : n \geq m\}$ for $m \in \mathbf{N}$, and if $\rightarrow$ designates convergence in the interval topology, i.e., $\mathcal{F} \rightarrow q$ means that $\mathcal{F}$ refines (or contains) the neighborhood filter at q, then the Cauchy Criterion is simply

$$\mathcal{F} - \mathcal{F} \rightarrow 0,$$

where $\mathcal{F} - \mathcal{F}$ is the filter with base sets $F - F = \{f_1 - f_2 : f_i \in F\}$ for $F \in \mathcal{F}$.

The definition of a Cauchy filter $\mathcal{F}$ on an arbitrary ℓ-group is the nonabelian counterpart of the foregoing:

$$\mathcal{F}\mathcal{F}^{-1} \rightarrow e \text{ and } \mathcal{F}^{-1}\mathcal{F} \rightarrow e,$$

where $\mathcal{F}\mathcal{F}^{-1}$ is the filter with base sets $FF^{-1} = \{f_1 f_2^{-1} : f_i \in F\}$ for $F \in \mathcal{F}$, and likewise for $\mathcal{F}^{-1}\mathcal{F}$. Of critical importance, however, is a suitable notion of convergence. Fortunately, lattice-ordered groups come equipped with numerous interesting and useful intrinsically defined convergences. These convergences need not be (and, in the cases we take up, are not) topological, but need satisfy only the following minimal requirements (plus two more which will be listed later). An ℓ-convergence on an ℓ-group G is determined by deciding which filters $\mathcal{F}$ on G converge to which elements $g \in G$, written $\mathcal{F} \rightarrow g$, subject to the following requirements.

(a) $g = \{X \subseteq G : g \in X\} \rightarrow g$ for all $g \in G$.

(b) $\mathcal{F} \supset \mathcal{M}$ and $\mathcal{M} \rightarrow m$ imply $\mathcal{F} \rightarrow m$.

(c) $\mathcal{F} \rightarrow g$ and $\mathcal{M} \rightarrow g$ imply $\mathcal{F} \cap \mathcal{M} \rightarrow g$.

(d) $\mathcal{F} \to f$ and $\mathcal{M} \to m$ imply $\mathcal{FM} \to fm$ and $\mathcal{F}^{-1} \to f^{-1}$.

(e) $\mathcal{F} \to f$ and $\mathcal{M} \to m$ imply $\mathcal{F} \vee \mathcal{M} \to f \vee m$.

The reader is hereby warned that, whereas $\mathcal{F} \cap \mathcal{M}$ is a literal set intersection and has a base of sets of the form $F \cup M$ for $F \in \mathcal{F}$ and $M \in \mathcal{M}$, the filters $\mathcal{FM}$, $\mathcal{F} \wedge \mathcal{M}$, and $\mathcal{F}^{-1}$ have bases $FM = \{xy : x \in F, y \in M\}$, $F \vee M = \{x \vee y : x \in F, y \in M\}$, and $F^{-1} = \{x^{-1} : x \in F\}$ for $F \in \mathcal{F}$, $M \in \mathcal{M}$.

This approach to the construction of completions has been profitably exploited in numerous contexts. For instance, if the convergence is topological then the filters Cauchy in the above sense are precisely those which are Cauchy in the two-sided uniformity of the topology ([Bourbaki, 1966]). The study of convergences is intertwined with the development of abstract topology, and in recent years generalizations of uniform completions have received considerable attention; see, for example, [Kent and Richardson, 1974] and [Reed, 1971]. Applications of these ideas to ordered sets can be found in [Nachbin, 1965], [Kent and Vaino, 1985], [Redfield, 1974] and [Ball, 1984]; applications to lattice-ordered groups can be found in [Conrad, 1974a], [Kenny, 1975], [Holland, 1963b], [Ball and Davis, 1983], [Ball, 1982], [Dashiell, Hager and Henricksen, 1980], and [Papangelou, 1965]. The specific development which we follow here is that of [Ball, 1980b]. Before further elaborating the Cauchy completion machinery, we introduce three of the four ℓ-convergences which we shall investigate in some detail.

The first ℓ-convergence is order convergence. A filter $\mathcal{F}$ *order converges* to an element f, written $\mathcal{F} \overset{o}{\to} f$, provided that $\wedge \mathcal{U}(\mathcal{F}) = \vee L(\mathcal{F}) = f$, where $\mathcal{U}(\mathcal{F}) = \cup \{\mathcal{U}(F) : F \in \mathcal{F}\}$ and dually for $L(\mathcal{F})$. As Birkhoff points out [1967, p. 294], order convergence is the lattice version of the familiar condition for sequences $\{f_n\}$

$$\lim \sup\{f_n\} = \lim \inf\{f_n\} = f.$$

That order convergence has the properties of an ℓ-convergence is easy enough to check, with the possible exception of the first part of (d), which goes as follows. Given $\mathcal{F} \overset{o}{\to} f$ and $\mathcal{M} \overset{o}{\to} m$ let $L = L(\mathcal{F})L(\mathcal{M})$ and $U = \mathcal{U}(\mathcal{F})\mathcal{U}(\mathcal{M})$. Since $\mathcal{FM} \supset \mathcal{K}$, where $\mathcal{K}$ is the filter with base sets of the form $[g,u] = \{k : g \leq k \leq u\}$ for $g \in L$ and $u \in U$, and since $\mathcal{K} \overset{o}{\to} fm$ because $\vee L = (\vee L(\mathcal{F}))(\vee L(\mathcal{M})) = fm = (\wedge \mathcal{U}(\mathcal{F}))(\wedge \mathcal{U}(\mathcal{M})) = \wedge U$, we get $\mathcal{FM} \overset{o}{\to} fm$.

It happens that the completion associated with order convergence is the ℓ-group Dedekind-MacNeille completion G^{ded} (Theorem 7.2.11). Since this completion may be viewed as the "intersection" of the lattice Dedekind-MacNeille completion with G^{dist} (in a sense which is evident from the construction of G^{ded} given prior to Theorem 7.2.11 but which we refuse to make otherwise more precise), and in view of Proposition 7.1.6, we would expect join and meet density to have something to do with order convergence. This expectation is met in Proposition 7.2.1.

For any ℓ-convergence on G, the *closure* of a subset $Y \subseteq G$ is cl$(Y) =$ $\{g \in G : Y \in \mathcal{F} \to g$ for some filter $\mathcal{F}$ on $G\}$, and a subset $Y \subseteq G$ is *closed* if $Y =$ cl(Y). But note well, dear reader, that cl(Y) need not be closed when the convergence is nontopological. It is this fact which necessitates the iteration of the Cauchy completion process which we shall encounter anon. $Y \subseteq G$ is *dense* in G if cl$(Y) = G$. Here, too, the reader should exercise caution: density with respect to order convergence as just defined does not coincide with the notion of order density. (Recall that H is order dense in G if for every $e < g \in G$ there is some $h \in H$ with $e < h \leq g$.)

PROPOSITION 7.2.1. *If G is an ℓ-subgroup which is jamd in H then G is dense with respect to order convergence on H.*

Proof. If G is jamd in H then for any $h \in H$ the filter $\mathcal{F}$ order converges to h, where $\mathcal{F}$ has base $\{[g,u]_G : g \in \mathcal{L}_G(h), u \in \mathcal{U}_G(h)\}$. This is so because $\mathcal{L}_G(h) \subseteq L(\mathcal{F})$, $\mathcal{U}_G(h) \subseteq \mathcal{U}(\mathcal{F})$, and G is order dense in H so that suprema and infima in G and H agree.

The second ℓ-convergence of particular interest here is defined by means of the complete Boolean algebra $\mathcal{P}$ of polars of G (see [Bigard, Keimel and Wolfenstein, 1977, Chapter 3]).   The infimum and supremum of a set of polars are given by the formulas

$$\wedge_\lambda P_\lambda = \cap_\lambda P_\lambda \text{ and } \vee_\lambda P_\lambda = (\cup_\lambda P_\lambda)^{\perp\perp}.$$

We shall use the symbols $0_\mathcal{P}$ and $1_\mathcal{P}$ to designate the bottom element $\{e\}$ and top element G of $\mathcal{P}$, respectively. Then a filter $\mathcal{F}$ *polar converges* to an element f, written $\mathcal{F} \xrightarrow{P} f$, provided that

$$\wedge\{(Ff^{-1})^{\perp\perp} : F \in \mathcal{F}\} = 0_\mathcal{P},$$

or equivalently (conjugate by f)

$$\wedge\{(f^{-1}F)^{\perp\perp} : F \in \mathcal{F}\} = 0_\mathcal{P}.$$

To see that polar convergence satisfies requirement (c) of the definition of ℓ-convergence, observe that

$$(\wedge_\mathcal{F}(Fg^{-1})^{\perp\perp})\vee(\wedge_\mathcal{M}(Mg^{-1})^{\perp\perp}) = \wedge_{\mathcal{F},\mathcal{M}}((Fg^{-1})^{\perp\perp}\vee(Mg^{-1})^{\perp\perp})$$

by the distributivity of finite suprema over arbitrary infima in any Boolean algebra, and then observe that the latter expression is $\wedge_{\mathcal{F},\mathcal{M}}((F\cup M)g^{-1})^{\perp\perp}$. To verify (d), observe that $A^\perp \cap B^\perp \subseteq (AB)^\perp$ for any subsets $A,B \subseteq G$, which gives $(AB)^{\perp\perp} \subseteq A^{\perp\perp}\vee B^{\perp\perp}$ upon passage to the complement. Thus

$$\wedge_{\mathcal{F},\mathcal{M}}(f^{-1}FMm^{-1})^{\perp\perp} \subseteq \wedge_{\mathcal{F},\mathcal{M}}((f^{-1}F)^{\perp\perp}\vee(Mm^{-1})^{\perp\perp}) = 0_\mathcal{P},$$

hence $\wedge_{\mathcal{F},\mathcal{M}}(FMm^{-1}f^{-1})^{\perp\perp} = 0_{\mathcal{P}}$ by conjugation, meaning $\mathcal{F}\mathcal{M} \xrightarrow{P} fm$. To verify (e) let us inspect products of the form

$$(f_1\vee m_1)(f\vee m)^{-1} = (f_1\vee m_1)(f^{-1}\wedge m^{-1})$$

for $f_1\in F\in \mathcal{F}$ and $m_1\in M\in \mathcal{M}$. The latter product expands to either

$$(f_1 f^{-1}\wedge f_1 m^{-1})\vee(m_1 f^{-1}\wedge m_1 m^{-1}) \text{ or}$$

$$(f_1 f^{-1}\vee m_1 f^{-1})\wedge(f_1 m^{-1}\vee m_1 m^{-1}),$$

and so is bounded above by $f_1 f^{-1}\vee m_1 m^{-1}$ and below by $f_1 f^{-1}\wedge m_1 m^{-1}$. This shows that

$$(F\vee M)(f\vee m)^{-1} \subseteq (Ff^{-1}\cup Mm^{-1})^{\perp\perp} = (Ff^{-1})^{\perp\perp}\vee(Mm^{-1})^{\perp\perp},$$

with the result that

$$\wedge_{\mathcal{F},\mathcal{M}}(F\vee M)(f\vee m)^{-1} \leq (\wedge_{\mathcal{F}}(Ff^{-1})^{\perp\perp})\vee(\wedge_{\mathcal{M}}(Mm^{-1})^{\perp\perp}) = 0_{\mathcal{P}},$$

or $\mathcal{F}\vee\mathcal{M} \xrightarrow{P} f\vee m$.

Polar convergence has an intuitively appealing meaning in $C(\mathbf{R})$. $\mathcal{F} \xrightarrow{P} f$ means that for any nonempty interval $I \subseteq \mathbf{R}$ there is some nonempty subinterval $J \subseteq I$ and set $F\in \mathcal{F}$ such that $f(x) = f_1(x)$ for all $f_1\in F$ and $x\in J$. Polar convergence is the subject of Section 5 of [Ball, 1980b].

The most important ℓ-convergence which we take up here is α-convergence, a notion due to Papangelou [1965] and pursued in [Ellis, 1968], [Madell, 1980], [Ball and Davis, 1983], and [Ball, b]. To define α-convergence it is convenient to let $a\uparrow b = \{g\in G : g\vee a \geq b\}$ and $a\downarrow b = \{g\in G : g\wedge b \leq a\}$ for pairs $a \leq b$ in G. Then a filter $\mathcal{F}$ α-converges to an element f provided that for all $a < b$ in G for which $f\in a\uparrow b$ ($f\in a\downarrow b$) there is a pair $c < d$ in G distinguishing a from b such that $c\uparrow d\in \mathcal{F}$ ($c\downarrow d\in \mathcal{F}$). We write $\mathcal{F}\xrightarrow{\alpha} f$. To check that α-convergence satisfies part (c) of the definition of ℓ-convergence, consider $\mathcal{F},\mathcal{M} \xrightarrow{\alpha} g\in a\uparrow b$. First find $c < d$ distinguishing a from b such that $c\uparrow d\in \mathcal{F}$. Since $g\in a\uparrow b \subseteq c\uparrow d$ there are elements $k < f$ distinguishing c from d such that $k\uparrow f\in \mathcal{M}$. But then $k\uparrow f\in \mathcal{F}\cap\mathcal{M}$, which with the dual proves $\mathcal{F}\cap\mathcal{M} \xrightarrow{\alpha} g$. Part (d) holds, but for slightly more technical reasons; a proof can be found in [Ball and Davis, 1983]. To check (e) suppose $\mathcal{F}\xrightarrow{\alpha} f$, $\mathcal{M} \xrightarrow{\alpha} m$, and $f\vee m\in a\uparrow b$. Then f and m cannot both be in $a\downarrow b$ (lest $f\vee m\in a\downarrow b$), so assume $f\in a\downarrow b$. Since $a < b_1 = (f\vee a)\wedge b$ and since $f\in a\uparrow b_1$, there are $c < d$ distinguishing a from b_1 such that $c\uparrow d\in \mathcal{F}$. Since distinction is a transitive relation on pairs, c and d also distinguish a from b, and $c\uparrow d\in \mathcal{F}\vee\mathcal{M}$. Thus we see that α-convergence is an ℓ-convergence.

The special role played by α-convergence rests on two facts, the first of which is its coarseness. The next result is a special case of Proposition 1.15 of [Ball and Davis, 1983]; its proof makes use of the method of primes again. (See the comment prior to Proposition 7.1.13.)

PROPOSITION 7.2.2. *If* $\mathcal{F} \xrightarrow{o} f$ *or* $\mathcal{F} \xrightarrow{P} f$ *then* $\mathcal{F} \xrightarrow{\alpha} f$.

Proof. Suppose $\mathcal{F} \xrightarrow{o} f \in a\hat{\uparrow}b$ for some $a < b$ in G. By replacing a and b with $a \wedge f$ and $b \wedge f$ if necessary, we may assume $a < b \le f$. Then $\vee \mathcal{L}(\mathcal{F}) = f$ implies $\vee(\mathcal{L}(\mathcal{F}) \wedge b) = b$, so there is some $F \in \mathcal{F}$ and $g \in \mathcal{L}(\mathcal{F})$ for which $g \wedge b \not\le a$. By taking $c = a$ and $d = (g \wedge b) \vee a$, we get a pair $c < d$ distinguishing a from b, and $g \in \mathcal{L}(\mathcal{F}) \cap c\hat{\uparrow}d$ implies $F \subseteq c\hat{\uparrow}d$, meaning $c\hat{\uparrow}d \in \mathcal{F}$. This argument and its dual show $\mathcal{F} \xrightarrow{\alpha} f$. Now suppose $\mathcal{F} \xrightarrow{P} f \in a\hat{\uparrow}b$ for some $a < b$ in G. Since $\wedge_{\mathcal{F}}(Ff^{-1})^{\perp\perp} = 0_{\mathcal{P}}$, there is some $F \in \mathcal{F}$ for which $ba^{-1} \notin (Ff^{-1})^{\perp\perp}$, say $e < c \le ba^{-1}$ with $c \in (Ff^{-1})^{\perp}$. Then $a < ca$ distinguish a from b, and we claim $F \subseteq a\hat{\uparrow}ca$. To verify this claim consider $x \in F$ and an arbitrary prime P; we seek to show that $P(x \vee a) \ge Pca$. If $Pa < Pca$ then $c \notin P$, so $xf^{-1} \in c^{\perp} \subseteq P$ implies $Px = Pf$. But then $fa \ge b \ge ca$ implies $Pca \le P(f \vee a) = P(x \vee a)$. This proves the claim, which shows $a\hat{\uparrow}ca \in \mathcal{F}$; with the dual argument we then have $\mathcal{F} \xrightarrow{\alpha} f$.

Before returning to the subject of the special role of α-convergence, let us outline the construction of the completion of G associated with an arbitrary ℓ-convergence. We need three additional properties of the convergence. Order convergence, polar convergence, and α-convergence all have these properties, though some effort is required to prove (h) in each case. See Propositions 4.11 and 5.2 of [Ball, 1980b] and Lemma 1.7 of [Ball and Davis, 1983].

(f) An ℓ-convergence is *Hausdorff* if no filter converges to more than one point.

(g) An ℓ-convergence is *convex* if $\mathcal{F} \to f$ implies $\mathcal{F}^{\sim} \to f$, where $\mathcal{F}^{\sim}$ has base $\{F^{\sim} : F \in \mathcal{F}\}$ and $F^{\sim} = \{g \in G : f_1 \le g \le f_2$ for some $f_i \in F\}$.

(h) An ℓ-convergence is *strongly normal* provided that, for any filters $\mathcal{F}$ and $\mathcal{M}$ such that $\mathcal{F}^{-1}\mathcal{F}, \mathcal{F}\mathcal{F}^{-1}$ and $\mathcal{M}$ converge to e, it is true that $\mathcal{F}\mathcal{M}\mathcal{F}^{-1} \to e$.

Suppose x-convergence is a convex Hausdorff strongly normal ℓ-convergence on G. A filter $\mathcal{F}$ on G will be termed x-*Cauchy* provided that $\mathcal{F}\mathcal{F}^{-1} \xrightarrow{x} e$ and $\mathcal{F}^{-1}\mathcal{F} \xrightarrow{x} e$. For x-Cauchy filters $\mathcal{F}$ and $\mathcal{M}$ define $\mathcal{F} \approx \mathcal{M}$ provided that $\mathcal{F} \cap \mathcal{M}$ is x-Cauchy. Let

$$[\mathcal{F}] = \{\mathcal{M} : \mathcal{M} \text{ is an } x\text{-Cauchy filter such that } \mathcal{F} \approx \mathcal{M}\},$$

and let

$$G^x = \{[\mathcal{F}] : \mathcal{F} \text{ is } x\text{-Cauchy}\}.$$

The strong normality requirement on the convergence is precisely what is needed to insure that products and inverses of Cauchy filters are themselves Cauchy (Proposition 2.14 of [Ball, 1980b]), while convexity is sufficient to insure that suprema and infima of Cauchy filters are themselves Cauchy ([Redfield, 1974], and Proposition 2.19 of [Ball, 1980b]). Therefore we can define the ℓ-group operations on G^x in the obvious way: $[\mathcal{F}][\mathcal{M}] = [\mathcal{FM}]$, $[\mathcal{F}]^{-1} = [\mathcal{F}^{-1}]$, $[\mathcal{F}] \vee [\mathcal{M}] = [\mathcal{F} \vee \mathcal{M}]$, and dually. Each element $g \in G$ is identified in this construction with $[\dot{g}] \in G^x$; the Hausdorff property insures that this identification is one-to-one.

Under this identification the operations on G^x certainly restrict to those on G, but a much closer semantic link between G and G^x exists. Let $c_1, c_2, ..., c_m$ denote constants from G, and let $w_1, w_2, ..., w_n$ be words built up form the c_i's and from variables $v_1, v_2, ..., v_k$ using group and lattice operations. A sentence is called *positive universal* if it is of the form

$$\forall v_1, v_2, ..., v_k \, \Psi(v_1, v_2, ..., v_k),$$

where Ψ is obtained by connecting the atomic formulas $w_i = e$ by logical conjunction or disjunction. For example, the defining laws for ℓ-groups are all equations (i.e., the atomic formulas are connected by conjunction) and are thus positive universal. For another example, the law defining total order in an ℓ-group is

$$\forall x, y (e \vee xy^{-1} = e \text{ or } e \vee yx^{-1} = e);$$

this law is positive universal but is not an equation. The following result has essentially nothing to do with ℓ-groups, but is instead an instance of a general principle ([Ball, c]).

THEOREM 7.2.3. *If x-convergence is a convex Hausdorff strongly normal ℓ-convergence on G, and if θ is any positive universal sentence with constants from G, then θ holds in G if and only if θ holds in G^x.*

Proof. θ clearly holds in G whenever it holds in G^x, so assume θ fails in G^x, where θ is $\forall v_1, v_2, ..., v_k (w_1 = e \text{ or } w_2 = e \text{ or } ... \text{ or } w_n = e)$. This means that there are elements $h_i \in G^x$, $1 \leq i \leq k$, such that $w_j(h_1, h_2, ..., h_k) \neq e$ for $1 \leq j \leq n$. Fix x-Cauchy filters $\mathcal{F}_i$ such that $h_i = [\mathcal{F}_i]$. Since the operations are all defined "filterwise", each $w_j(\mathcal{F}_1, \mathcal{F}_2, ..., \mathcal{F}_k)$ is an x-Cauchy filter inequivalent to e; more to the point, there must be sets $F_i \in \mathcal{F}_i$ such that $e \notin w_j(F_1, F_2, ..., F_k)$ for all j. But, choosing any element $f_j \in F_j$, we get $w_j(f_1, f_2, ..., f_k) \neq e$ for $1 \leq j \leq n$. That is, θ fails in G.

The immediate relevance of Theorem 7.2.3 to Cauchy constructions is that it shows that G^x satisfies the same equations as G does, and in particular satisfies the identities which define ℓ-groups. We summarize the general features of the construction in the next result ([Ball, 1980b]).

THEOREM 7.2.4. *Suppose that x-convergence is a convex Hausdorff strongly normal ℓ-convergence. Then G^x is an ℓ-supergroup of G which generates the same variety as G does.*

The construction of the completion up to this point will be quite familiar to the reader conversant with uniform completions. There is, however, an important difference. A uniform completion inherits a canonical uniformity from the underlying uniform space, whereas there is no canonical method of extending the underlying convergence or Cauchy structure of G to G^x in the construction above (see [Ball, 1980b] for some comments). This fact poses no obstacle, however, because the convergences we use are intrinsically defined from the ℓ-group structure, which is successfully extended to G^x above. What is necessary, however, is to check that the convergence intrinsically defined on the ℓ-supergroup G^x of G is consistent with the same convergence defined from the ℓ-group structure on G.

THEOREM 7.2.5. *Let $\rightarrow$ designate convergence with respect to order, polar, or α-convergence, and let H represent the corresponding completion. Then the convergence on H meshes nicely with the convergence on G in the following sense. For any $h \in H$ and filter $\mathcal{F}$ such that $G \in \mathcal{F}$, $\mathcal{F} \rightarrow h$ if and only if $h = [\mathcal{F}]$. In particular, it follows that G is dense in H with respect to the convergence, and that the convergence on H restricts to the same convergence on G in the following sense. If $\mathcal{F}$ is a filter on G and g is an element of G, then $\mathcal{F} \rightarrow g$ in H if and only if $\mathcal{F} \rightarrow g$ in G.*

Proof. The assertion for order convergence is Proposition 4.13 of [Ball, 1980b], for polar convergence it is Proposition 5.5 of [Ball, 1980b], and for α-convergence it is Proposition 1.11 of [Ball and Davis, 1983].

We are now free to investigate G^o, G^p, and G^α in more detail. The first objective will be to show that G^o and G^p can be taken to be ℓ-subgroups of G^α containing G. This fact, essentially a consequence of the coarseness of α-convergence, is one of the reasons that α-convergence is important. First we must prove a lemma.

LEMMA 7.2.6. *G is order dense in G^o, G^p, and G^α.*

Proof. We prove here that G is order dense in G^o; similar arguments can be made for G^p and G^α. The relevant results in the literature are Propositions 1.10, 4.3, and 5.1 of [Ball, 1980b]. Suppose $e < f = [\mathcal{F}] \in G^o$ for some order Cauchy filter $\mathcal{F}$ on G. Since $\mathcal{F}^{-1}\mathcal{F} \xrightarrow{o} e$, the filter $\mathcal{K}$ with base

$\{[g,u] : g\in L_G(F^{-1}F), u\in U_G(F^{-1}F), F\in \mathcal{F}\}$ also order converges to e. For each $F\in \mathcal{F}$ fix $f_F\in F$ and let $\mathcal{M}$ be the filter with base $\{f_F K : K\in \mathcal{K}\}$. Now for $x\in F\in \mathcal{F}$, $g\in L_G(F^{-1}F)$, and $u\in U_G(F^{-1}F)$ we have $f_F g \leq x \leq f_F u$; hence $F[g,u]_G \supset f_F[g,u]_G = \{f_F g, f_F u\}_G \supset F$, and $\mathcal{F}\supset \mathcal{M}\supset \mathcal{F}\mathcal{K}\overset{o}{\rightarrow} f$. In particular, $\mathcal{M}\vee e\overset{o}{\rightarrow} f$ in G^o means that f is the supremum in G^o of $\{f_F g\vee e : g\in L_G(F^{-1}F), F\in \mathcal{F}\}$. This shows the existence of at least one element $g\in G$ satisfying $e < g \leq f$.

In connection with the following proposition it is convenient to observe (and easy to show) that, for x-Cauchy filters $\mathcal{F}$ and $\mathcal{M}$, $\mathcal{F}\approx\mathcal{M}$ (i.e., $\mathcal{F}\cap\mathcal{M}$ is x-Cauchy) if and only if $\mathcal{F}\mathcal{M}^{-1}$ and $\mathcal{F}^{-1}\mathcal{M}$ x-converge to e.

PROPOSITION 7.2.7. *There are unique ℓ-injections* $\theta_1 : G^o \rightarrow G^\alpha$ *and* $\theta_2 : G^p \rightarrow G^\alpha$ *over* G.

Proof. Let $[\mathcal{F}]_x$ denote the equivalence class of $\mathcal{F}$ in the set of all x-Cauchy filters. Now if $\mathcal{F}$ is order Cauchy then $\mathcal{F}^{-1}\mathcal{F}\overset{o}{\rightarrow} e$ and so $\mathcal{F}^{-1}\mathcal{F}\overset{\alpha}{\rightarrow} e$ by Proposition 7.2.2, and likewise for $\mathcal{F}\mathcal{F}^{-1}$, meaning that $\mathcal{F}$ is α-Cauchy also. Furthermore $[\mathcal{F}]_o = [\mathcal{M}]_o$ implies $\mathcal{F}^{-1}\mathcal{M}\overset{o}{\rightarrow} e$, thus $\mathcal{F}^{-1}\mathcal{M}\overset{\alpha}{\rightarrow} e$, and likewise for $\mathcal{F}\mathcal{M}^{-1}$, meaning that $[\mathcal{F}]_\alpha = [\mathcal{M}]_\alpha$. Therefore define $\theta_1[\mathcal{F}]_o = [\mathcal{F}]_\alpha$. θ_1 is clearly an ℓ-homomorphism which is the identity map on G, and since G is order dense in G^o, θ_1 is one-to-one on G^o also. The uniqueness of this injection follows from Proposition 7.2.8 and the uniqueness of lattice injections into B over G (Theorem 7.1.11).

That all three Cauchy completions are in fact ℓ-subgroups of G^{dist} is the point of the next result. The proof uses Papangelou's original definition of α-convergence: $\mathcal{F}\overset{\alpha}{\rightarrow} f$ provided that

$$\vee(F\wedge x) = x = \wedge(F\vee x) \text{ for all } F\in \mathcal{F} \text{ if and only if } x = f.$$

The equivalence of the various definitions of α-convergence is Proposition 1.1 of [Ball, b].

PROPOSITION 7.2.8. *There is a unique ℓ-injection* $\theta : G^\alpha \rightarrow G^{dist}$ *over* G.

Proof. By Theorems 7.1.15 & 7.1.11, we need only show that G^α is a distinguished extension of G. Consider $[\mathcal{F}_1] = h_1 < h_2 = [\mathcal{F}_2]$ in G^α. Since $\mathcal{F}_2\mathcal{F}_1^{-1}$ α-converges in Papangelou's sense to $h_2 h_1^{-1} > e$, there are sets $F_i\in \mathcal{F}_i$ such that $\wedge(F_2 F_1^{-1}\vee e) \neq e$, say $F_2 F_1^{-1}\vee e \geq a > e$ for some $a\in G$. Because $\mathcal{F}_2\mathcal{F}_2^{-1}$ α-converges to $e\in e\downarrow a$ in the usual sense, there are $c < d$ in G distinguishing e from a and $K\in \mathcal{F}_2$ such that $KK^{-1}\wedge d \leq c$. By exchanging c and d for $(c\vee e)\wedge a$ and

$(d \vee e) \wedge a$, we may assume $e \leq c < d \leq a$. Fix $k \in F_2 \cap K \in \mathcal{F}_2$, and let $g_1 = d^{-1}k$ and $g_2 = c^{-1}k$. Then $kK^{-1} \wedge d \leq c$ implies $K \vee g_1 = K \vee d^{-1}k \geq c^{-1}k = g_2$, so $\dot{g}_2 = (\mathcal{F}_2 \vee g_1) \wedge g_2 \overset{\alpha}{\to} (h_2 \vee g_1) \wedge g_2$. Since α-convergence is Hausdorff, we get $g_2 = (h_2 \vee g_1) \wedge g_2$ or $h_2 \vee g_1 \geq g_2$. Also $kF_1^{-1} \vee e \geq a$ implies $kF_1^{-1} \vee c \geq d$, i.e., $F_1 \wedge g_2 \leq g_1$, from which follows $\dot{g}_1 = (\mathcal{F}_1 \wedge g_2) \vee g_1 \overset{\alpha}{\to} (h_1 \wedge g_2) \vee g_1$, and this yields $h_1 \wedge g_2 \leq g_1$. That is, g_1 and g_2 distinguish h_1 from h_2.

Having put G^o, G^p, and G^α in G^{dist} where they belong, we shall occasionally forget how they got there. That is, when convenient we identify each completion with its image under the appropriate embedding, and refer to it as an ℓ-subgroup of G^{dist}. From this perspective it is natural to wonder how these completions can be identified among the ℓ-subgroups of G^{dist}, and the most reasonable conjecture is that each completion is the closure of G in G^{dist} with respect to the relevant convergence. It happens that this is false in the case of G^o, as shown by the example following Theorem 7.2.11, but true for G^p and G^α, as shown by the next proposition.

COROLLARY 7.2.9. G^p and G^α are the closures of G in G^{dist} with respect to polar and α-convergence, respectively.

Proof. We prove this for G^α by making use of a key fact about α-convergence which the interested reader may prove: for any filter $\mathcal{F}$ on G^{dist} containing G and any element $g \in G$, $\mathcal{F} \overset{\alpha}{\to} g$ in G^{dist} if and only if $\mathcal{F} \overset{\alpha}{\to} g$ in G ([Ball and Davis, 1983], Proposition 1.10). Therefore, since G is α-dense in G^α by Theorem 7.2.5, the α-closure of G in G^{dist} contains G^α. Now suppose h lies in the α-closure of G in G^{dist}, say $G \in \mathcal{F} \overset{\alpha}{\to} h$. Then $\mathcal{F}^{-1}\mathcal{F} \overset{\alpha}{\to} h^{-1}h = e$ in G^{dist}, and likewise for $\mathcal{F}\mathcal{F}^{-1}$. Consequently these filters also α-converge to e in G. (It is the falsity of the corresponding step for order convergence that, loosely speaking, prevents G^o from being the closure of G in G^{dist} with respect to order convergence.) That is, $\mathcal{F}$ is an α-Cauchy filter on G, say $[\mathcal{F}] = k \in G^\alpha$. But Theorem 7.2.5 asserts that $\mathcal{F} \overset{\alpha}{\to} k$, which, considering the Hausdorff nature of α-convergence, shows that $h \in G^\alpha$.

Let us now turn to the completeness properties of G^o, G^p, G^α, and G^{dist}. The most immediate sort of completeness is Cauchy completeness: if x-convergence is a convex Hausdorff strongly normal ℓ-convergence then we say that G is x-Cauchy complete provided $G = G^x$. Note that, in the case of polar or α-convergence, this property simply expresses the closure of G in G^{dist} with respect to the relevant convergence. Since the closure operator of an ℓ-convergence is not in general idempotent, we do not expect G^{xx} to equal G^x in all cases. In fact, there are ℓ-groups for which $G^{\alpha\alpha} \neq G^\alpha$ ([Ball, b]), and it is not known whether there are any ℓ-groups for which $G^{pp} \neq G^p$. However, G^o is order Cauchy complete, a fact which we now prove in Theorem 7.2.11. First observe the Cauchy completeness of distinguished ℓ-groups.

PROPOSITION 7.2.10. *A distinguished ℓ-group is order, polar, and α-Cauchy complete.*

Proof. Let H represent the completion of G with respect to any one of the mentioned convergences. Then there is an ℓ-injection $\theta : H \to G^{dist} = G$ over G by Proposition 7.2.8. It follows that $G = H$.

The next result makes reference to the ℓ-group Dedekind-MacNeille completion G^{ded} of G ([Banaschewski, 1957]). Recall that the Dedekind-MacNeille completion of lattice G is $DM(G) = \{X \subseteq G : X = L\mathcal{U}(X)\}$, ordered by inclusion (see Proposition 7.1.6). For $X_1, X_2 \in DM(G)$ define the product $X_1 X_2$ to be $L\mathcal{U}\{x_1 x_2 : x_i \in X_i\}$. This definition makes $DM(G)$ a lattice-ordered semigroup with identity $L(e)$, and its invertible elements form the ℓ-group *Dedekind-MacNeille completion* of G, written G^{ded}. Since jamd extensions are distinguished (Lemma 7.1.7), it should come as no surprise that an element $X \in DM(G)$ is invertible if and only if for all $e < g \in G$, both $XL(g)$ and $L(g)X$ strictly exceed X in $DM(G)$, since this is the condition for membership in G^{dist}. It is light work to modify the proof of Proposition 7.1.6 to show that for any ℓ-supergroup $K \geq G$, G is jamd in K if and only if there is an ℓ-injection from K into G^{ded} over G.

THEOREM 7.2.11. *For any ℓ-group G there is an ℓ-supergroup H unique up to (unique) ℓ-isomorphism of G with respect to any one of the following equivalent properties.*
  (a) H is ℓ-isomorphic to G^o over G;
  (b) H is ℓ-isomorphic to G^{ded} over G;
  (c) For all ℓ-supergroups $K \geq G$, G is jamd in K if and only if there is an ℓ-injection of K into H over G;
  (d) G is jamd in H, and H is not jamd in any strictly larger ℓ-supergroup.

Proof. To show the equivalence of (a) and (b), define maps $\theta : G^{ded} \to G^o$ and $\psi : G^o \to G^{ded}$ as follows. $\psi[\mathcal{F}]$ is $L\mathcal{U}L(\mathcal{F})$, and $\theta(X)$ is $[\mathcal{F}(X)]$, where $\mathcal{F}(X)$ designates the filter with base $\{[x,u] : x \in X, u \in \mathcal{U}_G(X)\}$. We leave it as an exercise for the reader to show that the maps are well defined ℓ-homomorphisms which are inverses of one another. The argument used to establish Proposition 7.1.6 can then be used to prove the equivalence of (b) and (c).

Assume (c), and consider an ℓ-supergroup K in which H is jamd. Then there is an ℓ-injection $\tau: K \to H$ over G. Now τ restricts to the identity map on H by the uniqueness clause of Theorem 7.1.11, from which we see that K must be H. Thus part (d) holds. On the other hand, if (d) holds then at least H is (ℓ-isomorphic to) an ℓ-subgroup of G^{ded} (we are using the fact that (c) implies (b) here). We have roughly $G \leq H \leq G^{ded}$. But since G is jamd in G^{ded}, it follows that H is jamd in G^{ded}, whereupon the maximality of H gives $H = G^{ded}$. This proves that (b) holds and completes the proof of the theorem.

Here is a simple example which points out that G^o is generally smaller than the closure of G in G^{dist} with respect to order convergence; in this example the latter is G^{dist} while the former turns out to be G. Let H be the product $\Pi_I Z_i$ of copies Z_i of the integers over an infinite index set I, and let G consist of those members of H which are zero at all but finitely many indices. We first claim that G^{dist} is ℓ-isomorphic to H over G. To see this observe that H is a distinguished extension of G because H and G clearly satisfy the condition of Proposition 7.1.2. By referring to Theorem 7.3.11 we can see that H is distinguished because it is orthocomplete (i.e., projectable and laterally complete) and Dedekind-MacNeille complete. We next claim that G is dense with respect to order convergence on H. This is apparent from the observation that the filter with base $\{U_G(g) \cap L_G(h) : g \in L_G(h)\}$ order converges in H to h for arbitrary $0 < h \in H$. We finally claim that $G^o = G$. This is clear from the previous result, since G cannot be jamd in any ℓ-subgroup of H which contains G properly.

The polar Cauchy completion G^p is a somewhat more mysterious object than G^o, in part because it is not at present known whether G^p is polar Cauchy complete. However, there is an absolute polar completion in the following sense.

THEOREM 7.2.12. *For any ℓ-group G there is an ℓ-group G^{ip} unique up to (unique) ℓ-isomorphism over G with respect to the following properties. G^{ip} is a distinguished extension of G, G^{ip} is polar Cauchy complete, and $G \leq K < G^{ip}$ implies K is not polar Cauchy complete.*

Proof. Take G^{ip} to be

$$\cap\{K : G \leq K \leq G^{dist} \text{ and } K \text{ is polar-Cauchy complete}\}.$$

The initials *ip* in the notation above stand for *iterated polar* Cauchy completion. Though we shall not prove it here, the first of the three properties characterizing G^{ip} may be weakened to require only that G be order dense in G^{ip} ([Ball, 1980b]). Thus G^{ip} can be defined by applying the Theorem Prototype of the introduction to the property of polar Cauchy completeness. We may alternatively construct G^{ip} as follows. Define

$$G^{0p} = G,$$

$$G^{(\gamma+1)p} = (G^{\gamma p})^p, \text{ and}$$

$$G^{\beta p} = \cup_\beta G^{\gamma p} \text{ for limit ordinals } \beta.$$

Since all these ℓ-groups are distinguished extensions of G (and in fact can be taken to be ℓ-subgroups of G^{dist}), there is some ordinal γ for which $G^{\gamma p} = G^{(\gamma+1)p}$. Then G^{ip} is $G^{\gamma p}$. G^{ip} has the following interesting and frequently useful property.

PROPOSTION 7.2.13 G^{ip} *is laterally complete; that is, every pairwise disjoint subset of* G^{ip} *has a supremum in* G^{ip}.

Proof. In view of the construction of G^{ip} just outlined, it is enough to show that any pairwise disjoint subset $D \subseteq G$ has a supremum in G^p. Let $\mathcal{F}$ be the filter with base $\{F(C) : \text{finite } C \subseteq D\}$, where $F(C)$ denotes $\{\vee E : E \text{ finite and } C \subseteq E \subseteq D\}$. We claim $\mathcal{F}$ is polar Cauchy; let us show $\cap\{(F(C)F(C)^{-1})^{\perp\perp} : F \in \mathcal{F}\} = \{e\}$. Consider $e \leq g \in (F(C)F(C)^{-1})^{\perp\perp}$. Since $g \in D^{\perp\perp}$, if $g \neq e$ then there is some $d \in D$ with $g \wedge d > e$. The elements of D commute, so $(\vee E)(\vee E)^{-1} \in d^\perp$ for all finite $E \subseteq D$ containing d. Therefore $g \notin (F(\{d\})F(\{d\})^{-1})^{\perp\perp}$, proving $g = e$ or $\mathcal{F}\mathcal{F}^{-1} \xrightarrow{P} e$. An analogous argument shows $\mathcal{F}^{-1}\mathcal{F} \xrightarrow{P} e$, and we conclude that $\mathcal{F}$ is polar Cauchy. We next claim that $f = [\mathcal{F}] \in G^p$ is the supremum of D in G^{dist}. First note that for any $d \in D$, $d = \mathcal{F} \wedge d \xrightarrow{P} f \wedge d$, proving $f \geq d$. Now suppose there is some $k \in G^{dist}$ such that $f > k \geq D$. Then there are $a < b$ in G distinguishing k from f. But $k \wedge b \leq a$ and $k \geq D$ imply $(\vee E \wedge b) \vee a = a$ for all finite $E \subseteq D$, so $a = (\mathcal{F} \wedge b) \vee a \xrightarrow{P} (f \wedge b) \vee a$, contrary to assumption. Thus no such k can exist, meaning $f = \vee D$. This completes the proof of the claim and of the proposition.

COROLLARY 7.2.14. *A distinguished ℓ-group is laterally complete.*

The previous result suggests a simple construction of the *lateral completion* G^{lat} of G. (G^{lat} is defined by applying the Theorem Prototype of the introduction to the property of lateral completeness.) In a brilliant tour de force, Bernau [1975] showed the existence and uniqueness of the lateral completion of an arbitrary ℓ-group. We can recover Bernau's result in the present context just by taking

$$G^{lat} = \cap\{K : G \le K \le G^{dist} \text{ and } K \text{ is laterally complete}\}.$$

The latter approach avoids the combinatorial complexities of Bernau's construction, and thereby handily exemplifies the advantages of doing completions "from the top down" (i.e., inside G^{dist}). It must be admitted, however, that a more satisfying construction would result from isolating an ℓ-convergence whose completion is precisely G^{lat}. Can this be done?

The Cauchy completion which is perhaps most interesting is G^{α}. It is known ([Ball, b]) that $G^{\alpha\alpha}$ is in general strictly larger than G^{α}, but that G^{α} is α-Cauchy complete in case G is completely distributive, archimedean, or projectable (i.e., polars are cardinal summands). By arguing as in Theorem 7.2.12 we get an absolute α-Cauchy completion $G^{i\alpha}$.

THEOREM 7.2.15. *For any ℓ-group G there is an ℓ-group $G^{i\alpha}$ unique up to (unique) ℓ-isomorphism over G with respect to the following properties. $G^{i\alpha}$ is a distinguished extension of G, $G^{i\alpha}$ is α-Cauchy complete, and $G \le K < G^{i\alpha}$ implies K is not α-Cauchy complete.*

As in the case of the polar Cauchy completion, the initials $i\alpha$ stand for *iterated α*-Cauchy completion, and the conditions defining $G^{i\alpha}$ can be weakened to require only that G be order dense in $G^{i\alpha}$ ([Ball and Davis, 1983]). Therefore $G^{i\alpha}$ is the completion obtained by applying the Theorem Prototype of the introduction to the property of α-Cauchy completeness.

$G^{i\alpha}$ has a quite satisfying kind of completeness called cut completeness, which we shall define shortly. The motivation is like that for the formation of the lateral completion or of the ℓ-group Dedekind-MacNeille completion inasmuch as added elements are suprema or infima of subsets of G. These subsets, usually called *cuts*, are of quite different sorts in the two examples just mentioned, so one is led to wonder just what a cut ought to be. In reply to this speculation we offer the following result, whose proof, by the way, stands as an example of how computations which might be done in G^{α} can frequently be simplified in G^{dist}.

THEOREM 7.2.16. *The following are equivalent for a subset $X \subseteq G$.*
(a) $X = \mathcal{L}_G(h)$ *for some $h \in G^{\alpha}$.*

(b) *There is an element h in an ℓ-supergroup $H \geq G$ in which suprema and infima (in G and H) agree such that $X = L_G(h)$.*

(c) *X is an order closed (i.e., $g = \vee S$ for $S \subseteq X$ implies $g \in X$) lattice ideal of G such that $gX \neq X \neq Xg$ for any $e < g \in G$.*

Proof. (a) clearly implies (b), so assume $X = L_G(h)$ for an element h in some ℓ-supergroup H in which suprema and infima agree. X is order closed, for if g is the supremum in G of some subset $S \subseteq X$ then $g = \vee S$ in H also, which implies $g \leq h$. And if $e < g \in G$ then $Xg \neq X$ because $\vee(Xg) = (\vee X)g = hg > h = \vee X$, and $gX \neq X$ similarly. Hence (b) implies (c).

To show that (c) implies (a) let h be the supremum of X in B, the complete Boolean algebra which essentially extends G (see Section 7.1). We claim that $h \in G^{dist}$, for which we must show that, for any $e < g \in G$, both hg and gh strictly exceed h. To that end find $x \in X$ such that $xg \in Xg \setminus X$. Since X is order closed we know $xg \neq \vee(xg \wedge X)$, say $xg \wedge x_1 \leq y < xg$ for all $x_1 \in X$. Then $X \subseteq y \downarrow xg$ and $\vee X = h$ imply $h \in y \downarrow xg$ and (since we are in the Boolean algebra B) $h \wedge (xg \wedge y^\perp) = 0$. But since $hg \geq xg \geq (xg \wedge y^\perp) > 0$, we conclude that $hg > h$. A similar argument shows that $gh > h$, and this completes the proof the $h \in G^{dist}$.

To show $h \in G^\alpha$ it is enough by Corollary 7.2.9 to show that $\mathcal{F} \xrightarrow{\alpha} h$ in G^{dist}, where $\mathcal{F}$ is the filter with base $\{U_X(x) : x \in X\}$. To that end consider $a < b$ in G^{dist} for which $h \vee a \geq b$. Now there must be some $x \in X$ such that $x \wedge (b \wedge a^\perp) \neq 0$, for otherwise we get the contradiction $0 < b \wedge a^\perp = h \wedge b \wedge a^\perp = \vee X \wedge b \wedge a^\perp = \vee(X \wedge b \wedge a^\perp) = 0$. Therefore $x \notin a \downarrow b$; let $b' = (x \vee a) \wedge b > a$. Then $a \uparrow b' \supset U_X(x) \in \mathcal{F}$. On the other hand, $h \wedge b \leq a$ implies $a \downarrow b \supset X \in \mathcal{F}$. We have proved that $\mathcal{F} \xrightarrow{\alpha} h \in G^\alpha$ and therefore that $h \in G^\alpha$; it remains only to show $L_G(h) = X$. By construction $X \subseteq L_G(h)$, so consider $y \in L_G(h)$. Then $y = y \wedge h = y \wedge \vee X = \vee(y \wedge X)$ and X order closed imply $y \in X$. This completes the proof.

We shall term a subset satisfying the conditions of Theorem 7.2.16 a *cut*. An ℓ-group G will be said to be *cut complete* provided that every cut of G has a supremum in G.

PROPOSTION 7.2.17. *An α-Cauchy complete ℓ-group is cut complete. Consequently, a distinguished ℓ-group is cut complete.*

The *cut completion* G^{cut} of G is defined by applying the Theorem Prototype of the introduction to the property of cut completeness. The existence of a cut completion can be shown by simply intersecting those cut complete ℓ-supergroups of G within G^{dist} in the fashion by now familiar to the reader. The uniqueness of the cut completion can be shown with just a little more effort ([Ball, b]).

THEOREM 7.2.18. *Every ℓ-group G has a cut completion G^{cut} which is unique up to (unique) ℓ-isomorphism over G. In fact, $G \leq G^{cut} \leq G^{i\alpha}$.*

It is not at present known whether the converse of Proposition 7.2.17 is true, i.e., whether $G^{cut} = G^{i\alpha}$. An affirmative answer to this question is provided by Theorem 4.5 of [Ball, b] in the cases in which G is archimedean, projectable, or normal valued and completely distributive. In work in progress at the time of this writing, Steve McCleary and the author have shown that the answer is affirmative for any completely distributive ℓ-group. This result depends on a natural and interesting ℓ-permutation characterization of $G^{i\alpha}$. All other instances of the question are open.

The Cauchy completions we have discussed to this point have, when iteratively applied to G, eventually stabilized to produce a completion. We close this section by mentioning a Cauchy completion whose iterative application rarely stabilizes, but which readily proves a result of independent algebraic (i.e., nontopological) interest.

Among normal-valued ℓ-groups, a particularly tractable subclass consists of those in which every positive element is the supremum of a pairwise disjoint set of special elements. Recall that an element $e \neq g \in G$ is said to be *special* if g has just one value P (Section 5.1) [Bigard, Keimel & Wolfenstein, 1977, 2.5.2 & 2.5.11], and P is also called special in this case. It is easy to show that a special value is order closed and normal in its cover, so that an ℓ-group in which each positive element is supremum of disjoint special elements is normal valued and completely distributive. Furthermore, a given positive element can be written as the supremum of a set of disjoint special elements in at most one way; thus the special elements are the building blocks for all other elements in such ℓ-groups. If the ℓ-group is also laterally complete, then all possible combinations of these building blocks are already present.

Given element $g \in G$, define $\mathcal{F} \to g$ to mean that $\mathcal{F}$ contains the coset Pg for each prime P of G. It is straightforward to check that this is a convex Hausdorff strongly normal ℓ-convergence on G, and in fact is the convergence of a topology. (See [Ball, a] and Proposition 1.1 of [Ball, 1980b].) Furthermore, a filter $\mathcal{K}$ is readily seen to be Cauchy if and only if $\mathcal{K}$ is a proper filter (meaning $\emptyset \notin \mathcal{K}$) which contains both a right and a left coset of each prime of G. Let H designate the corresponding Cauchy completion. A close relative of the following theorem appears in [Ball, Conrad and Darnel, 1986]; other related results can be found in [Bixler and Darnel, a].

THEOREM 7.2.19. *An ℓ-group is normal valued if and only if it is an ℓ-subgroup of a laterally complete ℓ-group, every element of which is a supremum of disjoint special elements.*

Proof. Given a normal-valued ℓ-group G, construct H as above. Consider $h = [\mathcal{F}] \in H$, and let P be any prime of G for which there is some $e < g \in G$ such that $P \neq Pg \in \mathcal{F}$. Let S be the value of g containing P. We will construct a special element $h(S) = [\mathcal{K}] \in H$. Let $\mathcal{K}_1$ be the filter with base

$$\{\cap_1^n Q_i : Q_i \text{ prime and } Q_i \not\subseteq S\},$$

and let $\mathcal{K}_2$ be the filter with base

$$\{(\cap_1^m R_j f_j) \cap (\cap_1^m g_j R_j) : R_j \text{ prime}, R_j \subseteq S, R_j f_j \in \mathcal{F}, \text{ and } g_j R_j \in \mathcal{F}\}.$$

We claim $\mathcal{K}_1 \cup \mathcal{K}_2$ has the finite intersection property. To show this, consider $\cap_1^n Q_i \in \mathcal{K}_1$ and $R = (\cap_1^m R_j f_j) \cap (\cap_1^m g_j R_j) \in \mathcal{K}_2$. Now $R \in \mathcal{F}$ implies $R \neq \emptyset$, so the f_j's and g_j's can all be chosen to coincide, say $f_j = g_j = f$ for $1 \leq i \leq m$. Next choose $e < q_i \in Q_i \backslash S$ for $1 \leq i \leq n$. By replacing q_i by a sufficiently lofty power of itself, we can be sure that $Sq_i > Sg$ (remember that $\hat{S}/S$ is ℓ-isomorphic to a subgroup of $\mathbf{R}$, where $\hat{S}$ is the cover of S); then by replacing that q_i by $q_i \wedge f$ we can ensure that $R_j q_i = R_j f$ and $q_i R_j = f R_j$ for all i and j. Then

$$q = \wedge_1^n q_i \in (\cap_1^n Q_i) \cap ((\cap_1^m R_j f_j) \cap (\cap_1^m g_j R_j)) \neq \emptyset,$$

proving the claim. Let $h(S)$ be $[\mathcal{K}] \in H$, where $\mathcal{K}$ is the (clearly Cauchy) filter with base $\mathcal{K}_1 \cup \mathcal{K}_2$. We claim $h(S)$ is special with value $V = cl(S) = \{[L] \in H : S \in L\}$. Observe that $h(S) = [\mathcal{K}] \notin V$ because S is disjoint from $Sg \in \mathcal{K}$. To prove this claim it is sufficient to show that for any $e < m \in H \backslash V$ there is some positive integer n such that $m^n \geq h(S)$. Therefore consider $e < m = [\mathcal{M}] \in H \backslash V$. Without loss of generality assume that $G^+ = \{g \in G : g > e\} \in \mathcal{M}$. Now $St \in \mathcal{M}$ for some $t \in G^+ \backslash S$. If $t \notin \hat{S}$ then $St > Sg$ and we take n to be 1. If $t \in \hat{S}$ then we take n to be a positive integer large enough so that $St^n > Sg$, and then replace t^n by t. In either case we have a positive integer n and element $t \in G^+ \backslash S$ such that $St > Sg$ and $St \in \mathcal{M}^n$. We will now establish that $m^n = [\mathcal{M}^n] \geq [\mathcal{K}] = h(S)$ by showing that for any $L \in \mathcal{M}^n$ and $K \in \mathcal{K}$ there are elements $x \in L$ and $y \in K$ such that $x \geq y$. (This will show that $(L \vee K) \cap L \neq \emptyset$, which shows that $(\mathcal{M}^n \vee \mathcal{K}) \cup \mathcal{M}^n$ is a proper filter which is simultaneously finer than both $\mathcal{M}^n \vee \mathcal{K}$ and $\mathcal{M}^n$, and is therefore equivalent to both. This proves $m^n = [\mathcal{M}^n] = [\mathcal{M}^n \vee \mathcal{K}] = m^n \vee h(S)$.) Let Q be a prime of G such that $Q \not\subseteq S$. Then $Q \in \mathcal{K}_1 \subseteq \mathcal{K}$ while $Qv \cap vQ \in \mathcal{M}^n$ for some $e \leq v \in G$; in this case the x and y can be taken to be v and e, respectively. Let R be a prime of G such that $R \subseteq S$. Then $Rf \cap fR \in \mathcal{K}_2 \subseteq \mathcal{K}$ for some $f \in G^+$ while $Rv \cap vR \in \mathcal{M}^n$ for some $e \leq v \in G$. $Rf < Rv$ because $Sf = Sg < St = Sv$, and $fR < vR$ likewise. In this case the x and y can be taken to be v and $f \wedge v$, respectively. This completes the proof of the claim that V is the only value of $h(S)$. We leave it to the reader to show that

$\{h(S) : S$ a prime of G such that $S \neq Sg \in \mathcal{F}$ for some $g\}$

is a pairwise disjoint set whose supremum is h.

To prove H is laterally complete consider a pairwise disjoint set $D \subseteq H$; the discussion above allows us to assume that the elements of D are special. For each $d \in D$ let $\mathcal{K}(d)$ be a Cauchy filter such that $d = [\mathcal{K}(d)]$. Fix a prime P of G. Since $cl(P) = \{[\mathcal{K}] : P \in \mathcal{K}\}$ is prime in H and since D is pairwise disjoint in H, there is at most one $d \in D$ such that $d \notin cl(P)$, i.e., such that $Pg \cap gP \in \mathcal{K}(d)$ for some $g \notin P$. Label with $C(P)$ the set $Pg \cap gP$, where Pg is the greatest element in the chain of right cosets of P to be found in $\cup \{\mathcal{K}(d) : d \in D\}$. (It follows that gP will be the greatest element in the chain of left cosets of P to be found in $\cup \{\mathcal{K}(d) : d \in D\}$.) Let $\mathcal{D}$ be the filter with base $\{C(P) : P$ a prime of $G\}$. That $\mathcal{D}$ has the finite intersection property is clear upon reflection that, for any finite collection Q of primes of G, $\cap \{C(P) : P \in Q\}$ is an element of the Cauchy filter $\vee \{\mathcal{K}(d) : d \in D(Q)\}$, where $D(Q)$ is the finite set $\{d \in D : d \notin cl(P)$ for some $P \in Q\}$. $\mathcal{D}$ is Cauchy by design, and it is light work to show that $[\mathcal{D}] = \vee D$ in H.

Suppose that in the construction of H above the convergence is replaced by the following coarser one: $\mathcal{F} \to g$ is to mean that $\mathcal{F}$ contains Pg for each order closed prime P of G. This convergence is also convex and strongly normal, but is Hausdorff if and only if G is completely distributive. In this case it is easy to show that G is order dense in H, so that the argument above proves the theorem below.

THEOREM 7.2.20. *An ℓ-group is normal valued and completely distributive if and only if it is order dense in an ℓ-supergroup, every positive element of which is a supremum of disjoint special elements.*

7.3. Piecewise Approximation and Projectability

In this section we study generalizations of (and variations on) the projectable hull of a representable ℓ-group G. Though we show that such completions cannot be obtained as Cauchy completions, they nevertheless fit in G^{dist}. In fact, we conclude by showing that the property of being distinguished is the join of the properties of being projectable and of being α-Cauchy complete in the sense that G^{dist} is the cut completion of the projectable hull of G whenever G is representable.

We shall approach projectability not from the classical definition (that polars are cardinal summands), but from the auxiliary notion of piecewise approximation. Given an element g and subset X in an ℓ-group G, we say that g is *piecewise approximated by elements of X* provided that

$$\vee\{(gx^{-1})^{\perp} : x \in X\} = 1_{\mathcal{P}}.$$

The supremum is computed in the complete Boolean algebra $\mathcal{P}$ of polars; alternative equivalent formulations can be gotten by conjugation (e.g., $\vee\{(x^{-1}g)^{\perp} : x \in X\} = 1_{\mathcal{P}}$) or by passage to the Boolean complement (e.g., $\wedge\{(gx^{-1})^{\perp\perp} : x \in X\} = 0_{\mathcal{P}}$). An ℓ-supergroup $H \geq G$ is said to be a *(finitely) piecewise approximated extension* of G provided that each $h \in H$ is piecewise approximated by elements of (some finite subset of) G, and in addition provided that the polars of H and G are in one-to-one correspondence by intersection. The latter requirement holds if and only if every nontrivial polar of H meets G nontrivially. (Proposition 5.3 of [Ball, 1980b]), in which case the intersection map is actually a Boolean isomorphism from the polar algebra of H onto that of G. It is for this reason that we can (and do) use $\mathcal{P}$ to denote either algebra.

The intuitive motivation underlying piecewise approximation is very like the motivation for polar convergence; indeed, we show in Proposition 7.3.7 that the iterated polar Cauchy completion $G^{i\mathcal{P}}$ is a piecewise approximated extension of G. An example of a fundamentally different piecewise approximated extension, to which we return following Corollary 7.3.8, is the additive group H of 2×2 matrices with integral entries, ordered

by declaring $\begin{bmatrix} m_1 & m_2 \\ n_1 & n_2 \end{bmatrix} \geq 0$ if and only if the following condition

holds both for $i = 1$ and $i = 2$; either $m_i > 0$ or $m_i = 0$ and $n_i \geq 0$. Let G be the ℓ-subgroup consisting of those elements for which $m_1 = m_2$. Then H is a finitely piecewise approximated extension of G because an arbitrary

$\begin{bmatrix} m_1 & m_2 \\ n_1 & n_2 \end{bmatrix} \in H$ is piecewise approximated by elements $\begin{bmatrix} m_1 & m_1 \\ n_1 & n_2 \end{bmatrix}$ and

$\begin{bmatrix} m_2 & m_2 \\ n_1 & n_2 \end{bmatrix}$ of G.

Piecewise approximated extensions have tractable properties which directly suggest the existence of a maximal piecewise approximated extension for an arbitrary ℓ-group G. These properties are the subject of the next lemma, whose proof is easy.

LEMMA 7.3.1. *If $G \leq H \leq K$, then K is a (finitely) piecewise approximated extension of G if and only if K is a (finitely) piecewise approximated extension of H and H is a (finitely) piecewise approximated extension of G. If C is a set of ℓ-groups totally ordered by the (finitely) piecewise approximation relation then $\cup C$ is a (finitely) piecewise approximated extension of any $C \in C$.*

We establish the existence of maximal (finitely) piecewise approximated extensions in Theorem 7.3.4, whose proof requires the following two lemmas.

LEMMA 7.3.2. *G is order dense in any (finitely) piecewise approximated extension H.*

Proof. Given $e < h \in H$, find $g \in G$ such that $Q = (hg^{-1})^{\perp} \cap h^{\perp\perp} \neq 0_{\mathcal{P}}$, then find $e < q \in Q \cap G$, and let $f = (q \wedge g) \vee e \in G$. We claim $e < f \leq h$. To show $f \leq h$ assume for contradiction that $Pf > Ph$ for some prime P. Since $Pf = (Pq \wedge Pg) \vee P$, this means that $Pg > Ph$; so $hg^{-1} \notin P$. But then $q \wedge |hg^{-1}| = e$ forces $q \in P$, producing the contradiction $Pf \leq Pq = P \leq Ph$. To show $f \neq e$ find a prime P omitting $q \wedge h$; such a prime must exist because $0_{\mathcal{P}} \neq q^{\perp\perp} \subseteq Q \subseteq h^{\perp\perp}$ implies that $q \wedge h \neq e$. Then $h, q \notin P$ and $q \wedge |hg^{-1}| = e$ imply $Pg = Ph > P$, so that $Pf = (Pq \wedge Pg) \vee P > P$. This completes the proof of the claim and the lemma.

LEMMA 7.3.3. *A (finitely) piecewise approximated extension of G is a distinguished extension.*

Proof. Given $h_1 < h_2$ in H, let $Q = (h_2 h_1^{-1})^{\perp\perp}$, find $g_2 \in G$ such that $(h_2 g_2^{-1})^{\perp} \cap Q = R \neq 0_{\mathcal{P}}$, and then use the order density to find $d \in G \cap R$ such that $e < d \leq h_2 h_1^{-1}$. Let $g_1 = d^{-1} g_2$. We claim g_1 and g_2 distinguish h_1 from h_2. To show $h_2 \vee g_1 \geq g_2$, consider any prime P such that $Pg_1 = Pd^{-1} g_2 < Pg_2$. Then $P < Pd$, so $h_2 g_2^{-1} \in d^{\perp} \subseteq P$, hence $Ph_2 = Pg_2$. To show $h_1 \wedge g_2 \leq g_1$, consider any prime P for which $Pg_2 > Pg_1 = Pd^{-1} g_2$. Then $Pd^{-1} < P$ implies $(dPd^{-1}) < (dPd^{-1})d$, so $h_2 g_2^{-1} \in d^{\perp} \subseteq dPd^{-1}$, $(dPd^{-1})h_2 = (dPd^{-1})g_2$, and $Pd^{-1}h_2 = Pd^{-1}g_2$. We have $Ph_1 = Ph_1 h_2^{-1} h_2 \leq Pd^{-1}h_2 = Pd^{-1}g_2 = Pg_1$.

THEOREM 7.3.4. *Any ℓ-group G has a (finitely) piecewise approximated extension $\bar{G}$ ($\bar{G}^{\,\omega}$) which itself has no proper (finitely) piecewise approximated extensions.*

Proof. One need only check that the set of elements of G^{dist} which are piecewise approximated by elements of (some finite subset of) G constitutes an ℓ-subgroup.

The succinctness of the previous proof contrasts sharply with the 14 pages of detailed argument necessary to prove this theorem in [Ball, 1982]. Here, then, is another demonstration of the utility of G^{dist} in the study of completions of G.

COROLLARY 7.3.5. *A distinguished ℓ-group has no proper piecewise approximated extensions.*

By taking advantage of the unique ℓ-injection of any distinguished extension into G^{dist}, one can say slightly more.

THEOREM 7.3.6. *H is a (finitely) piecewise approximated extension of G if and only if there is an ℓ-injection $\theta: H \to \bar{G}$ ($\theta: H \to \bar{G}^{\,\omega}$) over G. Thus $\bar{G}$ ($\bar{G}^{\,\omega}$) is unique up to (unique) ℓ-isomorphism over G.*

We show now that the maximal piecewise approximated extension $\bar{G}$ contains the iterated polar Cauchy completion G^{ip}.

PROPOSITION 7.3.7. $G^{ip} \leq \bar{G}$.

Proof. It is enough to show that G^p is a piecewise approximated extension of G. To that end consider $e < f = [\mathcal{F}] \in G^p$. Now $\mathcal{F} \xrightarrow{P} f$ by Theorem 7.2.5, meaning

$$\wedge\{(Ff^{-1})^{\perp\perp} : F \in \mathcal{F}\} = 0_{\mathcal{P}}, \text{ or } \vee\{(Ff^{-1})^{\perp} : F \in \mathcal{F}\} = 1_{\mathcal{P}}.$$

We claim that h is piecewise approximated by elements from any $K \in \mathcal{F}$. To see this, fix $K \in \mathcal{F}$ and for each $F \in \mathcal{F}$ choose an element $k_F \in K \cap F \in \mathcal{F}$. Then $(Ff^{-1})^{\perp} \subseteq (k_F f^{-1})^{\perp}$, so $\vee\{(kf^{-1})^{\perp} : k \in K\} = 1_{\mathcal{P}}$. This completes the proof of the claim and of the proposition.

COROLLARY 7.3.8. *$\bar{G}$ is polar Cauchy complete and therefore laterally complete.*

$\bar{G}$ is in general properly larger then G^{ip}, and in fact we now show that $\bar{G}$ cannot be obtained by any (even iterated) Cauchy completion. Let G and H have the meaning of the example preceding Lemma 7.3.1. If H were contained in any iterated Cauchy completion of G, then a minor modification of the proof of Theorem 7.2.3 (Theorem 3.1 of [Ball, 1984]) would assert that a positive universal sentence θ with constants from G which holds in G would necessarily also hold in H. But if we denote by g, g_1, and g_2 the respective constants $\begin{bmatrix} 1 & 1 \\ 0 & 0 \end{bmatrix}, \begin{bmatrix} 0 & 0 \\ 1 & 0 \end{bmatrix}$, and $\begin{bmatrix} 0 & 0 \\ 0 & 1 \end{bmatrix}$ from G,

then the sentence

$$\forall z,w(\,|z|\wedge|w| >0 \text{ or } |z| \le g \text{ or } w = 0))$$

holds in G but not in H. (To violate this sentence in H take w to be g_1 and z to be $\begin{bmatrix} 0 & 2 \\ 0 & 0 \end{bmatrix}$.) This sentence hardly appears to be positive universal, but it is so because $|z|\wedge|w| > 0$ is expressible by

$$|z|\wedge|w| \ge g_1 \text{ or } |z|\wedge|w| \ge g_2,$$

and because any condition of the form $x \ge y$ reduces to the atomic formula

$$(x-y)\wedge 0 = 0.$$

Thus we see that the ideas involved in the formation of $\bar{G}$ fundamentally diverge from those of Section 7.2.

An ℓ-group G is said to be *projectable* if each of its polars is a cardinal summand (page 16 & Section 5.2). (The term strongly projectable is often used in the literature, but we do not consider weak projectability here and so drop the qualifying adjective.) Such an ℓ-group has normal polars and so is representable.

PROPOSITION 7.3.9. *G is projectable if and only if G is representable and G* $=\bar{G}^{\omega}$.

Proof. Suppose that G is representable and that $G = \bar{G}^{\omega}$, and consider a polar P of G. P and $P^{\perp}$ are then normal, and the natural ℓ-injection of G into the cardinal sum $H = G/P \boxplus G/P^{\perp}$ gives a finitely piecewise approximated extension: $h = (Pg_1, P^{\perp}g_2)$ is piecewise approximated by $(Pg_1, P^{\perp}g_1)$ and $(Pg_2, P^{\perp}g_2)$ from G. Since $G = \bar{G}^{\omega}$ has no proper finitely piecewise approximated extensions, $G = H$, and P is a summand of G.

Now suppose G is projectable. G is surely representable (Theorem 1.2.1), and so it remains to show that G has no proper finitely piecewise approximated extension H. So consider such an extension H and element $e < h \in H$, say $\vee_1^n (hg_i^{-1})^\perp = 1_{\mathcal{P}}$ for $\{g_1, g_2, ..., g_n\} \subseteq G$. Let P_i be $(hg_i^{-1})^\perp \cap G$, and use the projectability to find $p_i \in P_i$ such that $g_i p_i^{-1} \in P_i^\perp$. Note that $hp_i^{-1} = (hg_i^{-1})(g_i p_i^{-1}) \in P_i^\perp$, so that p_i is the projection of h on P_i. We claim that $h = \vee_1^n p_i \in G$, for $(h(\vee_1^n p_i)^{-1})^{\perp\perp} = (\wedge_1^n hp_i^{-1})^{\perp\perp} = \cap_1^n (hp_i^{-1})^{\perp\perp}$, the latter equality holding because $hp_i^{-1} > e$. And since $(hp_i^{-1})^{\perp\perp} \subseteq P_i^\perp$, we get $(h(\vee_1^n p_i)^{-1})^{\perp\perp} \subseteq \cap_1^n P_i^\perp = 0_{\mathcal{P}}$, proving the claim and proposition.

An ℓ-group is said to be *orthocomplete* if it is projectable and laterally complete.

PROPOSITION 7.3.10. G is orthocomplete if and only if G is representable and $G = \bar{G}$.

Proof. If G is representable and $G = \bar{G}$ then clearly $G = \bar{G}^\omega$, hence G is projectable, and G is laterally complete by Corollary 7.3.8. Now suppose G is orthocomplete. Since G is necessarily projectable and so representable, it remains to show $G = \bar{G}$. Consider a piecewise approximated extension $H \geq G$ and element $e < h \in H$. We have $\vee \mathcal{R} = 1_{\mathcal{P}}$, where $\mathcal{R} = \{P$ a polar $: P \subseteq (hg^{-1})^\perp$ for some $g \in G\}$. Let S be a maximal pairwise disjoint subset of $\mathcal{R}$; evidently $\vee S = 1_{\mathcal{P}}$ also. For each $R \in S$ choose $g(R) \in G$ such that $R \subseteq (hg(R)^{-1})^\perp$. By the projectability of G we may assume $g(R) \in R$ for each $R \in S$. Now $X = \{g(R) : R \in S\}$ is a pairwise disjoint subset of G, and so $\vee X = g$ exists in G by virtue of lateral completeness. Then $h = g$, for

$$(hg^{-1})^{\perp\perp} = (h(\vee_S g(R))^{-1})^{\perp\perp} = (h \wedge_S g(R)^{-1})^{\perp\perp} = (\wedge_S hg(R)^{-1})^{\perp\perp} \subseteq$$
$$\wedge_S (hg(R)^{-1})^{\perp\perp} \subseteq \wedge_S R^\perp = 0_{\mathcal{P}}.$$

We close by showing that G^{dist} is the cut completion of the projectable hull $\bar{G}^\omega$ of a representable ℓ-group G. This will require the following lemma, and this lemma uses two well-known facts about minimal primes: first, that every prime contains a minimal prime, and second, that if P is a minimal prime containing an element p then $p^{\perp\perp} \subseteq P$ (Lemma 1.1.14)

LEMMA 7.3.11.   *Let $e \leq a < b$ in G and $Q = (ba^{-1})^{\perp\perp}$. Suppose $a_1, b_1 \in Q$ satisfy $aa_1^{-1}, bb_1^{-1} \in Q^{\perp}$. Then $e \leq a_1 < b_1$, $a \uparrow b = a_1 \uparrow b_1$, $a \downarrow b = a_1 \downarrow b_1$, and $\mathcal{U}_G(e) \cap a \uparrow b = \mathcal{U}_G(b_1)$.*

Proof. We first claim that for any prime P, $Pa < Pb$ implies $Pa_1 = Pa$ and $Pb_1 = Pb$. For $ba^{-1} \notin P$ implies $aa_1^{-1}, bb_1^{-1} \in Q^{\perp} = (ba^{-1})^{\perp} \subseteq P$. We next claim that for any prime P, $Pa_1 \neq Pb_1$ implies $Pa = Pa_1$ and $Pb = Pb_1$ (and therefore $Pa_1 < Pb_1$). For $b_1a_1^{-1} \notin P$ implies $b_1 \notin P$ or $a_1 \notin P$, and in either case $aa_1^{-1}, bb_1^{-1} \in Q^{\perp} \subseteq P$. Simple arguments from these two claims give $e \leq a_1 < b_1$, $a \uparrow b = a_1 \uparrow b_1$, $a \downarrow b = a_1 \downarrow b_1$, and consequently $\mathcal{U}_G(b_1) \subseteq \mathcal{U}_G(e) \cap a \uparrow b$. Now let $x \in \mathcal{U}_G(e) \cap a \uparrow b$, and suppose for contradiction that $x \not\geq b_1$. Then there is some prime P such that $P \leq Px < Pb_1$, and without loss of generality P may be taken to be minimal. But in this case $b_1 \notin P$ and $b_1 \in (ba^{-1})^{\perp\perp}$ imply $ba^{-1} \notin P$, so $Pa < Pb$. And since $x \vee a \geq b$, we get $Px \geq Pb$. But by the first claim we are led to the contradiction $Pb = Pb_1$. We can only conclude that $x \geq b_1$. This completes the proof of the proposition.

THEOREM 7.3.12. $G^{dist} = (\overline{G}^{\omega})^{cut}$ *for a representable ℓ-group G.*

Proof. It is enough to show that $G^{dist} = G^{cut}$ for a projectable ℓ-group G. We show that $h = \vee L_G(h)$ for any $e < h \in G^{dist}$. Consider $k \in G^{dist}$ such that $e \leq k < h$. Since G^{dist} is a distinguished extension of G, there are $a < b$ in G distinguishing k from h. The elements a_1 and b_1 of the previous lemma exist in G by virtue of its projectability, and a_1 and b_1 also distinguish k from h. But then $h \in \mathcal{U}_H(e) \cap a \uparrow b = \mathcal{U}_H(b_1)$, but $k \notin \mathcal{U}_H(e) \cap a \uparrow b = \mathcal{U}_H(b_1)$, meaning that $b_1 \in L_G(h) \setminus L_G(k)$. This shows $\vee L_G(h) = h$.

By a very similar argument one can show that $G^{dist} = (\overline{G})^{ded} = \overline{G}^{0}$.

7.4. Questions

The topic of completions of ℓ-groups, even delimited by the Theorem Prototype of the introduction, has hardly been exhausted by the results mentioned in this survey. Here are several areas of inquiry of particular appeal to the author.

Much remains to be understood about G^{dist}. For example, is a distinguished ℓ-group nothing more than an α-Cauchy complete one without proper piecewise approximated extensions? More particularly, if one begins with G and applies these two closure operators (perhaps numerous times) does one obtain G^{dist}? That is, does something like Theorem 7.3.12 hold in general?

Distinguished ℓ-groups ought to be identifiable by means of concrete structural characteristics. That is, there ought to be a method of building all the distinguished ℓ-groups in some broad class from readily understood constituents using straightforward techniques. Failing that, one would like to have some descriptive structural results. For example, among representable distinguished ℓ-groups there are few which are completely distributive; in fact, results in [Ball, b] and [Ball, 1982] can be used to show that such an ℓ-group is completely distributive if and only if it is a product of Dedekind-MacNeille complete totally ordered groups.

Interesting completions are still being found in G^{dist} (e.g., [Ball, d]), and hard questions remain to be answered about those that have been found. We have already mentioned the very important question of whether every element of $G^{i\alpha}$ can be obtained by taking (iteratively) suprema and infima of elements of G; in short, whether $G^{i\alpha} = G^{cut}$. Here is another quite interesting question of the same ilk. Consider an ℓ-group G with polar P; let N be the normalizer of P (an ℓ-subgroup of G), and let $C(P)$ designate the *convex normalizer* of P, i.e., the largest ℓ-subgroup of N which is convex in G. For example, if G is representable then $G = N = C(P)$, and the condition that every polar is a cardinal summand of its convex normalizer is equivalent to projectability. Theorem 7.3.9 asserts that a representable ℓ-group G has no proper finitely piecewise approximated extensions if and only if every polar of G is a cardinal summand of its convex normalizer. We believe that, in this form, the theorem is true of all ℓ-groups.

Finally, what does it mean to say that ℓ-groups G and H are "codistinguished"? That is, if G^{dist} is ℓ-isomorphic to H^{dist} then G and H are cousins and ought to bear a family resemblance to one another. Is it possible to characterize this resemblance structurally?

There is much to be done.

Richard N. Ball
University of Denver
Denver, Colorado 80208
U. S. A.

Richard N. Ball and Anthony W. Hager

CHAPTER 8

CHARACTERIZATION OF EPIMORPHISMS IN ARCHIMEDEAN LATTICE-ORDERED GROUPS AND VECTOR LATTICES

8.1. Introduction[1]

This chapter is the first paper in a sequence of undetermined length, and this introduction is also a brief introduction to the sequence. An initial segment of that sequence is this chapter, followed by the essentially completed items listed in the bibliography as [Ball and Hager, a,b,c,d]. These papers treat various aspects of the theory of epimorphisms in the category $\mathcal{Arch}$ of archimedean ℓ-groups with ℓ-homomorphisms, and the category $\mathcal{Wu}$ of archimedean ℓ-groups with distinguished weak unit and unit-preserving ℓ-homomorphisms (and in the corresponding categories of vector lattices, which for all papers in the sequence can be disposed of with the comment: all this is true in vector lattices and the proofs require no change). $\mathcal{Wu}$ is itself interesting as a natural generalization of rings of continuous functions and as a setting for some functional analysis, and in any event seems to be a necessary bridge to $\mathcal{Arch}$.

In an abstract category C, a morphism ε is an *epimorphism*, or is *epi*, if whenever α,β have $\alpha\varepsilon = \beta\varepsilon$, then $\alpha = \beta$; it is *monic* if whenever α,β have $\varepsilon\alpha$

---

[1]The authors are much in debt to Donald G. Johnson for reading the manuscript with great care, eliminating many errors, and suggesting a number of improvements.

175

A. M. W. Glass and W. C. Holland (eds.), Lattice-Ordered Groups, 175–205.
© 1989 by Kluwer Academic Publishers.

$= \varepsilon\beta$, then $\alpha = \beta$. (Our convention is that $\alpha\varepsilon$ has α acting after ε.) And, an object G is *epicomplete* if $G \xrightarrow{\varepsilon} H$ epi and monic implies ε is an isomorphism, or, we can say, G has no proper epi extension; an *epicompletion* of G is an epicomplete object containing G epically.

In concrete situations, where one has surjections and embeddings, surjections are typically epi though not conversely (hence the present project); and one can study epimorphisms by studying epi embeddings (which we shall do), because $A \xrightarrow{\varepsilon} B$ is epi iff $\varepsilon(A) \leq B$ is epi.

In the category of abelian ℓ-groups, say $\ell A b$, the epi embeddings are known from the Anderson-Conrad work [1981]: $G \leq H$ is epi iff $H \leq dG$ over G, where dG is the divisible hull; the proof of this is less simple than the result. (Related, though seemingly not relevant to the present project, is Darnel's interesting paper [1985].) Thus, in $\ell A b$, epicomplete means divisible, and if $G \leq H$ is epi, the cardinalities satisfy $|H| = |G|$. In many other ways, the theory of epimorphisms is a model of propriety: In $\ell A b$,

1) If $G \leq H$ is epi, then this embedding is essential (whenever $H \xrightarrow{\tau} K$ is a homomorphism, $\tau | G$ one-to-one $\Rightarrow \tau$ one-to-one) or equivalently, kernel-distinguishing (distinct ℓ-ideals I, J of H have $I \cap G \neq J \cap G$).

2) If $G \leq H$ is epi, then this embedding is majorizing ($\forall h \in H \; \exists g \in G$ with $|h| \leq g$) and restrictably epi ($G \leq K \leq H \Rightarrow G \leq K$ epi).

3) Each G has a unique epicompletion.

4) Each G has a reflective epicompletion (an epicompletion over which each homomorphism of G to an epicomplete object lifts).

5) G is epicomplete iff G is epi-injective ($H \leq K$ epi and $H \xrightarrow{\varphi} G$ a morphism imply an extension $K \xrightarrow{\overline{\varphi}} G$).

Of course, in view of the Anderson-Conrad theorem, all of this is completely obvious, and the only reason for compiling the list is to emphasize the complexity of the theory of epimorphisms in $Arch$ and Wu: there, 4) is true (and that is involved), while *everything else in the list is completely false!*

We elaborate this slightly. The following refers to $Arch$ and Wu.

1) An epi embedding need not be essential (8.6.5 below), but there is unique essential-epicompletion [Ball and Hager, b]. In connection with this, we have considered "Archimedean-kernel-distinguishing" embeddings: $G \leq H$ is Akd if whenever I, J are distinct ℓ-ideals of H for which H/I and H/J are archimedean, then $I \cap G \neq J \cap G$. In [Ball and Hager, c] we show that Akd's are essential and epi, but not conversely, and there is unique Akd-completion (in W at least).

2) An epi embedding need not be majorizing or restrictably epi (8.6.5 below), the notions agree in $\mathcal{A}rch$ and differ in $\mathcal{W}u$, but the theory of majorizing-epicompletion is the theory of "algebraic closure" [Ball and Hager, d].

3) $C[0,1]$ has 2^C epicompletions (8.6.5 below shows $\geq 2^{\aleph_0}$; [Ball and Hager, b] shows $\leq 2^C$; scrutiny of that method reveals exactly 2^C.)

4) Madden and Vermeer [a] first showed that in $\mathcal{W}u$ there are reflective epicompletions, using locales. Then, using our characterization of epicompleteness in $\mathcal{W}u$ (of the form $D(X)$, X basically disconnected), we verified this fact by our methods, and proved it for $\mathcal{A}rch$ [Ball and Hager, a], and then gave a concrete construction of the reflective epicompletion and of all other epicompletions as the quotients over G of the reflective epicompletion [Ball and Hager, b].

5) There are no epi-injectives in $\mathcal{A}rch$ or $\mathcal{W}u$ [Ball and Hager, d].

So the theory of epimorphisms and epicompletions is, in $\mathcal{W}u$ and $\mathcal{A}rch$, much more complicated than the corresponding theory in $\ell\mathcal{A}b$. Before despairing, however, one might consider the corresponding theory in Tychonoff spaces. Here, epi means the image is dense, epicomplete means compact, and epicompletion means compactification. Regarding 1)-5) in the list above, in Tych, 2) makes no sense, 4) is true (the Stone-Čech compactification βX), and the rest are false; also, every epicompletion of X is a quotient over X of βX - all that just like $\mathcal{A}rch$ and $\mathcal{W}u$.

However, the present chapter merely *begins* the program of study of epimorphisms in $\mathcal{W}u$ and $\mathcal{A}rch$, and many of the features discussed above will not be reached here. The central point of this chapter is, quite simply, a description of the epimorphisms in $\mathcal{W}u$ and $\mathcal{A}rch$. The description/characterization is a bit technical, and is in somewhat unfamiliar terms, but it has the following features and consequences: (a) Epi in $\mathcal{W}u$ can be recognized (sections 8.6 and 8.7 here, and [Ball and Hager, a,b]). (b) Whereas in principle, epi in $\mathcal{A}rch$ can be recognized, this is more difficult. The problem is partly overcome by the curious theorem (8.5.2 below) that a $\mathcal{W}u$-epimorphism into an *algebra* is $\mathcal{A}rch$-epi. This has the simplifying consequence that a $\mathcal{W}u$-object is epicomplete in $\mathcal{A}rch$ iff it is epicomplete merely in $\mathcal{W}u$. (c) The description of epi in $\mathcal{W}u$ yields some conceptual insight into a new view of the action of the Yosida functor; see the remarks after 8.3.2. (d) The description of epi yields cardinal bounds on epi extensions in both $\mathcal{W}u$ and $\mathcal{A}rch$. Thus these categories are co-well-powered (see 6.27 of [Herrlich and Strecker, 1973] for the definition.) This fact for $\mathcal{W}u$ is the central one; another proof using locales is given in [Madden and Vermeer, a]. These "special" proofs may be viewed as analogous to the proof from the Anderson-Conrad theorem that $\ell\mathcal{A}b$ is co-well-powered. By contrast, $\ell\mathcal{A}b$ is co-well-powered also by "general principles", as the category of models of a first-order theory ([Freyd, 1964]; p. 93); see also [Isbell, 1966] for a sharper result applicable to $\ell \mathcal{A} \mathcal{b}$. We don't know of such a proof for $\mathcal{W}u$ or $\mathcal{A}rch$.

8.2. Archimedean ℓ-groups and the Yosida Functor

For "ℓ-homomorphism" and "ℓ-ideal" we say "homomorphism" and "ideal"; and "$G \leq H$" means that G is an ℓ-subgroup of H. We shall write the group operation in an ℓ-group additively, because we shall only be concerned with archimedean ℓ-groups, and each archimedean ℓ-group is abelian (Theorem 1.2.6). We note that any ℓ-subgroup of a power of the ℓ-group of real numbers $\mathbf{R}$ is archimedean.

It is a remarkable fact that the converse is almost true. i.e., any archimedean ℓ-group is "almost" an ℓ-subgroup of a power of $\mathbf{R}$. The precise statement of this is a version of what we call the Yosida Theorem, which shall be explained below, which can be used to bring some aspects of the subject quite close to the theory of $C(X)$ (as in [Gillman and Jerison, 1966], say), and without which this paper would surely not exist.

$\mathcal{A}rch$ stands for the class (or category) of archimedean ℓ-groups (with their homomorphisms, i.e., ℓ-homomorphisms). We shall frequently write things like "$G \in \mathcal{A}rch$", "$G \leq H \in \mathcal{A}rch$", "$G \leq H$ is $\mathcal{A}rch$-epi", etc.

Now let X be a topological space, which we always assume to be completely regular Hausdorff, and frequently compact. $C(X)$ is the ℓ-group of real-valued continuous functions on X with pointwise addition and order. $D(X)$ is the *set* of continuous $f: X \to [-\infty, +\infty]$ for which $f^{-1}(\mathbf{R})$ is dense in X ("almost" real-valued). For $f \in D(X)$, $\infty(f)$ stands for $f^{-1}(\pm\infty)$; also, $Z(f) = \{x \in X \mid f(x) = 0\}$ is the zero-set, and $\operatorname{coz} f = \{x \in X \mid f(x) \neq 0\}$ is the *cozero set*. $D(X)$ is a lattice (in the pointwise order), but it usually fails to be a group. That is, for $f, g, h \in D(X)$, we say "$f+g = h$ in $D(X)$" if $f(x)+g(x) = h(x)$ $\forall x \in f^{-1}(\mathbf{R}) \cap g^{-1}(\mathbf{R}) \cap h^{-1}(\mathbf{R})$; it may well happen that for particular $f, g \in D(X)$ there is *no* $h \in D(X)$ with "$f+g = h$ in $D(X)$". (For example, take $X = [-\infty, +\infty]$, f the obvious extension of $x+\sin x$, g the extension of $-x$.) But, it may well happen that a subset $G \subseteq D(X)$ has the property that $\forall f, g \in G$, $\exists h \in G$ with $f+g = h$ in $D(X)$; if also $f, g \in G \Rightarrow f \vee g, f \wedge g, -f \in G$, then we call G an "ℓ-group in $D(X)$". (E.g., $C(X)$ is always an ℓ-group in $D(X)$.) One sees easily that such a $G \in \mathcal{A}rch$. The converse of this is the Yosida Theorem mentioned above. We now sketch the theory leading to this result.

For $u \in G^+$, $u^\perp = \{g \in G \mid |g| \wedge u = 0\}$ is an ideal, and, importantly, $G \in \mathcal{A}rch \Rightarrow G/u^\perp \in \mathcal{A}rch$. The set of values of $u \in G$ is usually denoted $\operatorname{Val}_G u$, though we will almost always denote it $Y(G,u)$, because when given the hull-kernel topology, it is a "Yosida space".

THEOREM 8.2.1. *Let* $G \in \mathcal{A}rch$ *and* $u \in G^+$. *Then* $Y(G,u)$ *is a compact Hausdorff space, and there is an isomorphism* $G/u^\perp \ni g+u^\perp \mapsto \hat{g} \in \hat{G}_u \subseteq D(Y(G,u))$ *onto an ℓ-group in* $D(Y(G,u))$, *with* $\hat{u}$ *the constant function* 1, *and with* $\hat{G}_u$ *separating the points of* $Y(G,u)$).

8.2.1 has its origins in [Yosida, 1942], and is discussed in [Bigard, Keimel, and Wolfenstein, 1977], Chapter 13 (with historical remarks on p. 280). We indicate the construction of the functions $\hat{g}$ following [Bigard, Keimel, and Wolfenstein, 1977]: Let G,u be as in 8.2.1. For $P \in Y(G,u)$, there is a prime ideal P^* of G with P maximal in P^*; there results an isomorphism $P^*/P \xrightarrow{\varphi} \mathbf{R}$, chosen so that $u+P \mapsto 1$. The isomorphism φ can then be extended to a function $G/P \to \bar{\mathbf{R}}$ by setting $\varphi(g+P) = +\infty$ or $-\infty$ according as $g+P^* > P$ or $< P$ (since G/P^* is totally ordered). (See 2.5 and 2.6 of [Bigard, Keimel, and Wolfenstein, 1977].) Now set $\hat{g}(P) = \varphi(g+P)$. One verifies that $\hat{g} \in D(Y(G,u))$, and that $g \mapsto \hat{g}$ is a homomorphism (13.2 of [Bigard, Keimel, and Wolfenstein, 1977]). Finally, the kernel of the homomorphism is $\bigcap \{P \mid P \in Y(G,u)\} = u^\perp$.

Although we don't need it, we note that 8.2.1 yields the "full" Yosida Theorem, that any $G \in \mathcal{A}rch$ "is" an ℓ-group in some $D(X)$: By Zorn's Lemma, choose a maximal pairwise disjoint subset $U \subseteq G^+$. Then $U^\perp = \bigcap \{u^\perp \mid u \in U\} = (0)$, so $G \leq \prod \{G/u^\perp \mid u \in U\} \cong \prod \{\hat{G}_u \mid u \in U\} \subseteq \prod \{D(Y(G,u)) \mid u \in U\} \cong D(\Sigma(Y(G,u) \mid u \in U))$ (where the first two $\prod$'s are ℓ-group products, the last $\prod$ is the set-theoretic product, and Σ means topological sum (or disjoint union).)

We return to the setting of 8.2.1. The following says that the Yosida space is unique:

THEOREM 8.2.2. *Let* $G \in \mathcal{A}rch$, $u \in G^+$, *let* X *be compact, and let there be an isomorphism* $G/u^\perp \ni g+u^\perp \mapsto \bar{g} \in \bar{G} \subseteq D(X)$ *onto an ℓ-group in* $D(X)$, *with* $\bar{u} = 1$ *and with* $\bar{G}$ *separating the points of* X. *Then there is a homeomorphism* $\tau : X \to Y(G,u)$ *for which* $\bar{g} = \hat{g} \circ \tau$, $\forall g \in G$.

Further, the following quickly yields that $Y(\;\;)$ is, in a sense, functorial:

THEOREM 8.2.3. *Let $G \xrightarrow{\varphi} H$ be a homomorphism with $G, H \in \mathit{Arch}$; let $u \in G^+$, $v \in H^+$, and suppose that $\varphi(u) = v$. Then φ "drops" to a homomorphism $G/u^\perp \xrightarrow{\varphi_0} H/v^\perp$ via $\varphi_0(g+u^\perp) = \varphi(g)+v^\perp$, and there is a unique continuous $Y(G,u) \xleftarrow{\tau} Y(H,v)$ (given by $\tau(P) = \varphi^{-1}(P)$) which "realizes" φ_0 in that $\varphi_0(g)^\smallfrown = \hat{g} \circ \tau \;\; \forall g \in G$. Moreover, φ_0 is one-to-one iff τ is onto.*

The degree of generality in 8.2.1-8.2.3 is a bit unusual; most of the above is written out in [Hager and Robertson, 1977]. We shall need this generality at a few spots in the sequel, in connection with 8.4.4(b), in sections 8.5 and 8.7. But usually, 8.2.1-8.2.3 will be used in cases where the u and v are weak units.

Recall that u is a weak unit in G if $u^\perp = (0)$. (That makes $G \cong \hat{G}$ in 8.2.1.) The category Wu has objects (G, e_G), where $G \in \mathit{Arch}$ and e_G is a designated weak unit of G, and morphisms $(G, e_G) \to (H, e_H)$ are homomorphisms $G \xrightarrow{\varphi} H$ with $\varphi(e_G) = e_H$.

When possible, we shall suppress explicit mention of the weak unit of a Wu-object, and write things like: "$G \in \mathit{Wu}$", "$G \xrightarrow{\varphi} H$ is a Wu-morphism", "$G \le H \in \mathit{Wu}$", "$G \le H$ is Wu-epi", etc. And, when it is understood that $G \in \mathit{Wu}$, meaning there is a designated weak unit e_G, we shall usually write $Y(G)$ for $Y(G, e_G)$.

Also, we agree that whenever an ℓ-group G in some $D(X)$ is called a Wu-object, it means that the constant function 1 belongs to G and is the designated weak unit.

With these conventions, we can restate 8.2.1-8.2.3 in the category Wu, as follows:

THEOREM ON THE YOSIDA FUNCTOR 8.2.4. (a) *If $G \in \mathit{Wu}$, then G is Wu-isomorphic to an ℓ-group $\hat{G} \le D(Y(G))$ which separates the points. The representation is unique (in the sense of 8.2.2). (b) If $G \xrightarrow{\varphi} H$ is a Wu-morphism, then there is unique continuous $Y(G) \xleftarrow{\tau} Y(H)$ realizing φ in that $\varphi(g)^\smallfrown = \hat{g} \circ \tau \; \forall g \in G$. φ is one-to-one iff τ is onto.*

8.2.4 (or 8.2.2) can be used for, among other things, recognizing Yosida representations. We indicate two particular instances of this which will be referred to frequently in the sequel.

First, given $G \in \mathit{Wu}$, with weak unit e_G, let $G^* = \{g \in G \mid |g| \le n e_G$ for some $n \in N = \{0,1,2,...\}\}$, the principal ideal generated by e_G; giving G^* the weak unit e_G, we have $G^* \in \mathit{Wu}$. In the Yosida representation for G, $\hat{e_G} = 1$, so that $G^* = \{g \in G \mid \hat{g}$ is bounded on $Y(G)\}$. (So the notation G^* follows

[Gillman and Jerison, 1976; page 11] in the use of $C^*(X)$ for $\{f \in C(X) \mid f$ is bounded$\}$.) The representation for G yields a representation for G^* which satisfies 8.2.2, and we conclude

COROLLARY 8.2.5. *For* $G \in \mathcal{W}u$, $Y(G^*) = Y(G)$.

Second, consider $C(X)$ with weak unit 1. This presentation of the abstract $\mathcal{W}u$-object is "part of" the Yosida representation: let βX be the Stone-Čech compactification, and for $f \in C(X)$, let $\beta f : \beta X \to [-\infty, +\infty]$ be the extension of f. (See [Gillman and Jerison, 1976; page 86]). Since X is dense in βX, $\beta f \in D(\beta X)$. The conditions in 8.2.2 are seen to be met, so that

COROLLARY 8.2.6. *The map* $C(X) \ni f \mapsto \beta f \in D(\beta X)$ *is the Yosida representation of* $C(X) \in \mathcal{W}u$; $Y(C(X)) = \beta X$.

(Actually, it is easy to show that 8.2.1 implies the *existence* of Stone-Čech compactification.)

The τ in 8.2.4 ought to be called $Y\varphi$. If we do this, we see that $Y : \mathcal{W} \to \text{Comp}$ is a faithful functor, which we call the *Yosida functor*. It is not half of a duality: not every $Y(G) \xleftarrow{\tau} Y(H)$ is of the form $Y\varphi$ for some $G \xrightarrow{\varphi} H$; first, each $\tau^{-1}(\infty(\hat{g}))$ has to be nowhere dense, so that $\hat{g} \circ \tau \in D(Y(H))$, and that doesn't mean $\hat{g} \circ \tau \in H$. In spite of this, the functor Y can be *extremely* useful in studying $\mathcal{W}u$, particularly when something categorical is at issue, indeed, nearly as useful as Stone Duality is in studying Boolean Algebras. In fact, it is quite reasonable and useful to view Yosida pseudo-duality as an extension of Stone Duality in this simple way: given a Boolean Algebra A, let $G(A) \in \mathcal{W}u$ be the linear span of the characteristic functions of clopen sets on the Stone space $S(A)$. By 8.2.4, $Y(G(A)) = S(A)$. This point of view will be highly visible in [Ball and Hager, b]. See also the remarks after 8.3.2.

For the rest of this paper, we will almost always identify $\mathcal{W}u$-objects with their Yosida representations, writing G for $\hat{G}$, g for $\hat{g}$, etc. When we depart from this (in parts of sections 8.5, 8.6) the reader will be warned.

8.3. Epimorphisms in $\mathcal{W}u$

We now characterize epimorphisms in $\mathcal{W}u$, and deduce a bound on the cardinality of epi extensions of a given object.

8.3.1. *The setting.* Let $G \leq H \in \mathcal{W}u$, and consider the Yosida representation of $H \subseteq D(Y(H))$. This induces a representation of G on $Y(H)$.

Let $p, q \in Y(H)$. If $g(p) = g(q)$ for each $g \in G$, we write $p \sim_G q$. Since p and q are values of the weak unit e_H, for $p, q \subseteq H$, then $p \sim_G q$ iff $p \cap G = q \cap G$, and these are values of e_G. Or, if $Y(G) \xleftarrow{\tau} Y(H)$ is the Yosida surjection induced by the embedding of $G \leq H$, (defined by $\tau(p) = p \cap G$), then $p \sim_G q$ iff $\tau(p) = \tau(q)$.

Let $h \in H$. The set "split by h over G" is defined as $\Sigma(h) \equiv \{(p,q) \mid p, q \in Y(H), p \sim_G q, h(p) \neq h(q)\}$. One can think of $(p,q) \in \Sigma(h)$ as trying to prevent $G \leq H$ from being $\mathcal{W}u$-epi: If it were the case that all elements of H were real valued at both p and q, then evaluating at p and q would define homomorphisms $e_p, e_q : H \to \mathbf{R}$ which agree on G (since $p \sim_G q$.) but disagree at h (since $h(p) \neq h(q)$). Generally, however, this is not the case.

The following is the main theorem of the paper.

THEOREM 8.3.2. $G \leq H \in \mathcal{W}u$ is $\mathcal{W}u$-epi iff for each $h \in H$, there is a countable set $\{k_n \mid n \in N\} \subseteq H$ such that if $(p,q) \in \Sigma(h)$, then there is n for which $k_n(p) = \infty$ or $k_n(q) = \infty$.

Before we prove the theorem, we first make some remarks.

a) The set $\{k_n \mid n \in N\}$ in the theorem will be called an *epi-indicator* for h. Another way to put the condition is that $\Sigma(h) \cap \bigcap_n (k_n^{-1}(\mathbf{R}) \times k_n^{-1}(\mathbf{R})) = \emptyset$.

b) In the remainder of this paper, we will show the theorem to be useful.

c) We can shed some light on the meaning of theorem in the following way:

We have the functor $Y : \mathcal{W}u \to Cpt$, where Cpt is the category of compact Hausdorff spaces and continuous maps. An easy general argument shows that since Y is faithful, for $\varphi \in \mathcal{W}u$, if $Y(\varphi)$ is monic (and so one-to-one in this case), then φ is epi. The converse fails. However:

Let $\mathcal{W}u^*$ be the full subcategory of $\mathcal{W}u$ of those G whose weak unit is strong. Now, $G \in \mathcal{W}u^*$ iff the Yosida representation has $\hat{G} \subseteq C(Y(G))$. The little argument in the paragraph before 8.3.2 provides a proof that in $\mathcal{W}u^*$, $Y(\varphi)$ is one-to-one (i.e., monic) if φ is epi. That is, with $Y^* : \mathcal{W}u^* \to Cpt$ the restriction $Y \mid \mathcal{W}u^*$, we have: for $\varphi \in \mathcal{W}u^*$, φ epi iff $Y^*(\varphi)$ monic.

It turns out that 8.3.2 implies, and is virtually equivalent to, a statement of the same form, structurally just as simple but more complex in that a newly-viewed Yosida functor is permitted to range in a more complex category of spaces. We describe this briefly:

Consider the category of spaces with filters, $SpFi$, an object of which is a pair $(X, \mathcal{F})$, $X \in Cpt$ and $\mathcal{F}$ a filter base of dense sets in X, a morphism $(X, \mathcal{F}) \overset{f}{\leftarrow} (Y, \mathcal{K})$ being a continuous $X \overset{f}{\leftarrow} Y$ satisfying $f^{-1}(F) \in \mathcal{K}$ for each $F \in \mathcal{F}$. Whenever $G \in \mathcal{W}u$, we view $G \subseteq D(Y(G))$, and we have the naturally associated filter $G^{-1}(\mathbf{R}) = \{g^{-1}(\mathbf{R}) \mid g \in G\}$. With G we associate $(Y(G), G^{-1}(\mathbf{R})) \in SpFi$, and 8.2.4 implies that for $G \overset{\varphi}{\to} H \in \mathcal{W}u$, we have $(Y(G), G^{-1}(\mathbf{R})) \overset{Y(\varphi)}{\leftarrow} (Y(H), H^{-1}(\mathbf{R})) \in SpFi$. Thus we make an "enriched" Yosida functor $Y_e : \mathcal{W}u \to SpFi$, "enriched" in the sense that, clearly, Y_e forgets less that Y does. Then, 8.3.2 and an analysis of $SpFi$-monic, yield: for $\varphi \in \mathcal{W}u$, φ is epic iff $Y_e \varphi$ is monic in $SpFi$.

Of course this says $SpFi$-monomorphisms are just as complicated as $\mathcal{W}u$-epimorphisms, and thus does not provide immediate further insight into $\mathcal{W}u$-epimorphisms. It does, though, provide a point of view from which, we hope, 8.3.2 looks less wierd.

In due course, we hope to publish an account of $SpFi$ and the enriched Yosida functor. For now, we return to the business at hand.

Proof of Theorem 8.3.2. Suppose that $G \leq H$ is not epi. Then there are $H \overset{\alpha_i}{\to} K$, $i = 1, 2$, with $\alpha_1 \mid G = \alpha_2 \mid G$ but $\alpha_1 \neq \alpha_2$. Take $h \in H$ with $\alpha_1(h) \neq \alpha_2(h)$, and define

$$U = \{y \in Y(K) \mid \mid \alpha_1(h)(y) - \alpha_2(h)(y) \mid > 0\}.$$

U is open in $Y(K)$ since the functions are continuous, and $U \neq \emptyset$ since $\alpha_1(h) \neq \alpha_2(h)$.

Now, a point $y \in Y(K)$ is a value of e_K, so $\alpha_i^{-1}(y)$ is a value of e_H, i.e., a point of $Y(H)$, and for any $f \in H$, $f(\alpha_i^{-1}(y)) = \alpha_i(f)(y)$, from the Yosida representation. It follows that

$$\{(\alpha_1^{-1}(y), \alpha_2^{-1}(y)) \mid y \in U\} \subseteq \Sigma(h)$$

since $y \in U$ implies $h(\alpha_1^{-1}(y)) = \alpha_1(h)(y) \neq \alpha_2(h)(y) = h(\alpha_2^{-1}(y))$, and $\alpha_1 \mid G = \alpha_2 \mid G$ implies $g(\alpha_1^{-1}(y)) = \alpha_1(g)(y) = \alpha_2(g)(y) = g(\alpha_2^{-1}(y))$ for $g \in G$ and any y.

Seeking a contradiction, suppose there were $\{k_n \mid n \in N\} \subseteq H$ for which $(p,q) \in \Sigma(h)$ implies there is n with $k_n(p) = \infty$ or $k_n(q) = \infty$. Then by the above, for each $y \in U$, there is n with $k_n(\alpha_1^{-1}(y)) = \infty$ or $k_n(\alpha_2^{-1}(y)) = \infty$; so $\alpha_1(k_n(y)) = \infty$ or $\alpha_2(k_n(y)) = \infty$, i.e., $(\alpha_1(k_n) \vee \alpha_2(k_n))(y) = \infty$.

We have shown that

$$U \subseteq \bigcup_n \{y \in Y(K) \mid (\alpha_1(k_n) \vee \alpha_2(k_n))(y) = \infty\}.$$

Thus the open set U is contained in the union of a sequence of nowhere dense sets; that contradicts the Baire Category Theorem in the space $Y(K)$.

For the converse direction, we prove the contrapositive. We first construct some targets for testing whether a morphism is epi.

CONSTRUCTION 8.3.3. Let X be a set, and $\mathcal{F}$ a filter base of subsets of X. An inclusion $F_1 \subseteq F_2$ induces the restriction homomorphism $\mathbf{R}^{F_2} \to \mathbf{R}^{F_1}$ of the ℓ-groups of real-valued functions, and $\{\mathbf{R}^F \mid F \in \mathcal{F}\}$ is a directed system of abelian ℓ-groups. Let $\phi(X,\mathcal{F})$ be the direct limit of this system in the category of abelian ℓ-groups. More concretely, $\phi(X,\mathcal{F}) = \cup \{\mathbf{R}^F \mid F \in \mathcal{F}\}/\sim$, where $f_i \in \mathbf{R}^{F_i}$ have $f_1 \sim f_2$ if $f_1 \mid F = f_2 \mid F$ for some $F \in \mathcal{F}$ with $F \subseteq F_1 \cap F_2$; the definitions of $+$ and $\leq$ are the usual ones. Equivalently, $\phi(X,\mathcal{F})$ is the quotient $\mathbf{R}^X/I$, where $I = \{f \in \mathbf{R}^X \mid f \mid F = 0$ for some $F \in \mathcal{F}\}$. One can easily prove that

LEMMA 8.3.4. *The equivalence class* [1] *of the constant function 1 is a weak unit in* $\phi(X,\mathcal{F})$. $\phi(X,\mathcal{F})$ *is archimedean iff for each countable* $\{F_n \mid n \in N\} \subseteq \mathcal{F}$ *there is* $F \in \mathcal{F}$ *with* $F \subseteq \cap_n F_n$.

The expression "$\phi(X,\mathcal{F}) \in \mathcal{W}u$" is short for: $\phi(X,\mathcal{F})$ is archimedean and [1] is the designated weak unit.

Now suppose the condition in 8.3.2 fails; i.e., there is $h \in H$ such that no $\{k_n\}$ "works". This means that for any $\{k_n\}$ there is $(p,q) \in \Sigma(h)$ such that for each n, both $k_n(p)$ and $k_n(q)$ are real. Hence, for any $\{k_n\}$, $\Sigma(h) \cap \cap_n (k_n^{-1}(\mathbf{R}) \times k_n^{-1}(\mathbf{R})) \neq \varnothing$; this is a statement about subsets of $Y(H) \times Y(H)$.

We use 8.3.3 and 8.3.4 with $X = \Sigma(h)$ and the filter base

$$\mathcal{F} = \{\Sigma(h) \cap \bigcap_n (k_n^{-1}(\mathbf{R}) \times k_n^{-1}(\mathbf{R})) \mid \{k_n \mid n \in N\} \subseteq H\}.$$

Clearly, $\mathcal{F}$ is closed under countable intersection, so by 8.3.4, $\phi(\Sigma(h), \mathcal{F}) \in \mathcal{W}u$.

Since $\Sigma(h) \subseteq Y(H) \times Y(H)$, we have the two projections $\pi_i : \Sigma(h) \rightarrow Y(h)$ $(i = 1,2)$ and these induce $\mathcal{W}u$-homomorphisms $\psi_i : H \rightarrow \phi(\Sigma(h), \mathcal{F})$ by the formulas $\psi_i(f) = f \circ \pi_i$ interpreted as follows: for $f \in H$, f is real on $f^{-1}(\mathbf{R})$, so $f \circ \pi_i$ is real on $\pi_i^{-1}(f^{-1}(\mathbf{R}))$, whereas (for $i = 1$)

$$\pi_1^{-1}(f^{-1}(\mathbf{R}) = \{(p,q) \in \Sigma(h) \mid f(p) \in \mathbf{R}\}$$
$$\supseteq \{(p,q) \in \Sigma(h) \mid f(p), f(q) \in \mathbf{R}\}$$
$$= \Sigma(h) \cap (f^{-1}(\mathbf{R}) \times f^{-1}(\mathbf{R})) \in \mathcal{F}.$$

With $F = \Sigma(h) \cap (f^{-1}(\mathbf{R}) \times f^{-1}(\mathbf{R}))$, we thus have $f \circ \pi_1 " \in "\mathbf{R}^F$, and the equivalence class $[f \circ \pi_1] \in \phi(\Sigma(h), \mathcal{F})$ is, by definition, $\psi_1(f)$. Likewise for $i = 2$. It is easy to see that the ψ_i preserve the operations and the units (the constant functions 1).

If $g \in G$ and $(p,q) \in \Sigma(h)$, we have $p \sim_G q$; so $g(\pi_1(p,q)) = g(p) = g(q) = g(\pi_2(p,q))$, whence certainly $\psi_1(g) = \psi_2(g)$. Therefore $\psi_1 \mid G = \psi_2 \mid G$.

However, $\psi_1(h) \neq \psi_2(h)$ since for each $F \in \mathcal{F}$, $h \circ \pi_1 \mid F \neq h \circ \pi_2 \mid F$: indeed, for $(p,q) \in \Sigma(h)$, $h(\pi_1(p,q)) = h(p) \neq h(q) = h(\pi_2(p,q))$.

Thus, $\psi_1 \neq \psi_2$, and $G \leq H$ is not epi, proving the theorem.

We turn to the issue of cardinal bounds. Let $|S|$ denote the cardinal number of the set S.

COROLLARY 8.3.5. *Let $G \leq H$ be epi. Then*

(a) *There is a filter base $\mathcal{F}$ of dense sets in $Y(G)$ which satisfies the condition of 8.3.4, and a $\mathcal{W}u$-embedding $H \leq \phi(Y(G), \mathcal{F})$;*

(b) $|H| \leq 2^{2^{|G|}}$;

(c) *There is a $\mathcal{W}u$-epicomplete $K \in \mathcal{W}u$, and a $\mathcal{W}u$-epi embedding $H \leq K$.*

REMARKS. It can be shown (using [Ball and Hager, a]) that $\phi(Y(G), \mathcal{F})$ in (a) is $\mathcal{W}u$-epicomplete; but the embedding $H \leq \phi(Y(G), \mathcal{F})$ is not asserted to be, and need not be, $\mathcal{W}u$-epi. So $\phi(Y(G), \mathcal{F})$ cannot be used for K in (c).

The cardinal bound in (b) is not the best possible. We show in [Ball and Hager, b] that $|H| \leq |G|^{\aleph_0}$ holds; the method uses [Ball and Hager, a] (which uses 8.3.2), and considerable further argument.

Proof of 8.3.5. (a) As we saw in section 8.2, there is a continuous $Y(G) \xleftarrow{\tau} Y(H)$ for which $g \mapsto g \circ \tau$ is the embedding $G \leq H$. Let

$$\mathcal{F} = \{\tau(\cap_n k_n^{-1}(\mathbf{R}) \mid \{k_n \mid n \in N\} \subseteq H\}.$$

Each $k_n^{-1}(\mathbf{R})$ is dense and open, so each $\cap_n k_n^{-1}(\mathbf{R})$ is dense (by the Baire Category Theorem), and the continuous image of a dense set is dense. Thus $\mathcal{F}$ consists of dense sets.

The inclusion $\tau(\cap_\alpha S_\alpha) \subseteq \cap_\alpha \tau(S_\alpha)$ is true for any function τ. Thus $\tau(\cap_m \cap_n k_{nm}^{-1}(\mathbf{R})) \subseteq \cap_m \tau(\cap_n k_{nm}^{-1}(\mathbf{R}))$ always holds. Therefore $\mathcal{F}$ satisfies the condition in 8.3.4, whence $\phi(Y(G),\mathcal{F}) \in \mathcal{W}u$.

We define a $\mathcal{W}u$-homomorphism $H \ni h \mapsto h' \in \phi(Y(G),\mathcal{F})$: Given $h \in H$, let $\{k_n \mid n \in N\}$ be an epi-indicator for h; by definition, it follows that $p, q \in \cap_n k_n^{-1}(\mathbf{R})$ and $\tau(p) = \tau(q)$ (i.e., $p \sim_G q$) imply $h(p) = h(q)$. Thus we define: for $p \in h^{-1}(\mathbf{R}) \cap \cap_n k_n^{-1}(\mathbf{R})$, $h''(\tau(p)) = h(p)$; the equivalence class $[h''] = h' \in \phi(Y(G),\mathcal{F})$. One verifies without trouble that h' is independent of the choice of $\{k_n\}$, and that $h \mapsto h'$ preserves operations and units.

Finally, $h \mapsto h'$ is one-to-one because each $\cap_n k_n^{-1}(\mathbf{R})$ is dense: $h_1 \neq h_2$ implies h_1 and h_2 differ on any dense set, which implies $h_1' \neq h_2'$.

(b) From (a) and 8.3.5, $H \leq \phi(Y(G),\mathcal{F}) = \mathbf{R}^{Y(G)}/I$ for a certain ideal I. Thus $|H| \leq |\mathbf{R}^{Y(G)}| = 2^{(\aleph_0 \cdot |Y(G)|)} \leq 2^{2^{|G|}}$ since $|Y(G)| \leq 2^{|G|}$ (each point of $Y(G)$ is an ideal in G), and $|G|$ is infinite.

(c) From (b), there is, up to isomorphism over G, only a set of epi extensions of H, and we can find a maximal member K by the usual argument using Zorn's Lemma.

8.4. Epimorphisms in $\mathcal{A}rch$

We reduce, more or less, to $\mathcal{W}u$ by showing that $G \leq H$ is $\mathcal{A}rch$-epi iff $G/u^{\perp G} \leq H/u^{\perp H}$ is $\mathcal{W}u$-epi for all $u \in G^+$ and $G \leq H$ has an additional feature called "coessential" (defined below). This, coupled with the characterization of $\mathcal{W}u$-epimorphisms in section 8.3, constitutes a characterization of $\mathcal{A}rch$-epimorphisms, and permits derivation of a cardinal bound on $\mathcal{A}rch$-epi extensions.

8.4.1. *Induced $\mathcal{W}u$-embeddings.* An embedding $G \leq H$ in $\mathcal{A}rch$ induces a system of $\mathcal{W}u$-embeddings as follows: For $u \in G^+$, $u^{\perp G} = \{g \in G \mid |g| \wedge u = 0\}$ is an ideal in G; and also $u^{\perp H} = \{h \in H \mid |h| \wedge u = 0\}$ is an ideal in H. Since $u^{\perp G} = u^{\perp H} \cap G$, the map $g + u^{\perp G} \mapsto g + u^{\perp H}$ is a one-to-one homomorphism of $G/u^{\perp G}$ into $H/u^{\perp H}$. So we write $G/u^{\perp G} \leq H/u^{\perp H}$. Moreover, $G/u^{\perp G} \in \mathcal{A}rch$ by [Bigard, Keimel, and Wolfenstein, 1977, page 227] and also $H/u^{\perp H} \in \mathcal{A}rch$. Finally, $u + u^{\perp G}$ is clearly a weak unit in $G/u^{\perp G}$, and likewise $u + u^{\perp H}$ in $H/u^{\perp H}$. With these designations of weak units, we have $G/u^{\perp G} \leq H/u^{\perp H} \in \mathcal{W}u$.

8.4.2. *Coessential embeddings.* $G \leq H \in \mathcal{A}rch$ will be called *coessential* (in $\mathcal{A}rch$) if whenever $H \overset{\alpha}{\to} K \in \mathcal{A}rch$ and $\alpha(g) = 0$ for all $g \in G$, then $\alpha(h) = 0$ for all $h \in H$.

Denoting an identically zero homomorphism on H by $\bar{0}$, $G \leq H$ is coessential iff $\alpha | G = \bar{0} | G$ implies $\alpha = \bar{0}$; thus, any epi embedding is coessential. Note that $\mathcal{W}u$ does not have $\bar{0}$'s (weak units are strictly larger than 0) so we do not speak of $\mathcal{W}u$-coessentiality.

Given $H \in \mathcal{A}rch$, an ideal I of H will be called an *archimedean kernel* if $H/I \in \mathcal{A}rch$. Given $S \subseteq H$, there is the least archimedean kernel containing S (because $\mathcal{A}rch$ is closed in ℓAb under products and subobjects); we denote this kernel by $ak_H S$.

Clearly, $G \leq H$ is coessential iff $ak_H G = H$. Thus, what is $ak_H G$? For vector lattices, we have for general $S \subseteq H$, $ak_H S = \bigcup_{\alpha < \omega_1} S_\alpha$, where S_0 is the ideal generated by S, S_1 is the "relative uniform closure" in H of S_0, etc., over the countable ordinals. (See [Luxemburg and Zaanen, 1971, pages 427 and 85] and 8.5.4 below.) Thus there is more to coessentiality than one might first guess. For the purposes of this paper, though, calculation with relative uniform closures is confined to section 8.5.

Finally, we note that coessentiality is a strengthening of the condition that $G^{\perp\perp} = H$ (equivalently, $G^\perp = (0)$).

PROPOSITION 8.4.3. *(a)* If $G \leq H$ is coessential, then $G^{\perp\perp} = H$.

(b) Let D be an uncountable set, $H = \mathbf{R}^D$, and G the ℓ-subgroup of H comprising the functions of countable support in H. Then $G^{\perp\perp} = H$, while $G \leq H$ is not coessential.

Proof: (a) If $0 \neq h \in G^{\perp}$, then $H \overset{\alpha}{\to} H/h^{\perp}$ is not $\bar{0}$, but is 0 on G. ($H/h^{\perp} \in \mathit{Arch}$ by [Bigard, Keimel, and Wolfenstein, 1977, page 227].) Hence $G^{\perp\perp}$ is an archimedean kernel; so $ak_H G \subseteq G^{\perp\perp}$.

(b) If $h \neq 0$, then some $h(p) \neq 0$ and $h \wedge \chi_p \neq 0$ (where χ_p is the characteristic function of $\{p\}$); so $h \notin G^{\perp}$.

Indeed, G is a proper ideal with $H/G \in \mathit{Arch}$. We omit the verification.

THEOREM 8.4.4. $G \leq H \in \mathit{Arch}$ is Arch-*epi iff*

(a) $G \leq H$ *is coessential in* Arch, *and*

(b) *For each* $u \in G^{+}$, $G/u^{\perp G} \leq H/u^{\perp H}$ *is* Wu-*epi.*

Remarks. In (a), "coessential" cannot be replaced by the condition "$G^{\perp\perp} = H$". 8.4.3(b) is an example (in which $G/u^{\perp G} = H/u^{\perp H}$ for all $u \in G$).

In (b), it is not sufficient to have u range only over a maximal disjoint system. (See example 8.7.2.) However, "Wu-epi" can be replaced by "Arch-epi" as the proof will show.

Proof: Let $G \leq H$ be Arch-epi. We noted above that coessentiality follows. To get (b), note that we have the commuting diagram

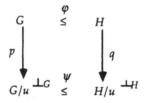

(φ, ψ are labels for the embeddings), with φ Arch-epi, and q a surjection, hence Arch-epi. Thus $q\varphi$ is Arch-epi. Since $\psi p = q\varphi$ is Arch-epi, so is ψ (as a second factor in an Arch-epi). But a Wu-map which is Arch-epi is Wu-epi.

The converse requires the following simple facts:

LEMMA 8.4.5. (a) *If* $G \leq H$ *is coessential, and* $H \overset{\beta}{\to} K$ *is a monomorphism, then* $\beta(G) \leq \beta(H)$ *is coessential.*

(b) *If* $A \leq B_i \leq K$ *for* $i = 1,2$, *with each* $A \leq B_i$ *coessential, and* B *is the* ℓ-*subgroup of* K *generated by* $B_1 \cup B_2$, *then* $A \leq B$ *is coessential.*

Proof. (a) If $\beta(G) \overset{\alpha}{\to} K$ has $\alpha | \beta(G) = 0$, then $\alpha\beta | G = 0$, whence $\alpha\beta = 0$; $\alpha = 0$ since β is onto its range, hence epi as a map onto its range.

(b) Let $B \xrightarrow{\alpha} K$ have $\alpha \mid A = 0$; thus $\alpha \mid B_i = 0$ $(i = 1,2)$. If $b \in B$, then $b = \underset{i}{\wedge} \underset{j}{\vee}(b_{1ij}+b_{2ij})$ over finite index sets, with all $b_{1ij} \in B_1$, $b_{2ij} \in B_2$ (see [Bigard, Keimel, and Wolfenstein, 1977, page 37]). Then $\alpha(b) = \underset{i}{\wedge} \underset{j}{\vee}(\alpha(b_{1ij})+\alpha(b_{2ij})) = 0$, completing the proof of the lemma.

We now show that if $G \leq H$ is not $\mathcal{A}rch$-epi, but is coessential, then there is $u \in G^+$ for which $G/u^{\perp G} \leq H/u^{\perp H}$ is not $\mathcal{W}u$-epi:

Let $H \underset{\beta_2}{\overset{\beta_1}{\rightrightarrows}} K \in \mathcal{A}rch$ have $\beta_1 \mid G = \beta_2 \mid G$, but $\beta_1(h_0) \neq \beta_2(h_0)$ for some $h_0 \in H$. By 8.4.5, $\beta_i(G)$ is coessential in the ℓ-subgroup, say B, of K generated by $\beta_1(H) \cup \beta_2(H)$. By 8.4.3, for each nonzero $b \in B$, there is $v \in \beta_1(G)$ with $b \wedge v \neq 0$. Thus there is $u \in G^+$ with $\mid \beta_1(h_0) - \beta_2(h_0) \mid \wedge \beta_i(u) \neq 0$ $(i = 1,2)$.

Define $H/u^{\perp H} \xrightarrow{\bar{\beta_i}} K/\beta_i(u)^{\perp K}$ $(i = 1,2)$ by $\bar{\beta_i}(h+u^{\perp H}) = \beta_i(h)+\beta_i(u)^{\perp K}$; this is valid because $\beta_i(u^{\perp H}) \subseteq \beta_i(u)^{\perp K}$. Moreover, upon giving $K/\beta_i(u)^{\perp K}$ the weak unit $\beta_i(u)+\beta_i(u)^{\perp K}$, we have $\bar{\beta_i} \in \mathcal{W}u$. Since $\beta_1(u) = \beta_2(u)$, $\bar{\beta_1}$ and $\bar{\beta_2}$ have the same codomain.

By construction, $\beta_1(h_0)-\beta_2(h_0) \notin \beta_i(u)^{\perp}$, whence $\bar{\beta_1}(h_0) \neq \bar{\beta_2}(h_0)$. Therefore, $\bar{\beta_1} \neq \bar{\beta_2}$. But for $g \in G$, $\beta_1(g) - \beta_2(g) = 0 \in \beta_i(u)^{\perp}$; so $\bar{\beta_1}$ and $\bar{\beta_2}$ agree on $G/u^{\perp G}$.

COROLLARY 8.4.6. *Let $G \leq H$ be $\mathcal{A}rch$-epi. Then*

        (a) For each $u \in G^+$, there is a filter base $\mathcal{F}_u$ of dense sets in $Y(G/u^{\perp})$ which satisfies the condition of 8.3.4, and a $\mathcal{W}u$-embedding $H/u^{\perp H} \leq \phi(Y(G/u^{\perp}),\mathcal{F}_u)$; thus $H \leq \prod_{u \in G^+} \phi(Y(G/u^{\perp}),\mathcal{F}_u)$.

        (b) $\mid H \mid \leq 2^{2^{|G|}}$.

        (c) There is $\mathcal{A}rch$-epicomplete K, and an $\mathcal{A}rch$-epi embedding $H \leq K$.

Remark. The bound in (b) can be reduced to $2^{|G|}$: See the remark after 8.3.5, and the proof below.

Proof. (a) 8.4.4(b) and 8.3.5(a) yield $H/u^{\perp}H \leq \phi(Y(G/u^{\perp}), \mathcal{F}_u)$ for each $u \in G^+$, and then there is an embedding $\prod \{H/u^{\perp}H \mid u \in G^+\} \leq \prod \{\phi(Y(G/u^{\perp})) \mid u \in G^+\}$. There is also the homomorphism $H \to \prod \{H/u^{\perp}H \mid u \in G^+\}$ given by $h \mapsto (h+u^{\perp}H)_{u \in G^+}$, which has kernel $\cap\{u^{\perp}H \mid u \in G^+\}$; this is (0), by 8.4.3.

(b) $|\phi(Y(G/u^{\perp}), \mathcal{F}_u)| \leq 2^{2^{|G/u^{\perp}|}} \leq 2^{2^{|G|}}$ by 8.3.5. Then, by (a),

$$|H| \leq 2^{2^{|G|}} |G| = 2^{2^{|G|}} .$$

(c) Apply Zorn's Lemma as in 8.3.5(c).

8.5. Practical Reduction of $\mathcal{A}rch$ to $\mathcal{W}u$

The reader is probably willing to believe (and should be convinced by section 8.6) that one can tell if a given $\mathcal{W}u$-embedding is $\mathcal{W}u$-epi or not by using our "Epi-indicator Theorem" 8.3.2. However, can one similarly recognize whether a given $\mathcal{A}rch$-embedding is epi from 8.4.4 and 8.3.2? It is daunting, at the very least, to contemplate, for a particular $G \leq H$, calculating all the Yosida representations for $G/u^{\perp}G \leq H/u^{\perp}H$ ($u \in G^+$), and then producing, or showing one can't produce, epi-indicators for all elements in these quotients; further, additional complications accompany coessentiality.

Thus most of this section is devoted to Theorem 8.5.2 below, which is the conducting of exactly that program in a setting of limited generality, still broad enough to cover many of the situations we want to look at. The thrust of this is that a $\mathcal{W}u$-epi-embedding $G \leq H$ will be $\mathcal{A}rch$-epi provided H is big enough. Before we get to this, however, we present a simple result (which still illustrates the problems).

PROPOSITION 8.5.1. *Let* $G \in \mathcal{W}u$. *If* G *is* $\mathcal{W}u$-*epicomplete, then* G *is* $\mathcal{A}rch$-*epicomplete. (And conversely, by 8.5.3.)*

Proof. Let $G \in \mathcal{W}u$, with the weak unit e_G. Suppose $G \leq H$ is $\mathcal{A}rch$-epi, and consider $G \overset{\varepsilon}{\leq} H \overset{\rho}{\to} H/e_G^{\perp}$ (with ε a label for the embedding, $e_G^{\perp}$ is the polar in H, and $\rho(h) = h+e^{\perp}$). We have $G \cap e_G^{\perp} = (0)$, of course, whence $\rho\varepsilon$ is one-to-one. Since $\varepsilon(e_G)+e^{\perp}$ is a weak unit in $H/e_G^{\perp}$ (as in 8.4), we construe

$\rho\varepsilon \in \mathcal{W}u$. Since ρ is a surjection, it is $\mathcal{A}rch$-epi, and thus so is $\rho\varepsilon$. Obviously, a $\mathcal{W}u$-morphism which is $\mathcal{A}rch$-epi is $\mathcal{W}u$-epi. Consequently, if G is $\mathcal{W}u$-epicomplete, then $\rho\varepsilon$ is an isomorphism in $\mathcal{W}u$, and hence also in $\mathcal{A}rch$. So ε is an epi "section", and therefore an isomorphism (all in $\mathcal{A}rch$). (See [Herrlich and Strecker, 1973, page 41] if need be.) It follows that G is $\mathcal{A}rch$-epicomplete.

We turn to the main topic of this section, a sufficient condition on the codomain that a $\mathcal{W}u$-epi be $\mathcal{A}rch$-epi.

Suppose that H is an archimedean vector lattice with weak units. Upon forgetting the scalar multiplication and designating a weak unit we obtain a $\mathcal{W}u$-object, again called H. Then its Yosida representation $\hat{H}$ has the property that for $r\in \mathbf{R}$ and $h\in H$, $(rh)^\smallfrown = r\hat{h}$ (the right side meant pointwise in $D(Y(H))$); see [Conrad, 1971a].

Similarly, if H is an archimedean f-ring with identity e_H then e_H is a weak unit, and upon forgetting the multiplication we have a $\mathcal{W}u$-object, again called H. Its Yosida representation $\hat{H}$ has the property that for $h_1, h_2 \in H$, $(h_1 h_2)^\smallfrown = \hat{h}_1 \hat{h}_2$ (the right side meant "pointwise" in $D(Y(H))$); see [Hager and Robertson, 1977] and [Bigard, Keimel, and Wolfenstein, 1977, section 13.3].

Thus we may and shall speak of $\mathcal{W}u$-objects which are vector lattices, rings (for short), and hence algebras (for short), and interpret the extra operations as pointwise in the appropriate $D(Y)$.

THEOREM 8.5.2. *Let $G \leq H\in \mathcal{W}u$, and suppose that H is an algebra. Then if $G \leq H$ is $\mathcal{W}u$-epi, it is also $\mathcal{A}rch$-epi.*

Before the lengthy proof of 8.5.2, we note a consequence:

COROLLARY 8.5.3. *Let $G\in \mathcal{W}u$. If G is $\mathcal{A}rch$-epicomplete, then G is $\mathcal{W}u$-epicomplete.*

Proof of Corollary 8.5.3. Let $G\in \mathcal{W}u$, and let $G \leq H$ be $\mathcal{W}u$-epi. There is the canonical $\mathcal{W}u$-extension of H to a ring K, due to Conrad [1974c] and Hager and Robertson [1977,1979]. Then there is the canonical $\mathcal{A}rch$-extension of K to a vector lattice L due to Conrad [1971a] and Bleier [1971], which is easily seen to be a $\mathcal{W}u$-extension. The multiplication on K naturally extends to multiplication on L. Thus L is an algebra.

Now both $H \leq K$ and $K \leq L$ are $\mathcal{W}u$-epi on the general grounds of [Herrlich and Strecker, 1973; 36.3]; this is shown also in [Hager, 1985;p. 181, p. 167, ex. 1,2]. Hence $H \leq L$ is $\mathcal{W}u$-epi, and so is $G \leq L$, each as a composition of $\mathcal{W}u$-epimorphisms. By 8.5.2, $G \leq L$ is $\mathcal{A}rch$-epi. So, if G is $\mathcal{A}rch$-epicomplete, then $G = L$, whence $G = H$. Consequently, G is $\mathcal{W}u$-epicomplete.

Remark. The proof of 8.5.3 shows that any $\mathcal{W}u$-epi-embedding $G \leq H$ can be "extended" to an $\mathcal{A}rch$-epi-embedding $G \leq K$; section 8.7 below contains descriptions of various $\mathcal{W}u$-epimorphisms which are not $\mathcal{A}rch$-epi.

We turn now to the proof of 8.5.2, which occupies the rest of this section. (We forgive the reader who wishes to skip these details.) By 8.4.4, it suffices to prove the following.

PROPOSITION 8.5.4. *Let $G \leq H \in \mathcal{W}u$, with H an algebra.*

(a) $G \leq H$ is coessential.

(b) If $G \leq H$ is $\mathcal{W}u$-epi, then for each $u \in G^+$, $G/u^{\perp G} \leq H/u^{\perp H}$ is $\mathcal{W}u$-epi.

Proof of 8.5.4(a). In the notation of 8.4.2, we are to show that $ak_H G = H$. Let e_H be the weak unit of H. Since $e_H \in G$, we have $ak_H(e_H) \subseteq ak_H G$. Notice that the ideal in H generated by e_H is H^* (the subobject of H of functions bounded in the Yosida representation); thus $ak_H H^* = ak_H(e_H)$. We shall show that when H is an algebra, $ak_H H^* = H$ in a very strong way. The result is easy, but we need to recall the description from [Luxemburg and Zaanen, 1971, pages 427 and 85] of $ak_H S$ in a vector lattice H:

$h_n \to h(u)$ means: given $\varepsilon > 0$, there is n_0 such that $n \geq n_0$ implies $|h_n - h| \leq \varepsilon u$.

Given $A \subseteq H$, let $A' \equiv \{h \in H \mid \exists h_1, h_2, \ldots \in A$ and $u \in H$ with $h_n \to h(u)\}$. Then $A = ak_H A$ iff A is an ideal with $A = A'$. It follows that $ak_H S = \bigcup_{\alpha < w_1} S_\alpha$, where S_0 is the ideal generated by S, $S_{\alpha+1} = S_\alpha'$, and for limit ordinals β, $S_\beta = \bigcup_{\alpha < \beta} S_\alpha$. Now the following shows (a).

LEMMA 8.5.5 [dePagter, 1981, p. 62]. *Let H be an algebra, and let $h \in H^+$. Then $h \wedge n \to h(h^2)$. Thus $(H^*)' = H$.*

Proof. Given $\varepsilon > 0$, let n_0 be the first positive integer $\geq 1/\varepsilon$, and let $n \geq n_0$. We show that $|(h(x) \wedge n) - h(x)| \leq \varepsilon h^2(x)$ for every x in the dense set $h^{-1}(\mathbf{R})$

$\subseteq Y(H)$, whence $|(h \wedge n) - h| \leq \varepsilon h^2$. So take $x \in h^{-1}(\mathbf{R})$. For $h(x) \leq n$, $|(h(x) \wedge n) - h(x)| = 0 < \varepsilon h^2(x)$. For $h(x) > n \geq n_0 \geq 1/\varepsilon$, we have $1/h(x) < \varepsilon$, and $|(h(x) \wedge n) - h(x)| = |n - h(x)| \leq h(x) = (1/h(x))h^2(x) < \varepsilon h^2(x)$.

Proof of 8.5.4(b). We are going to work with two Yosida representations at the same time, so we try to label these clearly.

Notation. Given $G \leq H \in \mathcal{W}u$, the common weak unit is e_G. With respect to $e_G \in H$, the Yosida space of H is $Y(H, e_G)$, and the Yosida representation is denoted $\hat{H} \subseteq D(Y(H, e_G))$. In this same representation, we have $\hat{G} \leq \hat{H}$.

Take and fix $u \in G^+$. With respect to $u \in H$, we have the Yosida representation $H/u^{\perp H} \cong \bar{H} \subseteq D(Y(H, u))$, and in this same representation, $G/u^{\perp G} \cong \bar{G} \leq \bar{H}$.

<u>Claim.</u> Suppose $G \leq H \in \mathcal{W}u$, and that H is a ring; let $h \in H$, and suppose that $\{k_n \mid n \in N\}$ is an epi-indicator for h. (All that is meant with respect to e_G.) This means that

(a) $\Sigma(\hat{h}) \cap \bigcap_n \hat{k}_n^{-1}(\mathbf{R}) \times \hat{k}_n^{-1}(\mathbf{R}) = \varnothing$ in $Y(H, e_G) \times Y(H, e_G)$.

With $u \in G^+$ (and fixed), since H is a ring we have an element $f \in H$ with $\hat{f}(x) = (\hat{u}(x))^2$ for all $x \in \hat{u}^{-1}(\mathbf{R})$. This f will just be called "u^2". Then, (we are claiming), $\{k_n \mid n \in N\} \cup \{e_G, u^2\}$ is an "epi-indicator for h with respect to u". This means that

(b) $\Sigma(\bar{h}) \cap \bigcap \{f^{-1}(\mathbf{R}) \times f^{-1}(\mathbf{R}) \mid f \in \{\bar{k}_n \mid n \in N\} \cup \{\bar{e}_G, \bar{u}^2\}\} = \varnothing$
      in $Y(H, u) \times Y(H, u)$.

The claim is not difficult to see once one recognizes the relationship between $\hat{H}$ and $\bar{H}$; so we set aside the claim for a moment, and write down that relationship.

$\underline{\hat{H} \text{ and } \bar{H}}$. We are in the situation of the paragraph "Notation", but the present paragraph is only about H and its two elements e_G and u. The following is from 2.2 of [Hager and Robertson, 1978], (but the proof is easy).

In $Y(H, e_G)$, let $\hat{U} = \operatorname{coz} \hat{u} \cap \hat{u}^{-1}(\mathbf{R})$. In $Y(H, u)$, let $\bar{U} = \operatorname{coz} \bar{e}_G \cap \bar{e}_G^{-1}(\mathbf{R})$; this is dense in $Y(H, u)$ since e_G is a weak unit. Then there is an onto homomorphism $\sigma : \hat{U} \to \bar{U}$ which relates the Yosida representation as

(c) $\forall f \in H, \bar{f}(\sigma(p)) = \dfrac{1}{\hat{u}(p)} \hat{f}(p), \forall p \in U.$

Note that (c) determines each $\bar{f}$, since $\sigma(\hat{U}) = \bar{U}$ is dense.

(Explanation: Algebraically, each of $\hat{U}$ and $\bar{U}$ is the same set of ideals $\mathrm{Val}_H(e_G) \cap \mathrm{Val}_H(u)$, σ is the obvious association, and (c) just says what is forced by $\hat{e}_G = 1$ and $\bar{u} = 1$.)

Finally, note that from the definition of $\bar{U}$, we have

(d) $Y(H,u) = \bar{U} \cup \infty(\bar{e}_G) \cup Z(\bar{e}_G)$, where $Z(\bar{e}_G) = \{x \in Y(H,u) \mid \bar{e}_G(x) = 0\}$.

Proof of the claim. We pass back to the setting of the claim.

(e) $Z(\bar{e}_G) \subseteq \infty((\bar{u}^2))$.

Proof. Insert $f = e_G$ and $f = u^2$, respectively, into (c), getting:

$$\bar{e}_G(\sigma(p)) = \frac{1}{\hat{u}(p)}\hat{e}_G(p) = \frac{1}{\hat{u}(p)}, \text{ and } (\bar{u}^2)(\sigma(p)) = \frac{1}{\hat{u}(p)}(\hat{u}^2)(p) = \hat{u}(p), \text{ for } p \in \hat{U}.$$

Now let $\bar{e}_G(x) = 0$, and choose a neighborhood O of x on which $\bar{e}_G$ is small. By density of $\sigma(\hat{U}) = \bar{U}$, there is $p \in \hat{U}$ with $\sigma(p) \in O$. For such a p, the first equation shows that $\hat{u}(p) = (\bar{u}^2)(\sigma(p))$ is large. By continuity, $(\bar{u}^2)(x) = \infty$.

(f) $Y(H,u) = \bar{U} \cup \infty(\bar{e}_G) \cup \infty(\bar{u}^2)$.

Proof. By (d) and (e).

We can now prove the claim, which is that (a) implies (b).

Let $(x,y) \in \Sigma(\bar{h})$. If one of x,y is in $\infty(\bar{e}_G) \cup \infty(\bar{u}^2)$ we are done. If not, then by (f), both $x,y \in \bar{U}$, so $x = \sigma(p)$ and $y = \sigma(q)$. Then $(p,q) \in \Sigma(\hat{h})$. (Since $(x,y) \in \Sigma(\bar{h})$, equation (c) with $f = h$ shows that $\hat{h}(p) \neq \hat{h}(q)$, and (c) for $f = g \in G$ shows that $\hat{g}(p) = \hat{g}(q)$ for all $g \in G$; thus $(p,q) \in \Sigma(\hat{h})$.) So, by (a), there is n for which $\hat{k}_n(p) = \infty$ or $\hat{k}_n(q) = \infty$. Now inserting k_n for f in equation (c), we see that $\bar{k}_n(x) = \infty$ or $\bar{k}_n(y) = \infty$. So (b) holds.

That completes the proof of 8.5.4(a), and hence of 8.5.2.

Finally, let us note that the proof of 8.5.4(b) yields something stronger, which we record for possible future use.

PROPOSITION 8.5.6. *Let $G \leq H$ in $\mathfrak{Arch}$. Let $e, u \in G^+$, and denote the Yosida representations for H with respect to e and u as $H/e^{\perp H} \cong \hat{H} = D(Y(H,e))$; $H/u^{\perp H} \cong \bar{H} \subseteq D(Y(H,u))$. Suppose that $u \in e^{\perp\perp}$ in H, and that there is $s \in H$ for which $\hat{s} = (\hat{u})^2$ (pointwise on the dense set $\hat{u}^{-1}(\mathbf{R}) \subseteq Y(H,u))$. Then:*

If $\{k_n \mid n \in N\} \subseteq H$ epi-indicates for $\hat{h}$ with respect to $\hat{G} \leq \hat{H}$ (i.e., $\Sigma(\hat{h}) \cap \bigcap_n \hat{k}_n^{-1}(\mathbf{R}) \times \hat{k}_n^{-1}(\mathbf{R}) = \varnothing$), then $\{e,s\} \cup \{k_n \mid n \in N\}$ epi-indicates for h with respect to $G \leq H$.

($u \in e^{\perp\perp}$ implies that $\text{Val}_H e \cap \text{Val}_H u$ is dense in $\text{Val}_H u = Y(H,u)$, and ensures that equation (a) determines each $\bar{f}$.)

8.6. Examples of Wu- and $Arch$-epimorphisms and Epicompletions

We now look at some relatively concrete situations in which there is, because of section 8.5, no difference between Wu and $Arch$. Examples illustrating differences will be presented in section 8.7.

We begin with what amounts to a very special case of the characterization of Wu-epicompleteness in [Ball and Hager, a]. Since the proof of the special case is attractive and close at hand, we present it.

PROPOSITION 8.6.1. *For any set E, the product* $\mathbf{R}^E$ *is* $Arch$-*epicomplete.*

Proof. Our weak unit in $\mathbf{R}^E$ is, of course, the constant function 1. Let βE be the Stone-Čech compactification of the discrete space E. Each $f \in \mathbf{R}^E$ has a unique extension in $D(\beta E)$. This provides an isomorphism $\mathbf{R}^E \cong D(\beta E)$; from section 8.2, we see that this is the Yosida representation for the unit $u = 1$. So $Y(\mathbf{R}^E) = \beta E$.

By section 8.5, it suffices to prove Wu-epicompleteness; so let $D(\beta E) \leq H$ be Wu-epi. There is, by 8.3.3, a filter base $\mathcal{F}$ of *dense* subsets of βE such that $H \leq \phi(\beta E, \mathcal{F})$. For each $F \in \mathcal{F}$, we have $E \subseteq F$ (since any dense set contains E), and we can define the restriction homomorphism $\mathbf{R}^F \to \mathbf{R}^E$; this produces a homomorphism $\phi(\beta E, \mathcal{F}) \to \mathbf{R}^E$. We thus have homomorphisms

$$D(\beta E)) \cong \mathbf{R}^E \overset{\alpha}{\leq} \overbrace{H \leq \phi(\beta E, \mathcal{F})}^{\beta} \to \mathbf{R}^E.$$

Here we have labelled the supposed epi embedding of $\mathbf{R}^E$ in H as α, and the composite from H into $\mathbf{R}^E$ as β, as shown. One sees easily that $\beta\alpha$ is the identity on $\mathbf{R}^E$. Thus α is a "section", also epi, hence an isomorphism. (See [Herrlich and Strecker, 1973, page 41].)

8.6.2. REMARKS. We now present a number of examples of epimorphisms (and non-epimorphisms) in Wu and $Arch$, and of epicompletions, all using the fundamental Theorem 8.3.2.

The block of examples 8.6.3-8.6.7 below can be described, roughly, as arranged in the order of *increasing complexity of epi-indicators*. That is, we have various Wu-epi-embeddings $G \leq H$, so that for each $h \in H$ there is an epi-indicator $E(h) = \{k_n \mid n \in N\}$ (by 8.3.2). In 8.6.3, every $E(h) = \emptyset$; in 8.6.5 there is a singleton $\{k\}$ serving as $E(h)$ for all h; in 8.6.7(b), there is a sequence $\{k_n \mid n \in N\}$ which serves as $E(h)$ for all h; in 8.6.7(c) and 8.6.8, in general, the $E(h)$ must be sequences and must depend on h; a remark in 8.6.8 shows how to arrange it that the $E(h)$'s can be singletons and must depend on h.

In a different direction, one can analyze epi-embeddings $G \le H$ in which the epi-indicators can be chosen out of G. It develops that this is the theory, in $\mathcal{W}u$, say, of algebraic extensions in the sense of model theory! We shall not get to that in this paper -- see [Ball and Hager, d].

EXAMPLE 8.6.3. Suppose that $G \le H \in \mathcal{W}u$ has the property that in the Yosida representaion of H, G separates the points of $Y(H)$ (i.e., that the realizing map $Y(G) \xleftarrow{\tau} Y(H)$ is one-to-one, thus a homeomorphism). Then $\Sigma(h) = \emptyset$ for all $h \in H$, and $G \le H$ is "vacuously" $\mathcal{W}u$-epi.

For example, for any $G \in \mathcal{W}u$, $G^* \le G$ is "vacuously" epi (by section 8.2). Also, for any $G \in \mathcal{W}u$, let uG denote the uniform closure of G within $D(Y(G))$. One sees easily that $uG \in \mathcal{W}u$, and $G \le uG$ is "vacuously" $\mathcal{W}u$-epi. (See [Hager and Robertson, 1977] for more about uG.)

"Vacuous" $\mathcal{W}u$-epimorphisms need not be $\mathcal{A}rch$-epi, as we will see in section 8.7 below.

EXAMPLE 8.6.4. We describe the setting for several examples. Let X be a topological space and $S \subseteq X$. Giving S the relative topology, we have a restriction homomorphism $C(X) \ni f \to (f \mid S) \in C(S)$ which is an embedding when S is dense. The Yosida representations are $C(X) \subseteq D(\beta X)$, $C(S) \subseteq D(\beta S)$, and the extension of the inclusion $X \supseteq S$ to a map $\beta X \xleftarrow{\tau_1} \beta S$ realizes the restriction homomorphism.

Now give S a finer topology, and call the resulting space S'. This produces an embedding $C(S) \le C(S')$, and the extension of the identity $S \leftarrow S'$ to a map $\beta S \xleftarrow{\tau_2} \beta S'$ realizes this embedding.

Assuming S dense and composing the two embeddings, we have an embedding $C(X) \le C(S')$ realized by the map $\beta X \xleftarrow{\tau} \beta S'$ ($\tau = \tau_1 \tau_2$) which extends the continuous inclusion $X \leftarrow S'$.

EXAMPLE 8.6.5. Let $S = \{x_n \mid n \in N\}$ be a countable dense subset of $[0,1]$, let S' be S with the discrete topology, and let $C[0,1] \le C(S')$ be the embedding described in 8.6.4.

(a) The single function $k_0(x_n) = n$ epi-indicates for every $h \in C(S')$.

(b) $C[0,1] \le C(S')$ is an epicompletion of $C[0,1]$ in $\mathcal{W}u$ and in $\mathcal{A}rch$.

(c) The embedding in (b) is not large, not majorizing, and not "restrictably epi": $C[0,1] \le C^*(S')$ is not epi.

Moreover, for S_1 and S_2 two countable dense subsets of $[0,1]$, with in-duced embeddings $C[0,1] \overset{\varepsilon_i}{\leq} C(S_i')$ $(i = 1,2)$ as above, the following are equivalent.

(1) $S_1 = S_2$.

(2) The pair $(\varepsilon_1,\varepsilon_2)$ of embeddings amalgamates in $\mathcal{A}rch$ (i.e., there are embeddings (δ_1,δ_2) with $\delta_1\varepsilon_1 = \delta_2\varepsilon_2$).

(3) The pair $(\varepsilon_1,\varepsilon_2)$ epically amalgamates in $\mathcal{A}rch$ (i.e., the δ_i can be chosen to be epi).

(4) $C(S_1')$ and $C(S_2')$ are isomorphic over $C[0,1]$.

Thus $C[0,1]$ has at least $2^{\aleph_0}$ "distinct" epicompletions.

Remarks. (i) Whereas the epicompletions $C(S')$ of $C[0,1]$ just described are pairwise non-isomorphic over $C[0,1]$, they are, of course, all isomorphic to $C(N)$, N being the discrete natural numbers. But we shall see in due course (8.6.8 below), there are epicompletions of $C[0,1]$ not isomorphic to $C(N)$ (or to any $C(X)$).

(ii) It is now clear that epicompletion theory is complicated. But it is by no means chaotic. See the introduction and [Ball and Hager, b].

(iii) In 8.6.5, the countability of S cannot be dropped. The next example shows this.

(iv) For all of 8.6.5 except (c), $[0,1]$ can be replaced by any compact space X. (In (c) there are complications involving isolated points. We won't pursue this subject here.)

(v) The implication $(2) \Rightarrow (3)$ is quite general. See the proof below.

(vi) The present situation illustrates the failure of the "amalgamation property" in $\mathcal{W}u$ and $\mathcal{A}rch$. See [Pierce, 1976] and compare [Hager and Madden, 1983].

Proof of 8.6.5. (a) Consider $[0,1]\overset{\varepsilon}{\leftarrow}\beta S'$ which realizes the embedding $C[0,1] \leq C(S') = D(\beta S')$, and note that $k_0(\beta S'-S') = \{\infty\}$. For $p,q \in S'$, $p \neq q$, we have $\tau(p) = p \neq q = \tau(q)$, and thus $(p,q)\notin \Sigma(h)$ for all $h \in C(S')$. Therefore, $(p,q)\in \Sigma(h)$ implies either p or q is a member of $\beta S'-S'$, so either $k_0(p) = \infty$ or $k_0(q) = \infty$.

(b) So the embedding is $\mathcal{W}u$-epi by 8.3.2. Since $C(S')$ is an "algebra", the embedding is $\mathcal{A}rch$-epi, by section 8.5. Since S' is discrete, $C(S') = \mathbf{R}^{S'}$, which is $\mathcal{A}rch$-epicomplete by 8.6.1.

(c) Let $h =$ the characteristic function of $x_1 \in S$. There is no $g \in C[0,1]$ with $0 < g \leq h$. So the embedding is not large. It is not majorizing, since no $g \in C[0,1]$ has $g \geq k$ (since g is bounded). The same $[0,1]\overset{\varepsilon}{\leftarrow}\beta S'$ realizes $C[0,1] \leq$

$C^*(S') = C(\beta S')$, τ is not one-to-one, and there are $\Sigma(h) \neq \varnothing$. (Take $p \neq q$ with $\tau(p) = \tau(q)$, then there exists $h \in C(\beta S')$ with $h(p) \neq h(q)$.) With no functions taking values ∞, there can't be epi-indicators. (Also, see the remarks preceding 8.3.2.)

Clearly, (1) implies (2).

For (2) $\Rightarrow$ (3): It is true in considerable generality that if two epi embeddings amalgamate then they epically amalgamate. This is visible in the following proof which, let us say, takes place in $\mathcal{A}rch$. (This is easily seen to be true, too, in any "(epi, extremal mono)-category" in the terminology of [Herrlich and Strecker, 1973, pages 250 and 260]. $\mathcal{A}rch$ is one of these, by [Ball and Hager, a]).

So let $G \xrightarrow{\varepsilon_i} H_i$ be epi embeddings and $H_i \xrightarrow{\delta_i} H$ embeddings with $\delta_1\varepsilon_1 = \delta_2\varepsilon_2$. Let K be the ℓ-subgroup of H generated by $\delta_1(H_1)\cup\delta_2(H_2)$; a typical element $k \in K$ is a finite join of finite meets of elements of the form $\delta_1(h_1)+\delta_2(h_2)$, $h_i \in H_i$, by Corollary 0.1.7, which fact we express by $k = \vee\wedge(\delta_1(h_1)+\delta_2(h_2))$. Let $H_i \xrightarrow{\gamma_i} K$ be the "range restriction" of δ_i; of course we have $\gamma_1\varepsilon_1 = \gamma_2\varepsilon_2$.

We show that γ_1 and γ_2 are epi. First note that given $K \xrightarrow[\beta]{\alpha} L$, $\alpha\gamma_1 = \beta\gamma_1 \Rightarrow \alpha\gamma_1\varepsilon_1 = \beta\gamma_1\varepsilon_1 \Rightarrow \alpha\gamma_2\varepsilon_2 = \beta\gamma_2\varepsilon_2 \Rightarrow \alpha\gamma_2 = \beta\gamma_2$, the last implication since ε_2 is epi. And conversely, $\alpha\gamma_2 = \beta\gamma_2 \Rightarrow \alpha\gamma_1 = \beta\gamma_1$, since ε_1 is epi. So, γ_1 is epi iff γ_2 is epi. Suppose that $\alpha\gamma_1 = \beta\gamma_1$. We show that $\alpha(k) = \beta(k)$ for all $k \in K$. Now, as noted above, $k = \vee\wedge(\delta_1(h_1)+\delta_2(h_2)) = \vee\wedge(\gamma_1(h_1)+\gamma_2(h_2))$; since α is a homomorphism, $\alpha(k) = \vee\wedge(\alpha\gamma_1(h_1)+\alpha\gamma_2(h_2)) = \vee\wedge(\beta\gamma_1(h_1)+\beta\gamma_2(h_2)) = \beta(\vee\wedge(\delta_1(h_1)+\delta_2(h_2))) = \beta(k)$. Thus $\alpha = \beta$, and the γ_i are epi.

(3) $\Rightarrow$ (4): If we have $\delta_1\varepsilon_1 = \delta_2\varepsilon_2$ with δ_i epi embeddings, then each δ_i is an isomorphism since $C(S_i')$ is epicomplete. Hence $\delta_1^{-1}\delta_2$ is an isomorphism over $C[0,1]$ of $C(S_2')$ onto $C(S_1')$.

(4) $\Rightarrow$ (1): If $C(S_2') \xrightarrow{\varphi} C(S_1')$ is an isomorphism over $C[0,1]$, i.e., $\varphi\varepsilon_2 = \varepsilon_1$, then $\varphi \in \mathcal{W}u$ and we have inducing Yosida maps $[0,1] \xleftarrow{\tau_i} \beta S_i'$ and $\beta S_2' \xrightarrow{\rho} \beta S_1'$ with $\tau_2\rho = \tau_1$, and ρ a homeomorphism (since φ is an isomorphism). Thus $\rho(S_1') = S_2'$ (since S_i' is exactly the set of isolated points in $\rho S_i'$). Therefore $S_1 = \tau_1(S_1') = \tau_2\rho(S_1') = \tau_2(S_2') = S_2$.

Finally, $[0,1]$ has $2^{\aleph_0}$ different countable dense sets, each of which yields a "distinct" epicompletion by (4) $\Rightarrow$ (1).

8.6.6. Let S be an uncountable dense subset of $[0,1]$, let S' be discrete S, and consider the embedding $C[0,1] \leq C(S')$.

(a) PROPOSITION. *If S has cardinal $2^{\aleph_0}$, then the embedding is not epi (in Wu or $Arch$).*

(b) CONJECTURE. *The statement "whenever S is uncountable, then $C[0,1] \leq C(S')$ is not epi" is undecidable in ZFC.*

By (a), under the Continuum Hypothesis, the statement in (b) is true. Thus the conjecture in (b) is: there is a model of ZFC in which there is an uncountable S (necessarily, by (a), of cardinal $< 2^{\aleph_0}$) with $C[0,1] \leq C(S')$ epi.

On the proof of (a). It is possible to grind this out, as a direct application of 8.3.2, along the lines of the proof of 8.6.9(c) below. On the other hand, it is shown in [Ball and Hager, a] that $|G|^{\aleph_0}$ is an upper bound on the cardinality of Wu-epi extensions of G so (a) follows since $|C[0,1]| = 2^{\aleph_0} = c$ and $|C(S')| = 2^c$. ([Ball and Hager, a] uses 8.3.2 also.)

8.6.7. Let X be a topological space and $S \subseteq X$. We call S a C_δ-set in X if $S = \bigcap_{n \in N} C_n$, with each C_n a cozero set in X. It is a theorem of topology that Z is a C_δ-set in βZ iff Z is a C_δ-set in some compact space iff Z is Lindelöf and Čech-complete.

Now consider an embedding $C(X) \leq C(S')$ of the type described in 8.6.4, and suppose that the topology of S' is Lindelöf and $S' = \beta S'$; so $S' = \bigcap \text{cozf}_n$, with $f_n \in C(\beta S')$ and $f_n \geq 0$. Then f_n is never 0 on S', so $k_n = 1/f_n \in C(S')$.

(a) The single sequence $\{k_n \mid n \in N\}$ epi-indicates (over $C(X)$) for *every* $h \in C(S')$.

(b) $C(X) \leq C(S')$ is Wu- and $Arch$-epi.

(c) Let X be compact (e.g., $[0,1]$), let $S \subseteq X$ be dense and Lindelöf in the relative topology (e.g., any dense subset of $[0,1]$). Then $C(X) \leq C(S)$ is Wu- and $Arch$-epi. Example 8.6.9 below shows "Lindelöf" cannot be dropped.

The proof of (a) is the obvious modification of the proof of 8.6.5(a), and (b) follows from 8.5.2 and 8.3.2.

We sketch the proof of (c). If $h \in C(S)$, then since S is Lindelöf, h extends continuously, say to $\bar{h}$, over some C_δ-set T in X, by [Hager, 1969]. As a C_δ-set in a compact space, T is Lindelöf and $T = \beta T$. By (a), there is $\{\bar{k}_n \mid n \in N\}$

epi-indicating over $C(X)$ for $\bar{h} \in C(T)$. Let $k_n = \bar{k}_n \mid S$. It is not hard to see that $\{k_n \mid n \in N\}$ epi-indicates for h.

8.6.8. Let $G \in Wu$, and let $C_\delta(G)$ be the filter base of dense C_δ-sets in $Y(G)$ (see 8.6.5). $C_\delta(G)$ is directed by set inclusion, and when $S_2 \subseteq S_1$, there is the restriction embedding $C(S_1) \leq C(S_2)$. With these as bonding maps, we have a direct system $\{C(S) \mid S \in C_\delta(G)\}$. By $C[Y(G), C_\delta]$ we mean the direct limit in Wu of that system.

Of course, $C[Y(G), C_\delta]$ is realized as $\cup\{C(S) \mid S \in C_\delta(G)\}$ modulo $\sim$, where $f \sim g$ means $f = g$ pointwise on the intersection of their codomains. The weak unit is, of course, the equivalence class of the constant function(s) 1. Note that $C[Y(G), C_\delta]$ is an "algebra".

We have a Wu-embedding $G \leq C[Y(G), C_\delta]$ given by $G \ni g \mapsto (g \mid g^{-1}(\mathbf{R})) \in C(g^{-1}(\mathbf{R}))(\bmod \sim)$.

(a) $G \leq C[Y(G), C_\delta]$ is Wu- and $Arch$-epi.

(b) Frequently, e.g., if $Y(G)$ is metrizable (e.g., $Y(C[0,1]) = [0,1]$), $C[Y(G), C_\delta]$ is Conrad's essential closure of G, thus is of the form $D(X)$ for compact extremally disconnected X (see [Conrad, 1971b]), and hence is Wu- and $Arch$-epicomplete (by [Ball and Hager, a]).

Remarks. (i) In 8.6.9 below, we exhibit G which is not epi in its essential closure, and for which $C[Y(G), C_\delta]$ is not epicomplete.

(ii) In [Ball and Hager, c], it is shown that $C[Y(G), C_\delta]$ is the (unique) maximum Archimedean-kernel-distinguishing extension of G in Wu.

(iii) Let $C = C(G)$ denote the collection of dense cozero sets of $Y(G)$, and consider $C[Y(G), C]$ (defined in the obvious way). Then (with reference to 8.6.2) $G \leq C[Y(G), C]$ is a Wu-epimorphism in which every epi-indicator $\mathcal{E}(h)$ can be a singleton, but the $\mathcal{E}(h)$'s must depend on h.

Proof. (a) We have $G^* \leq G \leq C[Y(G), C_\delta]$. It suffices to show that $G^* \leq C[Y(G), C_\delta]$ is epi, since a final factor of an epimorphism is epi.

We also have $G^* \leq C(Y(G)) \leq C[Y(G), C_\delta]$. The first embedding is epi by 8.6.3, so it suffices to show that the second embedding is epi, since the composition of epimorphisms is epi.

We have $C(Y(G)) < C[Y(G), C_\delta]$ as the direct limit of the restriction embeddings $C(Y(G)) \leq C(S)$, $(S \in C_\delta(G))$, so it suffices to show that each of these is epi, since the direct limit of epimorphisms is epi. But S is a C_δ in $Y(G)$, whence $C(Y(G)) \leq C(S)$ is epi by 8.6.7(b) (with $S' = S$) (or by 8.6.6(c)).

(b) It is known that for $G \in \mathcal{W}u$, Conrad's essential closure is $C[Y(G), \mathcal{G}_\delta]$, where $\mathcal{G}_\delta$ refers to the dense G_δ-sets of $Y(G)$. When $Y(G)$ is metrizable, each open set is cozero, so that $\mathcal{G}_\delta(G) = C_\delta(G)$.

8.6.9. Let X be an uncountable discrete space, p_α and p_λ points not in X, $\alpha X = X \cup \{p_\alpha\}$ the one-point compactification of X (in which neighborhoods of p_α have finite complement), $\lambda X = X \cup \{p_\lambda\}$ the (maximum, in a sense) one-point Lindelöfification of X (in which neighborhoods of p_λ have countable complement).

The obvious continuous map $\alpha X \leftarrow \lambda X$ ($x \leftarrow x$ and $p_\alpha \leftarrow p_\lambda$) induces an embedding $C(\alpha X) \leq C(\lambda X)$. Restriction to X defines an embedding $C(\lambda X) \leq C(X) = D(\beta X)$. (Recall that X is discrete.)

(a) αX has no proper dense cozero sets. Thus $C_\delta(C(\alpha X)) = \{\alpha X\}$, and $C(\alpha X) = C[\alpha X, C_\delta]$.

(b) $C(\alpha X) \leq C(\lambda X)$ is $\mathcal{W}u$- and $\mathcal{A}rch$-epi.

(c) $C(\alpha X) \leq C(X)$ is not epi, and thus neither is $C(\lambda X) \leq C(X)$.

(d) $C(X)$ is the essential closure of $C(\lambda X)$ and $C(\alpha X)$.

(This justifies the remark in 8.6.8(i), and shows that "Lindelöf" cannot be dropped in 8.6.7(c). Also, it follows from [Ball and Hager, a] that $C(\lambda X)$ is $\mathcal{W}u$- and $\mathcal{A}rch$-epicomplete, whence (c). But we prove (c) directly here.)

Proof. (a). Any cozero set of αX is Lindelöf (being F_σ in a compact space). The only dense subsets of αX are X and αX (since X is discrete), and X is not Lindelof (since X is uncountable).

(b) and (c). The embedding $C(\alpha X) \leq C(\lambda X)$ is realized by the map $\alpha X \xleftarrow{\tau} \beta \lambda X$ which extends the map $\alpha X \leftarrow \lambda X$ defined above. We need a few facts. Below, $\overline{(\ )}^\lambda$, $\overline{(\ )}^{\beta\lambda}$, and $\overline{(\ )}^\beta$ denote closure in λX, $\beta \lambda X$, and βX, respectively.

(1) If $U \subseteq X$ is uncountable, then $p_\lambda \in \bar{U}^\lambda$.

(2) $f : \lambda X \to \mathbf{R}$ is continuous iff $\{x \mid f(x) \neq f(p_\lambda)\}$ is countable.

(3) If $C \subseteq X$ is countable, then $\bar{C}^{\beta\lambda} \cap (\overline{X-C})^{\beta\lambda} = \emptyset$.

(4) Any dense cozero set in $\beta \lambda X$ contains λX.

(5) The Yosida representation of $C(\lambda X)$ is all of $D(\beta \lambda X)$.

Proofs of (1)-(5). (1) is obvious and (2) is easily verified. (3) Let f be the characteristic function in λX of C. By (2), f is continuous, and so extends to $f' \in C(\beta \lambda X)$. Then $\{p \in \beta \lambda X \mid f'(p) = 1\} = \bar{C}^{\beta\lambda}$ and $\{p \in \beta \lambda X \mid f'(p) = 1\} = (\overline{X-C})^{\beta\lambda}$. (4) Suppose $\text{coz} f$ is dense in $\beta \lambda X$. Any dense set contains X, since

X is discrete, so $\text{coz} f \supseteq X$. Also $p_\lambda \in \text{coz} f$, because otherwise (2) is violated. This proves (4). (5) is immediate from (4).

We now prove (b). By section 8.5, it suffices to verify the condition in 8.3.2. Let $h \in D(\beta\lambda X)$, and let $C = \{x \in X \mid h(x) \neq h(p_\lambda)\}$. By (2), C is countable, and by (3) we have $\bar{C}^{\beta\lambda} \cap (\overline{X-C})^{\beta\lambda} = \varnothing$. One easily sees that

$$\Sigma(h) \subseteq \{(p,q) \mid p \in \bar{C}^{\beta\lambda} - C \text{ and } q \in (\overline{X-C})^{\beta\lambda} - (X-C), \text{ or vice versa}\}.$$

If C is finite, $\Sigma(h) = \varnothing$ and there is nothing to prove. Otherwise, $C = \{x_1, x_2, \ldots\}$, and we define $k \in D(\beta\lambda X)$ by: $k(x_n) = n$ and $k \mid (\lambda X - C) = 0$; this is continuous on λX by (2), and we extend k to $\beta\lambda X$. This k has $\infty(k) = \bar{C}^{\beta\lambda} - C$, and thus epi-indicates for h.

We now prove (c). The embedding $C(\alpha X) \leq C(X)$ is realized by the map $\alpha X \overset{\sigma}{\leftharpoonup} \beta X$ which extends the inclusion $\alpha X \leftarrow X$; $D(\beta X)$ is the Yosida representation of $C(X)$. Write $X = X_0 \cup X_1$, the disjoint union of two uncountable sets, and let $h \in C(X)$ be the characteristic function of X_1. So $\bar{X}_0^\beta \cup \bar{X}_1^\beta$ is a disjoint union, and one sees easily that $\Sigma(h) = \{(p_0, p_1) \mid p_i \in X_i^\beta - X_i, \text{ or vice versa}\}$. Suppose there is an epi-indicator $\{k_n \mid n \in N\}$ for h over $C(\alpha X)$. With $F = \bigcap_n k_n^{-1}(\mathbf{R})$, we have $\sigma(\bar{X}_0^\beta \cap F) \cap \sigma(\bar{X}_1^\beta \cap F) = \varnothing$ (for if not, $\{k_n\}$ doesn't epi-indicate). But F is Lindelöf (8.6.7), so $\bar{X}_i^\beta \cap F$ is too (since it is closed in F), and so is the continuous image $\sigma(\bar{X}_i^\beta \cap F)$. But this contains X_i, and thus is an uncountable Lindelöf subset of αX which, therefore, must contain p_α. That is, $p_\alpha \in \sigma(\bar{X}_0^\beta \cap F) \cap \sigma(\bar{X}_1^\beta \cap F)$, a contradiction.

(d) By [Conrad, 1971b], the essential closure of (any) G is the unique $D(Z)$ with Z extremally disconnected in which G is large. Here, $C(X) = D(\beta X)$, and βX is extremally disconnected. Since characteristic functions of points of X are in $C(\alpha X)$, $C(\alpha X)$ and $C(\lambda X)$ are large in $C(X)$.

8.7 More Examples: Epimorphisms in 𝒲𝑢 and 𝒜𝑟𝑐ℎ Differ

Theorem 8.4.4 says that $G \leq H$ is 𝒜𝑟𝑐ℎ-epi iff the embedding is coessential, and $G/u^{\perp G} \leq H/u^{\perp H}$ is 𝒲𝑢-epi for all $u \in G^+$. We present two examples which show that 8.4.4 is sharp.

EXAMPLE 8.7.1. A 𝒲𝑢-epi-embedding $G \leq H$ which (a) fails to be coessential, and for which (b) $G/u^{\perp G} \leq H/u^{\perp H}$ is 𝒲𝑢-epi for all $u \in G^+$.

Let $G = C^*[1,+\infty)$. Note that the H can't be an algebra by 8.5.2. Let $v(x) = x$, and $H = \{f \in C[1,+\infty) \mid |f| \leq nv$ for some $n \in N\}$; the common weak unit is the constant 1. Since $Y(C^*[1,+\infty)) = \beta[1,+\infty) = Y(C[1,+\infty))$, we also have $Y(G) = \beta[1,+\infty) = Y(H)$; so every $\Sigma(h) = \varnothing$ and $G \leq H$ is $\mathcal{W}u$-epi "vacuously" (in the terminology of 8.6.2).

(a) We shall show that $G \leq H$ is not coessential, i.e., that $ak_H G \neq H$. We shall refer to the brief exposition about $ak_H G$ in the proof of 8.5.4(a) above. With $G' = \{f \in H \mid \exists \{g_n\} \subseteq G$ and $h \in H$ with $g_n \to f(h)\}$, it follows from (3) below that $ak_H G = G'$, and that $G' \neq H$.

The following very nice explanation of the situation was contributed by L. C. Robertson. We abbreviate $C^*[1,+\infty)$ to C^*, and let $C_0 = \{f \in C^* \mid \lim_{x \to \infty} f(x) = 0\}$. Define $\mu : H \to C^*$ by $\mu(f) = \frac{1}{v} f$. Then

(1) μ is an isomorphism onto.

(2) $\mu(G') = C_0$.

(3) $H/G' \cong C^*/C_0$.

(4) $C^*/C_0 \cong C(\beta X - X)$ $(X = [1,+\infty))$.

Proofs. (1) μ is clearly a one-to-one homomorphism. For $g \in C^*$, $g = \mu(vg)$, so μ is onto.

(2) If $g_n \to f(h)$ then $g_n \to f(v)$, since v is a strong unit. Then an inequality $|g_n - f| \leq \varepsilon v$ yields $\left| \frac{g_n}{v} - \frac{f}{v} \right| \leq \varepsilon$, i.e., $|\mu(g_n) - \mu(f)| \leq \varepsilon$. Therefore $\mu(g_n) \to \mu(f)$ uniformly on $[1,+\infty)$ (in the usual sense). Now, it is clear for bounded g that $\frac{g}{v} \in C_0$. Thus $\mu(G) \subseteq C_0$, and clearly C_0 is closed under this usual uniform convergence. Hence, $\tau(G') \subseteq C_0$.

Now let $k \in C_0$. We find $f \in G'$ with $\mu(f) = k$. Let C_{00} be the continuous functions on $[1,+\infty)$ of compact support. There is $\{k_n\} \subseteq C_{00}$ with $k_n \to k$ uniformly on $[1,+\infty)$. Then $vk_n \in C_{00} \subseteq C^* = G$, and evidently $vk_n \to vk(v)$. That is, $vk \in G'$; and of course, $\mu(vk) = k$.

(3) follows from (1) and (2).

(4) For $f \in C^*$, let $\hat{f}$ be the extension over βX. Define a homomorphism $\alpha : C^* \to C(\beta X - X)$ by: $\alpha(f) = \hat{f} | (\beta X - X)$. Clearly, the kernel of α is C_0, which proves (4).

(b) Let $u \in G^+$. We show that $G/u^{\perp G} \leq H/u^{\perp H}$ is $\mathcal{W}u$-epi. To see what the Yosida representation of $H/u^{\perp H}$ is, we look again at the paragraph "$\hat{H}$ and $\bar{H}$" in the proof of 8.5.4(b): the set $\mathrm{coz}u \subseteq [1,+\infty)$, denoted there by $\hat{U}$ in $Y(H,1) = \beta X$, becomes the dense set $\bar{U}$ in $Y(H,u)$, and the Yosida representation restricted to $\bar{U} = \text{cozu}$ is, by(c) of that paragraph, obtained by dividing the original representation by u; that is, for $h \in H$, and $x \in \text{cozu}$, $\frac{1}{u(x)}h(x)$ is the Yosida representation of h (or, $h+u^{\perp H}$) on $\text{cozu} = \bar{U} \subseteq Y(H,u)$.

We thus can express the embedding $G/u^{\perp G} \leq H/u^{\perp H}$, as $\frac{1}{u}G \leq \frac{1}{u}H$, and in fact we have $C^*(\text{cozu}) \leq \frac{1}{u}G \leq \frac{1}{u}H \leq C(\text{cozu})$. (The first inclusion says that each $f \in C^*(\text{cozu})$ is of the form $\frac{1}{u}g$ for some $g \in G = C^*[1,+\infty)$; take $g = uf$ on cozu, $g = 0$ off cozu.) Since $Y(C^*(\text{cozu}),1) = \beta\text{cozu} = Y(C(\text{cozu}),1)$, it follows that $Y(G,u) = Y(H,u) = \beta\text{cozu}$ also. Thus, $G/u^{\perp G} \leq H/u^{\perp H}$ is "vacuously" $\mathcal{W}u$-epi.

EXAMPLE 8.7.2. A $\mathcal{W}u$-epi-embedding $G \leq H$ (a) which is coessential, and (b) there is $v \in G^+$ with $G/v^{\perp G} \leq H/v^{\perp H}$ not $\mathcal{W}u$-epi, and for which (c) there is an order base $\mathcal{U}$ of G with $G/u^{\perp G} \leq H/u^{\perp H}$ $\mathcal{W}u$-epi for all $u \in \mathcal{U}$.

This is a modification of the previous example, again within $C[1,+\infty)$. Now let G be the ℓ-subgroup of $C[1+\infty)$ generated by $C^*[1,+\infty)$ together with v ($v(x) = x$), and as before, let $H = \{f \in C[1,+\infty) \mid |f| \leq nv \text{ for some } n \in N\}$. Using the weak unit 1, we have $G \leq H \in \mathcal{W}u$. Since $H^* \leq G$, we have $Y(G) = Y(H)$, and $G \leq H$ is "vacuously" $\mathcal{W}u$-epi. (a) Since v is a strong unit in H, the ideal which v, or G, generates in H is all of H, so $ak_H G = H$. (b) $G/v^{\perp G} \leq H/v^{\perp H}$ is not $\mathcal{W}u$-epi: This statement refers to the designation of weak unit as $v+v^{\perp}$, by virtue of which the embedding is in $\mathcal{W}u$. Of course, $v^{\perp G} = (0)$ and $v^{\perp H} = (0)$, so $G/v^{\perp G} \leq H/v^{\perp H}$ is an embedding of G in H different from the one presented to us. As in 8.7.1(b), this embedding can be expressed as $\frac{1}{v}G \leq \frac{1}{v}H$, which means: $[1,+\infty)$ is dense in $Y(H,v)$, and the representation $\bar{h}$ of h

on $Y(H,v)$ is an extension of $\frac{h}{v}$. A little thought reveals that all these functions are bounded, and that the extensions $\frac{g}{v}$ $(g \in G)$ are constant on $Y(H,v)$-$[1,+\infty)$. Thus it suffices to exhibit $h \in H$ with $\Sigma(\bar{h}) \neq \varnothing$, i.e., with the extension of $\frac{h}{v}$ *not* constant on $Y(H,v)$-$[1,+\infty)$. Note that any uniformly continuous function is in H, so it suffices to find a uniformly continuous function h such that $\frac{h(x)}{x}$ has no limit as $x \to \infty$; this is an exercise from calculus. (c) For each $n \in N$, choose $u_n \in G^+$ with $\mathrm{cozu}_n = (n,n+1)$. Then $\{u_n \mid n \in N\}$ is an order base (= maximal disjoint family) with, in fact, $G/u^{\perp G} = H/u^{\perp H}$ for all n.

Richard N. Ball
University of Denver
Denver, Colorado 80208
U. S. A.

Anthony W. Hager
Wesleyan University
Middletown, Connecticut 06457
U. S. A.

Stephen H. McCleary

CHAPTER 9

FREE LATTICE-ORDERED GROUPS

The definition of the free lattice-ordered group F_η of a given rank η is analogous to that of the free group G_η. Yet there are many questions which have the same "obvious" answers in both contexts, and for which the answers are very easy to confirm for free groups but not at all easy for free ℓ-groups. We assume $\eta > 1$. Some questions:

1. Does F_η have solvable word problem?

2. What is the center of F_η?

3. What is the centralizer of a free generator x?

4. Is F_η directly indecomposable?

5. Does F_η have a basic element?

6. Is F_η completely distributive?

Actually, the answer to Question 1 isn't quite so obvious, and of course the last two questions don't have group theoretic analogues. A *basic element* is an $e < w \in F_\eta$ such that $\{y \mid e < y < w\}$ is totally ordered. In several cases, the difficulty in confirming the answer[1] stems from the fact that Question 1 is certainly more complicated for ℓ-groups than for groups. For example, $(x \wedge e) \vee (x^{-1} \wedge e) = e$ for any free generator x. Thus in showing that any element w of the center of F_η must be e, any attempt to do this by showing that each of its constituent group words must be e is surely foredoomed. Of course, one of the reasons for the interest in Question 1 is that for any ℓ-

---

[1]The answers: yes, trivial, the ℓ-subgroup generated by x, yes, no, no.

A. M. W. Glass and W. C. Holland (eds.), Lattice-Ordered Groups, 206–227.
© 1989 by Kluwer Academic Publishers.

group word $w(x_1,...,x_n)$ in the free generators, $w = e$ in F_η iff "$\forall x_1,...,x_n$, $w(x_1,...,x_n) = e$" is a law in all ℓ-groups.

Aside from the solution of the word problem by Holland and McCleary [1979], virtually all progress in the understanding of free ℓ-groups has come from two quite different ways of representing them as ℓ-permutation groups, with each kind of representation yielding some results which cannot be obtained from the other.

First, Conrad [1970b] represented F_η by means of its actions on the various right orderings of the free group G_η (to be described in detail later). This approach has been exploited most intensively by Arora and McCleary [1986], who answered Questions 2 (previously answered by Medvedev [1981]), 3, and 4.

Later, Glass [1974] and McCleary [1985a] represented F_η as a 2-transitive ℓ-permutation group. (Also, with a different goal in mind, Kopytov [1983] obtained a similar representation of F_η which is easily seen to be 2-transitive.) From this it follows (Kopytov [1979]) that Conrad could have gotten by with just one right ordering of G_η. The 2-transitivity yields a spate of immediate corollaries (McCleary [1985a]), including the answers to Questions 2, 4, and 5, and an additional aspect of the representation answers Question 6.

Recently McCleary [1985b] constructed a representation giving the best of both worlds--a right ordering of G_η on which the action of F_η is both faithful and 2-transitive. The construction is explicit enough that variations of it can be utilized to get a great deal of information about the root system $\mathcal{P}_\eta$ of prime subgroups of F_η. All $\mathcal{P}_\eta$'s with $1 < \eta < \infty$ are isomorphic. This common root system $\mathcal{P}_f$ has only four kinds of branches (singleton, three-element, $\mathcal{P}_f$, and $\mathcal{P}_{\omega_0}$), each of which occurs 2^{ω_0} times. Each finite or countable chain having a largest element occurs as the chain of covering pairs of some root of $\mathcal{P}_f$.

The best-of-both-worlds representation gives all the earlier representations as corollaries, but it requires the Glass-McCleary representation as a lemma. Thus after solving the word problem, and looking at consequences of the Conrad-Kopytov representation, we shall establish and utilize the Glass-McCleary representation, and then use it to establish the best-of-both-worlds representation, obtaining the Conrad-Kopytov representation as a corollary. The proofs will be given only for $\eta \leq \omega_0$, but actually infinite rank is easier to handle than finite rank.

9.1. Background

Let x be a subset of an ℓ-group F. F is *free on x* if every function from x into an arbitrary ℓ-group H can be extended uniquely to an ℓ-homomorphism from F into H (in which case x is automatically a generating set for F). Any two ℓ-groups free on sets of the same cardinality η are ℓ-isomorphic.

F is *free* if it is free on some subset x. The *rank* of F means the cardinality of x.

PROPOSITION 9.1.1. *The rank of a free ℓ-group F is well defined.*

Proof. Let F be free on x_1 and on x_2. Then the abelianization F/F' of F is a free abelian ℓ-group on the set of cosets (necessarily distinct) having representatives in x_1 (resp. x_2). The rank of a free abelian ℓ-group is well defined [Weinberg, 1963, Theorem 2.13] (see Section 1.3)

Thus for each cardinality η, there is a unique free ℓ-group F_η of rank η. It is easy to see that F_1 is the cardinal sum $Z \boxplus Z$ of two copies of the integers, with $(1,-1)$ as a free generator; for in any ℓ-group, if w is incomparable with e, then the ℓ-subgroup $\ell(w)$ it generates is a copy of $Z \boxplus Z$. Almost all our results fail for $\eta = 1$, so we shall ordinarily assume that $\eta > 1$.

In contrast to the situation for free groups, an ℓ-subgroup of a free ℓ-group F_η need not be free. If $e < w \in F_\eta$, $\ell(w)$ is a copy of Z, which is certainly not free.

When dealing with F_η, we shall always envision a fixed x on which F_η is free. The subgroup generated by x is a free group on x, the free group of rank η, and it inherits from F_η the trivial order (no two elements comparable). For G_η can be totally ordered (Chapter 0 Example 9), and the reversal of this total order is another total order; and for each of these two orders, the function $x \to x$ from x into G_η can be extended to an ℓ-homomorphism. If $x' \subseteq x$, then $\ell(x')$ is the free ℓ-group on x'. Thus when $\eta' < \eta$, $F_{\eta'}$ is an ℓ-subgroup of F_η.

Each $w \in F_\eta$ can be put in a standard form $w = \vee_i \wedge_j w_{ij}$, a finite supremum of finite infima of elements of G_η, i.e., of reduced group words in the elements of x. As we saw in the introduction, this standard form is far from unique.

A *substitution* for F_η in $A(\Omega)$, Ω a chain and $A(\Omega)$ the ℓ-group of all automorphisms of Ω, is simply a function $x \to A(\Omega)$ which assigns to each free generator x an automorphism $\hat{x}$ of Ω; or equivalently, an ℓ-homomorphism $w \to \hat{w}$ from F_η into $A(\Omega)$, known also as an *action* of F_η on Ω. A *representation* is a faithful action. For $w = \vee_i \wedge_j w_{ij}$ $(w_{ij} \in G_\eta)$, we have $\alpha \hat{w} =$ $\max_i \min_j \alpha w_{ij}$ for each $\alpha \in \Omega$.

9.2. The word problem for F_η

The results in this section are due to Holland and McCleary [1979].

Our fundamental tool throughout the paper will be the notion of a *diagram* for an ℓ-group word $w = \vee_i \wedge_j w_{ij} \in F_\eta$, where each w_{ij} is a reduced group word $x_{i_1}^{\pm 1} \dots x_{i_n}^{\pm 1}$ ($n \geq 0$, and depending on i and j). As an example, consider $w = (x \wedge x^{-1} y x) \vee e$, and a substitution in $A(\mathbf{Q})$, $\mathbf{Q}$ the chain of rational numbers, having the indicated effect on 0:

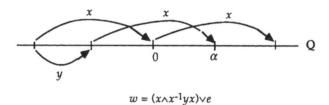

$$w = (x \wedge x^{-1} y x) \vee e$$

Note that $0w = \max_i \min_j 0 w_{ij} = \alpha \neq 0$. This diagram shows that $w \neq e$ in F_η. Of course, there are many other possible diagrams for this same word. The precise definition (making no reference to $\mathbf{Q}$) is as follows:

The *points* of the diagram are the formal initial segments $x_{i_1}^{\pm 1} \dots x_{i_k}^{\pm 1}$ ($k \geq 0$) of the w_{ij}'s. (Of course, a given point may arise from several w_{ij}'s.) For each ordered pair (α, β) of points such that $\alpha x_i^{\pm 1} = \beta$, the diagram includes an x_i-*arrow*, from α to β if the exponent on x_i is +1, otherwise from β to α. The remaining aspect of the diagram is a total order on the set D of points which is consistent with the arrows in that if there are x_i-arrows from α_1 to β_1 and from α_2 to β_2 (same i for both), then $\alpha_1 \leq \alpha_2$ iff $\beta_1 \leq \beta_2$.

An x_i-arrow from α to β may alternately be described as an x_i^{-1}-arrow from β to α. We emphasize that a diagram is necessarily *connected*, meaning that for all points α, β, there must be at least one sequence of arrows leading from α to β; and *loop-free*, meaning that there cannot be more than one such sequence.

By a *diagram on* **Q** we mean a diagram which arises from a substitution in $A(\mathbf{Q})$.

LEMMA 9.2.1. *Let* $w \in F_\eta$. *If* $w \neq e$, *there exists a diagram on* **Q** *showing this (by making* $0\hat{w} \neq 0$).

Proof. We use the Holland Representation Theorem [Theorem 2.2.5] to represent F_η on some chain Ω. Since $w \neq e$, there exists $\beta \in \Omega$ such that $\beta\hat{w} \neq \beta$. Then the points $\beta\hat{x}_{i_1}^{\pm 1}...\hat{x}_{i_k}^{\pm 1} \in \Omega$ (for $x_{i_1}^{\pm 1}...x_{i_k}^{\pm 1}$ an initial segment of some w_{ij}) together with the obvious arrows yield a diagram except that distinct initial segments may give identical points of Ω. We lay out this "diagram" on **Q** with 0 playing the role of β, and pick from each $x \in x$ an automorphism $\hat{x}$ of **Q** moving the points of the diagram in accordance with the x-arrows. (This can be done because all non-singleton closed intervals of **Q** are isomorphic.) Finally, the $\hat{x}$'s must be changed so that distinct initial segments send 0 to distinct points in **Q** (while retaining $0\hat{w} \neq 0$). Fairly obviously this can be done; for a rigorous proof see [Holland and McCleary, 1979]. Actually, the loop-freeness of diagrams won't be needed until §5.

THEOREM 9.2.2 [Holland and McCleary, 1979]. *The word problem for* F_η *is solvable.*

Proof. To decide whether or not $w = e$, draw on **Q** (with 0 as a base point) all of the finitely many diagrams for w (such as the diagram pictured above for $w = (x^{-1} \wedge x^{-1}yx) \vee e$). By the lemma, we have $w \neq e$ iff $0\hat{w} \neq 0$ for at least one diagram. (Of course, we could use the chain **R** of real numbers in place of **Q**.)

In fact, the decision procedure is quite practical as a real-life tool; see [Holland and McCleary, 1979]. The lemma also yields

THEOREM 9.2.3 [Holland, 1976]. *The ℓ-group $A(\mathbf{Q})$ generates the variety of all ℓ-groups.*

Proof. If the law "$\forall x_1,...,x_n, w(x_1,...,x_n) = e$" fails in some ℓ-group, $w \neq e$ in F_η, so this law fails in $A(\mathbf{Q})$.

9.3. The representation of F_η on a right ordering of G_η

Let G_η^λ be a right ordering of the free group G_η, i.e., a total ordering of the set G_η such that $g_1 \leq g_2$ implies $g_1 g_3 \leq g_2 g_3$. The right regular representation ϕ of G_η preserves the order and thus is a subgroup of $A(G_\eta^\lambda)$, but is not in general an ℓ-permutation group. By the freeness of F_η on the free generating set x (which, as mentioned above, freely generates G_η as a group), ϕ can be extended to a unique ℓ-homomorphism into $A(G_\eta^\lambda)$, (i.e., to a unique action on the chain G_η^λ), namely

$$w\bar{\phi} = (\vee_i \wedge_j w_{ij})\bar{\phi} = \vee_i \wedge_j w_{ij}\phi.$$

For $\alpha \in G_\eta^\lambda$, $\alpha(w\bar{\phi}) = \max_i \min_j \alpha w_{ij}$, the max and min being in the chain G_η^λ. We shall refer to this as the *natural action* of F_η on G_η^λ.

THEOREM 9.3.1 [Conrad, 1970b]. *The product of the natural actions of F_η on the collection of all right orderings G_η^λ of G_η gives a (faithful) representation of F_η.*

Surprisingly, a sharper result holds:

THEOREM 9.3.2 [Kopytov, 1979]. *There exists some one right ordering G_η^λ on which the natural action of F_η is a (faithful) representation.*

These results will be subsumed in the Best-of-Both-Worlds Theorem, which will be proved in §5.

Here we use Theorem 9.3.1 to answer Question 3:

THEOREM 9.3.3 [Arora and McCleary, 1986]. *The centralizer of a free generator x of F_η is just the ℓ-subgroup $\ell(x)$ generated by x (each element of which has the form $x^p \vee x^q$ or $x^p \wedge x^q$).*

The proof will take a while, and won't be completed until immediately before Corollary 9.3.9. Let G^λ be a right ordering of $G = G_\eta$. Of course, $ew^\lambda = \max_i \min_j ew_{ij} = \max_i \min_j w_{ij}$, but to retain our emphasis on permutations, we shall write ew_{ij} rather than just w_{ij}. The statement "$\alpha w_1^\lambda = \alpha w_2^\lambda$" will often be phrased as "$\alpha w_1 = \alpha w_2$ in G^λ".

LEMMA 9.3.4. *Let G^λ be a right ordering of G, and let $\alpha \in G^\lambda$. Then there exists another right ordering $G^{\lambda'}$ of G such that the order of $\{eg \mid g \in G\}$ in $G^{\lambda'}$ coincides with the order of $\{\alpha g \mid g \in G\}$ in G^λ (i.e., $eg \to \alpha g$ is an o-isomorphism from the chain $G^{\lambda'}$ onto the chain G^λ).*

Proof. The order of $\{\alpha g \mid g \in G\}$ in G^λ coincides with the order of $\{\alpha g \alpha^{-1} \mid g \in G\}$ in G^λ, so we take for $G^{\lambda'}$ the right ordering of G induced by the group automorphism $g \to \alpha g \alpha^{-1}$ (i.e., $g_1 \leq_{\lambda'} g_2$ iff $\alpha g_1 \alpha^{-1} \leq_\lambda \alpha g_2 \alpha^{-1}$).

PROPOSITION 9.3.5. *Let $w_1, w_2 \in F_\eta$. If $ew_1 = ew_2$ in every right ordering G_η^λ of G_η then $w_1 = w_2$ in F_η.*

Proof. If w_1 and w_2 differ at some α in some right ordering, then they differ at e in some other right ordering.

LEMMA 9.3.6. *Let $w_1, ..., w_n \in F_\eta$. Then there exists a right ordering G^λ of G in which $ew_1 < ew_2 < ... < ew_n$ if and only if there exists a substitution $w \to \hat{w}$ for F_η in $A(\mathbf{Q})$ (or equivalently in $A(\mathbf{R})$), with $0\hat{w}_1 < 0\hat{w}_2 < ... < 0\hat{w}_n$.*

Proof. The following are equivalent:

(1) There exists G^λ in which $ew_1 < ew_2 < ... < ew_n$.

(2) $w_2w_1^{-1} \wedge w_3w_2^{-1} \wedge ... \wedge w_nw_{n-1}^{-1} \not\leq e$ in F_η.

(3) There exists a substitution $w \to \hat{w}$ for F_η in $A(\mathbf{Q})$ (and thus also in $A(\mathbf{R})$) with $0(\hat{w}_2\hat{w}_1^{-1} \wedge \hat{w}_3\hat{w}_2^{-1} \wedge ... \wedge \hat{w}_n\hat{w}_{n-1}^{-1}) > 0$, i.e., $0\hat{w}_1 < 0\hat{w}_2 < ... < 0\hat{w}_n$.

To see that (2) implies (3), use Theorem 9.2.3.

Let $w = \vee_i \wedge_j w_{ij} \in F_\eta$. We shall say that w_{ij} is *determining* if there exists a right ordering G^λ in which

$$ew \ (= \max_{i'} \min_{j'} ew_{i'j'}) = ew_{ij}.$$

(Thus it is the group word, rather than a specific occurrence of it, that is "determining".) Equivalently (in view of Lemma 9.3.4), w_{ij} is determining if there exists a right ordering G^λ in which $\alpha w = \alpha w_{ij}$ for some α. Obviously at least one w_{ij} is determining.

Now we are ready to work directly on Theorem 9.3.3. Suppose $w = \vee_i \wedge_j w_{ij} \in F_\eta$ commutes with x. (A warning: Since $e = (y \wedge e) \vee (y^{-1} \wedge e)$ for any y, we cannot hope to show that every w_{ij} is a power of x.) Let w_{ij} be determining, with G^λ a right ordering of G in which $ew = ew_{ij}$. For any $n \in \mathbf{Z}$ we have $wx^n = x^nw$, so in G^λ we have

$$ew_{ij}x^n = ex^nw_{i'j'} \text{ for some } w_{i'j'}, \quad \text{i.e.,}$$
$$x^{-n}w_{ij}x^n = w_{i'j'} \text{ for some } w_{i'j'}.$$

Since there are only finitely many $w_{i'j'}$'s, we have

$$x^{-n_1}w_{ij}x^{n_1} = x^{-n_2}w_{ij}x^{n_2} \text{ for some } n_1 \neq n_2, \text{ and thus}$$
$$x^{n_2-n_1}w_{ij} = w_{ij}x^{n_2-n_1} \text{ for some } 0 \neq n_2-n_1.$$

Since w_{ij} commutes with the free generator x of the free group G, $w_{ij} = x^m$ for some $m \in \mathbf{Z}$.

We have shown that every determining w_{ij} is a power of x. That doesn't yet prove the theorem, but applying this result to two free generators is already enough to answer Question 2:

COROLLARY 9.3.7 [Medvedev, 1981]. F_η ($\eta > 1$) *has trivial center.*

We return to the proof of the theorem. If a *term* $\wedge_j w_{ij}$ of $w = \vee_i \wedge_j w_{ij}$, includes no determining w_{ij}, then deleting that term from w leaves each w^λ unchanged (because of the equivalent definition of "determining"), and so leaves w unchanged. Thus we may assume that each term of w includes at least one determining w_{ij}, which then must be a power of x. (Another warning: Deleting from one of the remaining terms a non-determining w_{ij} can in fact change w, by making the determining $w_{i'j'}$ applying in some G^λ come from that term instead of some other term -- so we may *not* assume that all w_{ij}'s are determining.) In any G^λ,

either $ex^m \le ex^n$ iff $m \le n$ (if $x > e$),

or else $ex^m \le ex^n$ iff $m \ge n$ (if $x < e$),

so we may assume that each term includes at most two w_{ij}'s which are powers of x (only the smallest and largest exponents being retained). Thus we may assume that w has the form

$$w = \vee_i(x^{p_i} \wedge x^{q_i} \wedge \wedge_j u_{ij}),$$

where $p_i \le q_i$ and where no u_{ij} is a power of x (and thus no u_{ij} is determining).

Let p be the largest p_i such that x^{p_i} is determining for some G^λ with $x > e$ (so that $x^m \le x^n$ iff $m \le n$). We claim that

(1) $ew = ex^p$ in *every* G^λ with $x > e$.

If (1) is false, then for some $G^{\lambda'}$ with $x > e$, and for some $p_0 < p$,

$$ex^{p_0} = ew = \max_i \min_j \{ex^{p_i}, ex^{q_i}, eu_{ij}\} = \max_i \min_j \{ex^{p_i}, eu_{ij}\},$$

so that for each $p_i \ge p$, there exists at least one u_{ij} and we have $\min_j eu_{ij} < ex^{p_0} \le ex^{p-1}$. Letting $u = \vee_{p_i \ge p} \wedge_j u_{ij}$, we have

(2) $eu = \max_{p_i \ge p} \min_j eu_{ij} < ex^{p-1}$ in $G^{\lambda'}$ (with $x > e$).

On the other hand, picking $G^{\lambda''}$ with $ew = ex^p$, we have

(3) $eu \geq \max_{p_i = p}\min_j eu_{ij} > ex^p$ in $G^{\lambda''}$ (with $x > e$).

To prove (1), it would suffice to show

(4) $ex^{p-1} < eu < ex^p$ for some $G^{\lambda'''}$ (with $x > e$).

For in $G^{\lambda'''}$, the determining w_{ij} $(= x^{p_i})$ of w could not have $p_i \geq p$ (since then $\min_j eu_{ij} \leq \max_{p_{i'} \geq p}\min_{j'} eu_{i'j'} = eu < ex^p \leq ex^{p_i}$) and also could not have $p_i < p$ (since then $eu < ex^{p_i} \leq ex^{p-1} < eu$), yielding a contradiction.

The following lemma guarantees that of the G^{λ}'s satisfying (2), at least one must satisfy (4).

LEMMA 9.3.8. *Let x be a free generator of F_η and let $u \in F_\eta$. Let $p \in \mathbf{Z}$. Suppose there exist two right orderings G^{λ_1} and G^{λ_2} of G with $x > e$, and with*

$$eu < x^{p-1} \text{ in } G^{\lambda_1} \text{ and } eu > x^p \text{ in } G^{\lambda_2}.$$

Then there exists a third right ordering G^{λ_3} with $x > e$, and with

$$x^{p-1} < eu < x^p \text{ in } G^{\lambda_3}.$$

Proof of Lemma 9.3.8. Multiplying by $(x^{p-1})^{-1}$, we may assume that $p = 1$. Thus we are given G^{λ_1} and G^{λ_2} with $x > e$, and with

$$eu < e \text{ in } G^{\lambda_1} \text{ and } eu > x \text{ in } G^{\lambda_2},$$

and we must produce G^{λ_3} with $x > e$, and with

$$e < eu < x \text{ in } G^{\lambda_3}.$$

Applying Lemma 9.3.6 to G^{λ_1}, we get a substitution $w \to \hat{w}$ for F_η in $A(\mathbf{R})$ making $0\hat{u} < 0 < 0\hat{x}$, and we may arrange that $0\hat{x} = 1$:

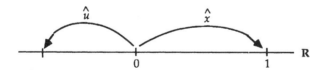

Applying Lemma 9.3.6 to G^{λ_2}, we get another substitution $w \to \bar{w}$ making $0 < 0\bar{x}$ $(= 1) < 0\bar{u}$:

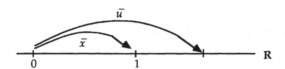

We continuously deform the automorphism $\hat{x}$ to x^-, arranging throughout that the intermediate automorphism $\dot{x}$ send 0 to 1. Then, for each other $y \in x$ involved in the spelling of u, we continuously deform $\hat{y}$ to $\bar{y}$ (having previously arranged that $\hat{z} = e = \bar{z}$ for all other $z \in x$). This process must move the point $0\hat{u} > 1$ continuously to the point $0\bar{u} < 0$, so there must exist an intermediate substitution $w \to \dot{w}$ such that $0 < 0\dot{u} < 1 = 0\dot{x}$. By Lemma 9.3.6 again, there exists G^{λ_3} with $e < u < x$.

We have proved the lemma and thus have proved (1). We also need to know that there exists $q \in Z$ such that

(5) $ew = ex^q$ in every G^λ with $x < e$.

But x^{-1} is also an element of a free generating set, and x^{-1} commutes with w, so applying (1) to x^{-1} and w, we get (5).

By (1) and (5), one of two cases must obtain:

(a) If $p \le q$, $ew = e(x^p \wedge x^q)$ in every G^λ whatsoever.

(b) If $p \ge q$, $ew = e(x^p \vee x^q)$ in every G^λ whatsoever.

By Proposition 9.3.5, either $w = x^p \wedge x^q \in \ell(x)$, or $w = x^p \vee x^q \in \ell(x)$. (We mention that the fact that every element of $\ell(x)$ has the form $x^p \vee x^q$ or $x^p \wedge x^q$ actually applies to any x in any ℓ-group [Arora and McCleary, 1986, Proposition 2].) This completes the proof of Theorem 9.3.3.

COROLLARY 9.3.9. *Let x be a free generator of F_η, and let $w \in F_\eta$. Then it can be decided whether or not $w \in \ell(x)$.*

Proof. By Theorem 9.2.2, it can be decided whether or not $wx = xw$.

Similar methods yield

THEOREM 9.3.10 [Arora and McCleary, 1986]. *It can be decided whether or not two elements of G_η are conjugate in F_η.*

However, the conjugacy problem in F_η remains open.

Now we try Question 5.

LEMMA 9.3.11. *Let $e < w \in F_\eta$. Then w is basic if and only if there exists a unique right ordering G^λ of $G = G_\eta$ in which $ew > e$.*

Proof. Suppose G^λ is the unique right ordering in which $ew > e$. Let $e \le h_i \le w$, $i = 1,2$, and suppose without loss of generality that $eh_1 \le eh_2$ in G^λ. Then $e(h_2 h_1^{-1} \wedge e) = e$ in G^λ, and certainly also in the other right orderings since there, $eh_i = e$. By Proposition 9.3.5, $h_2 h_1^{-1} \wedge e = e$ in F_η, i.e., $h_1 \le h_2$. Hence w is basic.

Conversely, let G^{λ_1} and G^{λ_2} be distinct right orderings in which $ew > e$. Pick $h_1 \in G$ such that $e < h_1$ in G^{λ_1} but not in G^{λ_2}. Let $k_1 = (h_1 \wedge w)^+ \le w$. Then $e < k_1$ in G^{λ_1} and $ek_1 = e$ in G^{λ_2}. Similarly, there exists $k_2 \in F_\eta$ such that $e \le k_2 \le w$, $e < k_2$ in G^{λ_2}, and $ek_2 = e$ in G^{λ_1}. Since k_1 and k_2 are incomparable, w is not basic.

PROPOSITION 9.3.12. *F_η has a basic element if and only if G_η has a finite subset S for which there is a unique right ordering of G_η making all elements of S positive.*

Proof. Suppose $w = \vee_i \wedge_j w_{ij}$ is basic. Then for some i_0, $(\wedge_j w_{i_0 j})^+$ is also basic. Let G^λ be the unique right ordering in which $e(\wedge_j w_{i_0 j})^+ > e$. Then G^λ is also the unique right ordering in which each $w_{i_0 j}$ is positive.

Conversely, if G^λ is the unique right ordering making all elements of S positive, the once e is deleted from S, $(\wedge S)^+$ is basic.

(Rhetorical) QUESTION. Does the free group G_η $(1 < \eta < \infty)$ have a finite subset S for which there is a unique right ordering of G_η making all elements of S positive?

(Open) QUESTION. Does G_η $(1 < \eta < \infty)$ have a finite subset S for which there is a unique (two-sided) total order of G_η making all elements of S positive?

For infinite η, both questions have easy negative answers. Some comments on finite η: The answer to the rhetorical question should be "no" since F_η "shouldn't" have a basic element; see Corollary 9.4.4. The open question can be shown to be equivalent to the question of whether the free representable ℓ-group of rank η has a basic element, so again the answer "should" be "no"--but is it?

4. The pathologically 2-transitive representation of F_η

An ℓ-permutation group (F,Ω) is 2-transitive if for all $\alpha < \beta$ and $\gamma < \delta$ in Ω, there exists $f \in F$ such that $\alpha f = \gamma$ and $\beta f = \delta$ (see Section 2.3). If in addition, no $e \neq f \in F$ has bounded support, (F,Ω) is called *pathological*. Supp(f) means $\{\omega \in \Omega \mid \omega f \neq \omega\}$. The following theorem was proved for most infinite ranks η by Glass [1974], and for finite η by McCleary [1985a] using the proof given here (which applies also for $\eta = \omega_0$).

THEOREM 9.4.1 (Glass-McCleary). *The free ℓ-group F_η ($\eta > 1$) has a (faithful) pathologically 2-transitive representation (on $\mathbf{Q}$, if $\eta \leq \omega_0$).*

Proof for $1 < \eta \leq \omega_0$. We shall represent F_η on $\mathbf{Q}$. For each $x \in x$, the action of its image $\hat{x}$ on $\mathbf{Q}$ will be specified at enough points to guarantee the desired results.

F_η is countable, and we enumerate its non-identity elements: $w_0, w_1, \ldots$. In the rational interval $[0,1]$, we lay out a copy of a diagram for $w_0 = \bigvee_i \bigwedge_j \prod_k x_{ijk}^{\pm 1}$ showing $e \neq w_0$, with the smallest point r_0 of the diagram taken to be 0, and the largest point t_0 taken to be 1. We specify about the $\hat{x}$'s that the point (corresponding to) $x_{ij1}^{\pm 1} \ldots x_{i,j,k-1}^{\pm 1}$ be sent by $\hat{x}_{ijk}^{\pm 1}$ to the point (corresponding to) $x_{ij1}^{\pm 1} \ldots x_{i,j,k-1}^{\pm 1} x_{ijk}^{\pm 1}$. Similarly, in each interval $[2n, 2n+1]$, we lay out such a diagram for w_n and make such specifications.

This is already enough to give a faithful action of F_η on $\mathbf{Q}$. Next we assure that all the points in the various diagrams lie in the same orbit of $\bar{F}_\eta$. For this it suffices to arrange that for each $n = 0, 1, \ldots$, the points $2n+1$ and $2(n+1)$ lie in the same orbit, and now we construct the appropriate "bridges".

We begin with the interval [1,2]. In the original diagram for w_0, t_0 ($\leftrightarrow 1$) must have been moved by at least one free generator, say x_{t_0}; and in the diagram for w_1, r_1 ($\leftrightarrow 2$) must have been moved by some free generator x_{r_1}. We decree that

(a_1) $1\hat{x}_{t_0} = \frac{4}{3}$ if t_0 was moved up by x_{t_0},

(a_2) $1\hat{x}_{t_0}^{-1} = \frac{4}{3}$ if t_0 was moved down by x_{t_0},

(b_1) $\frac{5}{3}\hat{x}_{r_1} = 2$ if r_1 was moved up by x_{r_1},

(b_2) $\frac{5}{3}\hat{x}_{r_1}^{-1} = 2$ if r_1 was moved down by x_{r_1}.

To connect $\frac{4}{3}$ and $\frac{5}{3}$, we further decree that $\frac{4}{3}\hat{x}_{t_0} = \frac{5}{3}$ if (a_1) obtains (or that $\frac{4}{3}\hat{x}_{t_0}^{-1} = \frac{5}{3}$ if (a_2) obtains); except that this may conflict with (b_2) or (b_1) if $x_{t_0} = x_{r_1}$, so in that case we pick any other $x \in x$ ($\eta > 1$) and decree that $\frac{4}{3}\hat{x} = \frac{5}{3}$. We build similar bridges in the other intervals $[2n+1,2n+2]$, $n = 1,2,\dots$.

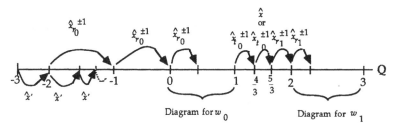

Diagram for w_0 Diagram for w_1

We decree also that $(-2)\hat{x}_{r_0}^{\pm 1} = -1$ and $(-1)\hat{x}_{r_0}^{\pm 1} = 0$; and picking any $x' \neq \hat{x}_{r_0}$, that $-(n+1)\hat{x}' = -n$, $n = 2,3,\dots$, and that $(-1-\frac{1}{n})\hat{x}' = -1-\frac{1}{n+1}$, $n = 1,2,\dots$. Now all diagram points, and all negative (as well as positive) integers, and all points of the form $-1-\frac{1}{n}$ do indeed lie in some one orbit Ω' of $\hat{F}_\eta$, and the action of $\hat{F}_\eta$ on Ω' is faithful. Because of $\hat{x}'$, the stabilizer subgroup of -1 has a (convex) orbit which includes all points in Ω' that are less than -1, forcing the action of $\hat{F}_\eta$ on Ω' to be 2-transitive. For each $e \neq w \in F_\eta$, $\text{supp}(w)$ is unbounded because the powers of w occur arbitrarily far out in the enumeration. Since Ω' is countable, dense in itself, and lacks endpoints, Ω' must be isomorphic to $\mathbf{Q}$.

Now we use Theorem 9.4.1 to answer Questions 2 (again), 4, 5, and 6. First, F_η ($\eta > 1$) has trivial center because this is true of every 2-transitive ℓ-permutation group (F, Ω). For let $e \neq f \in F$. Then $\alpha \neq \alpha f$ for some $\alpha \in \Omega$, and we pick $g \in F$ such that $\alpha g = \alpha$ but $(\alpha f)g \neq \alpha f$. Then $\alpha(gf) = \alpha f \neq \alpha(fg)$.

COROLLARY 9.4.2 [McCleary, 1985a]. F_η ($\eta > 1$) is finitely subdirectly irreducible (as an ℓ-group).

Proof. No transitive ℓ-permutation group can have two non-trivial ℓ-ideals whose intersection is trivial: If H and K are ℓ-ideals, and if $e < h \in H$ and $e < k \in K$, then some conjugate $f^{-1}kf$ of k fails to be disjoint from h, and $e < h \wedge f^{-1}kf \in H \cap K$.

Recently A. M. W. Glass and J. S. Wilson have sharpened this result by observing that every 2-transitive ℓ-permutation group is finitely subdirectly irreducible as a group. For 2-transitivity implies 4-transitivity (Theorem 2.3.1), so that non-trivial normal subgroups must be 2-transitive and thus no two of them can centralize each other (see the above proof that 2-transitive ℓ-permutation groups have trivial center.)

COROLLARY 9.4.3 [Glass and Wilson, a]. F_η ($\eta > 1$) is finitely subdirectly irreducible as a group.

COROLLARY 9.4.4 [McCleary, 1985a]. F_η ($\eta > 1$) has no basic elements.

Proof. No 2-transitive ℓ-permutation group (F, Ω) has basic elements. For let $e < f \in F$, and pick $\alpha \in \Omega$ such that $\alpha < \alpha f$. Since Ω is dense in itself, we may pick $\beta_1, \beta_2 \in \Omega$ such that $\alpha < \beta_1 < \beta_2 < \alpha f$. Now pick $g_1, g_2 \in G$ such that $\beta_1 g_1 = \beta_1$ but $\beta_2 g_1 > \beta_2$, and $\beta_2 g_2 = \beta_2$ but $\beta_1 g_2 > \beta_1$. Let $k_i = (g_i \wedge f) \vee e$, so that $e < k_i < f$. Then k_1 and k_2 are incomparable, so f is not basic.

COROLLARY 9.4.5 [McCleary, 1985a]. *The free group G_η ($\eta > 1$) has no finite subset S for which there is a unique right ordering of G_η making all elements of S positive.*

Proof. Proposition 9.3.12 and Corollary 9.4.4.

COROLLARY 9.4.6 [McCleary, 1985a]. F_η ($\eta > 1$) is not completely distributive.

Proof. No pathologically 2-transitive ℓ-permutation group is completely distributive [McCleary, 1973b, Theorem 1].

COROLLARY 9.4.7 [Kopytov,1983]. *The lattice order on F_η ($1 \le \eta \le \omega_0$) can be extended to a total order (respected by the group operation).*

Proof. In the proof of Theorem 9.4.1, it can be arranged that the graph of every $\hat{x}$ ($x \in x$) and thus of every $\hat{w}$ ($w \in F_\eta$) consist of a collection of linear pieces indexed by the natural numbers (with the domain of the $n+1^{st}$ piece to the right of the domain of the nth piece). To extend the lattice order, decree that $w > e$ iff the first piece of $\hat{w}$ not lying along the diagonal has slope $m > 1$.

COROLLARY 9.4.8 [Kopytov, 1983]. *F_η (any η) has unique root extraction.*

5. The Best-of-Both-Worlds Theorem

A cardinal number η is *regular* if, as an initial ordinal number, it has no cofinal subset of cardinality less than η. For the definition of α-*set* as used in the following theorem, see [Glass 1981b, page 187]. For regular cardinals $\eta > \omega_0$, our results require the Generalized Continuum Hypothesis.

THEOREM 9.5.1. *Let $1 < \eta \le \omega_0$, or (with G.C.H.) let η be regular. Then there exists a right ordering ($G_\eta, \le$) on which the natural action of F_η is both faithful and pathologically 2-transitive. ($G_\eta, \le$) must be isomorphic (as an ordered set) to $\mathbf{Q}$ if $\eta \le \omega_0$, and may be taken as an α-set ($\eta = \omega_\alpha$) if η is regular.*

Proof for $1 < \eta \le \omega_0$. We shall modify the proof of Theorem 9.4.1 so that no $e \ne \hat{g} \in \hat{G}_\eta$ fixes any point. Having $\beta\hat{g} = \beta$ (g a reduced group word $\ne e$) would produce a *loop*, i.e., a non-empty sequence of $x^{\pm 1}$-arrows (reduced in the obvious sense) such that the head of the last coincides with the tail of the first. We want our specifications to be loop-free.

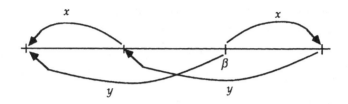

A loop for $g = xyxy^{-1}$

Once this is accomplished, the rest will be easy: Right order G_η by setting

$$g_1 \leq g_2 \Leftrightarrow 0\mathring{g}_1 \leq 0\mathring{g}_2,$$

and obviously the natural action of F_η on $(G_\eta, \leq)$ will coincide with the representation $(\mathring{F}_\eta, \Omega)$.

We retain the specifications made in the first part of the proof of Theorem 9.4.1, through the building of all the bridges, and including the specification that $(-1)\mathring{x}_{r_0}^{\pm 1} = 0$, but we change the specifications which guarantee 2-transitivity. For convenience, we reflect the figure used in the proof of Theorem 9.4.1 about the origin by negating all the numbers involved in the specifications, and then we translate the figure one unit to the left. Now $\mathring{x}_{r_0}^{\pm 1}$ moves -1 down. We denote $x_{r_0}^{\pm 1}$ by y, and specify that $0\mathring{y} = -1$.

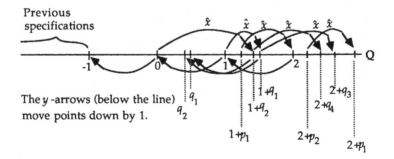

Next we make further specifications as shown in the positive half of the line in the above figure. Let $q_1, q_2, \ldots$ be a strictly decreasing sequence of rationals (all less than 1), with $q_n \downarrow 0$. Let $p_n = (q_{2n} + q_{2n+1})/2$; we have $p_n < q_{2n} < q_{2n+1} < p_{n-1}$. We pick any one $x \in x$ other than y, and specify that:

(c) $1\hat{x} = 1 + q_1$,

(d) $0\hat{x} = 1 + q_2$,

(e) $(n + q_{2n-1})\hat{x} = n + 1 + q_{2n+1}$, $n \geq 1$,

(f) $(n + q_{2n})\hat{x} = n + 1 + q_{2n+2}$, $n \geq 1$,

(g) $(1 + p_1)\hat{x} = 2$,

(h) $(n + p_n)\hat{x} = n + p_{n-1}$, $n \geq 2$,

(i) $\alpha\hat{y} = \alpha - 1$ for all $\alpha \geq 1$ which are integers, or which differ by an integer from some q_n or p_n.

Regardless of what further specifications we make, the resulting (transitive) action of F_η on the orbit $\Omega = 0\hat{F}_\eta$ will be faithful. We claim that $(\hat{F}_\eta, \Omega)$ will be 2-transitive. For specifications (c)-(f) and (i) make

$$0\hat{x}^n\hat{y}^n = q_{2n} \text{ and } 1\hat{x}^n\hat{y}^n = q_{2n-1},$$

so that

$$0(\hat{x}^n\hat{y}^n \wedge e) = 0 \text{ and } 1(\hat{x}^n\hat{y}^n \wedge e) = q_{2n-1}.$$

Thus the elements $\hat{x}^n\hat{y}^n \wedge e$ fix 0 and move 1 down arbitrarily close to 0. Moreover, specifications (g)-(i) make

$$0\hat{y}^{-1}\hat{x}^{-1} < 0 \text{ and } 1(\hat{y}^{-1}\hat{x}^{-1})^n = n + p_n,$$

so that

$$0((\hat{y}^{-1}\hat{x}^{-1})^n \vee e) = 0 \text{ and } 1((\hat{y}^{-1}\hat{x}^{-1})^n \vee e) = n + p_n.$$

Thus the elements $(\hat{y}^{-1}\hat{x}^{-1})^n \vee e$ fix 0 and move 1 arbitrarily far to the right. Therefore the representation of F_η on Ω will be 2-transitive (since the orbit containing 1 of the stabilizer subgroup $(\hat{F}_\eta)_0$, being convex, must be $(0, +\infty)$).[1]

For $e \neq w \in F_\eta$, supp(w) is unbounded below because the powers of w occur arbitrarily far out in the enumeration.

So far we have created no loops: Diagrams are loop-free, and no loops are created by (c)-(i) because the points in $\{q_1, q_2, \ldots\} \cup \{p_1, p_2, \ldots\}$ are distinct. To finish specifying the free generators without accidently forming any loops, we enumerate the set of ordered pairs (α, z), where $\alpha \in Q$ and z is a free generator or the inverse of one. Proceeding inductively, we specify a z-

---

[1]The author wishes to thank Mike Darnel for improving an earlier proof by adding conditions (g) and (h).

arrow with tail at α (unless one has already been specified, either earlier in the induction or prior to the induction).

The only limit points (rational or irrational) of the specified points are the non-negative integers; and if the y-arrows are deleted, there are no limit points at all. Thus of the tails of z-arrows below α, let β_L be the largest ($\beta_L = -\infty$ if there are no such tails); and of the tails of z-arrows above α, let β_U be the smallest ($\beta_U = +\infty$ if there are no such tails). This makes sense unless $z = y$ and $1 \le \alpha \in \mathbf{Z}$ or $z = y^{-1}$ and $0 \le \alpha \in \mathbf{Z}$, and in these cases a z-arrow with tail at α has already been specified. As the head of the z-arrow with tail at α, we choose any point $\delta \ne \alpha$ which has not previously been specified (as an end of any arrow at all), and which is greater than the head β'_L of the z-arrow with tail at β_L (no restriction here if $\beta_L = -\infty$) and less than the head β'_U of the z-arrow with tail at β_U. The above remark about limit points guarantees the existence of such a δ, and the choice of δ preserves consistency of z-arrows.

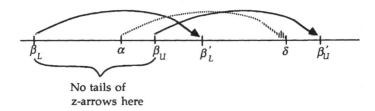

No tails of
z-arrows here

For each $z \in x$, this construction specifies an automorphism $\hat{z}$ of $\mathbf{Q}$ (onto $\mathbf{Q}$ because of the inclusion of z^{-1} in the induction). There were no loops before we began the induction, and the induction cannot produce a loop because the last arrow, which completed the loop, would have had as its head an already specified point, contrary to the above choice of δ. Restricting to $\Omega = 0\hat{F}_\eta$, we complete the construction. Of course, in any 2-transitive representation $(\hat{F}_\eta, \Omega)$, Ω must be countable and dense in itself and without endpoints, and thus isomorphic to $\mathbf{Q}$.

COROLLARY 9.5.2 [McCleary, 1985b]. *The free group G_η $(1 < \eta \le \omega_0)$ can be right ordered so as to have no proper convex subgroups.*

Proof. In the right ordering of Theorem 9.5.1, a proper convex subgroup would be a proper o-block (Section 2.3) for $\hat{G}_\eta$ and thus also for $\hat{F}_\eta$.

6. The root system of prime subgroups of F_η

For the representation of F_η given above, it can be shown [McCleary, 1985b] that the stabilizer subgroups of F_η are both maximal and minimal prime subgroups of F_η. The representation can be varied so as to provide the following detailed description of the root system $\mathcal{P}_\eta$ of prime subgroups of F_η (for proofs see [McCleary, 1985b]). Here we include $\eta = 1$ and $\eta = 0$. $F_1 = \mathbf{Z} \boxplus \mathbf{Z}$, so $\mathcal{P}_1$ has three elements, two minimal and one lying above them. $F_0 = \{e\}$, so $\mathcal{P}_0$ is a singleton.

$\mathcal{P}_\eta$ has F_η as its largest element. Its *branches* are the connected components of $\mathcal{P}_\eta \setminus \{F_\eta\}$. For $P \in \mathcal{P}_\eta$, $\mathcal{L}(P)$ will denote $\{Q \in \mathcal{P}_\eta \mid Q \leq P\}$.

The *roots* of $\mathcal{P}_\eta$ are the maximal subchains. Within each root the set of covering pairs is dense, and the bottom halves of these covering pairs are the *values* within that root. Given that for finite η every branch of $\mathcal{P}_\eta$ has a largest element [McCleary, 1985a, Corollary 16], part (5) of the following theorem says that every conceivable chain occurs as the chain of covering pairs (equivalently, of values) within some root of $\mathcal{P}_\eta$.

THEOREM 9.6.1 [McCleary, 1985b]. *For finite $\eta > 1$, all the root systems $\mathcal{P}_\eta$ are isomorphic to each other. For this common root system $\mathcal{P}_f$:*

(1) $Card(\mathcal{P}_f) = 2^{\omega_0}$.

(2) *Each branch contains a (unique) largest element.*

(3) *Every branch is isomorphic to $\mathcal{P}_0, \mathcal{P}_1, \mathcal{P}_f,$ or $\mathcal{P}_{\omega_0}$ (no two of which are isomorphic); and there are 2^{ω_0} branches of each type.*

(4) *For $P \in \mathcal{P}_f$, $\mathcal{L}(P)$ is isomorphic to $\mathcal{P}_0, \mathcal{P}_1, \mathcal{P}_f,$ or $\mathcal{P}_{\omega_0}$.*

(5) *The chains of covering pairs in the roots of $\mathcal{P}_f$ are precisely the finite and countable chains having largest elements.*

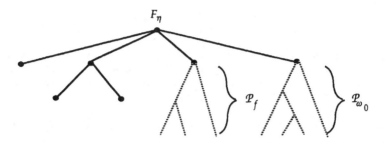

The branches of $\mathcal{P}_f$, each of which occurs 2^{ω_0} times.

THEOREM 9.6.2 [McCleary, 1985b]. *For* $\mathcal{P}_{\omega_0}$:

(1) $Card(\mathcal{P}_{\omega_0}) = 2^{\omega_0}$.

(2) *The isomorphism types of the branches are* $\mathcal{P}_0, \mathcal{P}_1, \mathcal{P}_f, \mathcal{P}_{\omega_0}$, *and* 2^{ω_0} *types (each of cardinality* 2^{ω_0}) *having no largest element. Each of these types occurs* 2^{ω_0} *times.*

(3) *For each* $P \in \mathcal{P}_{\omega_0}$, $L(P)$ *is isomorphic to* $\mathcal{P}_0$, $\mathcal{P}_1$, $\mathcal{P}_f$, *or* $\mathcal{P}_{\omega_0}$.

(4) *The chains of covering pairs in the roots of* $\mathcal{P}_{\omega_0}$ *are precisely the non-empty finite and countable chains.*

THEOREM 9.6.3 [McCleary, 1985b]. *Let* $\eta > \omega_0$ *be regular, and assume the Generalized Continuum Hypothesis.*[1]

(1) $Card(\mathcal{P}_\eta) = 2^\eta$.

(2) *The isomorphism types of the branches of* $\mathcal{P}_\eta$ *are*

    (a) $\mathcal{P}_0$, $\mathcal{P}_1$, $\mathcal{P}_f$, *and* $\mathcal{P}_\mu$ ($\omega_0 \le \mu \le \eta$),

    (b) 2^η *types of cardinality* 2^η *and cofinality* μ *(for each regular* μ *such that* $\omega_0 \le \mu \le \eta$).

    (c) *Perhaps some types of cardinality* η *and cofinality* η.

Each type occurs 2^η *times, except perhaps for the types in (b) with* $\mu < \eta$, *and even these occur at least* η *times.*

(3) *For each* $P \in \mathcal{P}_\eta$, $L(P)$ *is isomorphic to* $\mathcal{P}_0$, $\mathcal{P}_1$, $\mathcal{P}_f$, *or* $\mathcal{P}_\mu$ ($\omega_0 \le \mu \le \eta$).

---

[1]This theorem holds even without the GCH, except that in (2) no guarantees are offered about occurrences of $\mathcal{P}_0, \mathcal{P}_1, \mathcal{P}_f$ or $\mathcal{P}_\mu$ ($\omega_0 \le \mu < \eta$), though $\mathcal{P}_\mu$ is still guaranteed to occur 2^η times, and in (2c) the cardinality of a type is guaranteed merely to be at most 2^η. The author wishes to thank Manfred Droste for these observations.

(4) *Every chain of cardinality at most η occurs as an upper ray of the chain of covering pairs of some root of $\mathcal{P}_\eta$.*

7. Three open questions

For finite η, the free group G_η is Hopfian, i.e., every homomorphism from G_η onto itself is one-to-one. (There *are* monomorphisms from G_η into itself which are not onto, and similarly for F_η.)

QUESTION. *For finite η, is F_η Hopfian? I.e., is every ℓ-homomorphism from F_η onto itself one-to-one?*

For $\eta > 1$, F_η is non-abelian and thus non-Archimedean, so there must exist positive elements $w_1 \ll w_2$ (i.e., $w_1^n < w_2$ for all $n \in \mathbf{Z}^+$).

PROBLEM. *Give an example in F_η of $w_1 \ll w_2$.*

THE CONJUGACY PROBLEM. *Is it decideable whether or not two elements of F_η are conjugate?*

S. H. McCleary
University of Georgia
Athens, Georgia 30602
U. S. A

Norman R. Reilly

CHAPTER 10

VARIETIES OF LATTICE-ORDERED GROUPS

10.1. Generalities

We begin by recalling a fundamental theorem due to Birkhoff as shown in [Burris and Sankappanavar, 1981; Theorem II.11.9].

THEOREM 10.1.1. *For a class $\mathcal{K}$ of algebras of the same type, the following statements are equivalent.*

(1) *$\mathcal{K}$ is closed under the formation of products, homomorphic images and subalgebras.*

(2) *$\mathcal{K}$ is equationally defined (that is, for some family of equations or identities Σ, $\mathcal{K}$ is the class of all algebras of the specified type which satisfy all equations in Σ).*

A class of algebras which satisfies the equivalent conditions of Theorem 10.1.1 is called a *variety* or *equational class*. This chapter is devoted to the study of varieties of ℓ-groups, and the term "variety" will always mean "variety of ℓ-groups".

Since an equation involves only a finite number of variables, it is important to note that an ℓ-group G belongs to a variety $\mathcal{U}$ if and only if every finitely generated ℓ-subgroup of G lies in $\mathcal{U}$. If $w = w(x_1,...,x_n)$ is a word in the free ℓ-group on a set X with $x_i \in X$, we will, when

228

A. M. W. Glass and W. C. Holland (eds.), Lattice-Ordered Groups, 228–277.
© 1989 by Kluwer Academic Publishers.

convenient, write $w = w(x)$, where x represents the vector of variables $(x_1,...,x_n)$. Likewise, if G is an ℓ-group and $g_1,...,g_n \in G$ then we will write $w(g)$ or $w(g_1,...,g_n)$ for the element of G obtained by substituting the elements g_i for the x_i's in $w(x)$.

We will denote the variety of all ℓ-groups by L and the variety of trivial ℓ-groups by $\mathcal{E}$.

Varieties of ℓ-groups have some useful basic properties, the first of which is the following:

LEMMA 10.1.2. *If the variety $\mathcal{V}$ is finitely based, that is, if $\mathcal{V}$ is defined by a finite set of equations, then it is defined by a single equation.*

Proof: Let $\mathcal{V}$ be the class of all ℓ-groups satisfying the equations $u_i = v_i$ $(i = 1,...,n)$. The equation $u_i = v_i$ is clearly equivalent to the equation $u_i v_i^{-1} = e$. If we ensure that the variables in $u_i v_i^{-1}$ are distinct from those in $u_j v_j^{-1}$, for all $i \neq j$, then it is clear that the given equations hold in an ℓ-group G if and ony if the single equation $u_1 v_1^{-1} u_2 v_2^{-1}... u_n v_n^{-1} = e$ holds in G .

Another very useful observation about identities in ℓ-groups is that inequalities that hold universally in an ℓ-group are equivalent to identities. Conversely, certain identities are equivalent to inequalities. Suppose that, for $u(x)$, $v(x)$ in the free ℓ-group, we say that an ℓ-group G satisfies the inequality $u(x) \leq v(x)$ if and only if $u(g) \leq v(g)$ for all $g = (g_1,...,g_n)$ with $g_i \in G$. Then clearly G satisfies the inequality $u \leq v$ if and only if it satisfies the identity $u \vee v = v$. In like manner, the statement

$$u(x_1,...,x_i,...,x_j,...,x_n) = v(x_1,...,x_i,...,x_j,...,x_n) \quad \text{for } x_i \leq x_j$$

is equivalent to the identity

$$u(x_1,...,x_i,...,x_i \vee x_j,...,x_n) = v(x_1,...,x_i,...,x_i \vee x_j,...,x_n)$$

while

$$u(x_1,...,x_i,...,x_n) = v (x_1,...,x_i,...,x_n) \quad \text{for } e \leq x_i$$

is equivalent to the identity

$$u(x_1,...,e \vee x_i,...,x_n) = v (x_1,...,e \vee x_i,...,x_n) .$$

Consequently, we will use the inequalities between words and variables freely when defining varieties by 'identities'.

The set L of all varieties of ℓ-groups is a lattice where, for any varieties $\mathcal{U},\mathcal{V}$, the greatest lower bound, or meet, of $\mathcal{U}$ and $\mathcal{V}$ is their set intersection

$$\mathcal{U} \wedge \mathcal{V} = \mathcal{U} \cap \mathcal{V}$$

and their least upper bound, or join, is the intersection of all varieties containing both $\mathcal{U}$ and $\mathcal{V}$

$$\mathcal{U} \vee \mathcal{V} = \bigcap \{\mathcal{W} : \mathcal{U} \cup \mathcal{V} \subseteq \mathcal{W}\}.$$

For any variety $\mathcal{V}$, we will denote by $\mathcal{L}(\mathcal{V})$ the lattice of subvarieties of $\mathcal{V}$. For any class $\mathcal{K}$ of ℓ-groups let $H(\mathcal{K})$, $S(\mathcal{K})$ and $P(\mathcal{K})$ denote the classes of ℓ-groups which are homomorphic images, ℓ-subgroups or products, respectively, of elements of $\mathcal{K}$. Also, recall that an ℓ-group G is said to be *subdirectly irreducible* if, whenever G is a subdirect product of ℓ-groups G_i the projection of G onto G_i is an isomorphism for some i. Equivalently, G is subdirectly irreducible if and only if it has a smallest non-trivial ℓ-ideal. Then, from standard results in universal algebra (see [Burris and Sankappanavar, 1981; Theorems II.8.6, II.9.5 & Corollary II.9.7]) we have the following useful observations.

LEMMA 10.1.3.

(i) *Every ℓ-group is a subdirect product of subdirectly irreducible ℓ-groups.*

(ii) *If $\mathcal{K}$ is any class of ℓ-groups then the variety generated by $\mathcal{K}$ (that is, the smallest variety containing $\mathcal{K}$) is $HSP(\mathcal{K})$.*

(iii) *If $\mathcal{K}$ is the class of subdirectly irreducible elements of the variety $\mathcal{V}$, then $\mathcal{V} = HSP(\mathcal{K})$.*

(iv) *If $\mathcal{U}$ and $\mathcal{V}$ are varieties of ℓ-groups, then $\mathcal{U} \vee \mathcal{V} = HSP(\mathcal{U} \cup \mathcal{V})$.*

If $\mathcal{K} = \{G\}$, for some ℓ-group G, then we write $\mathcal{V}(G)$ for $HSP(\mathcal{K})$, that is, for the *variety generated* by G.

It is usually quite a difficult problem to characterize more precisely the join of two varieties. However, Martinez showed that varieties of ℓ-groups have an additional nice property in this regard.

LEMMA 10.1.4. [Martinez, 1974b]. *Let $\mathcal{U}, \mathcal{V}$ be varieties of ℓ-groups. Then $G \in \mathcal{U} \vee \mathcal{V}$ if and only if there exist ℓ-ideals M and N in G such that* (i) $M \cap N = \{e\}$ *and* (ii) $G/M \in \mathcal{U}$, $G/N \in \mathcal{V}$.

Proof: If there exist ℓ-ideals satisfying (i) and (ii), then it is clear that $G \in \mathcal{U} \vee \mathcal{V}$. So suppose that $G \in \mathcal{U} \vee \mathcal{V}$. By Lemma 10.1.3(ii), there exists an ℓ-group H with ℓ-ideals A, B such that $H/A \in \mathcal{U}$, $H/B \in \mathcal{V}$, $A \cap B = \{e\}$ and an ℓ-ideal C with $H/C \cong G$. Since the lattice of ℓ-ideals of H is distributive, Corollary 1.1.4, $AC \cap BC = (A \cap B)C = C$ so that if $M = AC/C$ and $N = BC/C$, then $M \cap N = C$, $G/M \cong H/AC \in \mathcal{U}$ and $G/N \cong H/BC \in \mathcal{V}$.

COROLLARY 10.1.5. *Let* $\mathcal{U}$ *and* $\mathcal{V}$ *be varieties of ℓ-groups and* $G \in \mathcal{U} \wedge \mathcal{V}$. *If* G *is either totally ordered or subdirectly irreducible, then* $G \in \mathcal{U} \cup \mathcal{V}$.

Proof. If G is subdirectly irreducible, then the result follows immediately from Lemma 10.1.4, so suppose that G is totally ordered and that M and N are ℓ-ideals in G satisfying (i) and (ii) of the conclusion in Lemma 10.1.3. Since G is totally ordered, M and N are comparable, say $M \subseteq N$. By (i) of Lemma 10.1.4, $M = \{e\}$ and so, by (ii) of Lemma 10.1.4, $G \in \mathcal{U}$. If $N \subseteq M$, then $G \in \mathcal{V}$.

The key property of ℓ-groups that makes the proof of Lemma 10.1.4 possible is the distributivity of the lattice of convex ℓ-subgroups and the same result holds in any congruence distributive variety if either G is subdirectly irreducible or the lattice of congruences on G is a chain.

Since ℓ-groups are also groups, congruences on an ℓ-group are also *permutable* (that is, for any ℓ-ideals M and N in an ℓ-group, $MN = NM$). A congruence distributive variety in which congruences are also permutable is called *arithmetical*. It is well known ([Burris and Sankappanavar, 1981; Theorem II.12.5]) that the property of being arithmetical for a variety of algebras is characterized by the existence of a Mal'cev term $w(x,y,z)$ with the property that

$$w(x,y,x) = w(x,y,y) = w(y,y,x) = x.$$

Such a term for the variety of ℓ-groups is

$$w(x,y,z) = x((x \vee y) \wedge (x \vee z) \wedge (y \vee z))^{-1}z.$$

Throughout, we will denote by X a countably infinite set and by F_X the free ℓ-group on X. Since F_X is a distributive lattice, every element w in F_X can be written in the form $w = \vee_i \wedge_j w_{ij}$ where each w_{ij} is a term formed from the generators using only group operations. We will denote by $FI(F_X)$ the lattice of fully invariant congruences on F_X. Standard arguments from universal algebra as shown in [Burris and Sankappanavar, 1981; Corollary II.14.10], show that L is anti-isomorphic to $FI(F_X)$. So, by [Bigard, Keimel, and Wolfenstein, 1977; Proposition 2.2.9], we have

THEOREM 10.1.6. L *is distributive. Moreover, for any varieties* $\mathcal{U}, \mathcal{V}_i$ ($i \in I$),

$$\mathcal{U} \vee (\bigwedge_{i \in I} \mathcal{V}_i) = \bigwedge_{i \in I} (\mathcal{U} \vee \mathcal{V}_i).$$

However, it has been shown by Kopytov and Medvedev [1975] and Smith [1980] that L is not Brouwerian, that is, it is not generally true that $\mathcal{U} \cap (\bigvee_{i \in I} \mathcal{V}_i) = \bigwedge_{i \in I}(\mathcal{U} \vee \mathcal{V}_i)$ (see also Section 10.13).

In addition to forming joins and meets of varieties, there is a third way of combining two varieties to form a third variety. For any varieties $\mathcal{U}, \mathcal{V}$ of lattice-ordered groups, let their *product* be

$\mathcal{U}\mathcal{V} = \{G \in L: G$ has an ℓ-ideal M such that $M \in \mathcal{U}$ and $G/M \in \mathcal{V}\}$.

It is easily verified that $\mathcal{U}\mathcal{V}$ is a variety. We define $\mathcal{U}^{n+1} = \mathcal{U}^n \mathcal{U}$ for any integer $n > 1$. Properties of this product will be discussed in detail in Section 10.9.

10.2. Abelian ℓ-groups

Let G be any non-trivial ℓ-group and let $a \in G$, $a \neq e$. Then either $a^+ \neq e$ or $a^- \neq e$. In either case, G contains a strictly positive element g, say, and $\langle g \rangle$ is then an ℓ-subgroup of G which is ℓ-isomorphic to $\mathbf{Z}$. On the other hand

LEMMA 10.2.1. [Weinberg, 1963]. *Every abelian ℓ-group is in the variety generated by* $\mathbf{Z}$.

Proof. The proof is taken from [Holland, Mekler and Reilly, 1986] and illustrates the usefulness of ultraproducts. Every abelian ℓ-group is a subdirect product of totally ordered abelian groups and therefore it suffices to show that $\mathcal{V}(\mathbf{Z})$ contains every finitely generated totally ordered abelian group. Let G be such a group. Since the divisible hull of G is a finite dimensional vector space, it follows that G has only finitely many convex subgroups. It follows from Hahn's theorem (Theorem 1.2.4 (o-group case)) that G can be embedded in a lexicographically ordered group $\Pi_{i<n}R_i$, where each R_i is a finitely generated additive ordered subgroup of the real numbers. It will suffice to prove the lemma for the full group $\Pi_{i<n}R_i$. Let the generators of R_i be $\{\alpha(i,j)\}$. By applying obvious automorphisms and exchanges, we may assume that $\alpha(i,0) = 1$ and $\alpha(i,j) > 0$. For each (i,j) we can choose a sequence of integers $(k(i,j,l)) \in \omega$ such that

(1) if $i < i'$, $\lim_{l \to \infty} k(i',j',l)/k(i,j,l) = \infty$ and

(2) $\lim_{l \to \infty} k(i,j,l)/k(i,0,l) = \alpha(i,j,)$.

Let $\mathcal{U}$ be a non-principal ultrafilter on ω. Then it is easy to see that the subgroup of Z^{ω}/U generated by the sequences $(k(i,j,l))(\mathrm{mod}\ U)$ is isomorphic as an ordered group to ΠR_i, where $\alpha(i,j) \longleftrightarrow (k(i,j,l))(\mathrm{mod}\ U)$.

THEOREM 10.2.2 [Weinberg, 1963]. *The smallest non-trivial variety of ℓ-groups is the variety $\mathcal{A}$ of all abelian ℓ-groups, that is, the variety defined by the single equation $xy = yx$.*

Proof. This is immediate from Lemma 10.2.1 and the remarks preceding it.

In looking beyond abelian ℓ-groups for other important varieties of ℓ-groups, it is natural to consider totally ordered groups. It is to be expected that the variety of ℓ-groups generated by the class of totally ordered groups will figure prominently in the study of the lattice of varieties of ℓ-groups. Of course the class of totally ordered groups is not a variety, since it is not closed with respect to products.

DEFINITION. We will say that an ℓ-group is *representable* if it is a subdirect product of totally ordered groups and denote the class of all representable ℓ-groups by $\mathcal{R}$. We recall Theorem 1.2.1:

THEOREM 10.2.3. *Let G be an ℓ-group. Then the following are equivalent:*
(i) G *is representable;*
(ii) *every polar of G is normal;*
(iii) *every minimal prime subgroup of G is normal;*
(iv) $a,b,c \in G$, $a \wedge b = e \Rightarrow a \wedge c^{-1}bc = e$;
(v) $x^+ \wedge y^{-1}(x^-)y = e$, *for all* $x,y \in G$;
(vi) $(x \wedge y)^2 = x^2 \wedge y^2$, *for all* $x,y \in G$.

It follows from Theorem 10.2.3 (v) or (vi) that the class $\mathcal{R}$ of representable ℓ-groups is a variety. Since $\mathcal{R}$ contains all totally ordered groups and must clearly be contained in any variety containing all totally ordered groups, $\mathcal{R}$ is, indeed, the variety generated by the class of all totally ordered groups.

Recently, Medvedev [1984] has shown that, like the lattice of all varieties of ℓ-groups, the lattice of subvarieties of $\mathcal{R}$ is not Brouwerian.

10.3. Mimicking

In this section we discuss a subtle but tremendously powerful tool in the study of varieties of ℓ-groups.

DEFINITION 10.3.1. We say that an ℓ-permutation group (G,Ω) *mimics* an ℓ-permutation group (H,Λ) (Glass, Holland and McCleary [1980]), if and only if $G \in \mathcal{V}$ and *whenever*

    (i) $\lambda \in \Lambda$

and

    (ii) $\{w_r(x)\}$ is a finite set of words in the variables $x_1,...,x_n$

and

    (iii) $h_1,...,h_n \in H$,

then

    (iv) there exist $\alpha \in \Omega$ and $g_1,...,g_n \in G$

such that

    (v) $\lambda w_i(h) < \lambda w_j(h)$ if and only if $\alpha w_i(g) < \alpha w_j(g)$.

We say that (G,Ω) *mimics a variety* $\mathcal{V}$ if and only if (G,Ω) mimics all transitive ℓ-permutation groups (H,Λ) with $H \in \mathcal{V}$ and that (G,Ω) *mimics an ℓ-group* H if and only if it mimics all ℓ-permutation groups of the form (H,Λ).

It is easy to see that if G and H are abelian o-groups, then the regular representation (G,G) mimics H if and only if it mimics the regular representation (H,H) and that, in order to verify this, the λ and α in Definition 10.3.1 may be chosen to be the identities of their respective groups. When this is done the λ and α essentially disappear from the calculation resulting in a useful simplification of the verification.

If (G,Ω) mimics the variety $\mathcal{V}$, then intuitively, in the notation of the definition, if we sketch the action of the words $w_i(h)$ on λ, then

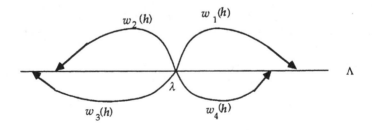

it is possible to copy or 'mimic' this action in (G,Ω)

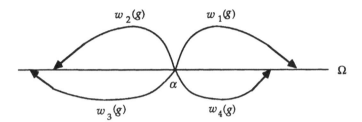

in terms of the relative positions of the points $\lambda w_i(h)$ and the points $\alpha w_i(g)$.

LEMMA 10.3.2. *If (G,Ω) mimics the variety $\mathcal{V}$, then G generates $\mathcal{V}$.*

Proof. Clearly $\mathcal{V}(G) \subseteq \mathcal{V}$. Therefore to establish equality it suffices to show that any equation which does not hold in $\mathcal{V}$ will not hold in G either. Let $w(x) = e$ not hold in $\mathcal{V}$. Then, for some subdirectly irreducible $H \in \mathcal{V}$ and some $h = (h_1,...,h_n) \in H^n$, $w(h) \neq e$. Let (H,Λ) be the Holland representation of H (Theorem 2.2.5). Then for some $\lambda \in \Lambda$, $\lambda w(h) \neq \lambda$. Since (G,Ω) mimics $\mathcal{V}$, there exists an $\alpha \in \Omega$ and $g = (g_1,...,g_n) \in G^n$ such that $\alpha w(g) \neq \alpha$. In particular, $w(g) \neq e$ in G, so that $w(x) = e$ does not hold in G.

The most frequently used example of an ℓ-permutation group which mimics a variety is given by the following (for a discussion of the related notion of universal equivalence of ordered abelian groups see Gurevich and Kokorin [1963]):

EXAMPLE 10.3.3 [Glass, Holland, and McCleary, 1980]. *The regular representation $(\mathbf{Z},\mathbf{Z})$ of the ℓ-group $\mathbf{Z}$ of integers mimics the variety $\mathcal{A}$ of all abelian ℓ-groups.*

Proof. Let (H,Λ) be a non-trivial transitive abelian ℓ-permutation group. Since H is abelian we will use additive notation. By Theorem 2.3.3 this is equivalent to a reprsentation $(H,H/P)$, for some prime subgroup P of H. Since the action of H on Λ is faithful, $P = \{0\}$, H is totally ordered and $(H,\Lambda) = (H,H)$. By Theorem 1.2.4 (o-group case) H can be embedded in a Hahn group $V = V(\Gamma,\mathbf{R})$ so it suffices to show that $(\mathbf{Z},\mathbf{Z})$ mimics V. The definition of mimicking only involves a finite number of elements from V

at a time and so we can assume that Γ is finite, say $|\Gamma| = m$. Then $V = \mathbf{R}^m$ ordered lexicographically from the right.

We first show that $\mathbf{R}$ mimics $\mathbf{R}^m$. Let $\{w_p(x)\}$ be a finite set of words, with $w_p(x) = \vee_i \wedge_j (w_{pij}(x))$ where each $w_{pij}(x)$ is a group word, and let $\{w_p(s)\}$ be a substitution in $\mathbf{R}^m$. Let D be the set of *positive* differences of the form $w_{pij}(s) - w_{qi'j'}(s)$. A typical element of D will be $d = (d_1, d_2, ..., d_m)$.

Let $1/c = \min\limits_{d \in D} \{|d_i| \mid d_i \neq 0\}$. Let $k_1 = 1$, and

for $2 \leq i \leq m$, let $k_i > \max\limits_{d \in D} |d_1 k_1 c + d_2 k_2 c + ... + d_{i-1} k_{i-1} c|$.

Define $\phi: \mathbf{R}^m \to \mathbf{R}$ by

$$(r_1, r_2, ..., r_m)\phi = \sum_{i=1}^{m} r_i k_i c .$$

Then ϕ is a group homomorphism, and ϕ preserves the order of the set $\{(w_p)_{ij}(s)\}$. For if $d = (d_1, d_2, ..., d_m)$ is an element of D with $d_j > 0$, then $d_j k_j c > d_j c | d_1 k_1 c + d_2 k_2 c + ... + d_{j-1} k_{j-1} c | \geq |d_1 k_1 c + ... + d_{j-1} k_{j-1} c|$, which implies that $d\phi > 0$. It follows that $w_p(s) > w_q(s)$ if and only if $w_p(s\phi) > w_q(s\phi)$, proving that $\mathbf{R}$ mimics $\mathbf{R}^m$.

Next we show that $\mathbf{Z}$ mimics $\mathbf{R}$. Let $\{w_p(s)\}$ be a substitution in $\mathbf{R}$. Since the set of s_{ijk} that occur in the various w_p's is a finite subset of $\mathbf{R}$, the subgroup A of $\mathbf{R}$ which it generates is a free abelian group with generators $\alpha_1, \alpha_2, ..., \alpha_n$, say. As before, let D be the set of positive differences $w_{pij}(s) - w_{qi'j'}(s)$. Each $d \in D$ has the form $d = \Sigma f_i(d)\alpha_i > 0$ with each $f_i(d)$ an integer. Thus $(\alpha_1, \alpha_2, ..., \alpha_n)$ is a solution of the system of inequalities $\Sigma f_i(d) z_i > 0$. By continuity, this system must also have a solution in rational numbers and thus must also have a solution in integers $(n_1, ..., n_m)$. That is, $\Sigma f_i(d) n_i > 0$. Let $\psi: A \to \mathbf{Z}$ be the homomorphism $(\Sigma m_i \alpha_i)\psi = \Sigma m_i n_i$. Then $w_p(s) > w_q(s)$ if and only if $w_p(s\psi) > w_q(s\psi)$, and the theorem is proved.

A second example of mimicking, one that will be used in the next section, concerns the variety of all ℓ-groups.

EXAMPLE 10.3.4. *If (G,Ω) is a doubly transitive ℓ-permutation group then (G,Ω) mimics L. In particular, $(A(\mathbf{Q}),\mathbf{Q})$ and $(A(\mathbf{R}),\mathbf{R})$ both mimic L.*

Proof. Let (H,Λ), λ, $\{w_r(x)\}$ and $h_1, ..., h_n$ be as in Definition 10.3.1. Every word $w_r(x)$ can be written in the form $w_r(x) = \vee_I \wedge_J w_{rij}(x)$, where I, J are finite and each w_{rij} is a group word. Let $\{u_t : t = 1, ..., m\}$ be the set of all the initial segments of the words w_{rij} when written in reduced form. Let $\lambda_0 = \lambda$ and $\lambda_t = \lambda u_t(h)$, $t = 1, ..., m$. Since (G,Ω) is doubly transitive, Ω is dense. Hence we can find a set of elements $\{\alpha_0,...,\alpha_m\} \subseteq \Omega$ such that $\lambda_t \to \alpha_t$ $(t = 0,...,m)$ is an order isomorphism of $\{\lambda_0,...,\lambda_m\}$ onto $\{\alpha_0,...,\alpha_m\}$.

Since (G,Ω) is doubly transitive, by Theorem 2.3.1, it is also m^2-transitive. Since Ω is also dense, it means that, for each i, we can find a $g_i \in G$ such that, for $s,t = 0,...,m$,

$$\lambda_s h_i \geq \lambda_t \Leftrightarrow \alpha_s g_i \geq \alpha_t.$$

Thus the g_i acting on $\{\alpha_0,...,\alpha_m\}$ simply copy or mimic the action of the h_i on $\{\lambda_0,...,\lambda_m\}$. Consequently, (v) of Definition 10.3.1 must hold and (G,Ω) mimics $\mathcal{L}$.

EXAMPLE 10.3.5. *If G is the relatively free ℓ-group of countable rank in a variety $\mathcal{V}$ and (G,Ω) is the Holland representation of G, then (G,Ω) mimics $\mathcal{V}$.*

Proof. The verification is straightforward and can be found in [Reilly and Wroblewski, 1981].

A variation on Definition 10.3.1 is obtained by replacing the single ℓ-permutation group (G,Ω) by a family $\{(G_t,\Omega_t): t \in T\}$ of ℓ-permutation groups with each $G_t \in \mathcal{V}$. In place of part (iv) of Definition 10.3.1 we substitute

(iv)* there exist $t \in T$, $\alpha \in \Omega_t$, $g_1,...,g_n \in G_t$ ... .

For the purposes of the general theory, one can work equally well with either definition since, for any family (G_t,Ω_t), $t \in T$, with $G_t \in \mathcal{V}$, there exists a single ℓ-permutation group (G,Ω) with $G \in \mathcal{V}$, with the sets Ω_t as the orbits of transitivity and G_t as the restriction of G to Ω_t. However, when considering specific varieties there are often natural families of ℓ-permutation groups that present themselves as candidates for mimicking families. In the examples which we list below, we write $(G,\mathcal{H}_G)$ for the Holland representation of the ℓ-group G. These examples are due to Huss [1984].

EXAMPLE 10.3.6. *The variety $\mathcal{R}$ of all representable ℓ-groups is mimicked by $\{(F,\mathcal{H}_F): F$ is a totally ordered free group$\}$.*

EXAMPLE 10.3.7. *Any variety $\mathcal{V}$ of representable ℓ-groups is mimicked by $\{(G,\mathcal{H}_G): G \in \mathcal{V}, G$ totally ordered$\}$.*

A variety that will be of some interest to us later in relation to varieties of nilpotent ℓ-groups is the variety $\mathcal{W}a$ of *weakly abelian* ℓ-groups introduced by Martinez [1972] and defined by the 'identity'

$$x^y \le x^2 \quad \text{for all } e \le x.$$

EXAMPLE 10.3.8. $\mathcal{W}a$ *is mimicked by* $\{(F, \mathcal{H}_F): F$ *is a free group with a weakly abelian order*$\}$.

EXAMPLE 10.3.9. *Any variety* $\mathcal{V}$ *of weakly abelian* ℓ*-groups is mimicked by* $\{(G,G): G$ *is a totally ordered member of* $\mathcal{V}$ *and* (G,G) *is the regular representation*$\}$.

Of course it is an easy extension of Lemma 10.3.2 that if the family $\{(G_t, \Omega_t): t \in T\}$ mimics the variety $\mathcal{V}$, then $\{G_t: t \in T\}$ generates $\mathcal{V}$. It might begin to appear that, conversely, if $\{G_t: t \in T\}$ generates $\mathcal{V}$ then $\{(G_t, \Omega_t): t \in T\}$ will mimic $\mathcal{V}$ for any reasonable representation (G_t, Ω_t). However, the next example due to Huss [1984] shows that this is not the case.

EXAMPLE 10.3.10. $\mathcal{W}a$ *is not mimicked by* $\{(F,F): F$ *is a free group with weakly abelian order and* (F,F) *the regular representation*$\}$. *By example 10.3.8,* $\mathcal{W}a$ *is, however, generated by* $\{F: F$ *is a free group with weakly abelian order*$\}$.

10.4. Normal-Valued ℓ-groups

DEFINITION 10.4.1. Recall that an ℓ-group G is said to be *normal valued* (Wolfenstein [1968]; see Section 1.1) if every regular subgroup M is normal in its cover M^*, that is, every value of every element is normal in its cover. We will denote the class of all normal-valued ℓ-groups by $\mathcal{N}$.

Recalling that $G(g)$ denotes the convex ℓ-subgroup of an ℓ-group generated by the element g, we say that an ℓ-permutation group (G, Ω) has *nested intervals of support* if and only if, for all $g, h \in G$, $\alpha \in \Omega$, the intervals $\alpha G(g)$, $\alpha G(h)$ are comparable, i.e., one contains the other.

LEMMA 10.4.2. *If* (G, Ω) *is a 2-transitive* ℓ*-permutation group then it does not have nested intervals of support.*

Proof. Let $\alpha,\beta,\gamma \in \Omega$ be such that $\alpha < \beta < \gamma$. Let $g,h \in G$ be such that $\alpha g = \beta$, $\gamma g = \gamma$, $\alpha h = \alpha$, $\beta h = \gamma$ (such elements exist since (G,Ω) is 2-transitive). Then $\alpha \in \beta G(g) \setminus \beta G(h)$ whereas $\gamma \in \beta G(h) \setminus \beta G(g)$ and the result follows.

We shall see in Corollary 10.4.4 that the class of normal-valued ℓ-groups is a very large class. This make the various characterizations of normal-valued ℓ-groups in the next theorem especially interesting. We recall (see Lemma 1.2.5):

THEOREM 10.4.3. *Let G be an ℓ-group. Then the following statements are equivalent.*

(i)   *G is normal valued.*

(ii)   *There is a plenary subset Δ of values in G each of which is normal in its cover.*

(iii)   *For all $x,y \in G$, $x^{-1}y^{-1}xy << |x| \vee |y|$.*

(iv)   *For all $x,y \in G^+$, $xy \leq y^2x^2$.*

(v)   *For all convex ℓ-subgroups A,B of G, $AB = BA$.*

(vi)   *All representations of G as an ℓ-permutation group of a totally ordered set have nested intervals of support.*

(vii)   *G is ℓ-isomorphic to an ℓ-subgroup of a product of wreath products of subgroups of the real numbers (acting on themselves by right regular translation).*

(viii) *G can be embedded in an ℓ-group in which the special values form a plenary subset of the set of all values.*

The proofs can be found in Chapter 1 and [Glass, 1981b; Chapter 11].

COROLLARY 10.4.4. *The class $\mathcal{N}$ of all normal-valued ℓ-groups is a variety.*

Proof. This follows directly from part (iv) of Theorem 10.4.3.

It is interesting to note, in connection with condition (ii) of Theorem 10.4.3, that it is not sufficient to assume that every element of G has a normal value:

EXAMPLE 10.4.5. *Let G denote the ℓ-group of all finitely piecewise linear order-preserving permutations of $\mathbf{R}$, the real line, of bounded support.* The prime subgroups of G have been completely characterized by Ball (see Kenoyer

[1984] or Section 5.4) and it is easily verified from that characterization that every non-trivial element has a value which is normal in its cover although this is not true for all values so that G is not normal valued.

The next result, due to Holland, establishes the important position of $\mathcal{N}$ in the lattice of varieties of ℓ-groups.

THEOREM 10.4.6. [Holland, 1976] *The variety $\mathcal{N}$ is the largest proper variety of ℓ-groups.*

Proof. Let $\mathcal{V}$ be any variety of ℓ-groups such that $\mathcal{V} \not\subseteq \mathcal{N}$. Let $G \in \mathcal{V} \setminus \mathcal{N}$. Then there is a value M in G such that M is not normal in its cover M^*. Let $K = \bigcap \{Mg : g \in M^*\}$. Then $M^*/K \in \mathcal{V}$ and the natural representation of M^* on $\mathcal{R}(M)$ induces a faithful representation of M^*/K as an ℓ-permutation group $(M^*/K, \mathcal{R}(M))$. Since M^*/K covers the stabilizer M/K of M and $M/K \neq \{K\}$ it follows from McCleary [1972a, 1973a] that $\mathcal{V}$ contains a 2-transitive ℓ-permutation group (H, Ω). By Example 10.3.4, (H, Ω) mimics L so that, by Lemma 10.3.2, $\mathcal{V}(H) = L$. Hence $\mathcal{V} = L$. Thus $\mathcal{N}$ contains all varieties properly contained in L. by Lemma 10.4.2 and Theorem 10.4.3. (vi), $\mathcal{N}$ contains no 2-transitive ℓ-permutation groups. Therefore $\mathcal{N} \neq L$ and the result follows.

An interesting consequence of Lemma 10.4.2, Theorem 10.4.3 and Theorem 10.4.6 is that any 2-transitive ℓ-group (G, Ω) generates the variety of all ℓ-groups. In particular

COROLLARY 10.4.7. [Holland 1976] $A(\mathbf{R})$ *generates the variety of all ℓ-groups.*

This is also an immediate consequence of Example 10.3.4.

10.5. The Size of L

Each variety is completely determined by the set of identities that hold in the variety. Since there are only a countably infinite number of identities there can be at most $2^{\aleph_0}$ distinct varieties.

That there are indeed $2^{\aleph_0}$ distinct varieties of ℓ- groups was first established by Kopytov and Medvedev [1976]. Olshanski [1970] showed that there is a continuum $\{X_\alpha : \alpha \in I\}$ of varieties of groups that are solvable of length 5. If N_α is the fully invariant subgroup of the free group G_X of rank $\aleph_0$ corresponding to X_α, then $N''_\alpha = [[N_\alpha, N_\alpha], [N_\alpha, N_\alpha]]$ is also fully invariant. Moreover, Smirnov [1965] showed that the groups F/N''_α can be totally ordered. These groups are distinguished by group laws and, therefore, the varieties of ℓ-groups generated by the totally ordered groups F/N''_α are distinct and provide $2^{\aleph_0}$ distinct varieties of ℓ-groups.

For any variety of groups $\mathcal{U}$, let $X(\mathcal{U}) = \{G \in L : G$ satisfies all identities of the form $z^+ \wedge u^{-1}(z^-)u = 1$ where u is a group word and $u = 1$ holds in $\mathcal{U}\}$. Reilly [1981] showed that X is a meet semilattice isomorphism of the lattice of varieties of groups into L. The varieties in the image are called *quasi-representable*. It follows from this theorem and Adyan [1973] that L contains a continuum of pairwise incomparable varieties. More recently, Gurchenkov [1982a], [1984c] ,[1985] has estab–lished the existence of $2^{\aleph_0}$ distinct varieties of nilpotent class 3 ℓ-groups. However, the method of establishing that the cardinality of L is $2^{\aleph_0}$ that we wish to detail here, on account of its simplicity, is due to Feil [1980,1981].

For any positive integers p and q, with $p \leq q$, let $\mathcal{U}_{p/q}$ denote the variety of representable ℓ-groups in $\mathcal{A}^2$ defined by the identity

$$| [x, | [x,y] |] |^p \geq | [x,y] |^q \text{ for } x \geq y \geq e. \tag{10.1}$$

In order to establish that these varieties are distinct, Feil introduced the following construction. For any real number t with $0 < t \leq 1$, let G_t denote $\mathbf{R} \times \mathbf{Z}$ with multiplication defined by

$$(r_1, n_1)(r_2, n_2) = \left(r_1 + \left(\frac{t}{t+1} \right)^{n_1} r_2, n_1 + n_2 \right). \tag{10.2}$$

Then G_t is a group with identity $e = (0,0)$ and it is a totally ordered group if we order it lexicographically from the right: $(r,n) > e$ if and only if (i) $n > 0$ or (ii) $n = 0$ and $r > 0$. Since $\mathbf{R} \times \{0\}$ is an abelian convex ℓ-subgroup of G_t and $G_t/(\mathbf{R} \times \{0\}) \cong \mathbf{Z}$, it follows that $G_t \in \mathcal{A}^2$.

LEMMA 10.5.1. *Let p and q be positive integers such that $p/q \leq 1$ and let t be a real number such that $0 < t < 1$. Then $G_t \in \mathcal{U}_{p/q}$ if and only if $t \leq p/q$.*

Proof. First suppose that $t \leq p/q$. We wish to show that G_t satisfies (10.1). Let $x = (r_1, n_1)$ and $y = (r_2, n_2) \in G_t$ where $n_1 \geq n_2 \geq 0$. The result is trivial if $n_1 = n_2 = 0$. So let $n_1 > 0$. A simple calculation shows that $|[x,y]| = (a,0)$, for some $a \in \mathbf{R}$ such that $a \geq 0$. The result again holds if $a = 0$. So suppose further that $a > 0$. Then

$$[x, |[x,y]|] = x^{-1} \, |[x,y]|^{-1} \, x \, |[x,y]|$$

$$= \left(-\left(\frac{t}{t+1} \right)^{-n_1} \cdot r_1, -n_1 \right) (-a,0)(r_1, n_1)(a,0))$$

$$= \left(-\left(\frac{t}{t+1} \right)^{-n_1} \cdot r_1 - \left(\frac{t}{t+1} \right)^{-n_1} \cdot a + \left(\frac{t}{t+1} \right)^{-n_1} \cdot r_1 + a, \, 0 \right)$$

$$= \left(-\left(\frac{t}{t+1} \right)^{-n_1} \cdot a + a, \, 0 \right)$$

$$= \left(\left(1 - \left(\frac{t+1}{t} \right)^{n_1} \right) a, \, 0 \right)$$

so that

$$||[x, |[x,y]|]|| = \left(\left(\left(\frac{t+1}{t} \right)^{n_1} - 1 \right) a, \, 0 \right)$$

$$\geq \left(\left(\left(\frac{t+1}{t} \right) - 1 \right) a, \, 0 \right) = \left(\frac{1}{t} a, \, 0 \right)$$

$$\geq \left(\frac{q}{p} a, \, 0 \right) \tag{10.3}$$

Therefore,

$$|[x, |[x,y]|]|^p \geq (qa, 0) = (a, 0)^q = |[x,y]|^q$$

and $G_t \in \mathcal{U}_{p/q}$, as required.

Now suppose that $t > p/q$. Let $x = (0,1)$, $y = (1,0)$. Then $|[x,y]| = (1/t, 0)$ so that, as above in (10.3) with $n_1 = 1$ and $a = 1/t$ we have

$$||[x, |[x,y]|]|| = (1/t^2, 0) < (q/pt, 0).$$

Hence

$$|[x, |[x,y]|]|^p < (q/t, 0) = (1/t, 0)^q = |[x,y]|^q$$

and $G_t \notin \mathcal{U}_{p/q}$.

We can now show that the varieties $\mathcal{U}_{p/q}$ form a dense infinite chain.

LEMMA 10.5.2. *If m,n,p,q are positive integers such that $0 < m/n < p/q \le 1$, then $\mathcal{U}_{m/n} \subset \mathcal{U}_{p/q}$.*

Proof. Let $u = |[x,|[x,y]|]|$ and $v = |[x,y]|$. In any totally ordered group and, therefore, in any representable ℓ-group, and for any positive integer m, $x^m \ge y^m$ implies that $x \ge y$. Hence

$$u^m \ge v^n \Rightarrow u^{mp} \ge v^{np} \ge v^{mq} \Rightarrow u^p \ge v^q$$

so that $\mathcal{U}_{m/n} \subseteq \mathcal{U}_{p/q}$. If we let $t = p/q$, then it follows from Lemma 10.5.1 that $G_t \in \mathcal{U}_{p/q}$ but $G_t \notin \mathcal{U}_{m/n}$. Thus $\mathcal{U}_{m/n} \subset \mathcal{U}_{p/q}$.

For $r \in \mathbf{R}$ and $0 < r \le 1$, we define a variety $\mathcal{U}_r$ by

$$\mathcal{U}_r = \cap \{\mathcal{U}_{p/q} \colon r \le p/q \le 1\}.$$

Clearly $G_r \in \mathcal{U}_r \setminus \mathcal{U}_{m/n}$ for $m/n < r$. Hence

THEOREM 10.5.3 [Feil, 1981]. *$\{\mathcal{U}_r \colon 0 < r \le 1\}$ is a tower of varieties of representable ℓ-groups in $\mathcal{A}^2$ whose order type is that of the continuum.*

COROLLARY 10.5.4 [Feil, 1981]. *There exists a continuum of varieties of representable ℓ-groups in $\mathcal{A}^2$ which are not finitely based.*

Proof. Any variety which is the intersection of an infinite properly descending chain of varieties cannot be finitely based. The varieties $\mathcal{U}_r$, with $r \in \mathbf{R} \setminus \mathbf{Q}$ and $0 < r < 1$ have the required properties.

In this connection, Gurchenkov [1982a] and [1984c], [1985] has established the existence of varieties of nilpotent class 3 ℓ-groups which are not finitely based, whereas Medvedev [1982a] has found varieties of non-nilpotent representable varieties which are not finitely based. In addition, Medvedev has shown that there exist infinitely many varieties of ℓ-groups with the property that any basis of identities contains a proper subset which is also a basis of identities.

An interesting related question is that of which varieties of lattice-ordered groups are generated by a single finitely presented lattice-ordered group. This has been considered by Glass [1986b], who has shown, in particular, that there exists a countably infinite set of distinct varieties of this type with solvable word and conjugacy problem. See also Smith [1980].

Returning to the work of Feil, Feil constructed a second tower of varieties similar to the tower $\{\mathcal{U}_r\}$. This time, for any positive integers p and q with $p \leq q$, let $\mathcal{W}_{p/q}$ denote the variety of representable ℓ-groups in $\mathcal{A}^2$ defined by the identity

$$| [x^{-1}, | [x,y] |] |^p \geq | [x,y] |^q \text{ for } x \geq y \geq e$$

and for any $r \in \mathbf{R} \backslash \mathbf{Q}$ with $0 < r < 1$, let

$$\mathcal{W}_r = \cap \{\mathcal{W}_{p/q} : 0 < r \leq p/q\}.$$

Totally ordered groups H_t $(0 < t \leq 1)$ can be constructed in a manner analogous to the groups G_t but using $\dfrac{t+1}{t}$ in the definition of the product in (R2) instead of $\dfrac{t}{t+1}$. These groups can then be used to distinguish the varieties $\mathcal{W}_r$ for different r. Then, analogous to Lemma 10.5.2, we have

LEMMA 10.5.5. *If m,n,p,q are positive integers such that $0 < m/n < p/q \leq 1$, then $\mathcal{W}_{m/n} \subset \mathcal{W}_{p/q}$.*

Feil also showed that these two towers contain only the variety of abelian ℓ-groups in common. This follows from the next lemma, since $\mathcal{U}_1$ and $\mathcal{W}_1$ are the largest varieties in the two towers.

LEMMA 10.5.6. $\mathcal{U}_1 \cap \mathcal{W}_1 = \mathcal{A}$.

Proof. Clearly $\mathcal{A} \subseteq \mathcal{U}_1 \cap \mathcal{W}_1$. So suppose that $G \in \mathcal{U}_1 \cap \mathcal{V}_1$. Then G satisfies the identities

$$| [x, | [x,y] |] | \geq | [x,y] |,$$
$$| [x^{-1}, | [x,y] |] | \geq | [x,y] |. \tag{10.4}$$

Since $\mathcal{U}_1$ and $\mathcal{W}_1$ are both varieties of representable ℓ-groups, we may assume that G is totally ordered.

Let $c = | [x,y] |$. Then

$$[x^{-1},c] = xc^{-1}x^{-1}c = x(c^{-1}x^{-1}cx)x^{-1} = x(x^{-1}c^{-1}xc)^{-1}x^{-1} = x[x,c]^{-1}x^{-1}$$

so that either $[x,c] \geq e$ or $[x^{-1},c] \geq e$. Suppose that $[x,c] \geq e$, the other case being similar. By (10.4), $[x,c] \geq c$ or $x^{-1}c^{-1}xc \geq c$. Thus $c \leq e$. But necessarily, by its definition $c \geq e$. Hence $c = e$ and therefore $[x,y] = e$ so that G is abelian, as required.

Now Theorem 10.5.3 can be interpreted as saying that $L(\mathcal{R})$ contains a sublattice isomorphic to the open unit interval $I = (0,1)$. Combining the $\mathcal{U}$-tower and the $\mathcal{W}$-tower, Huss showed that $L(\mathcal{R})$ contains a copy of $I{\times}I$.

THEOREM 10.5.7 [Huss, 1984]. *Let* $S = \{\mathcal{U}_r \vee \mathcal{W}_s : 0 < r,s < 1\}$. *Then* S *is a sublattice of* $L(\mathcal{R} \cap \mathcal{A}^2)$ *and is isomorphic to* $I{\times}I$.

Proof. Clearly $S \subseteq L(\mathcal{R} \cap \mathcal{A}^2)$. By Lemmas 10.5.2 and 10.5.4, for $0 < p,q,r,s < 1$,

$$(\mathcal{U}_p \vee \mathcal{W}_q) \vee (\mathcal{U}_r \vee \mathcal{W}_s) = \mathcal{U}_t \vee \mathcal{W}_u \in S$$

where $t = \max\{p,r\}$ and $u = \max\{q,s\}$. Since L is distributive

$$(\mathcal{U}_p \vee \mathcal{W}_q) \cap (\mathcal{U}_r \vee \mathcal{W}_s) = (\mathcal{U}_p \cap \mathcal{U}_r) \vee (\mathcal{U}_p \cap \mathcal{W}_s) \vee (\mathcal{W}_q \cap \mathcal{U}_r) \vee (\mathcal{W}_q \cap \mathcal{W}_s)$$
$$= \mathcal{U}_t \vee \mathcal{A} \vee \mathcal{A} \vee \mathcal{W}_u$$

by Lemmas 10.5.2, 10.5.5, and 10.5.6.

$$= \mathcal{U}_t \vee \mathcal{W}_u \in S$$

where $t = \min\{p,r\}$ and $u = \min\{q,s\}$. Thus S is a sublattice of $L(\mathcal{R} \cap \mathcal{A}^2)$.

Now define a mapping $\alpha{:}\ I{\times}I \to S$ by: $(r,s)\alpha = \mathcal{U}_r \vee \mathcal{W}_s$. Clearly α is a surjection. Let $(p,q),(r,s) \in I{\times}I$. If $(p,q) \leq (r,s)$, then $p \leq r$ and $q \leq s$ so that $\mathcal{U}_p \subseteq \mathcal{U}_r$, $\mathcal{W}_q \subseteq \mathcal{W}_s$ and $\mathcal{U}_p \vee \mathcal{W}_q \subseteq \mathcal{U}_r \vee \mathcal{W}_s$. Thus α is order preserving. If $(p,q) \neq (r,s)$, then either $p \neq r$ or $q \neq s$. Without loss of generality, let $p < r$. Then $G_r \in \mathcal{U}_r \subseteq \mathcal{U}_r \vee \mathcal{W}_s = (r,s)\alpha$ but $G_r \notin \mathcal{U}_p$, by Lemma 10.5.2, $G_r \notin \mathcal{W}_q$, by Lemma 10.5.5 so that $G_r \notin \mathcal{U}_p \vee \mathcal{W}_q = (p,q)\alpha$, by Corollary 10.1.4. Thus α is one-to-one and therefore an isomorphism.

One consequence of Theorem 10.5.7 is that there exist antichains (that is, sets of pairwise incomparable elements) in L of cardinality $2^{\aleph_0}$. This was established earlier by Reilly [1981]. A sharper result due to Medvedev [1983a] states the following: There exist $2^{\aleph_0}$ varieties of representable ℓ-groups each of which has $2^{\aleph_0}$ distinct covers in the lattice of varieties of representable ℓ-groups.

10.6. Torsion Classes

A concept similar to 'variety' is that of a torsion class. This notion was introduced by Martinez ([1975]; see also Chapter 6), who saw that many of the classes of ℓ-goups that had received attention in the past had, in fact, been torsion classes. Some of these were varieties and some (like the class of finite valued ℓ-groups or, as shown recently by Medvedev [1983b], the class of locally nilpotent ℓ-groups) were not.

In this short section we note some properties of torsion classes that are relevant to the study of varieties.

In many ways the most interesting condition in the definition of torsion class is that for any torsion class T and any ℓ-group G, there is a maximum convex ℓ-subgroup of G which lies in T (i.e., it contains all other convex ℓ-subgroups of G in T). We shall denote this subgroup by $T(G)$. The first important result relating varieties and torsion classes was due to Martinez.

LEMMA 10.6.1 [Martinez, 1976]; (Lemma 6.1.3). *The variety $\mathcal{N}$ of normal-valued ℓ-groups is a torsion class.*

This result was then used by Holland to prove the following generalization.

THEOREM 10.6.2 [Holland, 1979]. (Theorem 6.1.4). *Every variety of ℓ-groups is a torsion class.*

It is not hard to see that the family of torsion classes is a lattice when partially ordered by inclusion. Theorem 10.6.3 tells us that this lattice contains the lattice of varieties of ℓ-groups. However, Smith [1980] has shown that the lattice of varieties is not a sublattice of the lattice of torsion classses.

An important consequence of Theorem 10.6.2 for varieties is the following:

THEOREM 10.6.3 [Holland, 1979]. *For any ℓ-group G and any variety of ℓ-groups $\mathcal{U}$ there is a largest convex ℓ-subgroup $\mathcal{U}(G)$ of G such that $\mathcal{U}(G) \in \mathcal{U}$. Moreover, $\mathcal{U}(G)$ is an ℓ-ideal.*

The subgroups of the form $\mathcal{U}(G)$ appear to occupy no particular position in the lattice of convex ℓ-subgroups of G. To see this consider the examples of Kenoyer [1984] (Section 5.4). There we have ℓ-groups G and H with isomorphic lattices of convex ℓ-subgroups, but for any variety $\mathcal{U}$, with $\mathcal{A} \subseteq \mathcal{U} \subseteq \mathcal{N}$, we have $\mathcal{U}(G) = \{e\}$, , while $\mathcal{U}(H) = H$.

The subgroups of the form $\mathcal{U}(G)$ do have an interesting closure property which follows from the next result due to Bernau.

DEFINITION 10.6.4. An ℓ-subgroup H of an ℓ-group G is said to be *closed* (in G) if, for all subsets $\{x_i : i \in I\}$ of H such that $a = \vee \{x_i : i \in I\}$ exists in G we have $a \in H$. The *closure* $\bar{H}$ of H is the smallest ℓ-subgroup of G containing H which is closed. Clearly G is closed in G and $\bar{H}$ is the intersection of all closed ℓ-subgroups of G containing H.

THEOREM 10.6.5 [Bernau, 1977]. *Let $\mathcal{U}$ be a variety of ℓ-groups and H be an ℓ-subgroup of the ℓ-group G. If $H \in \mathcal{U}$, then also $\bar{H} \in \mathcal{U}$.*

COROLLARY 10.6.6. *For any variety of ℓ-groups $\mathcal{U}$ and any ℓ-group G, $\mathcal{U}(G)$ is closed in G.*

Of the various kinds of completions of ℓ-groups that have been studied, the lateral completion has received particular attention. G is the *lateral completion* (see page 26) of H provided: (i) if $a = \vee\{a_i : i \in I\}$ exists in H then a is the supremum of $\{a_i : i \in I\}$ in G, (ii) if M is a pairwise disjoint subset of H, then $\vee M$ exists in G, (iii) $e < g \in G$ implies that there exists $h \in H$ with $e < h \leq g$. From Corollary 10.6.6 we have, in particular:

COROLLARY 10.6.7. *Every variety of ℓ-groups is closed with respect to the formation of lateral completions.*

Torsion classes have attracted considerable attention in their own right and the interested reader can find more information on this topic in Chapter 6 and the references cited there, while Vosvi [1983] relates some of these ideas and results to more classical results in group theory.

10.7 Abelian Powers

In Section 10.1 we introduced the concept of the product $\mathcal{UV}$ of two varieties $\mathcal{U}$ and $\mathcal{V}$. It is easily verified that this product is associative (that is, $(\mathcal{UV})\mathcal{W} = \mathcal{U}(\mathcal{VW})$) so that the lattice L of all varieties is then endowed with a semigroup structure. In this and the next sections we explore the properties of this product. We begin by considering what happens when successive powers are formed $\mathcal{U}, \mathcal{U}^2, \mathcal{U}^3, \dots$.

In the study of the semigroup structure of L, the wreath product construction plays a vital role. This is also true for the theory of varieties of groups (see Neumann [1967]). But there is a difference here in that we must work with wreath products of permutation groups. The reason for this will become clear in Section 10.8.

We begin with some simple observations on wreath products (Section 2.3). If (H,Θ), (G,Ω) are ℓ-permutation groups, then the ℓ-groups $(H,\Theta)Wr(G,\Omega)$ and $(H,\Theta)wr(G,\Omega)$ are independent of Θ and so we can simply write $HWr(G,\Omega)$ or $Hwr(G,\Omega)$, respectively, for these ℓ-groups. Because of the Holland representation, we can have any ℓ-group as H in these constructions.

We will write $Wr^n(G,\Omega)$, $wr^n(G,\Omega)$ or simply Wr^nG, wr^nG, if the action is clear, for the group obtained by iterating the appropriate wreath product $(n-1)$ times with the same ℓ-permutation group (G,Ω).

LEMMA 10.7.1. *Let $\mathcal{U},\mathcal{V}$ be varieties, $H \in \mathcal{U}$, $G \in \mathcal{V}$ and (G,Ω) be an ℓ-permutation group. Then $W = HWr(G,\Omega) \in \mathcal{UV}$.*

Proof. The base group of W, $B = \Pi\{H_\alpha : \alpha \in \Omega\} \in \mathcal{U}$ and is an ℓ-ideal of W such that $W/B \cong G \in \mathcal{V}$.

Recall that $(\mathbf{Z},\mathbf{Z})$ denotes the regular representation of the integers and let $(\mathbf{R},\mathbf{R})$ denote the regular representation of the real numbers. From Lemma 10.7.1 we have:

LEMMA 10.7.2. $Wr^n\mathbf{Z}, Wr^n\mathbf{R} \in \mathcal{A}^n$, *for all positive integers n.*

LEMMA 10.7.3 [Martinez, 1974b]. $\mathcal{N}^2 = \mathcal{N}$.

Proof. Let $G \in \mathcal{N}^2$. Then there exists an ℓ-ideal K of G with $K, G/K \in \mathcal{N}$. Let M be any value in G. If $K \subseteq M$ then M/K is a value in G/K. Since $G/K \in \mathcal{N}$, (M/K) is normal in its cover $(M/K)^*$. Hence M is normal in its cover. On the other hand, if $K \not\subseteq M$ then, by Theorem 1.1.11, $M \cap K$ is a value in K. Since $K \in \mathcal{N}$, $M \cap K$ is normal in its cover. So, again by Theorem 1.1.11, M is normal in its cover. Thus M is always normal in its cover and so $G \in \mathcal{N}$. Consequently, $\mathcal{N}^2 \subseteq \mathcal{N}$, as required.

THEOREM 10.7.4 [Glass, Holland, and McCleary, 1980].

$$\mathcal{N} = \vee \{ \mathcal{A}^n : n = 1,2,... \}.$$

Proof: Since $\mathcal{A} \subseteq \mathcal{N}$ it follows from Lemma 10.7.3 by induction that $\mathcal{A}^n \subseteq \mathcal{N}$ and therefore that $\vee \{ \mathcal{A}^n : n = 1,2,... \} \subseteq \mathcal{N}$. Now let

$$w(x) = e \text{ where } w(x) = \bigvee_i \bigwedge_j \prod_k x_{ijk} = \bigvee_i \bigwedge_j w_{ij}$$

be a law that fails in $\mathcal{N}$ and therefore in some subdirectly irreducible ℓ-group $G \in \mathcal{N}$. It suffices to show that $w(x) = e$ does not hold in $\mathcal{A}^n$, for some n, and for that if suffices to show that it does not hold in $Wr^n Z$, for some n.

By Theorem 10.4.3 (vii) and the fact that G is subdirectly irreducible, we can assume that G is an ℓ-subgroup of a wreath product $(W, \Lambda) = Wr(R_\gamma : \gamma \in \Gamma)$ where each R_γ is the regular representation of the real numbers. Hence $w(x) = e$ fails in (W, Λ).

Therefore, for some $g_{ijk} \in W$, $\lambda \in \Lambda$

$$\lambda(\bigvee_i \bigwedge_j \prod_k g_{ijk}) \neq \lambda.$$

Let

$$\Theta = \{\lambda\} \cup \{\lambda u(g_1,...,g_n) : u(x_1,...,x_n) \text{ is an initial segment of some } w_{ij}\}.$$

Then Θ is finite. For each pair (θ_1, θ_2) of distinct elements of Θ, there exists a unique $\gamma \in \Gamma$ such that $\theta_1(\gamma) \neq \theta_2(\gamma)$ but $\theta_1(\delta) = \theta_2(\delta)$, for $\delta > \gamma$. Let Δ be the set of all such $\gamma \in \Gamma$. Then Δ is finite. Let $(W', \Lambda') = \Pi\{R_\delta : \delta \in \Delta\}$ and let $\pi : \mu \mapsto \mu'$ denote the projection of Λ onto Λ'. Clearly π is one-to-one and order preserving on Θ, and for $\theta \in \Theta$, let $\theta' = \theta\pi$. For each $g \in \{g_{ijk}\}$, define $g' = g'_{\delta,v} \in W' = Wr\{R_\delta : \delta \in \Delta\}$ by

$$g'_{\delta,v} = \begin{cases} g_{\delta,v} & \text{if there exist } \theta, \theta^* \in \Theta \text{ with } \theta \equiv^\delta v \text{ in } \Lambda' \text{ and } \theta g \equiv^\delta \theta^* \\ e & \text{otherwise.} \end{cases}$$

Then $g'_{\delta,v}$ is well defined and, if $g, h \in \{g_{ijk}\}$ are inverses then g' and h' will be also.

Now consider any $g = g_{ijk}$ and suppose that $w_{ij} = ux_{ijk}$. Then $\theta = \lambda u(g_1,...,g_n) \in \Theta$. Let $\delta \in \Delta$. Then

$$(\theta'g')(\delta) = \theta'(\delta)g'_{\delta,\theta'} = \theta(\delta)g_{\delta,\theta} = (\theta g)(\delta) = (\theta g)'(\delta)$$

since both θ and $\theta g = \lambda u(g)g_{ijk} \in \Theta$. Thus $\theta'g' = (\theta g)'$ and for each i, j, λ',

$$\lambda'w(g'_1,...,g'_n) = (\lambda w(g_1,...,g_n))'.$$

Since π is one-to-one and order preserving on Θ, a totally ordered set,

$$\lambda' \underset{i}{\vee} \underset{j}{\wedge} w_{ij}(g) \neq \lambda'.$$

Thus the identity $w(x) = e$ does not hold in W'. Since Δ is finite $W' = Wr^n R$, for some integer n. By Lemma 10.7.2, $W' \in \mathcal{A}^n$ so that $w(x) = e$ does not hold in $\mathcal{A}^n$ and the proof is complete.

The next result is an easy consequence of Lemma 10.7.3 and Theorem 10.7.4.

COROLLARY 10.7.5. *(i) If $\mathcal{V}$ is a variety with $\mathcal{A} \subseteq \mathcal{V} \subseteq \mathcal{N}$, then $\mathcal{N} = \vee \{\mathcal{V}^n: n = 1,2,3,...\}$.*

(ii) The only idempotents in the semigroup L *are* $\mathcal{E}, \mathcal{N}$ *and* $\mathcal{L}$.

10.8. Mimicking Revisited

In Section 10.3 we introduced the concept of an ℓ-permutation group mimicking a variety and saw there that such an ℓ-group always generates the variety. However, the real value of this concept is in relation to products of varieties and derives from the following result due to Glass, Holland and McCleary.

THEOREM 10.8.1 [Glass, Holland & McCleary, 1980]. *If $\mathcal{U} = \mathcal{V}(\{U_s : s \in S\})$ and (G,Ω) mimics $\mathcal{V}$ then $\mathcal{V}(\{U_s wr(G,\Omega): s \in S\}) = \mathcal{U}\mathcal{V}$.*

By Lemma 10.7.1, $U_s wr(G,\Omega) \in \mathcal{U}\mathcal{V}$, for all $s \in S$, so that it remains to show that $\mathcal{U}\mathcal{V}$ is the variety generated by these ℓ-groups. The technique is like that of Theorem 10.7.4, namely to consider any identity $w(x) = e$ which is not satisfied by $\mathcal{U}\mathcal{V}$ and, therefore, not satisfied by some $H \in \mathcal{U}\mathcal{V}$. Then there exists an ℓ-ideal M of H with $M \in \mathcal{U}$, $H/M \in \mathcal{V}$. For an appropriate representation $(H/M,\Omega)$ of H/M as an ℓ-group of permutations of the ordered set Ω, H is an ℓ-subgroup of $MWr(H/M,\Omega)$. Using the fact that the ℓ-permutation group (G,Ω) mimics $\mathcal{V}$ it is then possible to show that $w(x)$

$= e$ is not satisfied by one of the groups of the form $U_s wr(G,\Omega)$. Therefore $\mathcal{V}(\{U_s wr(G,\Omega) : s \in S\}) = \mathcal{UV}$. However, the proof is long and technical and so the reader is referred to [Glass, Holland and McCleary, 1980] for details.

Some interesting simple consequences flow from the above theorem.

COROLLARY 10.8.2. *For each positive integer* n, $\mathcal{A}^n = \mathcal{V}(wr^n \mathbf{Z})$.

Proof. By Example 10.3.3, $(\mathbf{Z},\mathbf{Z})$ mimics $\mathcal{A}$ and the result follows from Theorem 10.8.1.

COROLLARY 10.8.3. *The only varieties of ℓ-groups closed under the formation of wreath products with $\mathbf{Z}$ on either side are $\mathcal{N}$ and L.*

Proof. If $\mathcal{V}$ is such a variety, then, by Corollary 10.8.2, $\mathcal{A}^n \subseteq \mathcal{V}$, for all positive integers n. By Theorem 10.7.4, $\mathcal{N} \subseteq \mathcal{V}$.

THEOREM 10.8.4 [Glass, Holland, and McCleary, 1980].
(i) $wr^n \mathbf{Z}$ *is subdirectly irreducible.*
(ii) $\mathcal{A}^n \neq \mathcal{A}^{n+1}$, *for all positive integers* n.
(iii) $\mathcal{A}^n$ *is finitely join irreducible.*
(iv) *If* $\mathcal{U}$ *and* $\mathcal{V}$ *are proper subvarieties of* $\mathcal{N}$ *then so are* $\mathcal{UV}$ *and* $\mathcal{U} \vee \mathcal{V}$.
(v) $\mathcal{N}$ *is finitely join irreducible.*

Proof. (i) $wr^n \mathbf{Z}$ has a basic ℓ-ideal M which is a cardinal product of n copies of $\mathbf{Z}$. It is straighforward to show that this is the smallest non-trivial ℓ-ideal of $wr^n \mathbf{Z}$ which is, therefore, subdirectly irreducible.

(ii) Clearly $wr^{n+1}\mathbf{Z} \in \mathcal{A}^{n+1}$ whereas $wr^{n+1}\mathbf{Z} \notin \mathcal{A}^n$.

(iii) If $\mathcal{A}^n = \vee \{\mathcal{U}_i : i \in I\}$, where I is finite, then by Lemma 10.1.4 and (i), $wr^n \mathbf{Z} \in \mathcal{U}_i$ for some i so that, by Corollary 10.8.2, $\mathcal{A}^n \subseteq \mathcal{U}_i$.

(iv) Let $\mathcal{U}$ and $\mathcal{V}$ be proper subvarieties of $\mathcal{N}$. By Theorem 10.7.4 and Corollary 10.8.2, neither $\mathcal{U}$ nor $\mathcal{V}$ can contain $wr^n \mathbf{Z}$, for infinitely many positive integers n. So let m be the largest integer such that $wr^m \mathbf{Z} \in \mathcal{U} \cup \mathcal{V}$. By Corollary 10.1.4 and (i), $wr^{m+1}\mathbf{Z} \notin \mathcal{U} \vee \mathcal{V}$ and so $\mathcal{U} \vee \mathcal{V} \neq \mathcal{N}$. If $\mathcal{UV} = \mathcal{N}$, then $wr^n \mathbf{Z} \in \mathcal{UV}$ for all positive integers n. But the only ℓ-ideals of $wr^n \mathbf{Z}$ are the obvious ones consisting of products of $wr^m \mathbf{Z}$ for $m < n$ and with quotients $wr^{n+m}\mathbf{Z}$. Therefore either $\mathcal{U}$ or $\mathcal{V}$ contains $wr^n \mathbf{Z}$, for all positive integers n, a contradiction. Therefore, $\mathcal{UV}$ is a proper subvariety of $\mathcal{N}$.

(v) This follows immediately from (iv).

Among the most basic examples of ℓ-groups are the automorphism groups of chains and a natural question, with a surprising answer provided by Holland, is that of which varieties are generated by ℓ-groups of the form $A(\Omega)$, where Ω is a chain.

THEOREM 10.8.5 [Holland, 1985a]. *Let $\mathcal{V}$ be a variety of ℓ-groups. Then $\mathcal{V} = \mathcal{V}(A(\Omega))$ for some chain Ω if and only if $\mathcal{V} = \mathcal{A}^n$, for some $n = 0,1,2,...$, or $\mathcal{V} = \mathcal{N}$ or $\mathcal{V} = \mathcal{L}$.*

10.9. The Semigroup of Varieties

We have observed earlier that L is a semigroup with respect to the product of varieties. In this section, taken from Glass, Holland, and McCleary [1980], we will see that $L \setminus \{\mathcal{E}, \mathcal{N}, \mathcal{L}\}$ is a free semigroup. This requires some clever arguments for which the next two lemmas will prove useful.

We omit the proof of the first lemma which is fairly straightforward.

LEMMA 10.9.1 (HOURGLASS LEMMA [Glass, Holland and McCleary, 1980]). *Let H be an ℓ-group, (G,Ω) a transitive ℓ-permutation group and let N be an ℓ-ideal of $Hwr(G,\Omega)$. Then either N contains or is contained in the base ℓ-ideal $B = \Sigma\{H_\alpha: \alpha \in \Omega\}$ of $Hwr(G,\Omega)$. Moreover, if $N \subseteq B$, then*
$N = \Sigma\{H_\alpha \cap N: \alpha \in \Omega\}$ *and* $H_\alpha \cap N \cong H_\beta \cap N$ *for all* $\alpha,\beta \in \Omega$.

LEMMA 10.9.2 [Glass, Holland and McCleary, 1980]. *Let $\mathcal{U}_1, \mathcal{U}_2, \mathcal{V}_1, \mathcal{V}_2$ be varieties with $\mathcal{U}_1\mathcal{V}_1 \subseteq \mathcal{U}_2\mathcal{V}_2$. If $\mathcal{U}_1 \not\subseteq \mathcal{U}_2$ then $\mathcal{A}\mathcal{V}_1 \subseteq \mathcal{V}_2$. If $\mathcal{V}_1 \not\subseteq \mathcal{V}_2$, then $\mathcal{U}_1\mathcal{A} \subseteq \mathcal{U}_2$.*

Proof. We consider the case $\mathcal{V}_1 \not\subseteq \mathcal{V}_2$. The other case is similar and can be found in Glass, Holland and McCleary [1980]. Let $G \in \mathcal{V}_1 \setminus \mathcal{V}_2$ be subdirectly irreducible and let $H \in \mathcal{U}_1$. Then $W = Hwr(G,\Omega) \in \mathcal{U}_1\mathcal{V}_1 \subseteq \mathcal{U}_2\mathcal{V}_2$, for any faithful transitive representation (G,Ω) of G. Since $W \in \mathcal{U}_2\mathcal{V}_2$, there exists an ℓ-ideal M of W such that $M \in \mathcal{U}_2$ and $W/M \in \mathcal{V}_2$. By Lemma 10.9.1, M and $P = \Sigma\{H_\alpha: \alpha \in \Omega\}$ are comparable ℓ-ideals of W. If $M \subseteq P$, then G is a homomorphic image of W/M so that $G \in \mathcal{V}_2$, a contradiction. Hence $P \subset M$. Therefore if $K = M \cap \{(F,g): F$ is the identity function$\}$, then K is a non-trivial ℓ-subgroup of W. Moreover,

$Hwr(K,\Omega) \subseteq M$ so that $Hwr(K,\Omega) \in \mathcal{U}_2$. Now K is non-trivial, so $\mathbf{Z} \subseteq K$. Hence $Hwr(\mathbf{Z},\Omega) \subseteq \mathcal{U}_2$. But clearly $Hwr(\mathbf{Z},\mathbf{Z}) \subseteq Hwr(\mathbf{Z},\Omega)$, since $\mathbf{Z}$ acts faithfully on Ω, so that $Hwr(\mathbf{Z},\mathbf{Z}) \in \mathcal{U}_2$. This holds for all $H \in \mathcal{U}_1$. Hence from Example 10.3.3 and Theorem 10.8.1, $\mathcal{U}_1 \mathcal{A} \subseteq \mathcal{U}$.

We will say that a variety $\mathcal{W} \neq \mathcal{E}$ is *indecomposable* if $\mathcal{U}\mathcal{V} = \mathcal{W}$ implies that either $\mathcal{U} = \mathcal{E}$ or $\mathcal{V} = \mathcal{E}$. We are now ready for the main theorem.

THEOREM 10.9.3 [Glass, Holland and McCleary, 1980]. *The set* $\mathbf{L^*} = \mathbf{L} \setminus \{\mathcal{E}, \mathcal{N}, \mathcal{L}\}$ *is a free semigroup with respect to the product of varieties with the set of indecomposable varieties as a set of free generators.*

Proof. Let $\mathcal{U} \in \mathbf{L^*}$ and $\mathcal{U} = \mathcal{U}_1 ... \mathcal{U}_n$, $\mathcal{U}_i \in \mathbf{L^*}$. Then $\mathcal{A} \subseteq \mathcal{U}_i$, for $i = 1,...,n$, so that $\mathcal{A}^n \subseteq \mathcal{U}$. Since $\mathcal{U} \neq \mathcal{N}, \mathcal{L}$, it follows from Theorem 10.7.4 that there is a maximum integer n such that $\mathcal{U}$ can be written as a product of n elements from $\mathbf{L^*}$, say $\mathcal{U} = \mathcal{U}_1 ... \mathcal{U}_n$. By the maximality of n, each $\mathcal{U}_i$ must be indecomposable.

To complete the proof, it suffices to show that the representation of any $\mathcal{U} \in \mathbf{L^*}$ as a product of indecomposable varieties is unique. First suppose that $\mathcal{U} = \mathcal{U}_1\mathcal{V}_1 = \mathcal{U}_2\mathcal{V}_2$ where $\mathcal{U}_1$ and $\mathcal{U}_2$ are indecomposable. If $\mathcal{U}_1$ and $\mathcal{U}_2$ are incomparable, then by Lemma 10.9.2, $\mathcal{A}(\mathcal{A}\mathcal{V}_1) \subseteq \mathcal{A}\mathcal{V}_2 \subseteq \mathcal{V}_1$ which contradicts Corollary 10.8.3, since $\mathcal{V}_1 \neq \mathcal{N}, \mathcal{L}$. So we can assume that $\mathcal{U}_2 \subseteq \mathcal{U}_1$ and hence, by Lemma 10.9.2 that $\mathcal{V}_1 \subseteq \mathcal{V}_2$.

Let n be the largest positive integer such that $\mathcal{A}^n \subseteq \mathcal{V}_1$ and let

$$\mathcal{U}' = \{H : Hwr(G,\Omega) \in \mathcal{V}_2 \text{ for all transitive } \ell\text{-groups } (G,\Omega) \text{ with } G \in \mathcal{A}^n\}.$$

Since $\mathcal{A}^n \subseteq \mathcal{V}_1 \subseteq \mathcal{V}_2$, $\mathcal{U}'$ is non-empty and it is easily verified that $\mathcal{U}'$ is a variety. By Theorem 10.8.1 and the fact that the set of all transitive ℓ-permutation groups in $\mathcal{A}^n$ clearly mimics $\mathcal{A}^n$, we have that $\mathcal{U}'\mathcal{A}^n \subseteq \mathcal{V}_2$. Now let $\mathcal{U}^* = \mathcal{U}' \cap \mathcal{U}_1$. Then

$$\mathcal{U}_2\mathcal{U}^*\mathcal{A}^n \subseteq \mathcal{U}_2\mathcal{U}'\mathcal{A}^n \subseteq \mathcal{U}_2\mathcal{V}_2 = \mathcal{U}_1\mathcal{V}_1.$$

Since $\mathcal{A}^{n+1} \not\subseteq \mathcal{V}_1$, it follows from Lemma 10.9.2 that $\mathcal{U}_2\mathcal{U}^* \subseteq \mathcal{U}_1$. Now let $H \in \mathcal{U}_1 \setminus \mathcal{U}_2$ and (G,Ω) be a transitive ℓ-permutation group with $G \in \mathcal{A}^n$. Then $Hwr(G,\Omega) \in \mathcal{U}_1\mathcal{A}^n \subseteq \mathcal{U}_1\mathcal{V}_1 = \mathcal{U}_2\mathcal{V}_2$. Since $H \notin \mathcal{U}_2$, it follows from Lemma 10.9.1 that $\mathcal{U}_2(Hwr(G,\Omega)) = \Sigma\{\mathcal{U}_2(H_\alpha) : \alpha \in \Omega\}$ (recalling that $\mathcal{U}_2(H)$ means the largest convex ℓ-subgroup of H in $\mathcal{U}_2$). Also

$$(H/\mathcal{U}_2(H))wr(G,\Omega) \cong (Hwr(G,\Omega))/\mathcal{U}_2(Hwr(G,\Omega)) \in \mathcal{V}_2$$

so that $H/\mathcal{U}_2(H) \in \mathcal{U}^*$. Thus $H \in \mathcal{U}_2\mathcal{U}^*$, $\mathcal{U}_1 \subseteq \mathcal{U}_2\mathcal{U}^*$ and so $\mathcal{U}_1 = \mathcal{U}_2\mathcal{U}^*$. Since $\mathcal{U}_1$ and $\mathcal{U}_2$ are indecomposable, $\mathcal{U}_1 = \mathcal{U}_2$. By Lemma 10.9.2, we must also have $\mathcal{V}_1 = \mathcal{V}_2$. An induction argument on the number of factors will now show that the representation of $\mathcal{U}$ as a product of indecomposable factors in unique and therefore $\mathbf{L}^*$ is a free semigroup.

Indecomposable varieties in $\mathbf{L}^*$ are plentiful. For instance, any variety which does not contain $\mathcal{A}^2$ must be indecomposable. This includes the Feil varieties of Section 10.5 and the Medvedev varieties of Section 10.11. Indeed since $wr^2\mathbf{Z}$ is not representable, $\mathcal{A}^2 \not\subseteq \mathcal{R}$ and therefore every variety of representable ℓ-groups is indecomposable. In the same way, the Scrimger varieties of Section 10.12 are indecomposable.

For any varieties of ℓ-groups $\mathcal{U}$ and $\mathcal{V}$ it is possible to obtain a basis of identities for $\mathcal{U}\mathcal{V}$ in terms of any basis for $\mathcal{U}$ and $\mathcal{V}$ individually.

LEMMA 10.9.4. *Let* $\mathcal{U}$ *be the variety defined by the identities* $u_j(x) = e$, $j\in J$, *let* G *be an* ℓ-*group and* $g\in G$. *Then the convex* ℓ-*subgroup* $H = G(g)$ *lies in* $\mathcal{U}$ *if and only if*

$$u_j((y_1 \wedge |g|^k) \vee |g|^{-k}, (y_2 \wedge |g|^k) \vee |g|^{-k}, \dots, (y_n \wedge |g|^k) \vee |g|^{-k}) = e$$

for all $y_i \in G$ *and* $k = 1, 2, \dots$.

Proof. It suffices to observe that by Clifford's Lemma, every element of H can be written in the form $(y \wedge |g|^n) \vee |g|^{-n}$, for some element $y\in G$ and, conversely, every element of this form lies in H.

THEOREM 10.9.5 [Glass, Holland and McCleary, 1980]. *Let the variety of* ℓ-*groups* $\mathcal{U}$ *be defined by the identities* $\{u_j(x) = e : j\in J\}$ *and the variety of* ℓ-*groups* $\mathcal{V}$ *be defined by the laws* $\{v_i(x) = e : i\in I\}$. *Then* $\mathcal{U}\mathcal{V}$ *is defined by the laws*

$$u_j((y_1 \wedge |v_i(x)|^k) \vee |v_i(x)|^{-k}, (y_2 \wedge |v_i(x)|^k) \vee |v_i(x)|^{-k}, \dots) = e \quad (10.5)$$

where the variables y_r *do not appear in* $v_i(x)$, *for all choices of* $i\in I$, $j\in J$, *and* k *ranges over all positive integers.*

Proof. If $G\in \mathcal{U}\mathcal{V}$, then $G/\mathcal{U}(G)\in \mathcal{V}$ so that $v_i(x)\in \mathcal{U}(G)$, for all $x = (x_1,\dots,x_n)$ with $x_i\in G$. Consequently, $(y_r \wedge |v_i(x)|^k) \vee |v_i(x)|^{-k}\in \mathcal{U}(G)$ and (10.5) holds for all $x_r, y_r\in G$; that is, G satisfies the identity (10.5).

Conversely, let G satisfy (10.5). By Lemma 10.9.4, this says that $G(v_i(x))\in \mathcal{U}$, for all $x = (x_1,\dots,x_n)$ with $x_i\in G$, so that in particular, $v_i(x)\in \mathcal{U}(G)$. Hence $G/\mathcal{U}(G)$ must satisfy the identities $v_i(x) = e$, $i\in I$. Thus $G/\mathcal{U}(G)\in \mathcal{V}$ and $G\in \mathcal{U}\mathcal{V}$, as required.

When the first factor in a product is the variety $\mathcal{A}$ of abelian ℓ-groups then there is a simpler formulation of a basis for $\mathcal{A}\mathcal{V}$. This implies, in particular, that if $\mathcal{V}$ is finitely based, then so is $\mathcal{A}\mathcal{V}$.

THEOREM 10.9.6 [Glass, Holland and McCleary, 1980]. *If the variety $\mathcal{V}$ is defined by the identity $w(x) = e$ then $\mathcal{A}\mathcal{V}$ is defined by the law*

$$[|y| \wedge |w(x)|, \; |z| \wedge w(x)] = e$$

where y and z are variables not appearing in $w(x)$. Consequently $\mathcal{A}^n$, $n = 1,2,3,...$, is defined by the single identity $c_n(x,y,z) = e$ where $c_1(x) = [x_1,x_2]$ and, for $n > 1$,

$$c_n(x) = [|x_{2n-1}| \wedge |c_{n-1}(x)|, \; |x_{2n}| \wedge |c_{n-1}(x)|].$$

Proof. See Glass, Holland and McCleary [1980].

Now that **L** has both a lattice and a semigroup structure, it is natural to ask how they relate. The next result summarizes some observations in this regard. Detailed verification can be found in the cited reference.

THEOREM 10.9.7 [Glass, Holland and McCleary, 1980]. *The lattice-ordered semigroup of varieties satisfies*

(i) $\mathcal{U}(\mathcal{V}_1 \vee \mathcal{V}_2) = \mathcal{U}\mathcal{V}_1 \vee \mathcal{U}\mathcal{V}_2$,

(ii) $(\bigvee_i \mathcal{U}_i)\mathcal{V} = \bigvee_i (\mathcal{U}_i\mathcal{V})$,

(iii) $\mathcal{U}(\bigwedge_i \mathcal{V}_i) = \bigwedge_i (\mathcal{U}\mathcal{V}_i)$,

(iv) $(\bigwedge_i \mathcal{U}_i)\mathcal{V} = \bigwedge_i (\mathcal{U}_i\mathcal{V})$.

Although **L** has some nice properties (e.g., **L** is distributive, $\mathbf{L}^*$ is a free semigroup) it is clear already from Section 10.5 (see also Section 10.10) that **L** is an extremely complex lattice. It is therefore unrealistic to hope to determine its automorphism group. However, one rather curious automorphism has shown up.

If G is any ℓ-group and we reverse the order on G then we obtain another ℓ-group G^R. Thus $a \leq_R b$ in G^R if and only if $b \leq a$ in G. For any variety $\mathcal{V}$ let

$$\mathcal{V}^R = \{G^R : G \in \mathcal{V}\}.$$

It is easily verified that $\mathcal{V}^R$ is also a variety. So let $\theta: \mathbf{L} \to \mathbf{L}$ be the mapping defined by

$$\mathcal{V}\theta = \mathcal{V}^R \quad (\mathcal{V} \in L)$$

and let

$$\mathcal{F} = \{\mathcal{V} \in L: \ \mathcal{V}\theta = \mathcal{V}\}.$$

The mapping θ has some interesting properties.

THEOREM 10.9.8 [Huss and Reilly, 1984]. *The mapping* θ *is a lattice and semigroup automorphism of* L *with the following properties:*

(i) θ^2 *is the identity mapping;*

(ii) θ *preserves arbitrary joins and meets;*

(iii) $\mathcal{F}$ *is a complete sublattice of* L;

(iv)   *for any* $\mathcal{V} \in L$, $\mathcal{V} \vee \mathcal{V}^R$ *and* $\mathcal{V} \cap \mathcal{V}^R \in \mathcal{F}$.

Examples of varieties contained in $\mathcal{F}$ are: $\mathcal{A}^n$, $\mathcal{N}$, $\mathcal{W}a$, $\mathcal{R}$, Scrimger varieties (see Section 10.13), $\mathcal{V}(N)$ (see Section 10.11), all quasi-representable varieties (see Section 10.5) and all varieties defined by purely group-theoretic laws. Examples of varieties which are moved by θ include the Feil varieties (see Section 10.5) and the Medvedev varieties $\mathcal{V}(W^+)$, $\mathcal{V}(W^-)$ (see Section 10.11: in fact, $\theta: \mathcal{V}(W^+) \to \mathcal{V}(W^-)$). A full treatment of these observations can be found in Huss and Reilly [1984].

Another operation on varieties involves the formation of lexicographic products. For any ℓ-group G, let $G \overset{\leftarrow}{\times} Z$ denote the lexicographic extension of G by Z (that is $G \overset{\leftarrow}{\times} Z = G \times Z$ ordered lexicograpically from the right; Chapter 0 Example 3).

DEFINITION. For any variety of ℓ-groups $\mathcal{V}$, let $\mathcal{V}^L = \mathcal{V}(G \overset{\leftarrow}{\times} Z: G \in \mathcal{V})$ and say that $\mathcal{V}$ has the *lex-property* if $\mathcal{V} = \mathcal{V}^L$.

The formation of lexicographic extensions by Z seems a harmless enough construction and one might have expected that all varieties would have the lex property. Indeed, Smith [1980] showed that $\mathcal{N}$, $\mathcal{R}$, $\mathcal{A}^n$, the Scrimger varieties (see Section 10.13) and $\mathcal{W}a$ have the lex property while Huss [1984] showed that any variety of the form $\mathcal{V}^L$ and the quasi-representable varieties also have the lex property. On the other hand Huss [1984] showed that the Medvedev varieties $\mathcal{V}(N_0)$, $\mathcal{V}(W^+)$ and $\mathcal{V}(W^-)$ (see Section 10.11) and the Feil varieties do not have the lex-property.

The next result collects some general properties of the operator $\mathcal{V} \to \mathcal{V}^L$.

THEOREM 10.9.9 [Huss, 1984].

(i)   There exists a variety $\mathcal{V}$ such that the interval $[\mathcal{V}, \mathcal{V}^L]$ has cardinality $2^{\aleph_0}$.

(ii) The mapping $\mathcal{V} \to \mathcal{V}^L$ is not a lattice homomorphism.

(iii) $(\underset{i \in I}{\vee} \, \mathcal{V}_i)^L = \underset{i \in I}{\vee} \, \mathcal{V}_i^L$

(iv) $(\underset{i \in I}{\wedge} \, \mathcal{V}_i)^L \subseteq \underset{i \in I}{\wedge} \, \mathcal{V}_i^L$

(v) $(\mathcal{U}\mathcal{V})^L \subseteq \mathcal{U}(\mathcal{V}^L)$.

10.10. Locally Nilpotent ℓ-groups

Although one of the principal underlying structures of an ℓ-group is that of a group, the places in the theory of ℓ-groups where purely group theoretic conditions have had a large part to play are relatively and surprisingly few. Of course, the theory of *abelian* ℓ-groups is a major exception. Another is the remarkable result due to Kopytov [1975] which we present below, to the effect that any ℓ-group, which is locally nilpotent as a group, is weakly abelian. Recall that an ℓ-group G is said to be *weakly abelian* if $x^y \le x^2$ for all $x, y \in G$ with $x \ge e$ and that the variety of all weakly abelian ℓ-groups is denoted by $\mathcal{W}a$. Independently and by different methods, Hollister [1978] showed that every nilpotent ℓ-group is representable and Reilly [1983] extended Hollister's arguments to show that every nilpotent ℓ-group is weakly abelian.

We begin with some observations regarding weakly abelian ℓ-groups.

THEOREM 10.10.1 [Martinez, 1972]. *Let G be an ℓ-group. Then the following statements (i)-(v) are equivalent:*

(i) G *is weakly abelian;*

(ii) *There exists a positive integer $m \ge 2$ such that for all $x, y \in G$ with $x \ge e$, $x^m \ge x^y$;*

(iii) *For all $x, y \in G$ with $x \ge e$ and every value M of x, $xM = x^y M$;*

(iv) *For all $x, y \in G$ with $x \ge e$, $x \gg |[x, y]|$;*

(v)   *For all $x, y \in G$ with $x \ge e$, $x \ge |[x, y]|$.*

The statements (i)-(v) imply (vi) and (vii), which are equivalent.

(vi) *All regular subgroups of G are normal;*

(vii) *All convex ℓ-subgroups of G are normal.*

Finally, all these conditions imply that G is representable.

Proof. That (i) implies (ii) is trivial. So let (ii) hold.

We first note that the condition $x^y \leq x^m$, for $x,y \in G$, $x \geq e$, clearly implies that all convex ℓ-subgroups are normal in G , that is (ii) $\Rightarrow$ (vii). Now let M be any value of x . Then M is normal in its cover M^* and M^*/M is (isomorphic to) a subgroup of $\mathbf{R}$. Now conjugation by y induces an automorphism of M^*/M which, by Hion's Lemma, is simply multiplication by a scalar β . Replacing y be y^{-1} , if necessary, we may assume that $\beta \geq 1$. But $x^m \geq x^{y^n}$ for all integers n , where m is fixed, so that $(xM)^{y^n} \leq x^m M$, for all n . Hence $\beta = 1$, $(xM)^y = xM$ and (iii) holds.

(iii) $\Rightarrow$ (iv). Let $x,y \in G$, $x \geq e$ and M be any value of x , with cover M^* . By (iii), $[x,y] \in M$ so that $[M^*,G] \subseteq M \subseteq M^*$. Hence M^* is normal in G . Suppose $M^y \neq M$, for some $y \in G$. Since M^* covers M and M^* is normal in G , M and M^y must be incomparable. Consequently there exists $x \in M^* \backslash M$ with $x^y \in M$. Then $x^y M = M \neq xM$ which contradicts (iii). Hence M is normal in G . As every minimal prime is an intersection of values, we have that each minimal prime is normal in G . It now follows from Theorem 10.2.3 that G is representable. Without loss of generality, we may assume that G is totally ordered. Then (iii) implies that $[x,y] \in M$, the unique value of x so that (iv) must hold.

That (iv) implies (v) is trivial. If (v) holds, and $x,y \in G$ with $x \geq e$, then $x \geq |[x,y]| \geq x^{-1}y^{-1}xy$ so that $x^2 \geq x^y$ and (i) holds.

The equivalence of (vi) and (vii) is trivial and we have already seen that (i) implies (vii). That (vi) implies that G is representable follows from Theorem 10.2.3.

The missing implication in Theorem 10.10.1, namely (vi) implies (i), is not valid in general. An example is provided by Martinez [1972]. However, for varieties this implication does hold, as the following theorem demonstrates.

THEOREM 10.10.2 [Reilly, 1983]. *For a variety $\mathcal{V}$ of ℓ-groups, the following statements are equivalent:*

(i) $\mathcal{V}$ is weakly abelian;

(ii)   If $G \in \mathcal{V}$, then all regular subgroups of G are normal in G ;

(iii) If $G \in \mathcal{V}$, then all convex ℓ-subgroups of G are normal in G .

Proof. From Theorem 10.10.1, it only remains to show that (iii) implies (i). Let (iii) hold and suppose that (i) does not hold. Then there exist $x,y \in G \in \mathcal{V}$ with $x > e$ such that $x^y \not\leq x^2$. From the hypothesis that (iii) holds and Theorem 10.10.1, it follows that G is representable, and can therefore be considered as a subdirect product of totally ordered groups T_i ($i \in I$). For one of these groups, T say, the components a, b of x and y, respectively, must be such that $a > 1$ and $a^b > a^2$. Now, since T is a homomorphic image of G, $T \in \mathcal{V}$. Let N denote the set of natural numbers; for each $i \in N$ let $S_i = T$ and let $H = \Pi\{S_i : i \in N\}$. Then $H \in \mathcal{V}$. Let $f, g \in H$ be such that $f(i) = a$, $g(i) = b^i$ ($i \in N$). Then fg is not contained in the convex ℓ-subgroup of H generated by f, contradicting the hypothesis (iii). The result follows.

It is interesting to note that although the class of weakly abelian ℓ-groups is a variety, the class of ℓ-groups in which all convex ℓ-subgroups are normal is not. The last example in the paper by Kopytov and Medvedev [1976] illustrates this point.

In preparation for Kopytov's theorem, we require the following rather interesting observations about the centre of a subdirectly irreducible ℓ-group.

PROPOSITION 10.10.3 [Kopytov, 1975]. *Let G be a subdirectly irreducible ℓ-group with non-trivial centre C and smallest non-trivial ℓ-ideal N.*

   (i)   *Every non-identity element of C^+ is a weak unit in G.*
   (ii) *C is an abelian totally ordered group.*
   (iii) *If $N \cap C \neq \{e\}$, then $N = \{g \in G : \text{there exists } c \in N \cap C \text{ with } |g| \leq c\}$.*
   (iv) *$N \cap C$ is Archimedean.*

Proof. (i) Suppose that $e < a \in C$ and that a is not a weak unit. Then there exists $e \neq b \in G$ with $a \wedge b = e$. Since a is central in G, $a\backslash s^{\perp}$ and $a^{\perp\perp}$ are ℓ-ideals of G which are non-trivial ($a \in a^{\perp\perp}$, $b \in a^{\perp}$ and disjoint ($a^{\perp} \cap a^{\perp\perp} = \{e\}$). But this contradicts the minimality of N.

(ii) Clearly, if $a, b \in C$, then so do $a \vee b$ and $a \wedge b$. Thus C is an abelian ℓ-group in which every positive element different from e is a weak unit. Therefore C is totally ordered.

(iii) Clearly $K = \{g \in G: \text{there exists } c_g \in N \cap C \text{ with } |g| \leq c_g\}$ is contained in N. Since $N \cap C$ is normal in G, it follows easily that K is an ℓ-ideal of G. But $\{e\} \neq N \cap C \subseteq K \subseteq N$ so that we must have $K = N$, by the minimality of N.

(iv) Suppose that B is a non-trivial proper convex subgroup of $N \cap C$. Then $M = \{g \in G: \text{there exists } b_g \in B \text{ with } |g| \leq b_g\}$ is clearly an ℓ-ideal of G contained in N and such that $\{e\} \neq B = M \cap C \subset N \cap C$. Thus M is a non-trivial ℓ-ideal of G properly contained in N, which is a contradiction.

Hence $N \cap C$ has no non-trivial proper convex ℓ-subgroups and so is Archimedean.

We are now ready for Kopytov's Theorem. We emphasize that when we say that an ℓ-group G is *locally nilpotent* we mean that *as a group* it is locally nilpotent.

THEOREM 10.10.4 [Kopytov, 1975]. *If the ℓ-group G is a locally nilpotent group, then G is representable.*

Proof. Since G is a subdirect product of subdirectly irreducible ℓ-groups which must also be locally nilpotent, there is no loss in assuming that G is subdirectly irreducible. Since G is representable if and only if every finitely generated ℓ-subgroup is representable, we may also assume that G is generated as an ℓ-group by the elements $a_1,...,a_k$. Then every element of G can be expressed in the form $h = \underset{i}{\vee} \underset{j}{\wedge} h_{ij}$ $(1 \leq i \leq n, \; 1 \leq j \leq m)$ where each h_{ij} lies in the *subgroup* A generated by $a_1,...,a_k$. Since G is locally nilpotent, A is nilpotent and has non-trivial centre $Z(A)$. Since any element of $Z(A)$ will clearly commute with any element of the form of h, it follows that the centre C of G is non-trivial.

Let N denote the smallest non-trivial ℓ-ideal of G, which exists since G is subdirectly irreducible. Let $b \in N$, $b \neq e$ and H denote the subgroup generated by $b,a_1,...,a_k$. Then H is nilpotent and $N \cap H$ is normal in H. Hence, $(N \cap H) \cap Z(H)$ is non-trivial, where $Z(H)$ denotes the centre of H. But $Z(H) \subseteq C$ and so $N \cap C$ is non-trivial. By Proposition 10.10.3. (iv), $N \cap C$ is an Archimedean totally ordered group.

Claim: $N = N \cap C$. Suppose not. Then there exist $a \in N$, $x \in G$ with $e \neq [a,x] \in N \cap C$ (take $b \in N \backslash (N \cap C)$ and consider the second term of the upper central series of $H = \langle b,a_1,...,a_k \rangle$). Let $c = [a,x]$. Replacing x by x^{-1}, if necessary, we may assume that $c > e$ since $N \cap C$ is totally ordered and $c^{-1} = x^{-1}[a,x^{-1}]x$.

We have $x^{-1}ax = ac$ so that, for any integer k, $x^{-k}ax^k = ac^k$. But $a \in N$ and so, by Proposition 10.10.3.(iii) there exists a $z \in N \cap C$ such that $z^{-1} < a < z$. By Proposition 10.10.3.(iv), there exists an integer k such that $c^k > z$. Then $c^{-k} < a < c^k$ from which we get that $ac^k > e$, that is, $x^{-k}ax^k = ac^k > e$ so that $a > e$. But we also have, since $c = [a,x] \in C$,

$$[a^{-1},x^{-1}] = (ax)[a,x](ax)^{-1} = c$$

so that $e \neq [a^{-1},x^{-1}] \in N \cap C$. Thus we can apply exactly the same argument as above to conclude that $a^{-1} > e$. This is clearly a contradiction. Hence $N = N \cap C$, as claimed.

Now suppose that G is not representable. Then there exist $a,x \in G$ with $a \wedge a^x = e$. If $b \in N^+$ and $b \leq a$ then $e \leq b \wedge b^x \leq a \wedge a^x = e$ so that $b \wedge b^x = e$. But $N = N \cap C$ is totally ordered so that we must have $b = e$. Thus $a \wedge b = e$ for all $b \in N^+$ so that $a \in N^\perp$ and $\{e\} \neq N^\perp$. Since N is normal in G so is $N^\perp$. Thus $N^\perp$ is a non-trivial ℓ-ideal with $N^\perp \cap N = \{e\}$, which contradicts the assumption that N is the smallest proper ℓ-ideal in G. Therefore G must be representable.

Now that we know that every locally nilpotent ℓ-group is representable, we can go a step further.

THEOREM 10.10.5 [Kopytov, 1975]. *Every locally nilpotent ℓ-group is weakly abelian.*

Proof. Let G be a locally nilpotent ℓ-group. By Theorem 10.10.4, G is representable and so is a subdirect product of totally ordered groups G_i. To show that G is weakly abelian, it suffices to show that each G_i is weakly abelian and that will follow if every finitely generated subgroup of G_i is weakly abelian. Since each G_i is also (as a homomorphic image of G) locally nilpotent, we may assume that G is totally ordered and nilpotent of class n, say. Then, for all $a,b \in G$,

$$[a,\underbrace{b,...,b}_{n}] = e. \tag{10.6}$$

We argue by contradiction. Suppose that G is not weakly abelian. Then there must exist $x,y \in G$, with $e < x$ and $x^y > x^2$. Then $x^{y^2} = (x^y)^y > (x^2)^y = (x^y)^2 > x^4$, and by induction, $x^{y^m} > x^{2^m}$, for all positive integers m. We make an intermediate claim that

$$[x,\underbrace{y^m,...,y^m}_{k}]^{y^m} > [x,\underbrace{y^m,...,y^m}_{k}]^{2^{m-k}} > e \quad \text{for all } 1 \leq k < m. \tag{10.7}$$

For $k = 1$, we have

$$[x,y^m]^{y^m} = (x^{-1}x^{y^m})^{y^m} > (x^{-1}x^{2^m})^{y^m} > (x^{2^m-1})^{y^m}$$
$$= (x^{y^m})^{2^m-1} > (x^{-1}x^{y^m})^{2^m-1} = [x,y^m]^{2^m-1} > e$$

(since $x^{y^m} > x^{2^m}$). Thus (10.7) holds for $k = 1$. Now suppose that (10.7) holds for k, that $k+1 < m$ and let

$$u = [x,\underbrace{y^m,...,y^m}_{k}].$$

Then $u^{y^m} > u^{2^{m-k}} > e$ so that

$$[u,y^m]^{y^m} = (u^{-1}uy^m)^{y^m} > (u^{-1}u^{2^{m-k}})^{y^m} > (u^{2^{m-k-1}})^{y^m}$$

$$= (uy^m)^{2^{m-k-1}} > (u^{-1}uy^m)^{2^{m-k-1}} = [u,y^m]^{2^{m-k-1}} > e$$

where the final inequality follows from the fact that $uy^m > u^{2^{m-k}}$ and $k+1 < m$. Thus the claim holds for $k + 1$ and so (10.7) is verified. Hence, with $m = n + 1$ and $k = n$ we obtain

$$[x,\underbrace{y^{n+1},...,y^{n+1}}_{n}] \neq e$$

which contradicts (10.6). Therefore, G must be weakly abelian.

In this connection it is interesting to note the following observation due to Medvedev.

PROPOSITION 10.10.6 [Medvedev, 1983b]. *The class of locally nilpotent ℓ-groups is a torsion class.*

These results might give the impression that the lattice of varieties of nilpotent ℓ-groups might be in some sense simpler than the lattice L of all varieties and might even raise hopes of "describing" all or part of this lattice in some detail. However, all such hopes were dashed by the recent work of Gurchenkov. In both [1982a] and [1984c, 1985], Gurchenkov produced continua of varieties of nilpotent ℓ-groups. In particular,

THEOREM 10.10.7 [Gurchenkov, 1984c, 1985]. *For every integer $n \geq 3$, the lattice of varieties of ℓ-groups of nilpotent class n has the cardinality of the continuum.*

We say that a variety $\mathcal{V}$ has *infinite axiomatic rank* if there is no system of identites involving just a finite number of variables which defines $\mathcal{V}$.

THEOREM 10.10.8 [Gurchenkov, 1984c, 1985]. *There exists a variety of nilpotent class 3 metabelian ℓ-groups with infinite axiomatic rank.*

Another result indicating the complexity of varieties of nilpotent ℓ-groups is due to Kopytov.

THEOREM 10.10.9 [Kopytov, 1982]. *For every integer $k \geq 2$ the variety of nilpotent ℓ-groups of nilpotency class $\leq k$ is not generated by any finitely generated ℓ-group.*

The existence of varieties of ℓ-groups which are not generated by any finitely generated ℓ-groups was first established by Kopytov and Medvedev [1975].

10.11. Representable Covers for $\mathcal{A}$

In this and the next section we consider varieties that cover $\mathcal{A}$. In this section we consider representable covers and in the next non-representable covers. First we introduce some specific ordered groups.

The free nilpotent class two group on two generators can be described in terms of generators and relations as

$$N = \langle a,b,c: c = [a,b],[a,c] = [b,c] = e\rangle$$

where a and b are the two free generators. Every element of N can be written uniquely in the form $a^m b^n c^k$. Let N_0 denote N together with the total ordering defined by

$$a^m b^n c^k > e \Leftrightarrow \text{(i) } m > 0 \text{ or (ii) } m = 0 \text{ and } n > 0$$

$$\text{or (iii) } m = n = 0 \text{ and } k > 0.$$

It is straightforward to verify that N_0 is a totally ordered group.

Now let W^+ denote $\mathbf{Z}wr\mathbf{Z}$ totally ordered by

$$(F,k) > e \Leftrightarrow \text{either (i) } k > 0 \text{ or (ii) } k = 0 \text{ and } F(r) > 0$$
$$\text{where } r \text{ is the maximum element of the support of } F.$$

Let W^- denote $\mathbf{Z}wr\mathbf{Z}$ totally ordered by

$$(F,k) > e \text{ either (i) } k > 0 \text{ or (ii) } k = 0 \text{ and } F(r) > 0$$
$$\text{where } r \text{ is the minimum element in the support of } F.$$

Then W^+ and W^- are both totally ordered solvable groups.

LEMMA 10.11.1. *Let G be an o-group and $e < a, b \in G$.*

(i) If $a \ll a^b$, then $e < [a,b] \ll [a,b]^b$ and $[a,b]^{b^m} \ll b$, for all $m \in \mathbf{Z}$.

(ii) If $a^b \ll a$, then $e < [b,a]^b \ll [b,a]$ and $[b,a]^{b^m} \ll b$, for all $m \in \mathbf{Z}$.

Proof. (i) Since $a \ll a^b$ we have $a^2 < a^b = b^{-1}ab$, so that $e < a < a^{-1}b^{-1}ab = [a,b]$. Also $a \ll a^b$ implies that $a^b \ll a^{b^2}$. Hence, for any positive integer n,

$$e < [a,b]^n = (a^{-1}a^b)^n \le (a^b)^n = (a^b)^{-1}(a^b)^{n+1} \le (a^b)^{-1}a^{b^2}$$

$$= b^{-1}a^{-1}bb^{-2}ab^2 = b^{-1}(a^{-1}b^{-1}ab)b = [a,b]^b.$$

Thus $[a,b] \ll [a,b]^b$. Now, for any $e < g \in G$, if $g \ll g^b$, then $g^{n+1} \le bg^{n+1} \le bg^b = gb$ so that $g^n \le b$ and $g \ll b$. Since $[a,b] \ll [a,b]^b$, we have $[a,b]^{b^m} \ll [a,b]^{b^n}$ if (and only if) $m < n$. Hence $[a,b]^{b^m} \ll [a,b]^{b^{m+1}}$ so that $[a,b]^{b^m} \ll b$. The verificiation of (ii) is similar.

This prepares us for the first main theorem of this section due to Medvedev.

THEOREM 10.11.2 [Medvedev, 1977]. *The solvable representable covers of* $\mathcal{A}$ *are precisely* $\mathcal{V}(N_0)$, $\mathcal{V}(W^+)$ *and* $\mathcal{V}(W^-)$.

Proof. Let $\mathcal{V}$ be any solvable representable variety different from $\mathcal{A}$. Then there exists a solvable non-abelian o-group G, say, in $\mathcal{V}$. Replacing G by a member of its derived series, if necessary, we may assume that G is metabelian.

Case I. Suppose that there exist elements a,b with $e < a,b$ such that $a \ll a^b$. By Lemma 10.11.1, there exists a commutator c with $e < c \ll c^b$. Then we have

$$\ldots \ll c^{b^{-1}} \ll c << c^b \ll c^{b^2} \ll \ldots .$$

Moreover, $H = \langle c^{b^n} : n \in \mathbf{Z} \rangle$ is abelian, since G is metabelian. Again by Lemma 10.11.1, $H \ll b$. Therefore $K = \langle b, c^{b^n} : n \in \mathbf{Z} \rangle \cong W^+$ so that $W^+ \in \mathcal{V}$.

Case II. If there exist $e < a,b$ with $a^b \ll a$ then an argument similar to that in Case I gives $W^- \in \mathcal{V}$.

Case III. We can now assume that neither of Cases I and II arises in any metabelian o-group in $\mathcal{V}$. Therefore, for all, $e < a,b \in G \in \mathcal{V}$, where G is a metabelian o-group, we have that a and a^b are archimedean equivalent. Therefore all convex ℓ-subgroups are normal.

Now let $e < a \in G$ and $b \in G$ be such that $[a,b] \ne e$. Let M be a value of a. Then we claim that $[a,b] \in M$. Because, if not, then $x \to b^{-1}xb$ induces a non-trivial automorphism β of M^*/M, which is an archimedean o-group and therefore (isomorphic to) a subgroup of the real numbers. Hence

β is simply a scalar multiplication which can assume to be a scalar greater than 1. Therefore, for all positive integers n there exists an integer k_n such that

$$Mb^{-k_n} ab^{k_n} > Ma^n$$

so that $b^{-k_n} ab^{k_n} > a^n$. Let $f, g \in P = \Pi\{G_n: n = 1, 2, 3, ...\}$, $G_n = G$, be such that $f(n) = a$, $g(n) = b^{k_n}$, for all n. Then fg and f are not archimedeanly equivalent and if we take a quotient of P modulo any ultrafilter we are back in Case I or II.

Therefore $c = [a, b] \in M$ and also $[a, b^{-1}] \in M$. So we may assume that $e < a, b$. Taking $[a, b]$ or $[b, a]$, as appropriate, we can assume that $[a, b] > e$. Let $H = \langle a, b \rangle$ and V be a value of c in H. Then, as above, $[c, a], [c, b] \in V$. Therefore c is central in H/V and H/V is nilpotent class 2.

Let $H_n = H$, for all n, and $f, g, h \in Q = \Pi\{H_n: n = 1, 2, 3, ...\}$ be defined by $f(n) = ab^n$, $g(n) = b$, $h(n) = c$. Let U be any non-principal ultrafilter on $N = \{1, 2, 3, ...\}$ and f', g' and h' denote the images of f, g and h in the ultraproduct $S = \Pi_U H_n$. Then S is nilpotent class two and $h' \ll g' \ll f'$. Since $[f', g'] = h'$ and $[h', f'] = [h', g'] = e$ it follows that $\langle f', g', h' \rangle$ is isomorphic to N_0.

Finally $\mathcal{V}(N_0)$, $\mathcal{V}(W^+)$ and $\mathcal{V}(W^-)$ are all incomparable because:

$\mathcal{V}(N_0)$ satisfies $[[x, y], z] = e$

$\mathcal{V}(W^+)$ satisfies $|[x, y]|^2 \leq |z|^{-1} |[x, y]| |z|$ for $e \leq x \leq y \leq z$

$\mathcal{V}(W^-)$ satisfies $|[x, y]| \geq |z|^{-1} |[x, y]|^2 |z|$ for $e \leq x \leq y \leq z$

while in each case the other two varieties do not satisfy the given identity. This completes the proof.

That $\mathcal{V}(N_0)$ covers $\mathcal{A}$ was established independently by Martinez [1979].

Generators for two non-solvable representable varieties that cover $\mathcal{A}$ are also known.

LEMMA 10.11.3 [Feil, 1980]. *Let G be an o-group such that*

$$e < b \ll a \Rightarrow b \ll b^a. \tag{10.8}$$

Then G satisfies the identity

$$|[a, b]| \leq [|[a, b]|, |[a, b]|^a] \quad \text{for } e \leq b \leq a \tag{10.9}$$

and consequently $\mathcal{V}(G)$ contains none of $\mathcal{V}(N_0)$, $\mathcal{V}(W^+)$ and $\mathcal{V}(W^-)$.

Proof. Let G be an o-group satisfying (10.8) and $a,b \in G$ be such that $e \le b \le a$. Let M be the value of a and M^* its cover. Then M^*/M is abelian and therefore $[a,b] \in M$ so that $c = |[a,b]| \ll a$. By (10.8), $c \ll c^a$ and applying (10.8) again, we get $c \ll c^d$, where $d = c^a$. Hence $c^2 \le c^d$ so that $c \le c^{-1}c^d = [c,d]$ which is (10.9).

Now N_0, W^+ and W^- are all in $\mathcal{A}^2$ and therefore in any of these the right hand side of the inequality in (10.9) will always be e, whereas these groups are not abelian and so the left hand side will, for suitable a,b, differ from e. Thus the final claim holds.

Bergman and Kopytov provided an example of a group satisfying (10.8).

THEOREM 10.11.4 [Bergman, 1984], [Kopytov, 1985]. *Let X be a set with $|X| \ge 2$ and G be the free group on X. Then G can be ordered to give an ordered group satisfying (10.8).*

If G is as constructed in Theorem 10.11.4, then $\mathcal{V}(G)$ contains some cover C, say, of $\mathcal{A}$ since $\mathcal{A}$ is finitely based and cannot be the intersection of an infinite descending chain of varieties. By Lemma 10.11.3, $C \ne \mathcal{V}(N_0)$, $\mathcal{V}(W^+)$ or $\mathcal{V}(W^-)$. More specifically, we have:

THEOREM 10.11.5 [Powell and Tsinakis, a]. *Let F be the free group on two generators ordered so as to satisfy (10.8). Let F^R be the ordered group obtained from F by reversing the order. Then $\mathcal{V}(F)$ and $\mathcal{V}(F^R)$ are distinct representable non-solvable covers of $\mathcal{A}$.*

In fact, Powell and Tsinakis (private conversation) have shown that if G is the free group on X where $|X| = 2$ and ordered as in Theorem 10.11.6, then $\mathcal{V}(G)$ covers $\mathcal{A}$. If the order on G is reversed then we obtain a second ordered free group G' such that $\mathcal{V}(G')$ is distinct from $\mathcal{V}(G)$ and also covers $\mathcal{A}$. Thus there are at least five representable varieties that cover $\mathcal{A}$. It is conjectured that there may be a continuum of such covers.

10.12 Solvable Varieties

To this point, most of the examples of specific varieties considered have been varieties of representable ℓ-groups, with the notable exceptions of L, $\mathcal{N}$, and $\mathcal{A}^n$, $n = 2,3,\dots$. We have seen, especially in Sections 10.5, 10.10

and 10.11 just how rich and complicated the lattice of varieties of representable ℓ-groups is. Intuition might have led us to expect that varieties generated by totally ordered groups would be "easier" to deal with and one might have expected to be able to describe a significant part of the bottom of the lattice. However, not only has the determination of a complete list of representable covers for $\mathcal{A}$ proven elusive, but little progress has been made towards determining bases of identities for the known covers or what lies immediately above the covers. In contrast to this, a great deal has been discovered recently about small non-representable varieties and we will describe this work in the next two sections. Apart from Lemma 10.12.1, due to Martinez [1972], the main results of this section are due to Reilly [1986].

It turns out to be very useful to consider varieties that contain no representable ℓ-groups other than abelian ℓ-groups. The first varieties of this type were introduced by Martinez [1972]: for any positive integer n, let $\mathcal{L}_n$ denote the variety defined by $x^n y^n = y^n x^n$.

LEMMA 10.12.1 [Martinez, 1972]. $\mathcal{L}_n \cap \mathcal{R} = \mathcal{A}$.

Proof. That $\mathcal{A} \subseteq \mathcal{L}_n \cap \mathcal{R}$ is clear. Suppose that $\mathcal{L}_n \cap \mathcal{R}$ contains a non-abelian ℓ-group. Then it must contain a non-abelian o-group G, say. So there must be $a, x \in G$ with $e < a$ such that $a^x \neq a$. Replacing x by x^{-1}, if necessary, we may assume that $a < a^x$. Then, since conjugation by x is an order automorphism of G,

$$a < a^x < a^{x^2} < \dots a^{x^n}$$

so that $a^n < (a^{x^n})^n = (a^n)^{x^n} = a^n$, a contradiction. Thus $\mathcal{L}_n \cap \mathcal{R} \subseteq \mathcal{A}$, as required.

In case this suggests that the varieties $\mathcal{L}_n$ might be too restrictive to be interesting, we hasten to point out that $\vee \{\mathcal{L}_n : n = 1,2,3,\dots\} = \mathcal{N}$ (see Smith [1976, 1981] or Reilly and Wroblewski [1981]).

The next useful observation can be found in Darnel [1987] or Reilly [1986].

LEMMA 10.12.2. *If C is a convex ℓ-subgroup of the ℓ-group G, then the normalizer $N_G(C)$ of C in G is an ℓ-subgroup.*

Proof. It suffices to show that if $a, b \in N = N_G(C)$ then $a \vee b \in N$ and this will follow if, for all $e \leq g \in C$, we can show that $h = (a \vee b)^{-1} g(a \vee b) \in C$. But for $e \leq g \in C$,

$$e \le h = (a^{-1} \wedge b^{-1})g(a \vee b)$$
$$= (a^{-1}ga \wedge b^{-1}ga) \vee (a^{-1}gb \wedge b^{-1}gb)$$
$$\le a^{-1}ga \vee b^{-1}gb \in C$$

so that, by the convexity of C, $h \in C$, as required.

As seen in Lemma 10.12.1, if $G \in \mathcal{L}_n$ for some positive integer n, then $\mathcal{V}(G) \cap \mathcal{R} = \mathcal{A}$. The next result gives some useful characterizations of when the reverse implication holds.

THEOREM 10.12.3 [Reilly, 1986]. *Let G be an ℓ-group such that $\mathcal{V}(G) \cap \mathcal{R}$ = $\mathcal{A}$ and let n be a positive integer. Then the following statements are equivalent.*

(i) $G \in \mathcal{L}_n$.

(ii)   *For all convex ℓ-subgroups P of G and all $x \in G$, $x^{-n}Px^n = P$.*

(iii)   *There exists a family $\mathcal{P}$ of primes such that $\cap \mathcal{P} = \{e\}$ and for all $P \in \mathcal{P}, x \in G, x^{-n}Px^n = P$.*

Proof. First suppose that $G \in \mathcal{L}_n$ and let P be a convex ℓ-subgroup of G. Then, for all $a \in P$, $a > e$, $x^{-n}ax^n \le x^{-n}a^nx^n = a^n \in P$, so that $x^{-n}Px^n \subseteq P$. Applying the same argument to x^{-1}, we obtain $P \subseteq x^{-n}Px^n$ and equality follows. Thus (i) implies (ii). If (ii) holds then the family $\mathcal{P}$ of all minimal prime subgroups of G will clearly satisfy (iii).

Now let (iii) hold, $P \in \mathcal{P}$ and $x,y \in G$. By (iii), $x^n, y^n \in N = N_G(P)$. By Lemma 10.12.2, N is an ℓ-subgroup of G. Let H denote the ℓ-subgoup of N generated by P, x^n and y^n. Then P is a normal prime subgroup of H so that H/P is totally ordered. But then $H/P \in \mathcal{V}(G) \cap \mathcal{R} = \mathcal{A}$, so that H/P must be abelian. Thus $[x^n, y^n] \in P$ for all $P \in \mathcal{P}$, and so $[x^n, y^n] = e$. Therefore $G \in \mathcal{L}_n$ and (i) holds.

Recall that a prime subgroup P of an ℓ-group G is said to be a *representing* prime subgroup if $\cap \{P^g : g \in G\} = \{e\}$.

COROLLARY 10.12.4 [Reilly, 1986]. *Let G be an ℓ-group such that $\mathcal{V}(G) \cap \mathcal{R} = \mathcal{A}$. Let n be a positive integer and P be a representing prime. Then $G \in \mathcal{L}_n$ if and only if $x^{-n}Px^n = P$ for all $x \in G$.*

Proof. Let $G \in \mathcal{L}_n$. Then $x^{-n}Px^n = P$, for all $x \in G$, by Theorem 10.12.3 (ii). Conversely, let $x^{-n}Px^n = P$, for all $x \in G$. Let $\mathcal{P} = \{P^g : g \in G\}$. Then $\cap \mathcal{P} = \{e\}$, since P is a representing prime subgroup. Also, for any $Q = P^g$,

$x \in G$, $x^{-n}Qx^n = x^{-n}g^{-1}Pgx^n = g^{-1}(gxg^{-1})^{-n} P(gxg^{-1})^n g = g^{-1}Pg = Q$ so that by Theorem 2.3 (iii), $G \in L_n$.

One immediate and interesting consequence of these observations is that the varieties of the form L_n form a meet subsemilattice of the lattice of all varieties.

COROLLARY 10.12.5 [Reilly, 1986]. *For any positive integers m and n, $L_m \cap L_n = L_d$, where d is the greatest common divisor of m and n.*

Proof. Clearly $L_d \subseteq L_m \cap L_n$. So let G be any subdirectly irreducible ℓ-group in $L_m \cap L_n$. Then G has a representing prime P, say. Since $G \in L_m$, we have $x^{-m}Px^m = P$ and since $G \in L_n$, $x^{-n}Px^n = P$, for any $x \in G$. Since there exist integers a,b such that $d = am + bn$, it follows that $x^{-d}Px^d = P$. By Corollary 10.12.4, $G \in L_d$.

COROLLARY 10.12.6. *Let G be an ℓ-group and V be a variety such that $G \in L_n$ and $V \subseteq L_n$. If m is the smallest positive integer such that $G \in L_m$ (respectively, $V \subseteq L_m$) then m divides n.*

An important technique in the discussions below is a method of producing disjoint conjugate elements introduced by Scrimger [1975]. Towards this end we need some notation.

For any ℓ-group G, and $x,y \in G$ and integers n, p and r, let

$$d_0(n,p^r,x,y) = x \wedge (x^{-1}y^n x \wedge x^{-2}y^n x^2 \wedge \dots \wedge x^{p^r+1}y^n x^{p^r-1})$$

$$d_i(n,p^r,x,y) = x^{-i}d_0(n,p^r,x,y)x^i \qquad 0 \le i \le p^r - 1$$

$$e_i(n,p^r,x,y) = \bigvee_{\substack{j=0 \\ j \ne i}}^{p^r-1} (d_i \wedge d_j) \qquad 0 \le i \le p^r - 1$$

$$a_i(n,p^r,x,y) = d_i(n,p^r,x,y)e_i(n,p^r,x,y)^{-1}.$$

For any integers m,n let (m,n) denote the greatest common divisor of m and n.

LEMMA 10.12.7. *Let $G \in L_n$, where $n = p^r m$, for some prime number p, and $(p,m) = 1$. Let $y,z \in G$ and $x = z^m$. Let $a_i = a_i(n,p^r,x,y)$.*

(i) $x^{-1}a_i x = a_{i+1} \qquad 0 \le i < p^r - 1$

(ii) $x^{-1}a_{p^r-1}x = a_0$

(iii) $a_i \wedge a_j = e$ for $i \neq j$.

Proof. Straightforward.

Of course, for arbitrarily chosen elements in an ℓ-group G in L_n, there is no guarantee that $a_i(n, p^r, x, y)$ will not be the identity.

For any prime subgroup P of an ℓ-group G and any $g \in G$, g induces a permutation $g^\wedge$ on the set $\mathcal{P}$ of conjugates of P by conjugation $g^\wedge: P^a \to P^{ag}$. The set of all such actions forms a permutation group on $\mathcal{P}$. For any $g \in G$, we denote by $O(P, g)$ the orbit of $g^\wedge$ containing P. The statement "$x^{-n}Px^n = P$" appearing in Theorem 10.12.3 and Corollary 10.12.4 can then be replaced by the equivalent statement "$|O(P, x)|$ divides n".

LEMMA 10.12.8 [Reilly, 1986]. *Let $G \in L_n$, where $n = p^r m$ for some prime p, $r \geq 1$ and $(p, m) = 1$. Let P be a prime subgroup of G and $z \in G$ be such that $|O(P, z)| = p^r$. Let $x = z^m$. Then there exists an element $y \in G$ such that $a_i = a_i(n, p^r, x, y) > e$, for $i = 0, ..., p^r - 1$. In particular, $a_0 \notin P$ and all the elements a_i, $i = 0, ..., p^r - 1$, are distinct.*

Proof. Since $(p, m) = 1$, we also have $|O(P, x)| = p^r$. Let $P_i = x^{-i}Px^i$, for $i = 0, 1, ..., p^r - 1$. Let y be such that $e < y$ and

$$y \in P_0^* = P_0 \setminus \bigcup\{P_i: 1 \leq i \leq p^r - 1\}.$$

Then $y^n \in P_0^*$ and $x^{-i}y^nx^i \in P_i^* = P_i \setminus \bigcup\{P_j: j \neq i\}$. In particular, $x^{-i}y^nx^i \in P_i \setminus P$, $i \neq 0$, while $x \notin P$, since $P^x \neq P$. Hence $d_0(n, p^r, x, y) \in (\bigcap\{P_i: i \neq 0\}) \setminus P$. Similarly $d_i(n, p^r, x, y) \in (\bigcap\{P_j: j \neq i\}) \setminus P_i$. It follows that

$$e_i = e_i(n, p^r, x, y) \in \bigcap\{P_j\} \subseteq P_i.$$

Now it is clear that $e_i \leq d_i$, for all i. Since $e_i \in P_i$ but $d_i \notin P_i$, it follows that $e_i < d_i$ and therefore $a_i > e$, as required.

LEMMA 10.12.9 [Reilly, 1986]. *Let $G \in L_n$, where $n = p^r m$ for some prime p, $r \geq 1$ and $(p, m) = 1$. Let $x \in G$ and H be a convex ℓ-subgroup of G such that*

$$H \cap x^{-i}Hx^i = \{e\} \text{ for } 1 \leq i \leq p^r - 1 \tag{10.10}$$

and

$$H = x^{p^r}Hx^{p^r}.$$

Then $H \in L_m$. (Note that $L_1 = \mathcal{A}$.)

Proof. Suppose that $H \notin L_m$. Since $H \in L_n$, there exists a smallest integer s, say, such that $H \in L_s$. By Theorem 10.12.3 (ii) and since $G \in L_n$,

$x^{-n}Px^n = P$, for all minimal primes P in H and elements $x \in G$, whereas, by Theorem 10.12.3 (iii), s must be the smallest positive integer such that $x^{-s}Px^s = P$, for all minimal primes P in H and all elements $x \in H$. It follows that s divides n. Since $H \notin \mathcal{L}_m$, s does not divide m. Hence $s = p^k l$, for some integers $k,l \geq 1$. By Theorem 10.12.3 (iii), there must exist a prime subgroup P in H and $z \in H$ such that $|O(P,z)| = p^k t$, where $1 \leq t \leq l$. Replacing z by z^t, if necessary, we may assume that $|O(P,z)| = p^k$. By Lemma 10.12.8, there exist $u, v \in H$ such that $a_i = a_i(n, p^k, u, v) > e$, for $0 \leq i \leq p^k - 1$.

Now since $(p,m) = 1$, we can replace x by x^m without altering the hypothesis and so can assume that $x = g^m$, for some $g \in G$. Also, since $u \in H$, it follows from (10.10) that u commutes with $x^{-i}a_0 x^i$, for $1 \leq i \leq p^r - 1$. Therefore

$$a_0^n = (xu)^{-n}a_0^n(xu)^n$$
$$= (xu)^{-n+1}u^{-1}x^{-1}a_0^n xu(xu)^{n-1} = (xu)^{-n+1}x^{-1}a_0^n x(xu)^{-n+1}$$
$$= \ldots$$
$$= (xu)^{-n+p^r}u^{-1}x^{-p^r}a_0^n x^{p^r}u(xu)^{n-p^r}$$
$$= (xu)^{-n+p^r}u^{-1}a_0^n u(xu)^{n-p^r} \text{ since } x^{p^r} = g^n$$
$$= \ldots$$
$$= u^{-m}a_0^n u^m$$
$$= a_j^n \text{ where } j \equiv m \bmod p^r,$$

which is a contradiction, since $(p,m) = 1$. Hence $H \in \mathcal{L}_m$, as required.

Recall from Section 10.6 that for any ℓ-group G and any variety of ℓ-groups $\mathcal{U}$ there is a largest convex ℓ-subgroup $\mathcal{U}(G)$ of G such that $\mathcal{U}(G) \in \mathcal{U}$. Moreover, $\mathcal{U}(G)$ is an ℓ-ideal.

THEOREM 10.12.10 [Reilly, 1986]. *If the positive integer n can be written as a product of k prime numbers (not necessarily distinct) then $\mathcal{L}_n \subseteq \mathcal{A}^{k+1}$.*

Proof. The argument is by induction on k. If $k = 0$, then $n = 1$ and $\mathcal{L}_1 \subseteq \mathcal{A}$. So suppose that the claim holds for any integer that is a product of $k - 1$ or fewer prime numbers and let n be a product of k primes. Let $n = p_1^{r_1} \ldots p_l^{r_l}$, where the p_i, $i = 1,\ldots,l$ are distinct prime numbers and $r_i \geq 1$. Let $n_i = n/p_i$, $i = 1,\ldots,l$. Let $\mathcal{U} = \vee\{\mathcal{L}_{n_i} : i = 1,\ldots,l\}$ and $G \in \mathcal{L}_n$. By the induction aassumption $\mathcal{U} \subseteq \mathcal{A}^k$, so that if $G \in \mathcal{U}$ then the claim follows. So suppose that $G \notin \mathcal{U}$. Since $G \notin \mathcal{L}_{n_i}$, there must exist, by Theorem 10.12.3, a prime subgroup P of G and an element z of G such that $|O(P,z)| = p_1^{r_1}m$, for some integer $m \geq 1$. Replacing z by z^m, if necessary, we may assume that $|O(P,z)| = p_1^{r_1}$. By Lemma 10.12.8, there exist $x,y \in G$ such that $a_0 = a_0(n,p_1^{r_1},x,y) > e$. If we let H denote the convex ℓ-subgroup of G generated by a_0 then H is non-trivial and, by Lemma 10.12.9, $H \in \mathcal{L}_m$. By the induction hypothesis, $H \in \mathcal{A}^k$ and therefore has a non-trivial convex abelian ℓ-subgroup. Thus $\mathcal{A}(H) \neq \{e\}$ and $\mathcal{A}(G) \neq \{e\}$. By Theorem 10.6.3, $\mathcal{A}(G)$ is an ℓ-ideal.

Let $K = G/\mathcal{A}(G)$. If $K \in \mathcal{U}$, then $K \in \mathcal{A}^k$ by the induction hypothesis, so that $G \in \mathcal{A}^{k+1}$, as required. So suppose that $K \notin \mathcal{U}$. Then some subdirectly irreducible homomorphic image of K does not lie in $\mathcal{U}$ and so without loss of generality we can assume that K is subdirectly irreducible. Then K must have a representing minimal prime Q, say. Since, for all i, $K \notin \mathcal{L}_{n_i}$ it follows, by Corollary 10.12.4 and the argument in the first part of the proof for $i = 1$, that for each i there exists an element $z_i \in K$ such that $|O(Q,z_i)| = p_i^{r_i}$. Then, by Lemma 10.12.8, there exist g_i, $y_i' \in K$ such that for $x_i' = (g_i')^{m_i}$, where $m_i = n/p_i^{r_i}$, $b_i = a_0(n,p_i^{r_i},x_i',y_i') > e$ and $b_i \notin Q$. Let g_i and y_i be pre-images of g_i' and y_i' under the natural homomorphism of G onto $K = G/\mathcal{A}(G)$, and let $x_i = g_i^{m_i}$. Then $a_{0i} = a_0(n,p_i^{r_i},x_i,y_i)$ is a pre-image of b_i and so $a_{0i} > e$ and $a_{0i} \notin P$, where P is the pre-image of Q. By Lemma 10.12.9, the convex ℓ-subgroup H_i of G generated by a_{0i} must lie in $\mathcal{L}_{m_i}$. Since $a_{0i} \notin P$, for all i, $a = \wedge_i a_{0i} \notin P$. However, if H denotes the convex ℓ-subgroup of G generated by a, then $H \subseteq H_i$, for all i. Therefore $H \in \cap_i \mathcal{L}_{m_i}$ for all i. But the greatest common divisor of the m_i is 1. Hence $\cap_i \mathcal{L}_{m_i} = \mathcal{A}$ by Corollary 10.12.6. Thus $H \in \mathcal{A}$, yet $a \in H \setminus P \subseteq G \setminus \mathcal{A}(G)$, which is a contradiction. Therefore $K \in \mathcal{U}$ and the proof is complete.

The value of k in Theorem 10.12.4 is the best possible, since examples constructed by Smith [1981] and Reilly and Wroblewski [1981] show that if n is a product of k primes, then $\mathcal{L}_n \not\subseteq \mathcal{A}^k$.

An important special case of Theorem 10.12.10 was obtained earlier by Gurchenkov.

COROLLARY 10.12.11 [Gurchenkov, 1984a]. *For any prime* p, $\mathcal{L}_p \subseteq \mathcal{A}^2$.

It should be noted that the statement "$G \in \mathcal{A}^{k+1}$" is stronger than the statement "G is solvable class $k + 1$" since, for example, $N_0 \times N_0$ ordered lexicographically is solvable class two but does not lie in $\mathcal{A}^2$. The distinction between solvable and ℓ-solvable has been explored by Smith [1984].

A natural question to ask concerns the reverse inclusion: Does there exist for each integer k an integer n_k such that $\mathcal{A}^k \subseteq L_{n_k}$? One quickly sees that the ℓ-groups N_0, W^+ and W^- of Section 10.11 give a negative answer. Indeed, by Lemma 10.12.1, L_n contains no non-abelian o-group. If we sharpen the above question to make these necessary exclusions then, somewhat surprisingly, there is a positive answer.

The next result is crucial. However, the reader is referred to Reilly [1986] for the details which are somewhat technical.

PROPOSITION 10.12.12. *Let $\mathcal{V}$ be a variety such that for all positive integers n there exist an ℓ-group $G \in \mathcal{V}$, an element $x \in G$ and a minimal prime subgroup P of G such that $|O(P,x)| \geq n$. Then $\mathcal{A}^2 \subseteq \mathcal{V}$.*

With the aid of Proposition 10.12.12 it is a relatively simple matter to derive the 'converse' of Theorem 10.12.10.

THEOREM 10.12.13 [Reilly, 1986]. *If $\mathcal{V}$ is a variety of ℓ-groups such that $\mathcal{V} \cap \mathcal{R} = \mathcal{A}$, then $\mathcal{V} \subseteq L_n$, for some positive integer n.*

Proof. Let $\mathcal{V}$ be a variety of ℓ-groups with $\mathcal{V} \cap \mathcal{R} = \mathcal{A}$. Suppose that $\mathcal{V} \not\subseteq L_n$, for all n. For each positive integer n, let $G_n \in \mathcal{V} \setminus L_n$ and let $G = \Pi\{G_n : n = 1,2,3,...\}$. Then $G \in \mathcal{V}$ but $G \notin \cup L_n$. Consider any positive integer n. We claim that

there exists a minimal prime subgroup P in G
and an element $x \in G$ such that $|O(P,x)| \geq n$.   (10.11)

Supose that this is not the case and that for all minimal prime subgroups P in G and all $x \in G$, $|O(P,x)| \leq n$. Then $|O(P,x)|$ divides $m = n!$. So $x^{-m}Px^m = P$ for all $x \in G$, and, by Theorem 10.12.3, $G \in L_m$. Since this contradicts the hypothesis, the claim (10.11) must hold. By Proposition 10.12.12 $\mathcal{A}^2 \subseteq \mathcal{V}$ which again contradicts the hypothesis, since $\mathcal{A}^2$ contains non-abelian o-groups. Therefore $\mathcal{V}$ must be contained in L_n for some n.

COROLLARY 10.12.14 [Reilly, 1986]. *If $\mathcal{V}$ is any variety of ℓ-groups that contains no non-abelian o-groups, then $\mathcal{V} \subseteq \mathcal{A}^k$, for some positive integer k.*

Proof. This follows immediately from Theorems 10.12.10 and 10.12.13.

10.13. Covers of Varieties

From Section 10.11 we know that there are precisely three varieties of solvable representable ℓ-groups that cover $\mathcal{A}$. In this section we will consider (solvable) non-representable covers.

For any integer n, let $G_n = \mathbf{Z}^n \times \mathbf{Z}$ where multiplication is defined by

$$(F,r)(H,s) = (F + H^r, r + s)$$

where $H^r(t) = H(t')$ and t' is such that $1 \leq t' \leq n$, $t' \equiv t - r \pmod{n}$. Then G_n is a group with respect to this operation and becomes a lattice-ordered group if we introduce the order:

$$(F,r) \geq e \Leftrightarrow \text{ either (i) } r > 0 \text{ or (ii) } r = 0 \text{ and } F(t) \geq 0$$
$$\text{for all } 1 \leq t \leq n.$$

It is easily verified that $G_n \in \mathcal{L}_n$. For an alternate description of G_n in terms of generators and relations see Glass [1985] and Smith [1976, 1980]. The groups G_n, for n of the form 2^k, were introduced by Martinez [1972]. These groups and the varieties $S_n = \mathcal{V}(G_n)$ were studied by Scrimger and are now called *Scrimger groups* and *varieties*, respectively. His main result was the following.

THEOREM 10.13.1 [Scrimger, 1975]. *If p is a prime, then S_p covers $\mathcal{A}$.*

Rather than comment on Theorem 10.13.1, we will discuss the following more general result due to Gurchenkov.

THEOREM 10.13.2 [Gurchenkov, 1982b]. *The solvable varieties of ℓ-groups that cover $\mathcal{A}$ are precisely $\mathcal{V}(N_0)$, $\mathcal{V}(W^+)$, $\mathcal{V}(W^-)$ and S_p, p a prime.*

Proof. Let $\mathcal{V}$ be a variety of solvable ℓ-groups that covers $\mathcal{A}$. By Theorem 10.11.2, if $\mathcal{V} \cap \mathcal{R} \neq \mathcal{A}$ then $\mathcal{V}$ must be $\mathcal{V}(N_0)$, $\mathcal{V}(W^+)$ or $\mathcal{V}(W^-)$. So we may now assume that $\mathcal{V} \cap \mathcal{R} = \mathcal{A}$. Let $G \in \mathcal{V} \backslash \mathcal{A}$.

By Theorem 10.12.13, $G \in \mathcal{L}_n$ for some n. Choose n to be the smallest such integer and let $n = p^r m$ where p and m are relatively prime and p is prime. Now there must be a prime subgroup P and an element z in G such that $|O(P,z)| = p^r t$, for some integer $t \geq 1$. Otherwise, for all prime subgroups P and elements z in G, $|O(P,z)|$ divides $k = p^{r-1}m$ so that $z^{-k}Pz^k = P$. By Theorem 10.2.3, this contradicts the minimality of n. Relacing z by z^t, if necessary, we can assume that $|O(P,z)| = p^r$.

Let $\mathcal{P}$ denote the set of all conjugates of P and $\theta \colon G \to S_{\mathcal{P}}$ be the homomorphism of G into the group of permutations of $\mathcal{P}$ defined by $a \mapsto \theta(a)$ where $(g^{-1}Pg)\theta(a) = a^{-1}(g^{-1}Pg)a$. By Theorem 10.12.3, the order of $\theta(a)$ divides n for all $a \in G$. In particular, $\theta(z^-)$ has finte order and so $\theta((z^-)^{-1}) = \theta((z^-)^k)$, for some positive integer k. Hence $\theta(z) = \theta(z^+(z^-)^{-1}) = \theta(z^+)\theta((z^-)^k) = \theta(z^+(z^-)^k)$ where $z^+(z^-)^k > e$. Therefore, without loss of generality, we may assume that $z > e$.

Now let x and a_i, $i = 0,\dots,p^r - 1$ be as in Lemma 10.12.8. Let $P = \Pi\{H_i \colon H_i = G,\ i = 1,2,3,\dots\}$ and $S = \Sigma\{H_i \colon H_i = G,\ i = 1,2,3,\dots\}$. Let H be the ℓ-subgroup of P/S generated by the elements

$$\bar{x} = (x_1,x_2,\dots,x_n,\dots)S \text{ where } x_1 = x,\ x_{n+1} = x_n^{p^r+1}$$

$$\bar{a}_i = (a_i,a_i,\dots,a_i,\dots)S,\ i = 0,1,\dots,p^r - 1.$$

Clearly $\bar{x} \gg \bar{a}_i > e$ for all i, and the elements $\bar{x}, \bar{a}_i$ also satisfy (i), (ii) and (iii) of Lemma 10.12.7. Thus H must be isomorphic to the lexicographic extension of $\langle \bar{a_0} \rangle \times \langle \bar{a_1} \rangle \times \dots \times \langle \bar{a}_{p^r-1} \rangle$ by $\langle \bar{x} \rangle$ where conjugation by $\bar{x}$ "permutes" the components. This is clearly isomorphic to the Scrimger group G_{p^r}. Since G_p is an ℓ-subgroup of G_{p^r}, for $r \geq 1$, it follows that $S_p \subseteq \mathcal{V}$. Now $\mathcal{V}$ covers $\mathcal{A}$ while clearly $\mathcal{A} \subset S_p$. Hence $\mathcal{V} = S_p$ and this completes the proof.

Theorem 10.13.2 completely settles the question of non-representable covers of $\mathcal{A}$.

COROLLARY 10.13.3 [Kopytov and Gurchenkov, 1987], ([Darnel, 1987], [Reilly, 1986]). *The only non-representable covers of $\mathcal{A}$ are the Scrimger varieties S_p, p a prime.*

Proof. Let $\mathcal{V}$ be a non-representable cover of $\mathcal{A}$. Then clearly $\mathcal{V} \cap \mathcal{R} = \mathcal{A}$ so that, by Corollary 10.12.14, $\mathcal{V}$ is solvable. The result now follows from Theorem 10.13.2.

Two rather interesting general results have been obtained on the subject of covers in **L**. Gurchenkov [1984b] has shown that every proper subvariety of $\mathcal{L}$ has a cover in **L**. On the other hand, Medvedev [1984] has shown that this is not the case within $\mathcal{L}(\mathcal{R})$. He found a variety $\mathcal{V}$ of representable ℓ-groups with the following curious properties: (i) there is no variety of *representable* ℓ-groups covering $\mathcal{V}$, (ii) $\mathcal{V}$ contains every variety of representable ℓ-groups that covers $\mathcal{A}$, and (iii) $\mathcal{V}$ has no independent basis of identities.

While it is easily seen that $S_n \subseteq L_n$, it was not clear initially whether or not the containment is proper. However, in [1981], Smith established that S_n is properly contained in L_n for n a composite integer and then Fox [1983] showed that the containment is also proper when n is a prime.

For p a prime, the lattice of subvarieties of L_p is now completely known. The following remarkable result is due to Gurchenkov.

THEOREM 10.13.4 [Gurchenkov, 1984b]. *The lattice of subvarieties of L_p is isomorphic to the chain of natural numbers with a maximum element adjoined:*

- L_p
 ⋮
- S_p
- $\mathcal{A}$
- $\mathcal{E}$

Moreover, every subvariety of L_p is finitely based.

Independently, Holland and Reilly [1986] characterized the finitely generated subdirectly irreducible ℓ-groups in S_p and gave a basis of identities for S_p. This basis can be written very concisely with the aid of the following notational convention for conjugation. For any elements x, y, z, of a group G, let

$$x^{y+z} = x^y x^z \quad \text{and} \quad x^{-y} = (x^y)^{-1} = (x^{-1})^y.$$

THEOREM 10.13.5 [Holland and Reilly, 1986]. *For any prime p, the variety S_p is defined within the variety of all ℓ-groups by the identities:*

$$x^p y^p = y^p x^p, \quad |\, [x,y]^{1+z+z^2+...+z^{p-1}}\,| \wedge |\,[u,v]^{1-z}\,| = e.$$

This approach was continued by Holland, Mekler and Reilly [1986] where they characterized all the finitely generated subdirectly irreducible ℓ-groups in $L_p \cap \mathcal{A}^2$ (which is just L_p, by Theorem 10.12.11) and showed that, as in Theorem 10.13.4, the lattice of subvarieties of

$L_p \cap \mathcal{A}^2$ is a chain. It is also shown in Holland, Mekler and Reilly [1986] how to obtain a basis of identities for each of these varieties.

This work has now been extended to cover subvarieties of $L_n \cap \mathcal{A}^2$, where n is not necessarily a prime [Holland and Reilly, a].

The varieties S_p illustrate the point referred to in Section 10.1 that L is not Brouwerian.

THEOREM 10.13.6 ([Smith, 1980] and [Reilly and Wroblewski, 1981]). *If I is an infinite set of positive integers greater than* 1, *then*

$$\vee \{S_n \colon n \in I\} = \mathcal{A}^2.$$

Since $\mathcal{A}^2$ contains non-abelian o-groups whereas $\mathcal{R} \cap S_n = \mathcal{A}$, we have:

COROLLARY 10.13.7 ([Kopytov and Medvedev, 1975], [Smith, 1980]). L *is not Brouwerian.*

The construction used to obtain the Scrimger groups G_n can be iterated. For any finite sequence of positive integers $(n_1, n_2, ..., n_k)$ we define an ℓ-group $G(n_1, ..., n_k)$ inductively as follows: $G(n_1) = G_{n_1}$, the Scrimger group, while if $G = G(n_1, ..., n_{i-1})$ is defined then $G(n_1, ..., n_i) = G^{n_i} \times \mathbf{Z}$ with multiplication

$$(F, r)(H, s) = (FH^r, r + s)$$

where $H^r(t) = H(t')$ for $1 \le t' \le n$, $t' \equiv t - r \pmod{n}$, and ordered with the cardinal order on G^{n_i} and the lexicographic order on $G^{n_i} \times \mathbf{Z}$, from the right. The varieties generated by these ℓ-groups have been studied by Smith [1981] and Reilly and Wroblewski [1981]. It is shown, for instance, that for any infinite sequence $(n_1, n_2, ...)$ of integers greater than 1,

$$\underset{i}{\vee} \, \mathcal{V}(G(n_1, ..., n_i)) = \mathcal{N}.$$

Norman R. Reilly
Simon Fraser University
Burnaby, B. C. V5A 1S6
Canada

Wayne B. Powell and Constantine Tsinakis

CHAPTER 11

FREE PRODUCTS IN VARIETIES
OF LATTICE-ORDERED GROUPS

11.1 Introduction

The concept of a free product is fundamental to the study of any kind of algebraic system. Intuitively, a free product takes a family of algebras from a given class and combines it in the "loosest" or "freest" way possible. By this it is meant that any other algebra generated by the given family must be a homomorphic image of the free product.

Free products have been widely used in areas such as group theory and lattice theory to provide general methods of construction, establish various embedding theorems, and also produce pathological algebras.

Free products in various classes of ℓ-groups were not investigated outside of special cases until the early 1970's. A paper of Holland and Scrimger [1972] appeared which considered free products in the class of all ℓ-groups as well as free products of partially ordered groups. In the first case, the free product was described as a certain quotient of a free extension of a partially ordered group. About the same time as this work, Martinez [1972] and [1973d] began a study of free products in the classes of abelian ℓ-groups, of representable ℓ-groups, and of all ℓ-groups. His approach centered on determining when free products of ℓ-groups from one class were contained in another class. Beginning in the 1980's the work on free products

A. M. W. Glass and W. C. Holland (eds.), Lattice-Ordered Groups, 278–307.
© 1989 by Kluwer Academic Publishers.

concentrated more on representation theorems, structure theorems, and relationships to general embedding properties.

Our purpose here is to introduce some of the techniques available for examining the structure of free products in various varieties of ℓ-groups. No attempt is made to give all known constructions of these products or to prove all existing structure theorems. Rather we concentrate on the methods which to date have been the most fruitful and the most general. Only a few structural properties are addressed, and these are chosen because they demonstrate the power of the constructions given. More specifically section 11.2 deals with existence theorems for free products of ℓ-groups and free extensions of partially ordered groups (*po-groups*) while section 11.3 gives the details of the representation of these objects. Section 11.4 is devoted to showing that in most cases free products do not admit nontrivial cardinal decompositions, section 11.5 contains proofs of the failure of ℓ-group free products to limit the size of disjoint sets, and section 11.6 considers the relationship between free products in varieties of ℓ-groups and distributive-lattice free products. The final section briefly mentions other directions in which results have been achieved.

The varieties which will be most often discussed are given distinguished notation:

$\mathcal{L}$ = all ℓ-groups

$\mathcal{N}$ = normal-valued ℓ-groups

$\mathcal{R}$ = representable ℓ-groups

$\mathcal{N}_n$ = all ℓ-groups which are nilpotent of class n

$\mathcal{A}$ = abelian ℓ-groups.

Free products of ℓ-groups are closely tied with group free products and the various ways of ordering these latter products. It is frequently necessary to look at a group H with several different partial orders. To differentiate between these, the notation (H,P) will be used to signify a po-group with group structure H and positive cone P. If H is a po-group and if its positive cone has not been given a special label, then H^+ will refer to this cone.

Several different products are used on families of po-groups. For clarity we denote products involving only the group structure differently than those which also incorporate the order. The direct sum of groups is symbolized by $\oplus$ whereas $*$ refers to a free product in a class of groups without regard to order. When the order is also considered $\boxplus$ denotes the direct sum, Π denotes the direct product, and $\sqcup$ denotes the free product. The direct sum ($\boxplus$) and direct product (Π) of po-groups are usually called the cardinal sum and cardinal product, respectively. Thus, for example, if

G_1 and G_2 are po-groups, $G_1 \boxplus G_2$ denotes the po-group of all ordered pairs (g_1, g_2) where $g_i \in G_i$, and where both the group operation and the order relation are defined componentwise (compare Example 2 of Chapter 0).

11.2 Free Products and Free Extensions

In this section we consider two concepts which are generalizations of the notion of a free algebra. These are the free product of a family of algebras and the free extension of a partial algebra. Existence theorems are given for general situations and then specialized where possible to important classes of ℓ-groups. Further, it is shown that these two concepts agree for ℓ-groups in many situations beyond the base case of free ℓ-groups.

Let $\mathcal{U}$ be a class of algebras of the same similarity type and let $\{G_i \mid i \in I\}$ be a family of members of $\mathcal{U}$. The $\mathcal{U}$-*free product* of this family is an algebra $G \in \mathcal{U}$, denoted by $\mathcal{U}\bigsqcup_{i \in I} G_i$, together with a family of injective homomorphisms $\{\alpha_i : G_i \to G \mid i \in I\}$ such that

(i) $\bigcup_{i \in I} \alpha_i(G_i)$ generates G ;

(ii) if $H \in \mathcal{U}$ and $\{\beta_i : G_i \to H \mid i \in I\}$ is a family of homomorphisms, then there exists a (necessarily) unique homomorphism $\gamma : G \to H$ satisfying $\beta_i = \gamma \alpha_i$ for all $i \in I$.

We shall often find it convenient to identify each free factor G_i with its image $\alpha_i(G_i)$ in $\mathcal{U}\bigsqcup_{i \in I} G_i$ and thus view each α_i as the inclusion map.

It is easy to see that if the free product of a family exists, then it is unique up to isomorphism.

The interpretation of the concept of a free product as a solution of a universal-mapping problem is due to Sikorski [1952]. Of course, free products of groups had previously been studied extensively in the realm of combinatorial group theory.

We address first the question of existence of free products. The following result is quite adequate for our purposes.

THEOREM 11.2.1. ([Sikorski, 1952], [Christensen and Pierce, 1959]; see [Grätzer, 1979, p. 186] or [R. S. Pierce, 1968, p. 104]). *Let $\mathcal{U}$ be a class of algebras closed with respect to the formation of subalgebras and direct products. Let*

$\{G_i \mid i \in I\}$ *be a family in* $\mathcal{U}$ *for which there exists a* $K \in \mathcal{U}$ *and a family of injective homomorphisms* $\{\psi_i : G_i \to K \mid i \in I\}$. *Then* $\mathcal{U} \underset{i \in I}{\bigsqcup} G_i$ *exists.*

An instance of Theorem 11.2.1 occurs when all algebras G_i have one-element subalgebras. In this case one can let K be the direct product of the algebras G_i and let $\psi_i : G_i \to K$ be the canonical embedding for each i. In particular, if $\mathcal{U}$ is a class of groups or ℓ-groups closed with respect to the formation of subalgebras and direct products, then the $\mathcal{U}$-free product of any family in $\mathcal{U}$ exists.

Even though free products may be known to exist in certain classes, it is frequently difficult to give a suitable representation of these objects. Section 11.3 will be devoted almost entirely to constructing free products in certain varieties of ℓ-groups. Many of the results on the subject are based on these representations.

It is well known that the concept of a free product is a generalization of that of a free algebra. The exact relationship of the two notions can be stated in the context of ℓ-groups as: The $\mathcal{U}$-free ℓ-group on a nonempty set X is the $\mathcal{U}$-free product of $|X|$ copies of the $\mathcal{U}$-free ℓ-group on one generator. It is not difficult to show that this rank one $\mathcal{U}$-free ℓ-group is just $\mathbf{Z} \boxplus \mathbf{Z}$ independent of the class $\mathcal{U}$.

Another pertinent structure which is a generalization of the $\mathcal{U}$-free ℓ-group is the $\mathcal{U}$-free extension of a partially ordered group. This is a specialization of the general concept of the free extension of a partial algebra. We start with the definition of this concept.

Let $\mathcal{U}$ be a class of algebras of similarity type τ and let H be a partial algebra of type τ. The $\mathcal{U}$-*free extension of* H is an algebra $\mathcal{F}_{\mathcal{U}}(H) \in \mathcal{U}$ for which there exists an injective homomorphism $\alpha : H \to \mathcal{F}_{\mathcal{U}}(H)$ such that

(i) $\alpha(H)$ generates $\mathcal{F}_{\mathcal{U}}(H)$

(ii) if $K \in \mathcal{U}$ and $\beta : H \to K$ is a homomorphism, then there exists a (necessarily) unique homomorphism $\gamma : \mathcal{F}_{\mathcal{U}}(H) \to K$ such that $\gamma\alpha = \beta$.

Recall that if $A = (A, (f_k)_{k \in K})$ and $B = (B, (g_k)_{k \in K})$ are partial algebras of the same similarity type $\tau = (v_k)_{k \in K}$, a map $\phi : A \to B$ is called a homomorphism of A into B if whenever $f_k(a_1,...,a_{v_k})$ is defined, then $g_k(\phi(a_1),...,\phi(a_{v_k}))$ is defined and $\phi(f_k(a_1,...,a_{v_k})) = g_k(\phi(a_1),...,\phi(a_{v_k}))$.

In what follows we shall often identify H with $\alpha(H)$ and thus view α as the inclusion map. It is clear that if $\mathcal{F}_{\mathcal{U}}(H)$ exists, then it is unique up to isomorphism.

We specialize to the concept of a free extension of a po-group by viewing a po-group as a partial ℓ-group. The group operations are total and thus remain unaltered by the free extension process. The two partial lattice operations $\wedge$ and $\vee$ are given by

$$x \vee y = y \vee x = y \text{ iff } x \leq y, \text{ and}$$

$$x \wedge y = y \wedge x = x \text{ iff } x \leq y.$$

Thus, $x \vee y$ and $x \wedge y$ are defined if and only if x and y are comparable in the ordering of the po-group. Note that a po-group whose partial order is a lattice order, becomes via this definition a partial ℓ-group which is not an ℓ-group unless it is totally ordered. It should also be pointed out that a mapping between two po-groups considered as partial ℓ-groups is a homomorphism if and only if it is a po-group homomorphism. Thus, if $\mathcal{U}$ is a class of ℓ-groups and H is a po-group, the free extension specializes as follows: the $\mathcal{U}$-free extension of the po-group H is an ℓ-group $\mathcal{F}_{\mathcal{U}}(H) \in \mathcal{U}$ for which there exists an injective po-group homomorphism $\alpha : H \to \mathcal{F}_{\mathcal{U}}(H)$ such that

(i) $\alpha(H)$ generates $\mathcal{F}_{\mathcal{U}}(H)$ as an ℓ-group;

(ii) if $K \in \mathcal{U}$ and $\beta : H \to K$ is a po-group homomorphism, then there exists an ℓ-group homomorphism $\gamma : \mathcal{F}_{\mathcal{U}}(H) \to K$ such that $\gamma \alpha = \beta$.

If H is a po-group with positive cone P, we shall often write $\mathcal{F}_{\mathcal{U}}(H,P)$ for the $\mathcal{U}$-free extension of H.

This concept of the $\mathcal{U}$-free extension of a po-group is all that is necessary to the study of free products of ℓ-groups. Hence, in what follows we will consider no more general notion than this.

It is appropriate to mention that the only reference in the literature of lattice-ordered groups where the concept of free extension is considered is in [Bigard, Keimel, and Wolfenstein, 1977, page 302] . All other works on the subject deal with the more restricted concept of a $\mathcal{U}$-free ℓ-group over a po-group. The $\mathcal{U}$-free extension $\mathcal{F}_{\mathcal{U}}(H)$ of a po-group H is said to be the $\mathcal{U}$-free ℓ-group over the po-group H if the mapping α in the definition above is a po-group isomorphism between H and $\alpha(H)$. More specifically, if one views $\alpha(H)$ as a po-group with respect to the partial order inherited from $\mathcal{F}_{\mathcal{U}}(H)$, then $\alpha : H \to \alpha(H)$ and $\alpha^{-1} : \alpha(H) \to H$ are po-group homomorphisms. This concept specializes that of a free extension of a partial algebra as presented in [Grätzer, 1979, page 182].

The first concept of a free extension of a po-group described above is quite convenient for the description of the relationship between free extensions and free products of totally ordered groups (see Theorem 11.2.4 and Corollaries 11.2.5, 11.2.6, and 11.2.7 below).

The basic existence theorem for general $\mathcal{U}$-free extensions is similar to that for $\mathcal{U}$-free products.

THEOREM 11.2.2 [Grätzer and Schmidt, 1963]; see [Pierce, 1968, page 101]. *Let $\mathcal{U}$ be a class of algebras of type τ, closed with respect to the formation of subalgebras and direct products. Let H be a partial algebra of type τ such that there exists an injective homomorphism of H into some algebra in $\mathcal{U}$. Then the $\mathcal{U}$-free extension of H exists.*

If $\mathcal{U}$ is a class of ℓ-groups, we denote by $G(\mathcal{U})$ the class of all groups that can be embedded (as subgroups) in the members of $\mathcal{U}$. It is clear that if $\mathcal{U}$ is closed with respect to the formation of ℓ-subgroups and direct products, then $G(\mathcal{U})$ is closed under direct products and subgroups. It follows in this case that the $G(\mathcal{U})$-free product of any family in $G(\mathcal{U})$ exists. The results in the remainder of this section show that there is an intimate relationship between certain $\mathcal{U}$-free products, $\mathcal{U}$-free extensions of po-groups, and $G(\mathcal{U})$-free products.

We start with a straightforward consequence of Theorem 11.2.2. Part (c) can be found in [Bigard, Keimel, and Wolfenstein, 1977, page 302]. The corresponding results for the $\mathcal{U}$-free ℓ-group over a po-group are due to Weinberg [1963] and Conrad [1970b].

COROLLARY 11.2.3.

(a) Let $\mathcal{U}$ be $\mathcal{A}$ or one of the varieties $\mathcal{N}_n(n > 1)$. Then $\mathcal{F}_{\mathcal{U}}(H)$ exists for every po-group H with $H \in G(\mathcal{U})$.

(b) If H is a po-group, then $\mathcal{F}_{\mathcal{R}}(H)$ exists if and only if there exists a total order on H whose positive cone contains H^+.

(c) If H is a po-group, then $\mathcal{F}_{\mathcal{L}}(H)$ exists if and only if there exists a right order on H whose positive cone contains H^+.

Proof. (a) This follows from Theorem 11.2.2 and the result due to Mal'cev [1951] (see e.g., [Mura and Rhemtulla, 1977, page 56]) that states that every partial order of a torsion-free locally nilpotent group can be extended to a total order.

(b) Let H be a po-group. If the partial order of H can be extended to a total order, then there is an injective po-group homomorphism of H into some ℓ-group in $\mathcal{R}$ and hence, by Theorem 11.2.2, $\mathcal{F}_{\mathcal{R}}(H)$ exists. Conversely, suppose $\mathcal{F}_{\mathcal{R}}(H)$ exists. Then the inclusion map from H into $\mathcal{F}_{\mathcal{R}}(H)$ is a po-group homomorphism. Now the subdirectly irreducible ℓ-groups in $\mathcal{R}$ are totally ordered (see [Bigard, Keimel, and Wolfenstein, 1977, page 72] or Section 1.2) and hence $\mathcal{F}_{\mathcal{R}}(H)$ is an ℓ-subgroup of a product $K = \prod_{i \in I} K_i$ with each K_i totally ordered. Well-order I and totally order K lexicographically. That is, $x \in K$ is positive if its first component different from e (with respect to this well-ordering) is positive in the total order of the corresponding K_i. It is easily seen that this is a total order of K that contains the positive cone of $\mathcal{F}_{\mathcal{R}}(H)$. Thus its restriction to H is a total order that contains H^+.

(c) Let H be a po-group. It can be shown that there exists an injective po-group homomorphism of H into some ℓ-group if and only if there exists a right order on H whose positive cone contains H^+ (see [Bigard, Keimel, and Wolfenstein, 1977, page 84]). Hence this case follows from Theorem 11.2.2.

In order to relate free extensions of po-groups to free products, consider a family $\{G_i \mid i \in I\}$ of ℓ-groups. Let $\mathcal{U}$ be a class of ℓ-groups containing this family, and closed with respect to direct products and ℓ-subgroups. Write H for the $\mathcal{G}(\mathcal{U})$-free product of the G_i (viewed as groups) and Q for the set consisting of all products of conjugates in H of elements from $\bigcup_{i \in I} G_i^+$. We claim that Q is the positive cone of a partial order on H extending the orders of the G_i's. We proceed with the verification of this fact [Holland and Scrimger, 1972, proof of Theorem 2.2]. Q is clearly a normal subsemigroup of H, so if we can find a partial order on H whose positive cone contains Q it will follow that Q is the positive cone of a partial order on H. Let $\prod_{i \in I} G_i$ be the direct product in $\mathcal{U}$ of the family $\{G_i \mid i \in I\}$ and let $\phi : H \to \prod_{i \in I} G_i$ be the group homomorphism that extends the natural embeddings of the G_i's. Define

$$Q_1 = \{h \in H \mid \phi(h) > e \text{ in } \prod_{i \in I} G_i\} \cup \{e\}.$$

It is clear that Q_1 is the positive cone of a partial order on H and $G_i^+ \subseteq Q_1$ for all $i \in I$. Thus $Q \subseteq Q_1$ as was to be shown.

THEOREM 11.2.4. *Suppose that $\mathcal{U}$ is a class of ℓ-groups closed with respect to the formation of ℓ-subgroups and direct products. Let $\{G_i \mid i \in I\}$ be a family of* *totally ordered groups in* $\mathcal{U}$, *let* $H = \mathcal{G}^{(\mathcal{U})} \underset{i \in I}{*} G_i$,*and let* Q *be the set of all products of conjugates of* $\underset{i \in I}{\cup} G_i^+$ *in* H. *If* $\mathcal{F}_\mathcal{U}(H,Q)$ *exists, then* $\mathcal{U}\underset{i \in I}{\sqcup}G_i \cong \mathcal{F}_\mathcal{U}(H,Q)$.

Proof. In view of the preceding discussion Q is the positive cone of a partial order on H. Suppose now that $\mathcal{F}_\mathcal{U}(H,Q)$ exists and set

$$G = {}^\mathcal{U}\underset{i \in I}{\sqcup}G_i \ \text{ and } \ \bar{G} = \mathcal{F}_\mathcal{U}(H,Q).$$

Note that each G_i is an ℓ-subgroup of $\bar{G}$ and hence there exists an ℓ-homomorphism $\phi: G \to \bar{G}$ such that $\phi(g) = g$ for each $g \in G_i$ and each $i \in I$. Now the universal property of $G(\mathcal{U})$-free products yields a group homomorphism $\psi: H \to G$ such that $\psi(g) = g$ for all $g \in G_i$ and $i \in I$. A fortiori, $\psi: (H,Q) \to G$ is a po-group homomorphism and hence there exists an ℓ-group homomorphism $\bar{\psi}: \bar{G} \to G$ extending ψ. It is immediate that $\bar{\psi}\phi = id_G$ and $\phi\bar{\psi} = id_{\bar{G}}$. Thus, ϕ and $\bar{\psi}$ are inverses of each other and $G \cong \bar{G}$.

Note that the preceding proof shows that under the assumptions of Theorem 11.2.4, $\mathcal{G}^{(\mathcal{U})} \underset{i \in I}{*} G_i$ is isomorphic to the subgroup of ${}^\mathcal{U}\underset{i \in I}{\sqcup}G_i$ generated by $\underset{i \in I}{\cup}G_i$.

For several varieties this theorem has special significance.

COROLLARY 11.2.5. [Martinez, 1972]. *If* $\{G_i : i \in I\}$ *is a family of totally ordered abelian groups, then*

(i) $\mathcal{G}^{(\mathcal{A})} \underset{i \in I}{*} G_i \cong \underset{i \in I}{\oplus}G_i$

(ii) $\mathcal{F}_\mathcal{A}(\underset{i \in I}{\boxplus} G_i)$ *exists and is isomorphic to* ${}^\mathcal{A}\underset{i \in I}{\sqcup}G_i$.

Proof. The verification of (i) is straightforward. Next note that Q as in Theorem 11.2.4 is the positive cone of $\underset{i \in I}{\boxplus} G_i$. Thus (ii) follows from Theorem 11.2.4.

COROLLARY 11.2.6. *Let* $\{G_i : i \in I\}$ *be a family of totally ordered groups in* $\mathcal{N}_n$ $(n > 1)$, H *be the* $\mathcal{G}(\mathcal{N}_n)$-*free product of this family, and* Q *be the set of all products of conjugates of* $\underset{i \in I}{\cup}G_i^+$ *in* H. *Then* $\mathcal{F}_\mathcal{U}(H,Q)$ *exists and is isomorphic to* ${}^\mathcal{U}\underset{i \in I}{\sqcup}G_i$.

Proof. We remark that H is obtained by taking the free product of the family $\{G_i : i \in I\}$ in the class of all nilpotent class n groups and then taking the quotient of this group by its torsion subgroup. (The set of torsion elements of a nilpotent class n group is a subgroup; see, e.g., [D. J. S. Robinson, 1982; 5.2.7].) We have already observed that (H,Q) is a po-group (see the discussion preceding Theorem 11.2.4), and hence by Corollary 11.2.3, $\mathcal{F}_{\mathcal{N}_n}(H)$ exists. The conclusion now follows from Theorem 11.2.4.

COROLLARY 11.2.7. *Let* $\mathcal{U} \supseteq \mathcal{R}$ *be a class of ℓ-groups closed with respect to taking ℓ-subgroups and direct products.*

(a) *If* $\{G_i : i \in I\}$ *is a family in* $\mathcal{G}(\mathcal{R})$, *then* $\mathcal{G}^{(\mathcal{U})} \underset{i \in I}{*} G_i$ *is isomorphic to the free product of the G_i's in the class of all groups.*

(b) *If* $\{G_i : i \in I\}$ *is a family of totally ordered groups, H is the free product of the family, and Q is the set of all products of conjugates of elements of* $\underset{i \in I}{\cup} G_i^+$ *in H, then* $\mathcal{F}_{\mathcal{U}}(H,Q)$ *exists and is isomorphic to* $\underset{i \in I}{\overset{\mathcal{U}}{\sqcup}} G_i$.

Proof. It is well known ([Vinogradov, 1949]; see also [Bergman, a], and [Johnson, 1968]) that the free product of totally ordered groups can be totally ordered so that the given orders of the free factors are extended. The desired result is a consequence of this fact, of Corollary 11.2.3, and of Theorem 11.2.4.

The case $\mathcal{U} = L$ of the preceding corollary was first established in [Holland and Scrimger, 1972].

11.3. Constructions of Free Products and Free Extensions

In this section we shall present constructions for free extensions and free products in the varieties $\mathcal{A}$, $\mathcal{R}$, $\mathcal{N}_n$ $(n > 1)$, and L. These descriptions are invaluable both in intuitively understanding the role of free products in varieties of ℓ-groups and also in proving structure theorems about these products.

We start with the representations of free extensions in these varieties (Theorems 11.3.2 and 11.3.3).

LEMMA 11.3.1. *Let L and L' be ℓ-groups and let M be a subgroup of L which generates L as a lattice. Let $\phi : M \to L'$ be a group homomorphism such that for each finite subset* $\{x_{jk} \mid j \in J, K \in K\}$ *of M, the condition* $\underset{j \in J}{\vee} \underset{k \in K}{\wedge} x_{jk} = e$ *implies* $\underset{j \in J}{\vee} \underset{k \in K}{\wedge} \phi(x_{jk}) = e$. *Then ϕ can be extended to an ℓ-homomorphism $\phi' : L \to L'$.*

Proof. If $x = \underset{j \in J}{\vee} \underset{k \in K}{\wedge} x_{jk}$ with $\{x_{jk} \mid j \in J, K \in K\} \subseteq M$, we define $\phi'(x) = \underset{j \in J}{\vee} \underset{k \in K}{\wedge} \phi(x_{jk})$. The stated condition implies that this definition produces a function $\phi' : L \rightarrow L'$. A straightforward computation shows that ϕ' is an ℓ-group homomorphism.

Given a po-group H, we write $\mathcal{T}_H$ for the set consisting of all positive cones T of total orders on H with $T \supseteq H^+$. Let $\alpha : H \rightarrow \Pi\{(H,T) \mid T \in \mathcal{T}_H\}$ be the diagonal map $h \mapsto (...,h,h,h,...)$. In view of Corollary 11.2.3, if $\mathcal{U}$ is $\mathcal{A}$ or one of the varieties $\mathcal{N}_n$ and $H \in \mathcal{G}(\mathcal{U})$, then $\mathcal{T}_H \neq \emptyset$ and $\mathcal{F}_{\mathcal{U}}(H)$ exists. By the same corollary, $\mathcal{F}_{\mathcal{R}}(H)$ exists if and only if $\mathcal{T}_H \neq \emptyset$.

THEOREM 11.3.2. [Weinberg, 1963]. *Let $\mathcal{U}$ be $\mathcal{A}$, $\mathcal{R}$, or one of the varieties $\mathcal{N}_n$ and let H be a po-group such that $\mathcal{F}_{\mathcal{U}}(H)$ exists. Then $\mathcal{F}_{\mathcal{U}}(H)$ is isomorphic to the sublattice of $\Pi\{(H,T) \mid T \in \mathcal{T}_H\}$ generated by $\alpha(H)$.*

Proof. As was mentioned above, $\mathcal{T}_H \neq \emptyset$ in view of Corollary 11.2.3. Denote by G the sublattice of $\Pi\{(H,T) \mid T \in \mathcal{T}_H\}$ generated by $\alpha(H)$. It is easy to see that G is an ℓ-subgroup of the product and that $\alpha : H \rightarrow G$ is an injective po-group homomorphism. Now let $L \in \mathcal{U}$ and let $\phi : H \rightarrow L$ be a po-group homomorphism. Let $\bar{\phi} : \alpha(H) \rightarrow L$ be the group homomorphism defined by $\bar{\phi}(\alpha(h)) = \phi(h)$. It remains only to show that there exists an ℓ-homomorphism $\phi' : G \rightarrow L$ whose restriction to $\alpha(H)$ is $\bar{\phi}$. With the intention of using Lemma 11.3.1, we choose a finite subset $\{h_{jk} \mid j \in J, k \in K\}$ of H such that

$$\bigvee_{j \in J} \bigwedge_{k \in K} \bar{\phi}(\alpha(h_{jk})) = \bigvee_{j \in J} \bigwedge_{k \in K} \phi(h_{jk}) \neq e.$$

We need to show that

$$\bigvee_{j \in J} \bigwedge_{k \in K} \alpha(h_{jk}) \neq e.$$

Now L is a subdirect product of totally ordered groups and hence there exists a totally ordered group $L' \in \mathcal{U}$ and an ℓ-homomorphism $\psi : L \rightarrow L'$ such that

$$\bigvee_{j \in J} \bigwedge_{k \in K} \psi\bar{\phi}(\alpha(h_{jk})) = \bigvee_{j \in J} \bigwedge_{k \in K} \psi\phi(h_{jk}) \neq e.$$

We may therefore assume that L is totally ordered.

Suppose first that

$$\bigvee_{j \in J} \bigwedge_{k \in K} \phi(h_{jk}) > e.$$

Then there is some $j_0 \in J$ such that

$$\bigwedge_{k \in K} \phi(h_{j_0 k}) > e \quad \text{in } L.$$

Consider an arbitrary $T \in \mathcal{T}_H$ and let

$$T' = \{h \in H \mid \phi(h) \in L^+ \setminus \{e\} \text{ or both } \phi(h) = e \text{ and } h \in T\}.$$

It is easy to see that $T' \in \mathcal{T}_H$ and $h_{j_0 k} \in T'$ for all all $k \in K$. It follows that

$$\bigvee_{j \in J} \bigwedge_{k \in K} h_{jk} > e \quad \text{in } (H, T')$$

and thus

$$\bigvee_{j \in J} \bigwedge_{k \in K} \alpha(h_{jk}) \neq e \quad \text{in } G.$$

Assume next that

$$\bigvee_{j \in J} \bigwedge_{k \in K} \phi(h_{jk}) < e.$$

Then for each $j \in J$ there exists $k_j \in K$ such that $\phi(h_{jk_j}) < e$. Thus

$$\bigvee_{j \in J} \phi(h_{jk_j}) < e$$

and as above, there is $T'' \in \mathcal{T}_H$ such that

$$\bigvee_{j \in J} h_{jk_j} < e \quad \text{in } (H, T'').$$

But then

$$\bigvee_{j \in J} \bigwedge_{k \in K} \phi(h_{jk}) < e \quad \text{in } (H, T'')$$

and so

$$\bigvee_{j \in J} \bigwedge_{k \in K} \alpha(h_{jk}) \neq e \quad \text{in } G.$$

We next consider the construction for the L-free extension of a po-group H. Recall that in view of Corollary 11.2.3, $\mathcal{F}_L(H)$ exists if and only if there exists a right order on H whose positive cone contains H^+. Write $\mathcal{R}_H$ for the set of all positive cones T of right orders on H with $T \supseteq H^+$. For each $T \in \mathcal{R}_H$, we denote by $A(H, T)$ the ℓ-group of all order preserving permutations of H with respect to the right order induced by T. Let $\alpha_T : H \to A(H, T)$ denote the injective po-group homomorphism defined by $\alpha_T(h)(x) = xh$, for all $x, h \in H$. Finally, let $\alpha : H \to \Pi \{A(H, T) \mid T \in \mathcal{R}_H\}$ denote the diagonal map $h \mapsto (\alpha_T(h))_{T \in \mathcal{R}_H}$.

The proof of the next result is similar to the proof of Theorem 11.3.2. The original version of the result is concerned with the free ℓ-group over a po-group (see section 11.2).

THEOREM 11.3.3 [Conrad, 1970b]. $\mathcal{F}_L(H)$ exists if and only if $\mathcal{R}_H \neq \varnothing$. Furthermore in this case, $\mathcal{F}_L(H)$ is isomorphic to the sublattice of $\Pi \{A(H, T) \mid T \in \mathcal{R}_H\}$ generated by $\alpha(H)$.

We have already observed that free products of totally ordered groups in $\mathcal{A}$, $\mathcal{R}$, L, and $\mathcal{N}_n$ are free extensions of certain po-groups (see Corollaries 11.2.5, 11.2.6, and 11.2.7). Thus Theorems 11.3.2 and 11.3.3 provide useful representations for these products. The constructions for arbitrary free products in these varieties are more involved. The next theorem provides a very useful representation for a free product in $\mathcal{A}$, $\mathcal{N}_n$ ($n > 1$), and $\mathcal{R}$ in terms of suitable totally ordered groups. The underlying reason that makes this representation possible is the fact that the subdirectly irreducible ℓ-groups in these varieties, and in fact in every subvariety of $\mathcal{R}$, are totally ordered. It should also be kept in mind that if G is an ℓ-group and P is an ℓ-ideal (congruence relation) of G, then G/P is totally ordered if and only if P is a prime subgroup of G (Proposition 1.1.5). These remarks explain the choice of the sets Γ_i below.

Let $\mathcal{U}$ be one of the varieties $\mathcal{A}$, $\mathcal{R}$, or $\mathcal{N}_n$ ($n > 1$), and let $\{G_i \mid i \in I\}$ be a family of ℓ-groups in $\mathcal{U}$. In the ensuing discussion we will use the following notation:

$\Gamma_i = \{j \mid P_j$ is a normal prime subgroup of $G_i\}$, $i \in I$

$\Gamma = \cup_{i \in I} \Gamma_i$, the disjoint union of $\{\Gamma_i \mid i \in I\}$

Δ = the set of all choice functions $\delta : I \rightarrow \Gamma$.

For each $\delta \in \Delta$, let

H_δ = the $\mathcal{G}(\mathcal{U})$-free product of the family $\{G_i/P_{\delta(i)} \mid i \in I\}$

Q_δ = the set of all products of conjugates in H_δ of all elements of
$$\underset{i \in I}{\cup}(G_i/P_{\delta(i)})^+$$

$\mathcal{T}_\delta = \{T \mid T$ is the positive cone of a total order on H_δ, and $T \supseteq Q_\delta\}$

$A = \Pi\{(H_\delta T) \mid \delta \in \Delta, T \in \mathcal{T}_\delta\}$

$\rho_{\delta,T}$ = the projection map $A \rightarrow (H_\delta T)$, $\delta \in \Delta$, $T \in \mathcal{T}_\delta$

H = the $\mathcal{G}(\mathcal{U})$-free product of $\{G_i \mid i \in I\}$.

For each $i \in I$, there is a unique ℓ-homomorphism $\psi_i : G_i \rightarrow A$ satisfying $\rho_{\delta,T}\psi_i(g_i) = P_{\delta(i)}g_i$, for each $g_i \in G_i$, $\delta \in \Delta$, and $T \in \mathcal{T}_\delta$. Note that each ψ_i is injective. Let $\psi : H \rightarrow A$ be the unique group homomorphism extending the ℓ-homomorphisms ψ_i.

THEOREM 11.3.4 [Powell and Tsinakis, 1983a, 1984]. $^\mathcal{U}\underset{i \in I}{\bigsqcup}G_i$ is isomorphic to the sublattice of A generated by $\psi(H)$ whenever $\mathcal{U} = \mathcal{A}$, $\mathcal{R}$, or $\mathcal{N}_n$ ($n > 1$).

We omit the proof of this theorem as it is essentially a special case of the proof of Theorem 11.3.5 on L-free products given below.

To this date no satisfactory description of arbitrary L-free products has been obtained. Holland and Scrimger [1972] and Glass [1984] have given descriptions for special cases. The next theorem also considers a special case, this being when the free factors are from $\mathcal{R}$.

We start by establishing some notation. Let $\{G_i \mid i \in I\}$ be an arbitrary family of ℓ-groups in $\mathcal{R}$. For each $i \in I$ and $\delta \in \Delta$, let Γ_i, Γ, Δ, H_δ, Q_δ, and H be defined as above. Further let

$\mathcal{R}_\delta = \{T \mid T$ is the positive cone of a right order on H_δ, and $T \supseteq Q_\delta\}$

$A = \Pi\{A(H_\delta, T) \mid T \in \mathcal{R}_\delta\}$

$\rho_{\delta,T} = $ the projection map $A \to A(H_\delta, T)$.

As before, for each $i \in I$ there is a unique ℓ-homomorphism $\psi_i : G_i \to A$ satisfying $\rho_{\delta,T}\psi_i(g_i) = P_{\delta(i)}g_i$ for each $g_i \in G_i$, $\delta \in \Delta$, and $T \in \mathcal{T}_\delta$. (We identify here the coset $P_{\delta(i)}g_i$ with the order-preserving permutation in $A(H_\delta, T)$ induced by $P_{\delta(i)}g_i$.) Let $\psi : H \to A$ be the unique group homomorphism extending the ℓ-homomorphism ψ_i. With this notation established we get the following theorem.

THEOREM 11.3.5. $\mathop{{}^{L}\bigsqcup}\limits_{i \in I} G_i$ is isomorphic to the sublattice G of A generated by $\psi(H)$.

Proof. Let $L \in \mathcal{L}$ and let $\{\phi_i : G_i \to L \mid i \in L\}$ be a family of ℓ-group homomorphisms. There exists a group homomorphism $\phi : H \to L$ extending the homomorphisms ϕ_i.

Our first objective is to produce a group homomorphism $\bar{\phi} : \psi(H) \to L$ such that $\bar{\phi}\psi = \phi$. The obvious definition of $\bar{\phi}$ is $\psi(h) \mapsto \phi(h)$. This clearly is a group homomorphism if it is well-defined; i.e., if $\ker\psi \subseteq \ker\phi$.

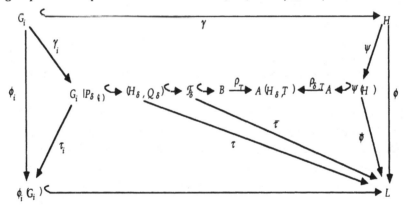

The family of ℓ-homomorphisms ϕ_i determines a $\delta \in \Delta$ such that $\phi_i(G_i) \cong G_i/P_{\delta(i)}$ under the natural isomorphism $\tau_i : G_i/P_{\delta(i)} \to \phi_i(G_i)$ defined by $\tau_i(P_{\delta(i)}g_i) = \phi_i(g_i)$. Let $\tau : H_\delta \to L$ be the group homomorphism extending the mappings τ_i. Let $\mathcal{F}_\delta = \mathcal{F}_L(H_\delta, T)$. In view of Corollary 11.2.7 and

Theorem 11.3.3, $\mathcal{F}_\delta$ is the sublattice of $B = \Pi\{A(H_\delta,T) \mid T \in \mathcal{R}_\delta\}$ generated by its image under the diagonal map. Now note that $\tau : (H_\delta, Q_\delta) \to L$ is a po-group homomorphism. Indeed, let $e \neq h_0 \in H$ and suppose that $\tau(h_0) \notin L^+$. Now L^+ is the intersection of right orders of L (see [Bigard, Keimel, and Wolfenstein, 1977, p. 84]), and hence there exists such an order with positive cone T_2 such that $\tau(h_0) \notin T_2$. Consider an arbitrary $T_1 \in \mathcal{R}_\delta$ and let

$$T = \{h \in H \mid \phi(h) \in T_2 \text{ or both } \phi(h) = e \text{ and } h \in T_1\}.$$

It is easy to see that $T \in \mathcal{R}_\delta$ and $h_0 \notin T \supseteq Q_\delta$. We have shown that τ is a po-group homomorphism. Thus, by the universal property of $\mathcal{F}_\delta$, there exists a unique ℓ-homomorphism $\bar{\tau} : \mathcal{F}_\delta \to L$ extending τ. Now let $\gamma : H \to \mathcal{F}_\delta$ be the group homomorphism extending the natural projections $\gamma_i : G_i \to G_i/P_{\delta(i)}$. It is straightforward to verify that $\phi = \bar{\tau}\gamma$. For each $T \in \mathcal{R}_\delta$ let ρ_T be the projection of B onto $A(H_\delta,T)$. One easily checks that $\rho_T\gamma = \rho_{\delta,T}\psi$ for each $T \in \mathcal{R}_\delta$. Consider now an element $h \in H$ with $\phi(h) \neq e$. The equality $\phi = \bar{\tau}\gamma$ yields that $\gamma(h) \neq e$ and so there is $T \in \mathcal{R}_\delta$ such that $\rho_T\gamma(h) = \rho_{\delta,T}\psi(h) \neq e$. It follows that $\psi(h) \neq e$ and thus $\ker\psi \subseteq \ker\phi$.

We will complete the proof by establishing the existence of an ℓ-homomorphism $\phi' : G \to L$ extending $\bar{\phi}$. With the intention of using Lemma 11.3.1, we choose a finite subset $\{h_{jk} \mid j \in J, k \in K\}$ of H such that

$$\bigvee_{j \in J} \bigwedge_{k \in K} \bar{\phi}(\psi(h_{jk})) = \bigvee_{j \in J} \bigwedge_{k \in K} \phi(h_{jk}) \neq e.$$

Suppose first that

$$(\bigvee_{j \in J} \bigwedge_{k \in K} \phi(h_{jk}))^+ = \bigvee_{j \in J} (\bigwedge_{k \in K} \phi(h_{jk}))^+ > e.$$

Then there is some $j \in J$ such that

$$(\bigwedge_{k \in K} \phi(h_{jk}))^+ > e.$$

But

$$(\bigwedge_{k \in K} \phi(h_{jk}))^+ = (\bigwedge_{k \in K} (\bar{\tau}\gamma)(h_{jk}))^+ = \bar{\tau}(\bigwedge_{k \in K} \gamma(h_{jk}))^+.$$

It follows that

$$(\bigwedge_{k \in K} \gamma(h_{jk}))^+ > e$$

and hence there must be $T \in \mathcal{R}_\delta$ with

$$\rho_T(\bigwedge_{k \in K} \gamma(h_{jk}))^+ > e.$$

But

$$\rho_T(\bigwedge_{k \in K} \gamma(h_{jk}))^+ = \bigwedge_{k \in K} (\rho_T\gamma(h_{jk}))^+ = \bigwedge_{k \in K} (\rho_{\delta,T}\psi(h_{jk}))^+.$$

It follows that

$$\bigvee_{j \in J} \bigwedge_{k \in K} (\rho_{\delta,T}\psi(h_{jk}))^+ = \rho_{\delta,T}(\bigvee_{j \in J} \bigwedge_{k \in K} \psi(h_{jk}))^+ > e$$

and so

$$\bigvee_{j\in J}\bigwedge_{k\in K}\psi(h_{jk}) \neq e$$

in this case.

Finally, assume that

$$(\bigvee_{j\in J}\bigwedge_{k\in K}\phi(h_{jk}))^+ = e.$$

Since

$$\bigvee_{j\in J}\bigwedge_{k\in K}\phi(h_{jk}) \neq e,$$

it follows that this element is strictly negative in L. Thus for each $j\in J$ there exists $k_j\in K$ such that $\phi(h_{jk_j}) < e$. It follows that

$$\bigvee_{j\in J}\phi(h_{jk_j}) < e$$

and, arguing as in the preceding case, we conclude that there is $T\in \mathcal{R}_\delta$ such that

$$\bigvee_{j\in J}(\rho_{\delta,T}\psi(h_{jk_j})) < e.$$

Hence

$$\bigvee_{j\in J}\bigwedge_{k\in K}(\rho_{\delta,T}\psi(h_{jk})) = \rho_{\delta,T}(\bigvee_{j\in J}\bigwedge_{k\in K}\psi(h_{jk})) < e.$$

It follows that in this case we also have

$$\bigvee_{j\in J}\bigwedge_{k\in K}\psi(h_{jk}) \neq e,$$

and the proof is complete.

As we pointed out earlier, free products of totally ordered groups in $\mathcal{A}$, $\mathcal{R}$, $\mathcal{N}_n$ ($n > 1$), and L are special cases of free extensions. For free products of arbitrary ℓ-groups in these varieties this is no longer the case. However, using Theorems 11.3.4 and 11.3.5 above we are able to establish an important embedding theorem (see section 11.4 for an application).

Let $\mathcal{U} = \mathcal{A}$, $\mathcal{R}$, $\mathcal{N}_n$ ($n > 1$), or L and consider an arbitrary family $\{G_i : i\in I\}$ in $\mathcal{U}\cap\mathcal{R}$. Define Γ_i, Γ, Δ, H_δ, Q_δ, and H as above, write

$$B = \Pi_{\delta\in\Delta}\mathcal{F}_\mathcal{U}(H_\delta, Q_\delta),$$

and denote by π_δ the projection of B onto $\mathcal{F}_\mathcal{U}(H_\delta, Q_\delta)$ for each $\delta\in \Delta$. For every $i\in I$, let $\chi_i : G_i\to B$ be the ℓ-homomorphism such that $\pi_\delta\chi_i(g) = P_{\delta(i)}g$ for all $g\in G$ and $\delta\in \Delta$. Finally, let $\chi : H\to B$ be the group homomorphism extending the ℓ-homomorphisms χ_i. The proof of the next theorem is a straightforward consequence of Theorems 11.3.2 - 11.3.5.

THEOREM 11.3.6. $\mathcal{U}\bigsqcup_{i\in I}G_i$ is isomorphic to the sublattice of B generated by $\chi(H)$.

11.4. Cardinal Decompositions

In this section we will look at a special property related to the internal structure of free products of ℓ-groups. In particular we will show that in most cases a free product of nontrivial ℓ-groups cannot be split into the cardinal product of two nontrivial ℓ-subgroups. We give complete proofs of the results for free products in all varieties except L. The case for L is merely sketched. Also, stated but not proved is an analogous result for free extensions.

An ℓ-group G will be called *cardinally indecomposable* if whenever $G = A \boxplus B$, then either $A = \{e\}$ or $B = \{e\}$. It is known that the $\mathcal{A}$-free ℓ-group is cardinally indecomposable if and only if its rank is greater than one. Further, a more general result for $\mathcal{A}$-free extensions exists which is applicable to free products.

THEOREM 11.4.1 [Bernau, 1969]. *The $\mathcal{A}$-free extension of any abelian po-group with nontrivial order is cardinally indecomposable.*

We remark that Bernau's original theorem is concerned with the free abelian ℓ-group over a po-group. His proof, which is technical, clearly applies without changes to the more general situation of Theorem 11.4.1. For free abelian ℓ-groups, the result is obvious by the first construction of the free abelian ℓ-group on a set X given in Chapter 1, section 3 of this book. We now outline Bernau's proof.

Firstly, let G be an abelian po-group with positive cone $P \neq \{e\}$ and let G_1 be the same group with the trivial order. Let H be the $\mathcal{A}$-free extension of G_1 and α the po-group injection of G_1 in H. Let K be the convex ℓ-subgroup of H generated by $\{\alpha(p)^- \mid p \in P\}$ where $x^- = x^{-1} \wedge e$ (the meet being in H). If $\alpha(g) \in K$ for some $g \in G \setminus \{e\}$, then $|\alpha(g)| \leq \prod_{i=1}^{n} \alpha(p_i)^-$ for some $p_1, \ldots, p_n \in P$. But $\{g, p_1, \ldots, p_n\}$ is positively independent in G_1 (replace g by g^- if necessary). Hence $\alpha(g)^+ \wedge \bigwedge_{i=1}^{n} \alpha(p_i)^+ \neq e$. But

$$\alpha(g)^+ \wedge \bigwedge_{i=1}^{n} \alpha(p_i)^+ \leq (\prod_{i=1}^{n} \alpha(p_i)^-) \wedge \bigwedge_{i=1}^{n} \alpha(p_i)^+ = e.$$

This contradiction implies that $\alpha(g) \in K$ only when $g = e$. It follows that the map α induces an order isomorphism of $\mathcal{F}_{\mathcal{A}}(G)$ into H/K. It is easy to lift any ℓ-homomorphism of $\mathcal{F}_{\mathcal{A}}(G)$ into an ℓ-group L to an ℓ-homomorphism of H/K into L. Consequently, $\mathcal{F}_{\mathcal{A}}(G) \cong H/K$. If $\mathcal{F}_{\mathcal{A}}(G)$ were cardinally decomposable, there would exist convex ℓ-subgroups H_1 and H_2 of H such that $H_1 H_2 = H$, $H_1 \cap H_2 = K$ and $H_1 \neq K \neq H_2$. Let $p \in P \setminus \{e\}$. Then $\alpha(p) = u_1 u_2$ for some $u_j \in H_j$ $(j = 1,2)$. But $|u_1| \wedge |u_2| \in K$ so

$$|u_1| \wedge |u_2| \leq \prod_{i=1}^{n} \alpha(p_i)^- \quad \text{for some } p_1, \ldots, p_n \in P.$$

Let

$$w = \alpha(p)^+ \wedge \bigwedge_{i=1}^{n} \alpha(p_i)^+.$$

By the Riesz Interpolation Theorem (Lemma 0.1.10) , $w = v_1 v_2$ for some $e \leq v_j \leq u_j^+$ $(j = 1,2)$. Now, as in the previous paragraph, $v_1 \wedge v_2 = e$. But $H = \mathcal{F}_{\mathcal{A}}(G_1)$ and so is a subdirect product of subgroups of $\mathbf{Q}$ (see section 1.3). It can be shown that the hypotheses force $v_1 = e$ or $v_2 = e$; i.e., $w \in H_1$ or $w \in H_2$. Without loss of generality, $w \in H_1$. Let $h \in H_2 \setminus K$. Then $Kw \wedge K|h| = K(w \wedge |h|) = K$ since $w \wedge |h| \in H_1 \cap H_2 = K$. If T is any total order on G extending P, then $w, p, p_1, \ldots, p_n \subset T \setminus \{e\}$. Because $H/K \cong \mathcal{F}_{\mathcal{A}}(G)$ is a subdirect product of o-groups (G, T) with $T \supseteq P$, Kw is a weak order unit in H/K. Therefore $|h| \in K$, the desired contradiction. This completes the outline of the proof of Theorem 11.4.1.

The ideas in the above proof lead to a more general result. Specifically we use Theorem 11.4.1 together with the results of section 11.3 to demonstrate the cardinal indecomposability of $\mathcal{A}$-free products with at least two nontrivial factors. This is the analog of a famous theorem of Baer and Levi [1936] which states that a group cannot be decomposed into both a nontrivial direct product and a nontrivial free product.

THEOREM 11.4.2. [Powell and Tsinakis, 1982]. *An abelian ℓ-group cannot be simultaneously decomposed into a nontrivial cardinal product and a nontrivial free product.*

Proof. Let $G \in \mathcal{A}$ and assume that $G = G_1 \mathcal{A} \sqcup G_2$ for nontrivial $G_1, G_2 \in \mathcal{A}$. It will be shown that G is cardinally indecomposable. To this end suppose $G = K_1 \boxplus K_2$.

In what follows we will make use of Theorem 11.3.6 and of the notation established in the paragraph preceding its statement. In addition, we let $G_i' = \chi(G_i) = \chi_i(G_i)$ $(i = 1,2)$, and write G' for the sublattice of $B =$

$\prod_{\delta\in\Delta} \mathcal{F}_{\mathcal{A}}(G_1/P_{\delta(1)} \boxplus G_2/P_{\delta(2)})$ generated by $\chi(H)$. Of course, by Theorem 11.3.6, $G \cong G'$. If we also denote this isomorphism by χ, then $G' = K_1' \boxplus K_2'$, where $K_i' = \chi(K_i)$ $(i = 1,2)$.

We proceed by first establishing the following properties.

(i) If $x,y\in G_1\cup G_2$, $x,y \neq e$, and $x\wedge y = e$, then $x,y\in G_1$ or $x,y\in G_2$.

(ii) If $x\in G_1$ and u,v are two disjoint elements of G such that $x = uv$, then $u,v\in G_1$.

In order to verify (i) suppose that $x\in G_1$, $y\in G_2$, and $x,y \neq e$. We will prove the equivalent statement that $\chi_1(x)\wedge\chi_2(y) \neq e$ in G'. There exist prime subgroups $P_{\delta(1)}$ of G_1 and $P_{\delta(2)}$ of G_2 such that $x\notin P_{\delta(1)}$ and $y\notin P_{\delta(2)}$. But then $(P_{\delta(1)}x)\wedge(P_{\delta(2)}y) \neq e$ in the lexicographic total order of $G_1/P_{\delta(1)}\oplus G_2/P_{\delta(2)}$. Thus, making use of Theorem 11.3.2 we conclude that $(P_{\delta(1)}x)\wedge(P_{\delta(2)}y) \neq e$ in $\mathcal{F}_{\mathcal{A}}(G_1/P_{\delta(1)} \boxplus G_2/P_{\delta(2)})$. It follows then by Theorem 11.3.6 that $\chi_1(x)\wedge\chi_2(y) \neq e$ in G'.

Next we proceed with the verification of (ii). Let $x\in G_1$ and u,v be two disjoint elements of G such that $x = uv$. Now for each $\delta\in\Delta$, $\mathcal{F}_{\mathcal{A}}(G_1/P_{\delta(1)} \boxplus G_2/P_{\delta(2)})$ is cardinally indecomposable by Theorem 11.4.1 and hence $\pi_\delta(u) = e$ or $\pi_\delta(v) = e$. Let $\pi : G' \to G_1'$ be the unique ℓ-epimorphism extending the identity on G_1' and the constant homomorphism $G_2' \to G_1'$. We will show that $u = \pi(u)$ and $v = \pi(v)$ which will clearly establish (ii). To begin with note that for each $z\in G'$ and each $\delta\in\Delta$, $\pi_\delta(z) = e$ implies $\pi_\delta(\pi(z)) = e$. Indeed, let $\delta\in\Delta$ and consider the ℓ-epimorphism

$$\sigma: \mathcal{F}_{\mathcal{A}}(G_1/P_{\delta(1)} \boxplus G_2/P_{\delta(2)}) \to G_1/P_{\delta(1)}$$

extending the identity on $G_1/P_{\delta(1)}$ and the constant homomorphism $G_2/P_{\delta(2)} \to G_1/P_{\delta(1)}$. The fact that σ is an ℓ-homomorphism implies that $\pi_\delta\pi = \sigma\pi_\delta$. But then if $z\in G'$ and $\pi_\delta(z) = e$, then $\sigma(\pi_\delta(z)) = e$ and hence $\pi_\delta(\pi(z)) = e$. Now let $\delta\in\Delta$. We have $\pi_\delta(x) = \pi_\delta(u)\pi_\delta(v)$ and $\pi_\delta(x) = \sigma(\pi_\delta(x)) = \pi_\delta(\pi(u))\pi_\delta(\pi(v))$. As was noted above, either $\pi_\delta(u) = e$ or $\pi_\delta(v) = e$. If, for instance, $\pi_\delta(u) = e$, then $\pi_\delta(\pi(u)) = e$ and so $\pi_\delta(v) = \pi_\delta(\pi(v))$. Thus we always have $\pi_\delta(u) = \pi_\delta(\pi(u))$ and $\pi_\delta(v) = \pi_\delta(\pi(v))$. Hence by Theorem 11.3.6, $u = \pi(u)$ and $v = \pi(v)$.

Now if $x\in G_1^+\backslash(K_1\cup K_2)$, then $x = x_1x_2$ for $x_1\in K_1^+\backslash\{e\}$ and $x_2\in K_2^+\backslash\{e\}$. By (ii) we have $x_1,x_2\in G_1^+$. Let $y\in G_2^+$. Then $y = y_1y_2$ where $y_1\in K_1^+$ and $y_2\in K_2^+$. Again by (ii) $y_1,y_2\in G_2^+$ and we can assume $y_2 \neq e$. But then $x_1\wedge y_2 = e$ in G contradicting (i). Thus, x cannot come from $G_1^+\backslash(K_1\cup K_2)$ and we get $G_1\cup G_2 \subseteq K_1\cup K_2$.

Finally suppose there is a nontrivial z in $G_1{}^+ \cap K_1$. Then for $y \in G_2{}^+ \setminus \{e\}$ we would have $x \wedge y \neq e$ by (i) which means that $y \notin K_2$. Hence $y \in K_1$. So in the end we must have $G_1 \cup G_2 \subseteq K_1$, showing K_2 is trivial.

Madden [1983, Proposition 11.10] subsequently gave another proof that the $\mathcal{A}$-free product of non-trivial abelian ℓ-groups is cardinally indecomposable. His proof mimics that of [Baker 1968] for free abelian ℓ-groups and free vector lattices. He first reduces to the case that there are two non-trivial factors in the $\mathcal{A}$-free product and each is a quotient of a free abelian ℓ-group of finite rank by the ℓ-ideal generated by a single element. As in section 1.3, this leads to a polyhedral cone with vertex the origin. Any cardinal decomposition of the $\mathcal{A}$-free product would lead to a "product" of polyhedral cones with vertex the origin which would not be path-connected if the origin were removed. But the product of such polyhedral cones is immediately seen to be path-connected even when the origin is removed. Consequently, the $\mathcal{A}$-free product of nontrivial abelian ℓ-groups is always cardinally indecomposable.

Turning to the non-abelian varieties we first consider those contained in $\mathcal{N}$ and finally take a look at L. For the proper varieties the indecomposability result is currently limited to free products where the free factors have strong order units although a proof eliminating this condition is almost certainly awaiting discovery. For any ℓ-group K we denote by K' the ℓ-ideal generated by the commutator subgroup of K.

THEOREM 11.4.3 [Powell and Tsinakis, 1984]. *If $\mathcal{U}$ is a nontrivial proper variety of ℓ-groups and $\{G_i : i \in I\} \subseteq \mathcal{U}$, then $\overset{\mathcal{U}}{\underset{i \in I}{\bigsqcup}} G_i$ is cardinally indecomposable if each G_i has a strong unit.*

Proof. Note first that for any nontrivial cardinal decomposition of $\overset{\mathcal{U}}{\underset{i \in I}{\bigsqcup}} G_i$ there is a finite subset $J \subseteq I$ such that the natural projection $\overset{\mathcal{U}}{\underset{i \in I}{\bigsqcup}} G_i \to \overset{\mathcal{U}}{\underset{i \in J}{\bigsqcup}} G_i$ causes a cardinal decomposition of $\overset{\mathcal{U}}{\underset{i \in J}{\bigsqcup}} G_i$. Thus we assume I is finite.

Suppose each of $G_1, \ldots, G_n \in \mathcal{U} \subseteq \mathcal{N}$ has a strong order unit and consider $G = G_1 \overset{\mathcal{U}}{\bigsqcup} \ldots \overset{\mathcal{U}}{\bigsqcup} G_n$. Then G has a strong order unit. Since $G \in \mathcal{N}$ it can easily be seen that $G' \neq G$. Further $G/G' \cong G_1/G_1' \overset{\mathcal{A}}{\bigsqcup} \ldots \overset{\mathcal{A}}{\bigsqcup} G_n/G_n'$ which is cardinally indecomposable by Theorem 11.4.2.

Now let $\{H_k \mid k \in K\} \subseteq \mathcal{U}$ be a set of subdirectly irreducibles of which G is a subdirect product. Then each H_k has a strong order unit as was the case for G we have $H_k' \neq H_k$.

If $G = L_1 \boxplus L_2$, then $G/G' = L_1/(G'\cap L_1) \boxplus L_2/(G'\cap L_2)$ so either $G' \supseteq L_1$ or $G' \supseteq L_2$. Assume that $G' \supseteq L_1$. For each $k \in K$ let $\pi_k : G \to H_k$ be the natural projection. Then $\pi_k(L_1) = \{e\}$ or $\pi_k(L_2) = \{e\}$ for each k since H_k is cardinally indecomposable. If $\pi_k(L_2) = \{e\}$ then $H_k = \pi_k(L_1) \subseteq \pi_k(G') \subseteq H_k'$ contradicting $H_k' \neq H_k$. Hence, $\pi_k(L_1) = \{e\}$ for each $k \in K$ and hence $L_1 = \{e\}$. This completes the proof.

To prove the indecomposability of free products in the variety L of all ℓ-groups a completely different approach is used. In this case countable ℓ-groups are first considered. If G and H a countably infinite ℓ-groups, then so is $G \, {}^L\!\!\sqcup \, H$ and we can list its non-identity elements $w_1, w_2, \dots$. Let

$$\Delta_n = (2n - \tfrac{1}{3}, 2n + \tfrac{4}{3}), \text{ and } \alpha_n \in (2n, 2n + 1)$$

be chosen so that for some ℓ-embedding ϕ_n of $G \, {}^L\!\!\sqcup \, H$ in $A(\Delta_n)$, $(\phi_n(w_n))(\alpha_n) \neq \alpha_n$, $2n = \min\{(\phi_n(u_n))(\alpha_n) : u$ is an initial subword of $w_n\}$ and $2n + 1 = \max\{(\phi_n(u_n))(\alpha_n) : u$ is an initial subword of $w_n\}$, $(n = 1,2,3,\dots)$. As in the proof of Theorem 9.4.1, it is possible to build bridges between $2n + 1$ and $2n + 2$ using elements from G and H so that if ϕ and ψ are the resulting ℓ-homomorphisms of G and H in $A(\mathbf{R})$, then $\phi(g)\,|\,(2n, 2n + 1) = \phi_n(g)\,|\,(2n, 2n + 1)$ and $\psi(h)\,|\,(2n, 2n + 1) = \psi_n(h)\,|\,(2n, 2n + 1)$ for all $g \in G$, $h \in H$ and $n \in \{1,2,3,\dots\}$. Hence if θ is the resulting ℓ-homomorphism of $G \, {}^L\!\!\sqcup \, H$ into $A(\mathbf{R})$, then $(\theta(w_n))(\alpha_n) = (\phi_n(w_n))(\alpha_n) \neq \alpha_n$ for each n; consequently, θ is an ℓ-embedding. Moreover, as in the proof of Theorem 9.4.1, $\theta(G \, {}^L\!\!\sqcup \, H)$ can be made to act doubly transitively (and faithfully) on the orbit of α_1. Thus $G \, {}^L\!\!\sqcup \, H$ *is* ℓ-isomorphic to a doubly transitive ℓ-subgroup of $A(\mathbf{Q})$. This immediately yields that such a free product is cardinally indecomposable. Indeed, $G \, {}^L\!\!\sqcup \, H$ is directly indecomposable even as a group [Glass and Wilson, a]. This is in sharp contrast to the $\mathcal{A}$-free product which can be directly decomposable as a group (let G and H be free abelian ℓ-groups of finite rank; then so is $G \, {}^{\mathcal{A}}\!\!\sqcup \, H$ and hence is a direct sum of $\aleph_0$ copies of $\mathbf{Z}$ as a group - see Section 1.3.) Finally, it is possible to generalize the above constructions to obtain a representation of the L-free product of two non-trivial ℓ-groups which is "locally" doubly transitive. This yields

THEOREM 11.4.4 [Glass, 1987]. *An ℓ-group cannot be simultaneously decomposed into a nontrivial L-free product and a nontrivial direct product. In fact, the L-free product of any two nontrivial ℓ-groups has trivial center and the variety it generates is L.*

11.5. Disjoint Sets in Free Products

By a *disjoint set* in an ℓ-group G we mean a subset S of $G \setminus \{e\}$ such that $x \wedge y = e$ for all distinct $x, y \in S$. Given an infinite cardinal m, an ℓ-group is said to satisfy the m-*disjointness condition* if it contains no disjoint subset of cardinality m.

We remark that $\mathcal{U}$-free ℓ-groups on any number of generators satisfy the $\aleph_1$-disjointness condition for every nontrivial variety of ℓ-groups ([Bleier, 1975], [Powell and Tsinakis, 1983a], [Baldwin, Berman, Glass, and Hodges, 1982]). Furthermore, it was shown in [Powell and Tsinakis, 1986] that any $\mathcal{U}$-free ℓ-group on two or more generators has a disjoint set of cardinality $\aleph_0$. (For more general results along these lines see [Powell and Tsinakis, b].) The situation for arbitrary free products is radically different. The main result of this section (Theorem 11.5.3) implies that for every infinite cardinal m, $\mathcal{U}$-free products fail to preserve the m-disjointness condition in the worst possible fashion. More specifically, it is shown that given an infinite cardinal m, there exist totally ordered abelian groups G and H each of cardinality m such that $G \overset{\mathcal{U}}{\sqcup} H$ has a disjoint set of cardinality m, for every nontrivial ℓ-group variety $\mathcal{U}$.

This last result is quite surprising when contrasted with the analogous theorem valid for lattice varieties. For example, Adams and Kelly [1977] showed that free products in the variety of all lattices preserve the m-disjointness condition for all m (where e is replaced by an arbitrary distinguished element). In the same paper (see also [Lakser, 1973]), Adams and Kelly established that free products in the variety of distributive lattices preserve the m-disjointness condition whenever m is singular or weakly compact. Further, they proved that the question of the preservation of the m-disjointness condition for these products cannot be decided in general on the basis of the standard axioms of set theory.

LEMMA 11.5.1. *Let* a_1, a_2, b_1, b_2 *be positive elements of an* ℓ-group G *such that* $a_2{}^2 < a_1$ *and* $b_1{}^2 < b_2$. *If* $x = [(a_i{}^2 b_i{}^{-1}) \wedge (b_i{}^2 a_i{}^{-1})]^+$ *for* $i = 1,2$, *then* $x_1 \wedge x_2 = e$.

Proof. In view of Theorem 2.2.5, we may view G as an ℓ-permutation group (G, Ω) for a suitable chain Ω. Since $a_1{}^2 b_1{}^{-1} \geq a_2{}^2 b_2{}^{-1}$ and $b_2{}^2 a_2{}^{-1} \geq b_1{}^2 a_1{}^{-1}$, it follows that

$$e \leq x_1 \wedge x_2 \leq (b_1{}^2 a_1{}^{-1})^+ \wedge (a_2{}^2 b_2{}^{-1})^+,$$

and hence it is sufficient to prove that the right-hand side of the inequality is equal to e. This is of course equivalent to showing that for all $\omega \in \Omega$,

$$\min\{b_1{}^2 a_1{}^{-1}(\omega), (a_2{}^2 b_2{}^{-1}(\omega)\} \leq \omega.$$

To this end, let $\omega \in \Omega$ and suppose that $b_1{}^2 a_1{}^{-1}(\omega) > \omega$, i.e., $a_1{}^{-1}(\omega) > b_1{}^{-2}(\omega)$. Now by hypothesis, $a_2{}^{-2}(\omega) \geq a_1{}^{-1}(\omega)$ and $b_1{}^{-2}(\omega) \geq b_2{}^{-1}(\omega)$. It follows that $a_2{}^{-2}(\omega) > b_1{}^{-2}(\omega)$ or $a_2{}^2 b_2{}^{-1}(\omega) < \omega$.

The next result shows that disjoint sets in a free product can be built from suitable chains in the free factors. Let G be an ℓ-group and let m be a cardinal number. A chain $\{a_\lambda : \lambda < m\}$ in G of length m is said to be an *admissible ascending (descending) chain* if for all $\lambda < \mu < m$, $e < a_\lambda{}^2 < a_\mu$ $(e < a_\mu{}^2 < a_\lambda)$.

LEMMA 11.5.2. *Let m be an infinite cardinal. Suppose that G_1 and G_2 are abelian ℓ-groups such that G_1 has an admissible descending chain of length m and G_2 has an admissible ascending chain of length m. Then for every nontrivial variety $\mathcal{U}$ of ℓ-groups, $G_1{}^{\mathcal{U}}\sqcup G_2$ has a disjoint set of cardinality m.*

Proof. Let $\{a_\lambda : \lambda < m\}$ be an admissible descending chain in G_1, and let $\{b_\lambda : \lambda < m\}$ be an admissible ascending chain in G_2. Consider the elements

$$x_\lambda = [(a_\lambda{}^2 b_\lambda{}^{-1})\wedge(b_\lambda{}^2 a_\lambda{}^{-1})]^+ \qquad (\lambda < m)$$

in $G_1{}^{L}\sqcup G_2$ and set $S_L = \{x_\lambda : \lambda < m\}$.

Recall that $\mathcal{A} \subseteq \mathcal{U} \subseteq L$ and hence the identity mappings in G_1 and G_2 induce the epimorphisms

$$\phi_{\mathcal{U}} : G_1{}^{L}\sqcup G_2 \rightarrow G_1{}^{\mathcal{U}}\sqcup G_2$$

and

$$\psi_{\mathcal{U}} : G_1{}^{\mathcal{U}}\sqcup G_2 \rightarrow G_1{}^{\mathcal{A}}\sqcup G_2.$$

The goal is to show that $S_{\mathcal{U}} = \{\phi_{\mathcal{U}}(x_\lambda) : \lambda < m\}$ is a disjoint set in $G_1{}^{\mathcal{U}}\sqcup G_2$ of cardinality m.

To begin with, in view of Lemma 11.5.1, it is sufficient to prove that $S_{\mathcal{U}}\cap\{e\} = \varnothing$. Further, since every $S_{\mathcal{U}}$ is projected onto $S_{\mathcal{A}}$ by $\psi_{\mathcal{U}}$, we only need to show that $S_{\mathcal{A}}\cap\{e\} = \varnothing$. To this end, we fix $\lambda < m$ and proceed to show that $\phi_{\mathcal{A}}(x_\lambda) \neq e$ in $G_1{}^{\mathcal{A}}\sqcup G_2$. Let P_i be a prime subgroup of G_i $(i = 1,2)$ such that $a_\lambda \notin P_1$ and $b_\lambda \notin P_2$. Consider the totally ordered groups $H_i = G_i/P_i$ $(i = 1,2)$ and let $\psi : G_1{}^{\mathcal{A}}\sqcup G_2 \rightarrow H_1{}^{\mathcal{A}}\sqcup H_2$ be the epimorphism extending the natural projections. It will be sufficient to verify that $\psi\phi_{\mathcal{A}}(x_\lambda) \neq e$ in $H_1{}^{\mathcal{A}}\sqcup H_2$. Set $\psi(a_\lambda) = a$ and $\psi(b_\lambda) = b$. In view of the universal property of $\mathcal{A}$-free products, it will be enough to to prove that the elements $a^2 b^{-1}$ and $b^2 a^{-1}$ in $H_1 \oplus H_2$ are positive with respect to some total order on $H_1 \oplus H_2$ extending the orders of H_1 and H_2. Now any partial order on a torsion-free abelian group can be extended to a total order (see Chapter 0 Example 5), and thus it will suffice to prove that there is a partial order on $H_1 \oplus H_2$ whose cone contains the set $(H_1 \boxplus H_2)^+ \cup \{a^2b^{-1}, b^2a^{-1}\}$. It is easy to see that such a partial order exists if and only if for all $n_1, n_2 \in \mathbf{Z}^+ \backslash \{0\}$, the combination $(a^2b^{-1})^{n_1}(a^{-1}b^2)^{n_2}$ is not negative in $H_1 \boxplus H_2$. Thus suppose that $(a^2b^{-1})^{n_1}(a^{-1}b^2)^{n_2} \leq e$ in H_1 and H_2, respectively. As $a > e$ in H_1 and $b > e$ in H_2, this implies $2n_1-n_2 \leq 0$ and $2n_2-n_1 \leq 0$ and so $n_1 = n_2 = 0$.

We can now use Theorem 11.5.2 to establish the main result of this section.

THEOREM 11.5.3. *Let m be an infinite cardinal. There exist totally ordered abelian groups G_1 and G_2 of cardinality m such that for any nontrivial variety $\mathcal{U}$ of ℓ-groups, $G_1{}^{\mathcal{U}} \sqcup G_2$ has a disjoint set of cardinality m.*

Proof. Let Λ be a totally ordered set of cardinality m. Define

$$G_1 = \underset{\lambda \in \Lambda}{\oplus} G_\lambda^{(1)}$$

and

$$G_2 = \underset{\lambda \in \Lambda}{\oplus} G_\lambda^{(2)},$$

where $G_1 \cap G_2 = \{e\}$, and for each $\lambda \in \Lambda$, $G_\lambda^{(1)}$ and $G_\lambda^{(2)}$ are isomorphic to the totally ordered group $\mathbf{Z}$ of integers. Endow G_1 and G_2 with the lexicographic and the dual lexicographic order respectively. Thus if $x = x_1 x_2 ... x_n$ with $x_i \in G_{\lambda_i}^{(1)} \backslash \{e\}$ (respectively, $x_i \in G_{\lambda_i}^{(2)} \backslash \{e\}$) and $\lambda_1 < \lambda_2 < ... < \lambda_n$, then $x > e$ in G_1 iff $x_1 > e$ in $G_{\lambda_1}^{(1)}$ (and $x > e$ in G_2 iff $x_n > e$ in $G_{\lambda_n}^{(2)}$).

Now for each $\lambda \in \Lambda$, pick $a_\lambda \in (G_\lambda^{(1)})^+ \backslash \{e\}$ and $b_\lambda \in (G_\lambda^{(2)})^+ \backslash \{e\}$. It is readily seen that $\{a_\lambda \mid \lambda \in \Lambda\}$ is an admissible descending chain in G_1 and $\{b_\lambda \mid \lambda \in \Lambda\}$ is an admissible ascending chain in G_2. The result now follows from Lemma 11.5.2 since these two chains have length m.

It would be of interest to determine whether Lemma 11.5.2. remains true when G_1 and G_2 are arbitrary ℓ-groups in $\mathcal{U}$. The following result shows that this is the case if $\mathcal{U} = L$.

THEOREM 11.5.4 [Powell and Tsinakis, b]. *Let m be an infinite cardinal and let $\mathcal{U} = \mathcal{A}$ or L. Suppose that G_1 and G_2 are ℓ-groups in $\mathcal{U}$ such that G_1 has an admissible descending chain of length m and G_2 has an admissible ascending chain of length m. Then $G_1{}^{\mathcal{U}} \sqcup G_2$ has a disjoint set of cardinality m.*

Proof. The case $\mathcal{U} = \mathcal{A}$ is found in Lemma 11.5.2. Thus suppose that $\mathcal{U} = L$. Let $\{a_\lambda \mid \lambda < m\}$ be an admissible descending chain in G_1 and let $\{b_\lambda \mid \lambda < m\}$ be an admissible ascending chain in G_2. Define the subset $S_L = \{x_\lambda \mid \lambda < m\}$ of $G = G_1{}^{\mathcal{U}}\sqcup G_2$ as in the proof of Lemma 11.5.2:

$$x_\lambda = [(a_\lambda{}^2 b_\lambda{}^{-1}) \wedge (b_\lambda{}^2 a_\lambda{}^{-1})]^+ \quad (\lambda < m)$$

In view of Lemma 11.5.1, we only need show that $S_L \cap \{e\} = \varnothing$. To this end, we fix $\lambda < m$ and proceed to show that $x_\lambda \neq e$. By [2.2.5 and 2.3.2], we may view G as an ℓ-permutation group (G,Ω), where $A(\Omega)$ is doubly transitive. Now $a_\lambda b_\lambda > e$ and hence there exist $\omega_1, \omega_2 \in \Omega$ such that $a_\lambda{}^{-1}(\omega_1) < \omega_1$ and $b_\lambda{}^{-1}(\omega_2) < \omega_2$. Thus, since Ω is doubly homogeneous, there exists $h \in A(\Omega)$ such that $h(\omega_2) = \omega_1$ and $h(b_\lambda{}^{-1}(\omega_2)) = a_\lambda{}^{-1}(\omega_1)$. We have $a_\lambda{}^{-1}(\omega_1) = c_\lambda{}^{-1}(\omega_1) < \omega_1$, where $c_\lambda = h b_\lambda h^{-1}$. It follows that

$$a_\lambda{}^2 c_\lambda{}^{-1}(\omega_1) = a_\lambda(\omega_1) > \omega_1$$

and

$$c_\lambda{}^2 a_\lambda{}^{-1}(\omega_1) = c_\lambda(\omega_1) > \omega_1.$$

This shows that

(*) $\qquad (a_\lambda{}^2 c_\lambda{}^{-1})^+ \wedge (c_\lambda{}^2 a_\lambda{}^{-1})^+ > e$

in $A(\Omega)$. Consider now the ℓ-homomorphism $f : G \to A(\Omega)$ extending the inclusion map $f_1 : G_1 \to A(\Omega)$ and the ℓ-homomorphism $f_2 : G_2 \to A(\Omega)$ defined by $f_2(g) = hgh^{-1}$ for all $g \in G_2$. In view of (*), $f(x_\lambda) > e$ in $f(G)$ and hence $x_\lambda > e$ in G. This completes the proof of the theorem.

11.6. Embedding Distributive Lattice Free Products in ℓ-group Free Products

We have seen in sections 11.2 and 11.3 that free products in varieties of ℓ-groups are intimately related to free products in classes of groups. The objective of this section is to show that there is also a significant link between free products of ℓ-groups and free products of distributive lattices.

Initially, recall that the lattice of any ℓ-group is distributive (Lemma 0.1.5) and hence any ℓ-group can be considered as a member of the variety $\mathcal{D}_e$ of all distributive lattices $(L, \wedge, \vee, e)$ with a distinguished element (nullary operation) e. The main result of this section is Theorem 11.6.4 which asserts that if $\mathcal{U}$ is $\mathcal{A}$ or L and $\{B_i \mid i \in I\}$ is a family in $\mathcal{U}$, then the sublattice of ${}^{\mathcal{U}}\underset{i \in I}{\sqcup} G_i$ generated by $\underset{i \in I}{\cup} G_i$ is the $\mathcal{D}_e$-free product of the family $\{G_i \mid i \in I\}$.

This result was established in [Franchello, 1978] for the case $\mathcal{U} = L$ and in [Powell and Tsinakis, 1983b] for the case $\mathcal{U} = \mathcal{A}$. The proof below for $\mathcal{U} = L$ borrows ideas from both papers.

We first consider $\mathcal{D}_e$-free products. A $\mathcal{D}_e$-homomorphism is, of course, a lattice homomorphism which preserves e. In Lemma 11.6.2 we give a characterization of $\mathcal{D}_e$-free products which will directly apply to ℓ-group free products. This characterization is a straightforward modification of the well known characterization of free products of bounded distributive lattices (see for example [Grätzer, 1978, page 131]).

For $L \in \mathcal{D}_e$, we use the notation $L^+ = \{x \in L \mid x \geq e\}$ and $L^- = \{x \in L \mid x \leq e\}$. If S is a finite non-empty subset of a set T, then we write $S \subseteqq T$. Note that for $S \subseteq L \in \mathcal{D}_e$, the subalgebra $\langle S \rangle$ of L generated by S is given by

$$\langle S \rangle = \bigvee_{j \in J} \bigwedge_{k \in K} \{s_{kj} \mid s_{kj} \in S \cup \{e\}; K \text{ and } J \text{ are finite}\}$$
$$= \bigwedge_{k \in K} \bigvee_{j \in J} \{s_{kj} \mid s_{kj} \in S \cup \{e\}; K \text{ and } J \text{ are finite}\} .$$

It will be convenient to isolate the following lemma whose straightforward proof can be found in [Balbes and Dwinger, 1974, page 86].

LEMMA 11.6.1. *Let* $L, L' \in \mathcal{D}_e$ *and* $e \in S \subseteq L$. *A map* $f : S \to L'$ *such that* $f(e) = e$ *can be extended to a* $\mathcal{D}_e$-*homomorphism* $\overline{f} : \langle S \rangle \to L'$ *if and only if*

$$\bigwedge_{k \in K} t_k \leq \bigvee_{j \in J} r_j \text{ implies } \bigwedge_{k \in K} f(t_k) \leq \bigvee_{j \in J} f(r_j)$$

whenever $\{t_k \mid k \in K\}, \{r_j \mid j \in J\} \subseteqq S$.

LEMMA 11.6.2. *Let* $L \in \mathcal{D}_e$, *and let* $\{L_i \mid i \in I\}$ *be a family in* $\mathcal{D}_e$ *such that each* L_i *is a subalgebra of* L. *Then* L *is the* $\mathcal{D}_e$-*free product of this family if and only if the following conditions are satisfied:*

(1) $\langle \bigcup_{i \in I} L_i \rangle = L$.

(2) *For all* $J, K \subseteqq I$ *and all families* $\{a_j \mid j \in J\}, \{b_k \mid k \in K\}$ *in* L *such that either* $a_j \in L_j^+ \backslash \{e\}$ *and* $b_k \in L_k^+$ *for all* $j \in J, k \in K$ *or* $a_j \in L_j^-$ *and* $b_k \in L_k^- \backslash \{e\}$ *for all* $j \in J, k \in K$, *the relation*

$$\bigwedge_{j \in J} a_j \leq \bigvee_{k \in K} b_k$$

implies the existence of $i \in J \cap K$ *with* $a_i \leq b_i$.

Proof. (i) Suppose first that L is the free product of $\{L_i \mid i \in I\}$. Condition (1) holds by the definition of the free product. To establish condition (2), let $J, K \subseteqq I$ indexing the sets $\{a_j \in L_j^+ \backslash \{e\} \mid j \in J\}$ and $\{b_k \in L_k^+ \mid k \in K\}$. Suppose that $\bigwedge_{j \in J} a_j \leq \bigvee_{k \in K} b_k$ and that the conclusion of (2) is not satisfied. Denote by C the two element chain $\{0,1\}$ with $e_C = 0$.

If $J \cap K \neq \emptyset$, then by assumption $a_i \not\leq b_i$ for all $i \in J \cap K$. Thus, for each such i there exists a lattice homomorphism $f_i : L_i \to C$ such that $f_i(a_i) = 1$ and $f_i(b_i) = 0$. Note that f_i is actually a $\mathcal{D}_e$-homomorphism since $f_i(e) \leq f_i(b_i) = 0 = e_C$.

If $i \in I \setminus K$, then $a_i > e$, so again there exists a $\mathcal{D}_e$-homomorphism $f_i : L_i \to C$ such that $f_i(a_i) = 1$. Finally, for $i \in I \setminus J$, let $f_i : L_i \to C$ be the zero map.

By the universal property of the free product, there is a $\mathcal{D}_e$-homomorphism $f : L \to C$ such that $f | L_i = f_i$ for all $i \in I$. But then $1 = \bigwedge_{j \in J} f_j(a_j) = \bigwedge_{j \in J} f(a_j) = f(\bigwedge_{j \in J} a_j) \leq f(\bigvee_{k \in K} b_k) = \bigvee_{k \in K} f_k(b_k) = 0$. The contradiction implies that the conclusion of (2) must indeed be satisfied.

A similar proof takes care of the other case for condition (2).

(ii) Conversely, suppose conditions (1) and (2) are satisfied, and consider a family $\{f_i : L_i \to L' \mid i \in I\}$ of $\mathcal{D}_e$-homomorphisms into $L' \in \mathcal{D}_e$. We first verify that the following implication holds.

(*) If $J, K \subseteq I$, $a_j \in L_j$ for each $j \in J$, $b_k \in L_k$ for each $k \in K$, and $\bigwedge_{j \in J} a_j \leq \bigvee_{k \in K} b_k$ in L, then $\bigwedge_{j \in J} f_j(a_j) \leq \bigvee_{k \in K} f_k(b_k)$.

For a_j, b_j as in (*) we get

$$\bigwedge_{j \in J}(a_j \vee e) \leq \bigvee_{k \in K}(b_k \vee e) \text{ and } \bigwedge_{j \in J}(a_j \wedge e) \leq \bigvee_{k \in K}(b_k \wedge e).$$

There are four cases which must be considered.

Case I. For all $j \in J$, $k \in K$, $a_j \vee e > e$ and $b_k \wedge e < e$.

Case II. For all $j \in J$, $a_j \vee e > e$, and there is $k \in K$ such that $b_k \wedge e = e$.

Case III. There is $j \in J$ such that $a_j \vee e = e$, and for all $k \in K$, $b_k \wedge e < e$.

Case IV. There are $j \in J$, $k \in K$ such that $a_j \vee e = e$ and $b_k \wedge e = e$.

We first consider case I. In view of condition (2), there exist $i_1, i_2 \in J \cap K$ such that

$$a_{i_1} \vee e \leq b_{i_1} \vee e \text{ and } a_{i_2} \wedge e \leq b_{i_2} \wedge e.$$

As each f_i is a $\mathcal{D}_e$-homomorphism, it follows that $f_{i_1}(a_{i_1}) \vee e \leq f_{i_1}(b_{i_1}) \vee e$ and $f_{i_2}(a_{i_2}) \wedge e \leq f_{i_2}(b_{i_2}) \wedge e$. An easy computation yields

$$f_{i_1}(a_{i_1}) \wedge f_{i_2}(a_{i_2}) \leq f_{i_1}(b_{i_1}) \vee f_{i_2}(b_{i_2}),$$

which clearly implies the conclusion in (*).

We next consider case II. By condition (2), there exists $i_1 \in J \cap K$ such that $a_{i_1} \vee e \leq b_{i_1} \vee e$. Again this implies that $f_{i_1}(a_{i_1}) \vee e \leq f_{i_1}(b_{i_1}) \vee e$. Now let $k \in K$ be such that $b_k \wedge e = e$, and consider an arbitrary $j \in J$. Then $a_j \wedge e \leq e \leq b_k \wedge e$ and hence $f_j(a_j) \wedge e \leq f_j(e) = e = f_k(b_k) \wedge e$. The relations $f_{i_1}(a_{i_1}) \vee e \leq f_{i_1}(b_{i_1}) \vee e$ and $f_j(a_j) \wedge e \leq f_k(b_k) \wedge e$ imply the inequality

$$f_{i_1}(a_{i_1}) \wedge f_j(a_j) \leq f_{i_1}(b_{i_1}) \vee f_k(b_k),$$

which implies the conclusion in (*).

Case III is treated as case II, whereas case IV immediately implies the conclusion in (*).

Now define $f : \cup_{i \in I} L_i \to L'$ by $f(a_i) = f_i(a_i)$ for each $a_i \in L_i$. In view of condition (I), (*), and Lemma 11.6.1, f can be extended to a $\mathcal{D}_e$-homomorphism $\bar{f} : L \to L'$. Evidently $\bar{f} | L_i = f_i$ for each $i \in I$, and thus L is the $\mathcal{D}_e$-free product of the family $\{L_i \mid i \in I\}$.

The next result is an immediate consequence of Lemma 11.6.2.

LEMMA 11.6.3 [Powell and Tsinakis, 1983b]. *Let G be an arbitrary ℓ-group and $\{G_i \mid i \in I\}$ be a family of ℓ-subgroups of G such that $\cup_{i \in I} G_i$ generates G as an ℓ-group. Let $L = \langle \cup_{i \in I} G_i \rangle$ be the sublattice of G generated by $\cup_{i \in I} G_i$. The following are equivalent:*

(1) *L is the $\mathcal{D}_e$-free product of $\{G_i \mid i \in I\}$.*

(2) *For $J, K \subseteq I$, $a_j \in G_j^+ \setminus \{e\}$ for all $j \in J$, and $b_k \in G_k^+$ for all $k \in K$, the relation*

$$\bigwedge_{j \in J} a_j \leq \bigvee_{k \in K} b_k$$

implies the existence of $i \in J \cap K$ with $a_i \leq b_i$.

We are now prepared to establish the main result of this section.

THEOREM 11.6.4 [Franchello, 1978]; [Powell and Tsinakis, 1983b]. *Let $\mathcal{U}$ be $\mathcal{A}$ or $\mathcal{L}$. Suppose $\{G_i \mid i \in I\}$ is a family of ℓ-groups in $\mathcal{U}$, and let $G = \mathcal{U}\bigsqcup_{i \in I} G_i$ be their $\mathcal{U}$-free product. Then the sublattice L of G generated by $\mathcal{U}\bigsqcup_{i \in I} G_i$ is the $\mathcal{D}_e$-free product of the family $\{G_i \mid i \in I\}$.*

Proof. We will verify that condition (2) of Lemma 11.6.3 is satisfied by L. To this end, let $J, K \subseteq I$, $a_j \in G_j^+ \setminus \{e\}$ for all $j \in J$, and $b_k \in G_k^+$ for all $k \in K$. Suppose

(*) $\bigwedge_{j \in J} a_j \leq \bigvee_{k \in K} b_k$.

We need to show that there is $i \in J \cap K$ such that $a_i \leq b_i$. Note that $^U\bigsqcup_{i \in J \cup K} G_i$ is an ℓ-subgroup of $^U\bigsqcup_{i \in I} G_i$, and hence we may assume without loss of generality that $I = J \cup K$.

We first prove the result when $\mathcal{U} = \mathcal{A}$ by invoking Theorem 11.3.4. (The reader is reminded that $\mathcal{G}(\mathcal{A})$-free products are direct sums.) The proof will be divided into several cases and the notation introduced in the paragraph preceding the statement of Theorem 11.3.4 will be used without any further explanations.

Case I. $J \cap K = \varnothing$. For each $j \in J$, let $P_{\delta(j)}$ be a prime subgroup of G_j such that $a_j \notin P_{\delta(j)}$, and for each $k \in K$, let $P_{\delta(k)}$ be an arbitrary prime subgroup of G_k. Note that for each $j \in J$ and $k \in K$, $P_{\delta(j)} a_j > P_{\delta(j)}$ in $G_j / P_{\delta(j)}$ and $P_{\delta(k)} b_k \geq P_{\delta(k)}$ in $G_k / P_{\delta(k)}$. Now let T be the positive cone of a lexicographic order on $H_\delta = \bigoplus_{i \in I}(G_i / P_{\delta(i)})$ such that for each $j \in J$ and $k \in K$, $G_j / P_{\delta(j)}$ dominates $G_k / P_{\delta(k)}$. It is clear that $T \in T_\delta$ and $\bigwedge_{j \in J}(P_{\delta(j)} a_j) > \bigwedge_{k \in K}(P_{\delta(k)} b_k)$ in (H_δ, T). In view of Theorem 11.3.4, the last inequality contradicts (*).

Case II. $J \cap K = \{i_1\}$. Suppose $a_{i_1} \not\leq b_{i_1}$. Then there is a prime subgroup $P_{\delta(i_1)}$ of G_{i_1} such that $P_{i_1} a_{i_1} > P_{i_1} b_{i_1}$ in $G_{i_1} / P_{\delta(i_1)}$. For $j \in J \setminus \{i_1\}$, $k \in K \setminus \{i_1\}$, pick $P_{\delta(j)}$ and $P_{\delta(k)}$ as in case I. Now let T be the positive cone of a lexicographic order on H_δ such that $G_j / P_{\delta(j)}$ dominates $G_{i_1} / P_{\delta(i_1)}$ for each $j \in J \setminus \{i_1\}$, and $G_{i_1} / P_{\delta(i_1)}$ dominates $G_k / P_{\delta(k)}$ for each $k \in K \setminus \{i_1\}$. Then $T \in T_\delta$ and $\bigwedge_{j \in J}(P_{\delta(j)} a_j) = P_{\delta(i_1)} a_{i_1} > P_{\delta(i_1)} b_{i_1} = \bigvee P_{\delta(k)} b_k$ in (H_δ, T). This again contradicts (*) by Theorem 11.3.4.

Case III. $J \cap K = \{i_1, i_2\}$. Suppose $a_{i_1} \not\leq b_{i_1}$ and $a_{i_2} \not\leq b_{i_2}$. Choose prime subgroups $P_{\delta(i_1)}$ of G_{i_1} and $P_{\delta(i_2)}$ of G_{i_2} such that $P_{\delta(i_1)} a_{i_1} > P_{\delta(i_1)} b_{i_1}$ and $P_{\delta(i_2)} a_{i_2} > P_{\delta(i_2)} b_{i_2}$. It is easy to see that there is a partial order of $(G_{i_1} / P_{\delta(i_1)}) \oplus (G_{i_2} / P_{\delta(i_2)})$ extending the cardinal order and whose positive cone contains the elements $(P_{\delta(i_1)} a_{i_1})(P_{\delta(i_2)} b_{i_2})^{-1}$ and $(P_{\delta(i_2)} a_{i_2})(P_{\delta(i_1)} b_{i_1})^{-1}$. Extend this partial order to a total order. That such a total order exists follows from Mal'cev's aforementioned result (see [Mura and Rhemtulla, 1977, page 56]). For $j \in J \setminus \{i_1, i_2\}$, $k \in K \setminus \{i_1, i_2\}$, pick $P_{\delta(j)}$ and $P_{\delta(k)}$ as in case I. Now define a total order on H_δ by lexing the ℓ-groups $G_j / P_{\delta(j)}$ $(j \in J \setminus \{i_1, i_2\})$ on top, then $((G_{i_1} / P_{\delta(i_1)}) \oplus (G_{i_2} / P_{\delta(i_2)}), T')$, and finally the ℓ-groups $G_k / P_{\delta(k)} (k \in K \setminus \{i_1, i_2\})$ on the bottom. If T is the positive cone of this order it is clear that $T \in T_\delta$. Furthermore, we have in (H_δ, T),

$$\bigwedge_{j \in J}(P_{\delta(j)} a_j) \geq (P_{\delta(i_1)} a_{i_1}) \wedge (P_{\delta(i_2)} a_{i_2}) > (P_{\delta(i_1)} b_{i_1}) \vee (P_{\delta(i_2)} b_{i_2}) \geq \bigvee_{k \in K} P_{\delta(k)} b_k,$$

which again contradicts (*).

Case IV. $J \cap K = \{i_1, ..., i_n\}$, $n > 2$. Suppose there is an element of $J \cap K$, say i_1, such that $a_{i_1} \not\leq b_{i_1}$. Now $G = G_{i_1} \bigsqcup (\bigsqcup_{i \neq i_1} G_i)$ and $\bigsqcup_{i \neq i_1} G_i$ is the ℓ-subgroup of G generated by $\bigcup_{i \neq i_1} G_i$. Consider the elements $a = \bigwedge_{j \neq i_1} a_j$ and $b = \bigvee_{k \neq i_1} b_k$ in $\bigsqcup_{i \neq i_1} G_i$. We have $a_{i_1} \wedge a \leq b_{i_1} \vee b$ and hence, by case III, $a \leq b$.

An easy inductive argument implies that there is $r \in \{2,...,n\}$ such that $a_{i_r} \leq b_{i_r}$.

We next consider the case $\mathcal{U} = L$. We retain our notation, established at the beginning of the proof and our assumption that $I = J \cup K$. Using Holland's representation theorem (Theorem 2.2.5) we may view $G = {}^{L}\bigsqcup_{i \in I} G_i$ as an ℓ-subgroup of the doubly transitive ℓ-group $A(\Omega)$. It is well known (Theorem 2.3.1) that such an ℓ-group is n-transitive for all $n \geq 1$. The following observation will be used repeatedly. Suppose that for each $i \in I$, there exists $h_i \in A(\Omega)$ such that for some $\omega \in \Omega$,

$$(**) \qquad h_k b_k h_k^{-1}(\omega) < h_j a_j h_j^{-1}(\omega) \text{ for all } (j,k) \in J \times K.$$

Then $\bigwedge_{j \in J} a_j \not\leq \bigvee_{k \in K} b_k$. Indeed, for each $i \in I$ define the ℓ-homomorphism $\phi_i : G_i \to A(\Omega)$ by $\phi_i(x) = h_i x h_i^{-1}$. Now consider the ℓ-homomorphism $\phi : G \to A(\Omega)$ extending the ϕ_i's. In view of $(**)$, $\bigwedge_{j \in J} \phi(a_j) \not\leq \bigvee_{k \in K} \phi(b_k)$ in $\phi(G)$, and hence $\bigwedge_{j \in J} a_j \not\leq \bigvee_{k \in K} b_k$ in G.

The proof will involve two cases.

Case I. $J \cap K = \emptyset$. For each $j \in J$, there exists $\omega_j \in \Omega$ such that $a_j(\omega_j) > \omega_j$. Fix $j_1 \in J$. By the 2-transitivity of $A(\Omega)$, for each $j \in J \setminus \{j_1\}$ there exists $h_j \in A(\Omega)$ such that $h_j(\omega_j) = \omega_{j_1}$, and $h_j(a_j(\omega_j)) = a_{j_1}(\omega_{j_1})$. Let $\omega \in \Omega$ be such that $\omega_{j_1} < \omega < a_{j_1}(\omega_{j_1})$. If there exists $k \in K$ such that $b_k(\omega_{j_1}) \geq a_{j_1}(\omega_{j_1})$, then by the 2-transitivity of $A(\Omega)$ we have that for each such $k \in K$ there exists $h_k \in A(\Omega)$ such that $h_k(\omega_{j_1}) = \omega_{j_1}$ and $h_k(b_k(\omega_{j_1})) = \omega$. For all other $k \in K$, let $h_k = id_\Omega \in A(\Omega)$. Now it is easy to see that for all $(j,k) \in J \times K$, $h_k b_k h_k^{-1}(\omega_{j_1}) < h_j a_j h_j^{-1}(\omega_{j_1})$. Thus, $(**)$ fails.

Case II. $J \cap K \neq \emptyset$. Suppose further that there is no $i \in J \cap K$ such that $a_i \leq b_i$. Then for each $i \in J \cap K$ there is $\omega_i \in \Omega$ such that $a_i(\omega_i) > b_i(\omega_i)$. Two possibilities arise; namely, (1) there is $i \in J \cap K$ with $b_i(\omega_i) > \omega_i$ or (2) for all $i \in J \cap K$, $b_i(\omega_i) = \omega_i$. For case (1) fix i_1 to be any such element $i \in J \cap K$ and let $\omega = b_{i_1}(\omega_{i_1})$. For case (2) fix i_1 to be any element of $J \cap K$ and let $\omega \in \Omega$ be any point such that $\omega_{i_1} < \omega < b_{i_1}(\omega_{i_1})$. Now the 3-transitivity of $A(\Omega)$ implies that for every $i \in J \cap K$ there exists an $h_i \in A(\Omega)$ such that $h_i(\omega_i) = \omega_{i_1}$, $h_i(b_i(\omega_i)) \leq \omega$, and $h_i(a_i(\omega_i)) = a_{i_1}(\omega_{i_1})$. Next note that for each $j \in J \setminus K$ there is $\omega_j \in \Omega$ such that $a_j(\omega_j) > \omega_j$. By the 2-transitivity of $A(\Omega)$, for each such j there is $h_j \in A(\Omega)$ such that $h_j(\omega_j) = \omega_{i_1}$ and $h_j(a_j(\omega_j)) = a_{i_1}(\omega_{i_1})$. If there is $k \in K \setminus J$ such that $b_k(\omega_{i_1}) \geq a_{i_1}(\omega_{i_1})$, then by the 2-transitivity of $A(\Omega)$, for each such k there is $h_k \in A(\Omega)$ such that $h_k(\omega_{i_1}) = \omega_{i_1}$ and $h_k(b_k(\omega_{i_1})) = \omega$. For all other $k \in K \setminus J$ let $h_k = id_\Omega$. It is again easy to see that for all $(j,k) \in J \times K$, $h_k b_k h_k^{-1}(\omega_{i_1}) < h_j a_j h_j^{-1}(\omega_{i_1})$. Thus, again $(**)$ fails.

11.7. Concluding Remarks

The results given in the preceding discussion only serve as an introduction to free products of ℓ-groups. We will mention here some other related results and some directions which future research might take.

In the class L of all ℓ-groups Glass [1986a, 1987] has successfully analyzed some aspects of the free product by finding a faithful doubly transitive representation of it. As was noted in Theorem 11.4.4, this approach yields that L-free products have trivial center. It is unlikely that this method can be used for smaller varieties, but the theorem on the centers of an ℓ-group free product should be sought for more general cases.

The results of section 11.6 have not yet been extended to other varieties but probably could be.

The amalgamation property is closely related to the study of free products. In fact many times amalgamations can be determined by using the properties of free products discussed in this chapter. The next chapter is devoted to the amalgamation property in various classes of ℓ-groups and its implication for other embedding properties.

Further problems on free products of ℓ-groups are given in section 10 of [Powell and Tsinakis, 1984].

Wayne B. Powell
Oklahoma State University
Stillwater, Oklahoma 74078
U. S. A.

Constantine Tsinakis
Vanderbilt University
Nashville, Tennessee 37235
U. S. A.

Constantine Tsinakis and Wayne B. Powell

CHAPTER 12

AMALGAMATIONS OF LATTICE-ORDERED GROUPS

12.1 Introduction

The word amalgamation refers to the process of combining a pair of algebras
in such a way as to preserve a common subalgebra. This is made precise in
the following definition:

A class $\mathcal{U}$ of algebras is said to satisfy the *amalgamation property* (*AP*)
if for any $A, B_1, B_2 \in \mathcal{U}$ and embeddings $\sigma_1 : A \to B_1$, $\sigma_2 : A \to B_2$, there exist
$C \in \mathcal{U}$ and embeddings $\tau_1 : B_1 \to C$ and $\tau_2 : B_2 \to C$ such that $\tau_1 \sigma_1 =$
$\tau_2 \sigma_2$. The situation is depicted by the following diagram:

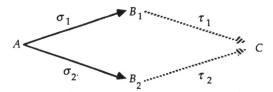

The quintuple $(A, B_1, B_2, \sigma_1, \sigma_2)$ described above is called a *V-formation*
in $\mathcal{U}$ and the triple (τ_1, τ_2, C) an *amalgamation in* $\mathcal{U}$ of this formation. An
algebra $A \in \mathcal{U}$ is said to be an *amalgamation base* for $\mathcal{U}$ if every V-
formation $(A, B_1, B_2, \sigma_1, \sigma_2)$ in $\mathcal{U}$ has an amalgamation in $\mathcal{U}$. The
amalgamation class of $\mathcal{U}$, $Amal(\mathcal{U})$, is the class consisting of all
amalgamation bases for $\mathcal{U}$. It is evident that AP holds in $\mathcal{U}$ if and only if
$Amal(\mathcal{U}) = \mathcal{U}$.

A. M. W. Glass and W. C. Holland (eds.), Lattice-Ordered Groups, 308–327.

Historically, amalgamations were first considered for groups by Schreier [1927] in the form of amalgamated free products. The general form of AP was first formulated by Fraïssé [1954] and the significance of this property to the study of algebraic systems was further demonstrated by Jónsson's pioneering work [1956], [1960], [1961] and [1962]. The development of AP in various branches of algebra is described in Jónsson [1965] and a comprehensive list of references is included in Kiss et al. [1983].

Our objective in this chapter is to introduce some of the main results in the theory of ℓ-groups which are directly or indirectly related to AP. Specifically, in the second section the satisfaction of this property will be investigated in a number of classes of ℓ-groups. The proof that AP holds or fails for a particular class is often instructive about the structure of the members of the class. In the third section we focus our attention on amalgamation classes and restricted types of V-formations. In the final section we outline an approach for establishing the existence of η_α-groups which is based on the fact that totally ordered abelian groups satisfy AP.

A number of classes, and in particular varieties, will be considered throughout this chapter. Those which need special recognition are:

$\mathcal{A}$ = abelian ℓ-groups;

$\mathcal{A}_T$ = totally ordered abelian groups;

$\mathcal{V}$ = vector lattices (Riesz spaces);

$\mathcal{V}_T$ = totally ordered vector lattices;

$\mathcal{R}$ = representable ℓ-groups;

$\mathcal{M}^+, \mathcal{M}^-$ = the solvable non-nilpotent covers of $\mathcal{A}$ below $\mathcal{R}$;

$\mathcal{W}a$ = weakly abelian ℓ-groups;

$\mathcal{N}_n$ = all ℓ-groups which are nilpotent groups of class n;

$\mathcal{L}$ = all ℓ-groups.

Many of these classes were discussed in detail in Chapter 10.

In addition to AP, we shall consider two other properties of significance.

A variety $\mathcal{U}$ is said to satisfy the *special amalgamation property* (SAP) if for any embeddings $\sigma_i : A_i \to B_i$ (i = 1,2) in $\mathcal{U}$ the resulting homomorphism of the $\mathcal{U}$-free product $A_1^{\mathcal{U}} \sqcup A_2$ into the $\mathcal{U}$-free product $B_1^{\mathcal{U}} \sqcup B_2$ is an embedding. In the literature of ℓ-groups, SAP is often called the *subalgebra property for free products*.

An algebra A is said to satisfy the *congruence extension property (CEP)* if every congruence of an arbitrary subalgebra is the restriction of a congruence of A. Of course, the congruences of an ℓ-group are in bijective correspondence with its ℓ-ideals. A variety $\mathcal{U}$ is said to satisfy CEP if every member of $\mathcal{U}$ satisfies this property. CEP is easily seen to hold in every subvariety of $\mathcal{Wa}$.

The next result describes an interesting connection among the three proerties.

THEOREM 12.1.1 (Jónsson [1961]; Grätzer and Lakser [1971]). *In a variety admitting free products, AP implies SAP. Conversely, SAP implies AP in the presence of CEP.*

We close this introductory section by recalling the concept of positive independence.

Let G be an abelian *po*-group with positive cone P. We say that a subset S of G is *positively independent* in G if for every finite subset $\{a_1,...,a_k\}$ of S and every set $\{n_1,...,n_k\}$ of nonnegative integers,

$$\prod_{i=1}^{K} a_i^{n_i} \in P^{-1}$$

only in case all integers n_i are zero. Note that S is positively independent if and only if there is a partial order on G whose positive cone contains $P \cup S$. This is equivalent to the existence of a total order on G whose positive cone contains $P \cup S$, since every partial order on a torsion-free abelian group can be extended to a total order (see Mura and Rhemtulla [1977, Chapter III]).

12.2. The Amalgamation Property for Classes of ℓ-Groups

Our goal in this section is to discuss results concerned with the satisfaction or failure of AP or SAP in particular classes of ℓ-groups and vector lattices. The first four theorems of this section establish the satisfaction of AP for the variety $\mathcal{A}$ of abelian ℓ-groups, the class $\mathcal{A}_T$ of totally ordered abelian groups, the variety $\mathcal{V}$ of vector lattices, and the class $\mathcal{V}_T$ of totally ordered vector lattices. AP is quite rare for general varieties. Among those which are known to satisfy this property are the classes of all groups, all lattices, and all distributive lattices. In the lattice setting it has been shown by Day and Jezek [1984] that these are the only nontrivial lattice varieties satisfying AP. The case for ℓ-groups appears to be somewhat similar although it is an important open question whether $\mathcal{A}$ is the only nontrivial variety of ℓ-groups satisfying AP. Theorems 12.2.5, 12.2.6, and 12.2.10 establish that a multitude of both representable and nonrepresentable varieties fail this property.

THEOREM 12.2.1 (Pierce [1972a]; see also Pierce [1972b], Powell and Tsinakis [1983]). *The variety $\mathcal{A}$ satisfies AP.*

Proof. Let $(A,B_1,B_2,\sigma_1,\sigma_2)$ be a V-formation in $\mathcal{A}$. Let N be the ℓ-ideal of the $\mathcal{A}$-free product $B = B_1^{\mathcal{A}} \sqcup B_2$ (see Chapter 11) generated by $\{\sigma_1(a)\sigma_2(a^{-1}) \mid a \in A\}$, and let $\pi.B \to B/N$ be the canonical projection. Finally, let $\tau_1 = \pi|_{B_1}$ and $\tau_2 = \pi|_{B_2}$.

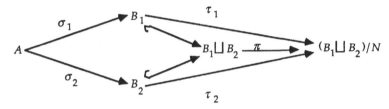

Now for each $a \in A$, $\tau_1\sigma_1(a) = \sigma_1(a)N = \sigma_2(a)N = \tau_2\sigma_2(a)$. Hence $\tau_1\sigma_1 = \tau_2\sigma_2$. It remains to be shown that τ_1 and τ_2 are embeddings or, what amounts to the same, that $B_1 \cap N = B_2 \cap N = \{e\}$.

Notice first that N is generated by the set $\{\sigma_1(a)\sigma_2(a^{-1}) \mid a \geq e \text{ in } A\}$. Indeed, if $a \in A$, then $\sigma_1(a)\sigma_2(a^{-1}) = [\sigma_1(a^+)\sigma_2((a^+)^{-1})][\sigma_1(a^-)\sigma_2((a^-)^{-1})]^{-1}$.

Let now $x \in B_1^+ \cap N$. We need to show that $x = e$. Assume, for *reductio ad absurdum*, that $x \neq e$. Then there is a finite family $(a_k \mid k \in K)$ of positive elements of A such that

$$x \leq \bigvee_{k \in K} [(\sigma_1(a_k)\sigma_2(a_k^{-1}))^+ \vee (\sigma_1(a_k^{-1})\sigma_2(a_k))^+]. \qquad (1)$$

Let P_1 be a prime in B_1 such that $x \notin P_1$. Let $Q_1 = \sigma_1(A) \cap P_1$, and $Q_2 = \sigma_2(\sigma_1^{-1}(\sigma_1(A) \cap P_1))$. Then Q_i is prime in $\sigma_i(A)$ for $i = 1,2$. Let P_2 be a prime in B_2 such that $P_2 \cap \sigma_2(A) = Q_2$. Note that

for each $a \in A$, $\quad \sigma_1(a) \in P_1 \quad$ iff $\quad \sigma_2(a) \in P_2$. $\qquad (2)$

Let N_1 be the kernel of the epimorphism from B to $(B_1/P_1)^{\mathcal{A}} \sqcup (B_2/P_2)$ extending the natural projections $B_i \to B_i/P_i$ $(i = 1,2)$. Condition (2) yields for all $k \in K$,

$$\sigma_1(a_k)N_1 > N_1 \quad \text{iff} \quad \sigma_2(a_k)N_1 > N_1. \qquad (3)$$

Note also that $xN_1 > N_1$. Let $K' = \{k \in K \mid \sigma_1(a_k)N_1 > N_1\} = \{k \in K \mid \sigma_2(a_k)N_1 > N_1\}$. It is clear that $K' \neq \varnothing$, and we have in $(B_1/P_1)^{\mathfrak{A}} \sqcup (B_2/P_2)$,

$$xN_1 \leq \bigvee_{k \in K} [(\,\sigma_1(a_k)\sigma_2(a_k^{-1})N_1)^+ \vee (\,\sigma_1(a_k^{-1})\,\sigma_2(a_k)N_1)^+]. \qquad (4)$$

We next claim that the elements $\{x\sigma_1(a_k^{-1})\sigma_2(a_k)N_1, x\sigma_1(a_k)\sigma_2(a_k^{-1})N_1 \mid k \in K'\}$ are positively independent in $B_1/P_1 \boxplus B_2/P_2$. To begin with, notice that all these elements are different from N_1. Next, let $(\lambda_k \mid k \in K')$ and $(\mu_k \mid k \in K')$ be two families of nonnegative integers such that

$$\prod_{k \in K'} [x\,\sigma_1(a_k^{-1})\sigma_2(a_k)N_1]^{\lambda_k} \prod_{k \in K'} [x\,\sigma_1(a_k)\,\sigma_2(a_k^{-1})N_1]^{\mu_k} \leq N_1.$$

Then

$$\prod_{k \in K'} (xN_1)^{\lambda_k + \mu_k} \leq \prod_{k \in K'} (\,\sigma_1(a_k)N_1)^{\lambda_k - \mu_k} \qquad (5)$$

and

$$\prod_{k \in K'} (\sigma_2(a_k)N_1)^{\lambda_k - \mu_k} \leq N_1. \qquad (6)$$

Now (6) can be written as $\sigma_2(\prod_{k \in K'} a_k^{\lambda_k - \mu_k})P_2 \leq P_2$, which is equivalent to $\sigma_2(\prod_{k \in K'} a_k^{\lambda_k - \mu_k})^+ \in P_2$. It follows from (2) that $\sigma_1(\prod_{k \in K'} a_k^{\lambda_k - \mu_k})^+ \in P_1$, and hence $\prod_{k \in K'} (\,\sigma_1(a_k)N_1)^{\lambda_k - \mu_k} \leq N_1$. But then (5) implies that

$$\prod_{k \in K'} (xN_1)^{\lambda_k + \mu_k} \leq N_1,$$

and so $\lambda_k = \mu_k = 0$ for all $k \in K'$. We have established the positive independence of the aforementioned elements.

Consider now a total order (with positive cone T) on $H = B_1/P_1 \boxplus B_2/P_2$ extending the cardinal order, such that for all $k \in K'$, $x\sigma_1(a_k^{-1})\sigma_2(a_k)N \in T$ and $x\sigma_1(a_k)\sigma_2(a_k^{-1}) \in T$. Now (H,T) is clearly a homomorphic image of $(B_1/P_1)^{\mathfrak{A}} \sqcup (B_2/P_2)$, and in it,

$$xN_1 > N_1, \quad xN_1 > \sigma_1(a_k)\,\sigma_2(a_k^{-1})N_1, \text{ and } xN_1 > \sigma_1(a_k^{-1})\,\sigma_2(a_k)N_1.$$

Hence in (H,T),

$$xN_1 > \bigvee_{k \in K} [(\,\sigma_1(a_k)\,\sigma_2(a_k^{-1})N_1)^+ \vee (\,\sigma_1(a_k^{-1})\,\sigma_2(a_k)N_1)^+].$$

The last inequality contradicts (4), and hence the proof of the theorem is complete.

Pierce's original proof of Theorem 12.2.1 made use of η_α-groups whose existence--assuming the Generalized Continuum Hypothesis--had previously been established by Alling [1960] and [1962]. The existence of these totally ordered groups will be obtained in Section 4 as a consequence of

Theorem 12.2.2 below. Other less direct proofs than the one given above are presented in Pierce [1972b] and Powell and Tsinakis [1983a]. The present proof was obtained by the authors in 1979.

THEOREM 12.2.2 (Pierce [1972a]; see also Pierce [1972b]). *The class $\mathcal{A}_T$, of totally ordered abelian groups satisfies AP.*

Proof. Let $(A, B_1, B_2, \sigma_1, \sigma_2)$ be a V-formation in $\mathcal{A}_T$, and let $\tau_1, \tau_2, B,$ and N be as in the proof of the preceding theorem. If T is a total order on B/N extending the lattice order and $C = (B/N, T)$, then (τ_1, τ_2, C) is an amalgamation in $\mathcal{A}_T$ of $(A, B_1, B_2, \sigma_1, \sigma_2)$.

The following two results can be proved similarly.

THEOREM 12.2.3. *The class $\mathcal{V}$ of vector lattices satisfies AP.*

THEOREM 12.2.4. *The class $\mathcal{V}_T$ of totally ordered vector lattices satisfies AP.*

We next turn our attention to varieties of ℓ-groups failing AP. The proof of the next theorem follows Pierce's approach but is simpler. Recall the definition of S_p from Section 10.13.

THEOREM 12.2.5 (Pierce [1972b]). *If $\mathcal{U}$ is a variety of ℓ-groups exceeding one of the abelian covers S_p (p prime), then $\mathcal{U}$ fails AP.*

Proof. We prove the result for S_2. The proof for the other covers S_p is similar.

Let I be the group homomorphism from $\mathbf{Z}$ into the group of o-automorphisms (i.e., group and order automorphisms) of $\mathbf{Z} \boxplus \mathbf{Z}$ defined by

$$I_n(k, \ell) = \begin{cases} (k, \ell) & \text{if } n \text{ is even} \\ (\ell, k) & \text{if } n \text{ is odd} \end{cases} .$$

Recall that S_2 is generated by the semidirect product $B_1 = (\mathbf{Z} \boxplus \mathbf{Z}) \overrightarrow{\times} \mathbf{Z}$. It will be convenient to use multiplicative notation for B_1 and write $x = (1,0,0)$, $y = (0,1,0)$, and $z = (0,0,1)$. Let A be the ℓ-subgroup of B_1 generated by $\{x, y, z^2\}$. It is clear that A is isomorphic to the abelian ℓ-group $(\mathbf{Z} \boxplus \mathbf{Z}) \overrightarrow{\times} \mathbf{Z}$. Denote by B_2 the abelian ℓ-group $(\mathbf{Z} \overrightarrow{\times} \mathbf{Z}) \boxplus (\mathbf{Z} \overrightarrow{\times} \mathbf{Z})$. We set $x_1 = (1,0,0,0)$, $y_1 = (0,0,1,0)$, $z_1 = (0,1,0,0)$, $z_2 = (0,0,0,1)$, and use the multiplicative notation for the group operation of B_2.

Let $\sigma_1: A \to B_1$ be the inclusion map, and let $\sigma_2: A \to B_2$ be the ℓ-homomorphism such that $\sigma_2(x) = x_1$, $\sigma_2(y) = y_1$, and $\sigma_2(z^2) = z_1 z_2$. We claim that the V-formation $(A_1, B_1, B_2, \sigma_1, \sigma_2)$ in S_2 cannot be amalgamated in L. Suppose, for a proof by contradiction, (τ_1, τ_2, C) amalgamates the above formation. We will view C as an ℓ-permutation group (C, Ω). The proof will be established by verifying four simple facts.

$$\tau_1(x)(\omega) > \omega \text{ and } \tau_2(z_2)(\omega) = \omega \quad \text{for some } \omega \in \Omega. \tag{1}$$

This is an immediate consequence of the fact that $\tau_1(x)$ and $\tau_2(z_2)$ are disjoint order automorphisms of Ω. To verify this fact, note that $e = x_1 \wedge z_2 = \sigma_2(x) \wedge z_2$, and hence $e = \tau_2 \sigma_2(x) \wedge \tau_2(z_2) = \tau_1 \sigma_1(x) \wedge \tau_2(z_2) = \tau_1(x) \wedge \tau_2(z_2)$.

For the remainder of the proof we fix $\omega \in \Omega$ satisfying (1).

$$\tau_1(z)(\omega) = \tau_2(z_1)\tau_1(z)(\omega). \tag{2}$$

Indeed, the equalities $e = y_2 \wedge z_1 = \sigma_2(y) \wedge z_1$ imply $e = \tau_2 \sigma_2(y) \wedge \tau_2(z_1) = \tau_1 \sigma_1(y) \wedge \tau_2(z_1)$ or $\tau_1(y) \wedge \tau_2(z_1) = e$. It follows that $\tau_1(y)\tau_1(z) \wedge \tau_2(z_1)\tau_1(z) = \tau_1(z)$ or

$$\tau_1(yz) \wedge \tau_2(z_1)\tau_1(z) = \tau_1(z). \tag{*}$$

Note now that $z^{-1}yz = x$. Since, by (1), $\tau_1(x)(\omega) > \omega$, we obtain the equivalent inequalities, $\tau_1(z^{-1}yz)(\omega) > \omega$, and $\tau_1(yz)(\omega) > \tau_1(z)(\omega)$. Combining the last inequality with (*), we obtain (2).

$$\tau_1(z)(\omega) > \tau_2(z_1)(\omega). \tag{3}$$

Note that $z > x$. Thus $\tau_1(z) > \tau_1(x)$ and so, by (1), $\tau_1(z)(\omega) > \omega$. Condition (3) is now a direct consequence of (2).

$$\tau_1(z)(\omega) > \tau_1(z^2)(\omega). \tag{4}$$

Indeed, by (3) and (1), $\tau_1(z)(\omega) > \tau_2(z_1)(\omega)) = \tau_2(\sigma_2(z^2)z_2^{-1})(\omega) = \tau_2 \sigma_2(z^2)\tau_2(z_2^{-1})(\omega) = \tau_2 \sigma_2(z^2)(\omega) = \tau_1(z^2)(\omega)$. Condition (4) contradicts our assumption that τ_1 is an ℓ-homomorphism. Thus, the formation $(A, B_1, B_2, \sigma_1, \sigma_2)$ cannot be amalgamated in L.

The proof of the corresponding result for the covers $\mathcal{M}^+ = \mathcal{V}(W^+)$ and $\mathcal{M}^- = \mathcal{V}(W^-)$ is more complicated and will require a few auxiliary lemmas. An important idea used in the proof is due to Glass et al. [1984].

THEOREM 12.2.6 (Powell and Tsinakis [a]). *If $\mathcal{U}$ is a variety of representable ℓ-groups exceeding one of the abelian covers $\mathcal{M}^+$ or $\mathcal{M}^-$, then $\mathcal{U}$ fails AP.*

Consider the small wreath product $GwrH$ of two groups G and H (see Section 2.3). The elements of $GwrH$ will be written in the form $((a_i),b)$ where $b \in H$ and (a_i) is a vector of elements of G indexed by H whose support, $\text{supp}(a_i) = \{i \in H \,|\, a_i \neq e\}$, is finite. The group operation for $GwrH$ is given by

$$((a_i),b)((a_i'),b) = (c_i),bb'),$$

where $c_i = a_i a_{ib}$. If G and H are totally ordered, then $GwrH$ admits two natural total orders. The positive cone of one of these orders is defined by $((a_i),b) > (e,e)$ if $b > e$ in H or $b = e$ and $a_j > e$ in G, where j is the largest index such that $a_j \neq e$. The other order just reverses the significance of the vectors from G. In this case $((a_i),b) > (e,e)$ if $b > e$ in H or $b = e$ and $a_j > e$ in G, where j is the smallest index such that $a_j \neq e$, $GwrH$ with the first order above will be denoted by $Gw\overset{\leftarrow}{r}H$ while under the second order it will be denoted by $Gw\vec{r}H$.

In particular we will concentrate on the wreath product $ZwrZ$. For notational purposes we will write $W^+ = Zw\overset{\leftarrow}{r}Z$ and $W^- = Zw\vec{r}Z$. Recall (see Chapter 10) that $\mathcal{M}^+$ and $\mathcal{M}^-$ are the varieties generated by W^+ and W^-, respectively. It was shown by Medvedev [1977] that $\mathcal{M}^+$ and $\mathcal{M}^-$ are the only solvable covers of $\mathcal{A}$ inside $\mathcal{R}$ which contain non-nilpotent, solvable groups.

Theorem 12.2.6 can be established by finding ℓ-groups H_1 and H_2 in $\mathcal{M}^+$ and an ℓ-subgroup G common to both H_1 and H_2 such that the formation (G,H_1,H_2) cannote be amalgamated in $\mathcal{R}$. The proof for the interval $[\mathcal{M}^-,\mathcal{R}]$ is completely analogous and so it will be omitted. The construction of the ℓ-groups in question uses an idea from Glass et al. [1984], which is based on the fact that members of $\mathcal{R}$ have unique extraction of roots. That is, they satisfy the implication

$$x^n = y^n \Rightarrow x = y$$

for all positive integers n. Given a positive integer n, we will construct totally ordered groups $G, H_1,$ and H_2 so that they contain elements $a, h_1,$ & h_2, respectively, such that $h_1^n = h_2^n \in G$ but $a^{h_1} = h_1^{-1}ah_1$ and $a^{h_2} = h_2^{-1}ah_2$ are distinct elements of G. Thus any amalgam of (G,H_1,H_2) will not be in $\mathcal{R}$ since it will satisfy $h_1^n = h_2^n$ and $h_1 \neq h_2$. The burden of proof consists in showing that the ℓ-groups H_1 and H_2 can be constructed in $\mathcal{M}^+$.

Let G be any totally ordered group and let α be an o-automorphism (i.e., a group and order automorphism) of G. The cyclic extension of G by α is the set $G \times \{\alpha^n \,|\, n \in Z\}$ with operation defined by

$$(g,\alpha^n)(h,\alpha^m) = (g\alpha^n(h),\alpha^{n+m})$$

and order defined by

$(g, \alpha^n) > (e, \alpha^0)$ if $(n > 0)$ or $(n = 0$ and $g > e)$.

This totally ordered group will be denoted by $G(\alpha)$.

Suppose now that $\alpha, \beta \gamma$, are o-automorphism of the totally ordered group G such that $\alpha = \beta^n = \gamma^n$. Clearly, $G(\alpha)$ is an ℓ-subgroup of $G(\beta)$ and $G(\gamma)$. Further, for all $g \in G$, $g^\beta = \beta(g)$ in $G(\beta)$ and $g^\gamma = \gamma(g)$ in $G(\gamma)$. Hence, if $\beta \neq \gamma$, then $g^\beta \neq g^\gamma$ in $G(\alpha)$. Thus, in view of the preceding discussion, we have the following result:

LEMMA 12.2.7. *Let G be totally ordered group with distinct o-automorphisms α, β, and γ such that $\alpha = \beta^n = \gamma^n$ for some positive integer n. then any subvariety of $\mathcal{R}$ which contains the totally ordered groups $G(\beta)$ and $G(\gamma)$ fails the amalgamation property.*

In view of the above result, our goal can be accomplished for $\mathcal{M}^+$ by finding and ℓ-group G with α, β, and γ as described such that $G(\beta)$ and $G(\gamma)$ are in $\mathcal{M}^+$. Initially, let I be a totally ordered set and $\bar{\alpha}$ an order automorphism of I. Let

$$G = \overset{\leftarrow}{\underset{i \in I}{\oplus}} Z_i, \quad (Z_i \cong Z)$$

be the lexicographically ordered direct sum of copies of Z indexed by I. Then $\bar{\alpha}$ induces a natural o-automorphism α of G defined by $(n_i) \mapsto (m_i)$ where $m_i = n_{\bar{\alpha}(i)}$.

In order to find appropriate α, β, and γ for G we need the following lemma:

LEMMA 12.2.8. *Given an integer $n \geq 2$, there exists a totally ordered set I and distinct order automorphisms $\bar{\alpha}, \bar{\beta}$, and $\bar{\gamma}$ of I such that*

(i) $\bar{\alpha} = \bar{\beta}^n = \bar{\gamma}^n$;

(ii) $\bar{\alpha}(i) > i$ *for all $i \in I$;*

and   (iii)   *if $i < j$ in I, then there exists $m \in Z^+$ with $j < \bar{\alpha}^m(i)$.*

Proof. Consider the set $I = Z \overset{\leftarrow}{\times} Z$ and define $\bar{\beta}, \bar{\gamma}$ by

$$\bar{\beta}(a, i) = \begin{cases} (a, i + 1) & \text{if } i \notin nZ \\ (a + 1, i + 1) & \text{if } i \in nZ \end{cases}$$

$$\bar{\gamma}(a, i) = \begin{cases} (a, i + 1) & \text{if } i \notin nZ + 1 \\ (a + 1, i + 1) & \text{if } i \in nZ + 1 \end{cases}.$$

Note that $\bar{\beta} \neq \bar{\gamma}$, but $\bar{\beta}^n(a, i) = (a + 1, i + n) = \bar{\gamma}^n(a, i)$. Further, conditions (ii) and (iii) can be easily verified for $\bar{\alpha} = \bar{\beta}^n = \bar{\gamma}^n$.

In working with $W^+ = Z w \bar{r} Z$ it is useful to understand the significance of certain elements. Let G be a totally ordered group generated as a group by two elements x and y where

1) $x \ll x^y \ll y$,

and 2) the set $\{x y^n \mid n \in Z\}$ is a free basis for the abelian subgroup of W^+ it generates.

Then $G \cong W^+$. This is easily verified via the correspondence $y \mapsto (0,1)$,

$$x \mapsto ((a_i),0), \text{ where } a_i = \begin{cases} 1 & \text{if } i = 0 \\ 0 & \text{if } i \neq 0. \end{cases}$$

To determine if an ℓ-group is in a variety it suffices to consider all of its finitely generated ℓ-subgroups. For the next two lemmas, we assume that $I, \bar{\alpha}, \bar{\beta}$, and $\bar{\gamma}$ are as in Lemma 12.2.8. As before, let $G = \overset{\leftarrow}{\underset{i \in I}{\oplus}} Z_i$, $Z_i \cong Z$. Further, for each $i \in Z$ and for $\delta = \alpha, \beta$, or γ let c_i denote the element $((b_j),1)$ of $G(\delta)$ where

$$b_j = \begin{cases} 1 & \text{if } j = i \\ 0 & \text{if } j \neq i. \end{cases}$$

Also, let c denote $(0,\delta) \in G(\delta)$.

LEMMA 12.2.9. *Let H be a finitely generated ℓ-subgroup of $G(\delta)$. Then there exists a finite subset $J \subset I$ such that*

(i) $max(J) < \delta(min(J))$,

and (ii)   *H is contained in the ℓ-subgroup generated by $\{c_j \mid j \in J\} \cup \{c\}$.*

Proof. Let $\{(f_1, \delta^{m_1}), \ldots, (f_r, \delta^{m_r})\}$ be a set of generators of H. Let

$$I' = \overset{r}{\underset{k=1}{\cup}} \text{supp}(f_k).$$

Then $\{c_j \mid j \in I'\} \cup \{c\}$ generates an ℓ-subgroup containing H.

The proof will be completed by showing that for every finite subset K of I, there exists a finite subset J of I such that $\max J < \delta(\min J)$ and the sets $\{c_j \mid j \in K\} \cup \{c\}$, $\{c_j \mid j \in J\} \cup \{c\}$ generate the same subgroup. We verify this fact by induction on $|K|$. For $|K| = 1$ the conclusion follows immediately from the fact that $i < \delta(i)$ for all $i \in I$. Suppose next that $|K| \geq 2$ and that the conclusion holds for all subsets of I with $|K| - 1$ elements. Pick $i_2 \in K$. By the induction hypothesis, there exists a finite subset J' of I such that $\max J' < \delta(\min J')$ and $\{c_j \mid j \in J'\} \cup \{c\}$ and $\{c_j \mid j \in K \setminus \{i_2\}\} \cup \{c\}$ generate the same subgroup. Let $i_1 = \min J'$. If $i_2 < \delta(i_1)$, then $J' \cup \{i_2\}$ satisfies the required properties; so suppose that $\delta(i_1) \leq i_2$. By condition (iii) of Lemma 12.2.8, there exists a positive integer ℓ such that

$\delta^\ell(i_1) \leq i_2 < \delta^{\ell+1}(i_1)$. Define $i = \delta^{-\ell}(i_2)$ and $J = J' \cup \{i\}$. It is clear that $i_1 \leq i$ $< \delta(i_1)$ and hence $\max(J') < \delta(\min J)$. Furthermore, $c_{i_2} = c^{-\delta^\ell} c_i c^{\delta^\ell}$ and hence the sets $\{c_j \mid j \in K\} \cup \{c\}$ and $\{c_j \mid j \in J\} \cup \{c\}$ generate the same subgroup. The proof is now complete.

Proof of Theorem 12.2.6. It suffices to assume that $\mathcal{M}^+ \subseteq \mathcal{U} \subseteq \mathcal{R}$. Let I be a totally ordered set with order automorphisms $\bar{\alpha}, \bar{\beta}$ and $\bar{\gamma}$ satisfying conditions (i)-(iii) of Lemma 12.2.8 for a fixed positive integer n. Let $G = \overset{\leftarrow}{\underset{i \in I}{\oplus}} \mathbf{Z}_i$ and $\alpha, \beta,$ and γ be the o-automorphisms of G induced by the automorphisms $\bar{\alpha}, \bar{\beta},$ and $\bar{\gamma}$, respectively. In view of Lemma 12.2.7, any subvariety of $\mathcal{R}$ containing $G(\beta)$ and $G(\gamma)$ fails the amalgamation property.

The proof will be completed by showing that $\mathcal{M}^+$ is the smallest ℓ-group variety containing $G(\alpha)$, $G(\beta)$, and $G(\gamma)$. This will be accomplished by verifying that every finitely generated ℓ-subgroup H of $G(\beta)$ (or $G(\gamma)$) can be embedded in W^+. By Lemma 12.2.9, we may assume that H is generated by a set of the form $\{c_{i_1}, \ldots, c_{i_n}, c\}$ with $i_0 < i_1 < \ldots < i_{n-1} < \beta(i_0)$ in I. Note that $X = \{\beta^k(c_{i_\mu}) \mid k \in \mathbf{Z}, \mu \in \{0,1,\ldots,n-1\}\}$ is a free basis for the abelian subgroup $\mathbf{Z}^{(X)}$ of $G(\beta)$. Similarly, $Y = \{c_k \mid k \in \mathbf{Z}\}$ is a free basis for the subgroup $\mathbf{Z}^{(Y)}$ of W^+. Thus the bijection $\phi : X \to Y$ defined by

$$\phi(\beta^k(c_{i_\mu})) = c_\mu^{(0,kn)} = c_{kn+\mu}, \quad (k \in \mathbf{Z}, \mu \in \{0,1,\ldots,n-1\}) \qquad (*)$$

induces a group isomorphism ϕ between $\mathbf{Z}^{(X)} \times \{1\}$ and $\mathbf{Z}^{(Y)} \times \{0\}$. Observe that (*) implies that for all $k \in \mathbf{Z}$ and $f \in \mathbf{Z}^{(X)}$,

$$(\phi(\beta^k(f)),1) = (\phi(f),0)^{(0,kn)}. \qquad (**)$$

The map ϕ can be extended to an injective map $\psi : G(\beta) \to W^+$ by

$$\psi(f,\beta^k) = (\phi(f),nk).$$

A routine computation, making use of (**), establishes that ϕ is a group homomorphism.

Finally note that for all $k, \ell \in \mathbf{Z}$ and $\mu, \upsilon \in \{0,1,\ldots,n-1\}$, $\beta^k(c_\mu) < \beta^\ell(c_\upsilon)$ if and only if $k < \ell$ or both $k = \ell$ and $\mu < \upsilon$. Thus an easy application of (*) shows that ψ preserves order.

We note that the collection of varieties considered above is uncountable since it clearly contains those discussed in Feil [1981].

The statement and the proof of the next theorem have been inspired by Berman's corresponding result for lattice varieties (see Berman [1981]).

THEOREM 12.2.10. *Every join-reducible variety of ℓ-groups fails SAP.*

Proof. Consider a join-reducible variety $\mathcal{U}$ of ℓ-groups. Let $\mathcal{U}_1$, $\mathcal{U}_2$ be varieties which are strictly contained in $\mathcal{U}$ and whose join is $\mathcal{U}$. It is clear that $\mathcal{U}_1$ and $\mathcal{U}_2$ are incomparable; hence there exist ℓ-group words $w_1(x_1,...,x_n)$, $w_2(y_1,...,y_m)$ such that $w_1 = e$ in $\mathcal{U}_1$, $w_1 > e$ in $\mathcal{U}_2$, $w_2 = e$ in $\mathcal{U}_2$ and $w_2 > e$ in $\mathcal{U}_1$. Let $F = \mathcal{F}_{\mathcal{U}}(x_1,...,x_n,y_1,...,y_m)$ be the $\mathcal{U}$-free ℓ-group on $n + m$ generators. It is clear that $w_1, w_2 > e$ in F. However, $w_1 \wedge w_2 = e$ in F. This can be seen by noting that $w_1 \wedge w_2 = e$ in $\mathcal{U}_i$ ($i = 1,2$) and by recalling Jónsson's result (Jónsson [1967]; see Corollary 10.1.5 above) which asserts that a subdirectly irreducible in a congruence distributive variety $\mathcal{U}$ = $\mathcal{U}_1 \vee \mathcal{U}_2$ belongs to either $\mathcal{U}_1$ or $\mathcal{U}_2$. Let A_i ($i = 1,2$) be the ℓ-subgroup of F generated by w_i. It is clear that $A_1 \cong A_2 \cong \mathbf{Z}$, A_1 is an ℓ-subgroup of $\mathcal{F}_{\mathcal{U}}(x_1,...,x_n)$, A_2 is an ℓ-subgroup of $\mathcal{F}_{\mathcal{U}}(y_1,...,y_m)$, and $\mathcal{F}_{\mathcal{U}}(x_1,...,x_n) \overset{\mathcal{U}}{\sqcup} \mathcal{F}_{\mathcal{U}}(y_1,...,y_m) = F$. We claim that the ℓ-subgroup $\langle A_1 \cup A_2 \rangle$ of F generated by $A_1 \cup A_2$ is not isomorphic to $A_1 \overset{\mathcal{U}}{\sqcup} A_2$. Indeed, as a special case of Theorem 11.6.4, we have that if $A_1, A_2 \in \mathcal{A}$, and $w_i > e$ in A_i ($i = 1,2$), then for every variety $\mathcal{U}$ of ℓ-groups, $w_1 \wedge w_2 > e$ in $A_1 \overset{\mathcal{U}}{\sqcup} A_2$. However, $w_1 \wedge w_2 = e$ in $\langle A_1 \cup A_2 \rangle$.

It is worth noting that AP fails in many representable varieties that do not exceed $\mathcal{M}^+$ or $\mathcal{M}^-$. For example, it is shown in Glass et al. [1984] that $\mathcal{W}a$ fails AP. Hence, by Theorem 12.1.1, it also fails SAP. The same conclusion is valid for all nilpotent varieties $\mathcal{N}_n$ (Powell and Tsinakis [1985]).

12.3. Amalgamation Classes and Particular V-Formations

The results reported in the preceding section provide strong evidence for the conjecture that $\mathcal{A}$ is the only nontrivial variety of ℓ-groups satisfying AP. Consequently, investigations in this area should focus on amalgamation classes of ℓ-group varieties and particular types of V-formations.

We begin by noting that the amalgamation class of any variety is a proper class. This striking fact along with some other results about the amalgamation class of an elementary class appear in Yashuhara [1974].

Call a subclass $\mathcal{U}_1$ of $\mathcal{U}$ *cofinal* in $\mathcal{U}$ if every member of $\mathcal{U}$ can be embedded in a member of $\mathcal{U}_1$.

THEOREM 12.3.1 (Yasuhara [1974]; see Grätzer [1978, p. 258]). *For any variety $\mathcal{U}$, Amal($\mathcal{U}$) is cofinal in $\mathcal{U}$.*

In general very little is known about the amalgamation class of a particular variety. It is important to note that Yasuhara's proof of Theorem 12.3.1 shows that the existentially complete algebras in $\mathcal{U}$ belong to Amal($\mathcal{U}$). Existentially complete ℓ-groups in $\mathcal{A}$ and L are examined in Glass and Pierce [1980a], [1980b] and [1980c]. The most significant result on amalgamation classes of ℓ-group varieties is concerned with Amal(L).

THEOREM 12.3.2 (Pierce [1972a]). *Every totally ordered archimedian ℓ-group belongs to Amal(L).*

We refer the reader to Glass [1982b; Chapter 10] for the proof of this theorem and other related theorems. We remark that AP is indirectly involved in some of these proofs, since η_α-sets are the α-universal-homogeneous members of the Jónsson class of totally ordered sets (see Section 12.4 below and Comfort and Negrepontis [1974; Section 5]). It is also worth mentioning that the proof of Theorem 12.2.5 shows that $\mathcal{A} \not\subseteq$ Amal(L).

Another result of interest concerns the varieties $\mathcal{N}_n$.

THEOREM 12.3.3 (Powell and Tsinakis [1985]). *No nontrivial totally ordered abelian group is in Amal($\mathcal{N}_n$), for all $n > 1$.*

We next turn our attention to the question of amalgamating particular V-formations.

It is well-known that if the embeddings of a V-formation are elementary, then the formation has an amalgamation whose embeddings are elementary. The corresponding result for V-formations of pure embeddings has been established by Jónsson [1984].

It has been already remarked that $\mathcal{A} \not\subseteq$ Amal(L). However, Pierce [1972a] showed that every V-formation $(A, B_1, B_2, \sigma_1, \sigma_2)$ with A an archimedean ℓ-group and $B_1, B_2 \in \mathcal{R}$ can be amalgamated in L.

Other interesting amalgamations are related to divisible embeddings. An ℓ-group variety $\mathcal{U}$ is said to satisfy the *divisible embedding property* (*DEP*) if the class of divisible members of $\mathcal{U}$ is cofinal in $\mathcal{U}$.

THEOREM 12.3.4. *For a variety $\mathcal{U}$ of ℓ-groups, the following statements are equivalent:*

(i) $\mathcal{U}$ *satisfies DEP;*

(ii) *Every V-formation in $\mathcal{U}$ of the form $(\mathbf{Z},\mathbf{Q},G,\sigma_1,\sigma_2)$ can be amalgamated in $\mathcal{U}$;*

(iii) *Every V-formation of the form $(\mathbf{Z},\mathbf{Z},G,\sigma_1,\sigma_2)$, where G is a finitely generated subdirectly irreducible ℓ-group in $\mathcal{U}$, can be amalgamated in $\mathcal{U}$;*

(iv) *Every finitely generated subdirectly irreducible ℓ-group in $\mathcal{U}$ can be embedded in a divisible ℓ-group in $\mathcal{U}$.*

The proof is straightforward. Evidently (i)$\Rightarrow$(iv), and a routine proof establishes the implication (iv)$\Rightarrow$(iii). In order to establish (iii)$\Rightarrow$(ii), note that (ii) holds if every formation in $\mathcal{U}$ of the form $(\mathbf{Z},\mathbf{Z},G,\sigma_1,\sigma_2)$, with G finitely generated, can be amalgamated in $\mathcal{U}$. This follows from the general fact that if $(A,B_1,B_2,\sigma_1,\sigma_2)$ is a V-formation in a variety $\mathcal{U}$ which cannot be amalgamated in $\mathcal{U}$, then there exist finitely generated subalgebras A' of A and B_i' of B_i $(i=1,2)$, such that $(A',B_1',B_2',\sigma_1|_{B_1'},\sigma_2|_{B_2'})$ cannot be amalgamated in $\mathcal{U}$ (see Grätzer [1978, p. 254]). Then the proof of (iii)$\Rightarrow$(ii) can be easily completed by making use of Birkhoff's subdirect product theorem for varieties. Finally, the proof of the implication (ii)$\Rightarrow$(i) is Neuman's [1943] well-known proof of embedding a group into a divisible group (see also Pierce [1972a]). Specifically, if $g > e$ is an element of an ℓ-group G in $\mathcal{U}$, since formations in (ii) can be amalgamated in $\mathcal{U}$, one can obtain an extension of G in which the equation $x^n = g$ has a solution for all positive integers n. Then a standard direct limit construction will yield a divisible ℓ-group containing G as an ℓ-subgroup.

The next result shows that DEP is also related to SAP. Its proof is a localized version of the proof of Theorem 12.1.1 (see Grätzer and Lakser [1971]) and will therefore be omitted.

THEOREM 12.3.5. *If $\mathcal{U}$ satisfies DEP, then $\mathcal{U}$ satisfies SAP for embeddings of the form $\mathbf{Z} \to \mathbf{Q}$, $id:G \to G$, whenever G is an arbitrary ℓ-group in $\mathcal{U}$. If $\mathcal{U}$ satisfies CEP, then the converse is true.*

We conclude this section by listing varieties for which the last two theorems apply. Whether $\mathcal{R}$ satisfies DEP is one of the most important open questions in the theory of l-groups.

All varieties $\mathcal{N}_n$ satisfy DEP. Inasmuch as all these varieties are contained in $\mathcal{R}$ (see Theorem 10.10.4), it suffices to verify DEP for totally ordered groups in $\mathcal{N}_\alpha$. The proof of this fact is essentially a consequence of Mal'cev's [1949] result which shows that every torsion-free nilpotent group G can be embedded in a divisible torsion-free nilpotent group of the same nilpotency class. Further, one can assume that H consists of all the roots of elements of G (see for example Kargapolov and Merzlyakov [1979, p. 126]). Lastly, if G and H are as in the preceding sentence, then every total order on G can be extended to a total order on H (see Mura and Rhemtulla [1977, p. 111]).

Another representable variety with DEP is that of all representable l-groups which are metabelian groups. We refer the reader to the original paper by Bludov and Medvedev [1974] for the simple but ingenious proof.

We finally mention that $\mathcal{L}$ satisfies DEP (Theorem 2.2.11). This result was first established by Holland [1963a], and alternative proofs were given in Weinberg [1967] and Pierce [1972a]. Still a simpler proof can be obtained by using Theorem 12.3.4 (ii)⇔(iv) and the fact that every finitely generated l-group can be embedded in the l-group of all order automorphisms of $\mathbf{Q}$ (see [Glass, 1981b; Corollary 2L]).

4. α-homogeneous-universal totally ordered abelian groups

We consider now an application of the amalgamation property to the existence of various universal structures. Further, we use the existence of these particular structures to show that sets with a special separation property can actually be endowed with a compatible group structure. The classes we will consider in this section are $\mathcal{A}, \mathcal{A}_T, \mathcal{V},$ and $\mathcal{V}_T$. Because of the nature of the classes considered, we will use additive notation throughout. Some of the preliminary definitions are reviewed first.

Throughout this section let $\aleph_\alpha$ be a cardinal number with $\alpha > 0$. $\aleph_\alpha$ is said to be *admissible* whenever $\sum_{\beta < \alpha} 2^{\aleph_\beta} < \aleph_\alpha$, and it is called *regular* if $|\bigcup_{\lambda < \Lambda} S_\lambda| < \aleph_\alpha$ whenever $|S_\lambda|, |\Lambda| < \aleph_\alpha$. The existence of arbitrarily large admissible regular cardinals is guaranteed by GCH.

A totally ordered set S is called an η_α-set if for all $A, B \subseteq S$ with $A <$ B and $|A|, |B| < \aleph_\alpha$, there exists $x \in S$ with $A < x < B$. A totally ordered abelian group which is also an η_α-set is called an η_α-group.

We will show the existence of η_α-groups of cardinality $\aleph_\alpha$ whenever $\aleph_\alpha$ is admissible and regular. Further, we prove that any two such η_α-groups which are divisible are isomorphic, and that the α-homogeneous-universal structures in $\mathcal{A}_T$ are precisely these groups. These results are for the most part contained in the classic papers of Alling [1960, 1962].

Ribenboim [1965] later gave alternate proofs using the theory of upper classes in abelian totally ordered groups. The proofs given in the following discussion are somewhat different in their approach and use the results on amalgamation from the second section along with the general theory of Jónsson classes.

A class $\mathcal{U}$ is called an α-*Jónsson class* if it satisfies each of the following properties:

(1) for any cardinal $\aleph_\alpha$ there exists $G \in \mathcal{U}$ with $|G| \geq \aleph_\alpha$;

(2) if $G \in \mathcal{U}$ and $H \cong G$, then $H \in \mathcal{U}$;

(3) if $G, H \in \mathcal{U}$, then there exist $K \in \mathcal{U}$ and embeddings
$\phi: G \to K$ and $\psi: H \to K$ (the joint embedding property);

(4) the amalgamation property holds for $\mathcal{U}$;

(5) if $\{G_i \mid i \in I\}$ is a chain of structures in $\mathcal{U}$, then $\cup \{G_i \mid i \in I\} \in \mathcal{U}$ (the compactness property);

and (6) if $G \in \mathcal{U}$ and $H \subseteq G$ with $|H| < \aleph_\alpha$, then there exists $K \in \mathcal{U}$ such that $H \subseteq K \subseteq G$ and $|K| < \aleph_\alpha$ (the downward Löwenheim-Skolem property for $\aleph_\alpha$).

Let $\mathcal{U}$ be a class of similar relational systems and let $G \in \mathcal{U}$. G is said to be α-*universal* for $\mathcal{U}$ if every $H \in \mathcal{U}$ with $|H| \leq \aleph_\alpha$ can be embedded in G. If for every $H \in \mathcal{U}$ with $|H| < \aleph_\alpha$ and for every pair of embeddings $\phi, \phi': H \to G$ there exists $\psi: G \to G$ such that $\psi \circ \phi = \phi'$, then G is said to be α-*homogeneous* for $\mathcal{U}$. A structure in $\mathcal{U}$ which is both α-universal and α-homogeneous is called an α-homogeneous-universal structure for $\mathcal{U}$.

A nice treatment of Jónsson classes, η_α-sets, and α-homogeneous-universal structures can be found in Comfort and Negrepontis [1974], and we direct the reader to this reference for details on these subjects.

In view of Theorems 12.2.1-12.2.4 the classes $\mathcal{A}, \mathcal{V}, \mathcal{A}_T$, and $\mathcal{V}_T$ can easily be seen to be α-Jónsson classes. Now, Theorem 4.10(b) in Comfort and Negrepontis [1974] asserts that such classes have unique α-homogeneous-universal structures for admissible regular $\aleph_\alpha$. Thus our first two theorems are immediate, and with the lemma that follows we are able to establish the promised results.

THEOREM 12.4.1. *The classes* $\mathcal{A}$, $\mathcal{V}$, $\mathcal{A}_T$, *and* $\mathcal{V}_T$, *are* α-*Jónsson classes for any* $\aleph_\alpha$ $(\alpha > 0)$.

THEOREM 12.4.2. *Let* $\aleph_\alpha$ *be admissible and regular. Then there exist unique* α-*homogeneous-universal structures of cardinality* $\aleph_\alpha$ *in each of the classes* $\mathcal{A}$, $\mathcal{V}$, $\mathcal{A}_T$, *and* $\mathcal{V}_T$.

LEMMA 12.4.3. *Let* $\mathcal{U}$ *be a class of torsion-free groups for which an* α-*homogeneous structure* $G \in \mathcal{U}$ *exists. Then* G *is divisible.*

Proof. Suppose $g_0 \in G$ and fix $n \in Z^+$. Consider the maps ϕ_n, ψ from G into G where $\phi_n(g) = ng$ and $\psi(g) = g$ for all $g \in G$. By the homogeneity of G there exists $\phi': G \to G$ such that $\phi' \circ \phi_n = \psi$. But then $n\phi'(g_0) = \phi'(ng_0) = \phi' \circ \phi_n(g_0) = \psi(g_0) = g_0$.

The next theorem is proved with the aid of the notion of positive independence defined in the chapter introduction.

THEOREM 12.4.4. (GCH) *If* $\aleph_\alpha$ *is admissible and regular, then the* α-*homogeneous-universal structures for* $\mathcal{A}_T$ *are divisible* η_α-*groups and for* $\mathcal{V}_T$ *are just the* η_α-*groups in* $\mathcal{V}$. *In particular* η_α-*groups exist for any cardinal* $\aleph_\alpha$.

Proof. If $\aleph_\alpha$ is an arbitrary cardinal $(\alpha > 0)$, then there exists an admissible regular cardinal $\aleph_\beta > \aleph_\alpha$ (GCH). As an η_β-group is always an η_α-group if $\beta \geq \alpha$, we may assume $\alpha = \beta$.

Let G be the (unique up to isomorphism) α-homogeneous-universal structure in $\mathcal{A}_T$, and suppose $A < B$ for some $A, B \leq G$ with $|A|, |B| < \aleph_\alpha$. Pick $x \notin G$ and consider the abelian group $\bar{G} = G \oplus Zx$ generated by G and $\{x\}$. Then $\bar{G}$ is a partially ordered abelian group with positive cone G^+. Consider the set

$$P = \{x - a \mid a \in A\} \cup \{b - x \mid b \in B\}.$$

We now show that this set is positively independent with respect to G^+.

Let $\{n_1, \ldots, n_k\}, \{m_1, \ldots, m_\ell\} \subseteq Z^+ \cup \{0\}$ and

$$\left(\sum_{i=1}^{k} n_i (x - a_i) + \sum_{j=1}^{\ell} m_j (b_j - x)\right) \in -G^+$$

for some $\{a_1,...,a_k\} \subseteq A$ and $\{b_1,...,b_\ell\} \subseteq B$.   Then

$$(\sum_{i=1}^{k} n_i - \sum_{j=1}^{\ell} m_j)x - \sum_{i=1}^{k} n_i a_i + \sum_{j=1}^{\ell} m_j b_j \in -G^+.$$

Hence $\sum_{j=1}^{\ell} m_j b_j \leq \sum_{i=1}^{k} m_i a_i$ and $(\sum_{i=1}^{k} n_i - \sum_{j=1}^{\ell} m_j)x = 0$; so $\sum_{i=1}^{k} n_i = \sum_{j=1}^{\ell} m_j$. Assume $a_1 \leq a_2 \leq ... \leq a_k$ and $b_1 \leq ... \leq b_\ell$. We then have

$$(\sum_{j=1}^{\ell} m_j)b_1 \leq \sum_{j=1}^{\ell} (m_j b_j) \leq \sum_{i=1}^{k} (n_i a_i) \leq (\sum_{i=1}^{k} n_i)a_k = (\sum_{j=1}^{\ell} m_j)a_k.$$

But $b_1 > a_k$ so $\sum_{j=1}^{\ell} m_j = 0 = \sum_{i=1}^{k} n_i$. Thus $n_i = 0 = m_j$ for each i,j and the set P is in fact positively independent.

Consequently, there exists a total order on $\bar{G}$, with positive cone $\bar{G}^+$, such that $A < \{x\} < B$ in this order.

Now consider the following diagram

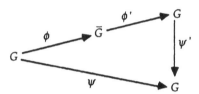

where ϕ is the inclusion map, ϕ' is the embedding guaranteed by the universality of G, ψ is the identity , and ψ' is the map with $\psi' \circ \phi' \circ \phi = \psi$ guaranteed by the homogeneity of G. Finally, this gives the required inequalities:

$$A = \psi(A) = (\psi' \circ \phi' \circ \phi)(A) < (\psi' \circ \phi')(x) < (\psi' \circ \phi' \circ \phi)(B) = \psi(B) = B.$$

In fact more can be shown about the relationship between α-homogeneous-universal totally ordered abelian groups and divisible η_α-groups. As we shall see for groups of cardinality $\aleph_\alpha$ these concepts coincide. The assumption below that an η_α-set has cardinality $\aleph_\alpha$ implies that $\aleph_\alpha$ is admissible and regular (see [Alling, 1960]; cf [Chang & Keisler, 1973; Section 5.1]).

THEOREM 12.4.5. *Any two divisible η_α-groups of cardinality $\aleph_\alpha$ are isomorphic.*

Proof. By Theorem 3.1.2, any two non-trivial divisible abelian totally ordered groups are elementarily equivalent. Consequently, the theorem follows at once from [Chang & Keisler, 1973; Theorems 5.1.13 and 5.1.14].

For those unfamiliar with model theory, we now give the above proof specifically in ℓ-group terms.

Let G and H be divisible η_α-groups of cardinality $\aleph_\alpha$, and denote by ω_α the least ordinal corresponding to $\aleph_\alpha$. Then G and H can be enumerated in the following way:

$$G = \{g_\beta \mid \beta < \omega_\alpha\}$$
$$H = \{h_\beta \mid \beta < \omega_\alpha\}.$$

Our objective is to construct ascending sequences of divisible subgroups $G_\beta \subseteq G$ and $H_\beta \subseteq H$ and o-isomorphisms $\phi_\beta \colon G_\beta \to H_\beta$ such that for each $\beta < \omega_\alpha$:

(1) $|G_\beta|, |H_\beta| \leq |\beta| \cdot \aleph_0$;

(2) $\{g_\gamma \mid \gamma < \beta\} \subseteq G_\beta$;

and   (3) $\{h_\gamma \mid \gamma < \beta\} \subseteq H_\beta$.

In doing so we will also get an o-isomorphism $\bigcup_{\beta<\omega} \phi_\beta$ between G and H. For $\beta = 0$ we need only set $G_0 = \{0\}$, $H_0 = \{0\}$, and $\phi_0 = \{(0,0)\}$. For a limit ordinal β, let $G_\beta = \bigcup_{\gamma<\beta} G_\gamma$, $H_\beta = \bigcup_{\gamma<\beta} H_\gamma$ and $\phi_\beta = \bigcup_{\gamma<\beta} \phi_\gamma$; the required properties are clearly transmitted. Finally, assume that properties (1), (2), and (3) hold for the ordinal β. Clearly, $|G_\beta| \leq |\beta| \cdot \aleph_0 < \aleph_\alpha$. Hence, since H is an η_α-set, there exists $x \in H$ with $\phi_\beta(\{g \in G_\beta \mid g < g_\beta\}) < \{x\} < \phi_\beta(\{g \in G_\beta \mid g > g_\beta\})$. If $g_\beta \in G_\beta$, then $g < g_\beta < g'$ implies $\phi_\beta(g) < \phi_\beta(g_\beta) < \phi_\beta(g')$ so in this case we can assume $x = \phi_\beta(g_\beta)$. We wish to extend ϕ_β to an o-isomorphism $\bar\phi_\beta \colon G_\beta + Qg_\beta \to H_\beta + Qx$. If $g_\beta \in G_\beta$ this is already achieved by our assumption that $x = \phi_\beta(g_\beta)$. On the other hand, if $g_\beta \notin G_\beta$ then define $\bar\phi_\beta(g + qg_\beta) = \phi_\beta(g) + qx$ for $g \in G_\beta$ and $q \in Q$. This is clearly a group homomorphism (recall that G_β and H_β are divisible). Suppose now that $u = g + qg_\beta > 0$ for some $g \in G_\beta$ and $q \in Q$. If $q = 0$, then $\bar\phi_\beta(u) = \phi_\beta(g) > 0$. If $q > 0$, then $g_\beta > (\frac{-1}{q})g$; so $\phi_\beta(-\frac{1}{q}g) < x$ whence $\bar\phi_\beta(u) = \phi_\beta(g) + qx > 0$. If $q < 0$, then let $q' = -q$ and proceed as before. Thus, in all cases $\bar\phi_\beta$ preserves the order structure and so is an o-isomorphism.

The group $G_\beta + Qg_\beta$ meets the requirements for $G_{\beta+1}$ but $H_\beta + Qx$ may not contain h_β. To remedy this situation we first apply the fact that G is an η_α-set to pick $y \in G$ such that

$$\bar{\phi}_\beta^{-1}(\{h \in H_\beta \mid h < h_\beta\}) < \{y\} < \bar{\phi}_\beta^{-1}(\{h \in H_\beta \mid h > h_\beta\}).$$

If in fact $h_\beta \in H_\beta$ then we can let $y = \bar{\phi}_\beta^{-1}(h_\beta)$. Following the same course as before we define an o-isomorphism $\bar{\bar{\phi}}_\beta: G_\beta + Qg_\beta + Qy \to H_\beta + Qx + Qh_\beta$ by $g + qy \mapsto \bar{\phi}_\beta(g) + qh_\beta$ for all $g \in G_\beta + Qg_\beta$ and $q \in Q$. Now let $G_{\beta+1} = G_\beta + Qg_\beta + Qy$, $H_{\beta+1} = H_\beta + Qx + Qh_\beta$ and $\phi_{\beta+1} = \bar{\bar{\phi}}_\beta$. These clearly satisfy conditions (1), (2), and (3) and the theorem is proved.

Together with theorem 12.4.4 the preceding theorem allows for a complete characterization of η_α-groups of cardinality $\aleph_\alpha$.

THEOREM 12.4.6. *If $\aleph_\alpha$ is an admissible regular cardinal, then the α-homogeneous-universal structures in $\mathcal{A}_T$ are precisely the divisible η_α-groups of cardinality $\aleph_\alpha$. Those in $\mathcal{V}_T$ are the η_α-groups in $\mathcal{V}$.*

Constantine Tsinakis
Vanderbilt University
Nashville, Tennessee 37235
U. S. A.

Wayne B. Powell
Oklahoma State University
Stillwater, Oklahoma 74078
U. S. A.

A. M. W. Glass[1]

CHAPTER 13

GENERATORS AND RELATIONS
IN LATTICE-ORDERED GROUPS:
DECISION PROBLEMS AND EMBEDDING THEOREMS

This chapter is devoted to proving that many problems in ℓ-groups are un-decidable yet many embedding theorems can be made effective (where, throughout this chapter, homomorphism, embedding etc., is in the category of ℓ-groups). Specifically,

THEOREM A [Glass and Gurevich, 1983]. *There is a finitely presented ℓ-group with insoluble group word problem.*

THEOREM B [Glass and Madden, 1984]. *The isomorphism problem for the ten-generator one-relator members of any recursively axiomatised variety of ℓ-groups (other than $\forall x \forall y (x=y)$) is insoluble.*

THEOREM C [Glass, 1983]. *Any countable ℓ-group can be embedded in a seven-generator ℓ-simple ℓ-group.*

[1]Research supported in part by NSF grant no. 8401745.

328

A. M. W. Glass and W. C. Holland (eds.), Lattice-Ordered Groups, 328–346.
© *1989 by Kluwer Academic Publishers.*

THEOREM D [Glass, 1983; 1986a]. *Every finitely presented ℓ-group can be constructively embedded in*

   (i) *a two-generator one-relator ℓ-group.*

   (ii) *a finitely presented ℓ-group with trivial centre.*

   (iii) *an eight-generator one-relator perfect ℓ-group.*

Note that as a consequence of Theorems A and D(i), we obtain

COROLLARY E. *There exists a two-generator one-relator ℓ-group with insoluble group word problem.*

Corollary E is in marked contrast to Magnus' result that every finitely generated one-relator group has soluble word problem (see [Lyndon and Schupp, 1977, §IV.5].) As we saw (Theorem 9.2.2), finitely generated no-relator ℓ-groups (i. e., finitely generated free ℓ-groups) have soluble word problem.

By Corollary 3.2.9, there is an algorithm which when given an arbitrary universal sentence in the language of abelian ℓ-groups, determines whether or not it holds in every abelian ℓ-group. Consequently, if $w(x)$ denotes an arbitrary element of the free abelian ℓ-group A on $x_1,...,x_m$ and G is the quotient of A by the ℓ-ideal generated by the set of elements $\{r_1(x),...,r_n(x)\}$ of A, then the algorithm will tell us whether or not $\forall x(\underset{i=1}{\overset{n}{\&}} r_i(x) = 0 \rightarrow w(x) = 0)$ holds in every abelian ℓ-group; i.e., whether or not $w(x) = 0$ in G. Hence there is a single algorithm which solves the word problem in every finitely presented abelian ℓ-group. We say that the variety has *uniformly soluble word problem* in this case. By Theorem B it follows that

COROLLARY F [Glass and Madden, 1984]. *The variety of abelian ℓ-groups has uniformly soluble word problem but the isomorphism problem for ten-generator one-relator abelian ℓ-groups is insoluble.*

The rest of the chapter is devoted to elucidating the techniques needed to prove these major theorems. With the exception of Theorems B and D(i), the proofs of all the results hinge on some technical results about conjugacy and roots in doubly transitive $A(\Omega)$ (see [Glass, 1981b,§2.2] or [Holland, 1963a].)

LEMMA 13.1. *Let Ω be doubly homogeneous and $e < f, g \in B(\Omega)$ have one bump with the suprema (infima) of their supports belonging to the same $A(\Omega)$ orbit. If $\alpha \in supp(f)$ and $\beta \in supp(g)$, then any ordermorphism between $[\alpha, \alpha f]$ and $[\beta, \beta g]$ can be extended to an element $h \in A(\Omega)$ such that $h^{-1}fh = g$ and $\alpha f^n h = \beta g^n$ $(n \in \mathbf{Z})$. Moreover, we may assume that $supp(h)$ is contained in any open interval that includes the closure of the interval generated by $supp(f) \cup supp(g)$.*

This lemma can be proved in the same manner as Lemma 2.5.2.

LEMMA 13.2. *Let Ω be doubly homogeneous and $e < f \in A(\Omega)$. If $\alpha_0 < \alpha_1 < \alpha_2 < \alpha_3 = \alpha_0 f$ in Ω and $h_i : [\alpha_i, \alpha_{i+1}] \cong [\alpha_{i+1}, \alpha_{i+2}]$ $(i = 0,1)$ are arbitrary, there exists $h \in A(\Omega)$ such that $h \mid [\alpha_i, \alpha_{i+1}] = h_i$ $(i = 0,1)$ and $h^3 = f$.*

This lemma, too, can be proved in the same manner as Lemma 2.5.1.

Of course, 3 is not sacred in Lemma 13.2. An analogous result holds with any integer $m \geq 2$ in place of 3.

Let F be the free ℓ-group on $\{x_i : i \in I\}$ and $\{r_j(x) : j \in J\} \subseteq F$. We write $G = (x_i ; r_j(x) = e)_{i \in I, j \in J}$ if G is the quotient of F by the ℓ-ideal generated by $\{r_j(x) : j \in J\}$. If I is finite we say that G is *finitely generated* and if J is finite we say that G is *finitely related*. A finitely generated finitely related ℓ-group is called a *finitely presented* ℓ-group. If J is finite, then as $\bigvee \{ \mid r_j(x) \mid : j \in J\} = e$ if and only if $r_j(x) = e$ for all $j \in J$ $(\mid a \mid = a \vee a^{-1} \geq e; \mid a \mid = e \Leftrightarrow a = e$ (Lemma 0.1.8(b)), any finitely related ℓ-group can be written in the form $(x_i ; r(x) = e)_{i \in I}$. Note that the passage from such $(x_i ; r_j(x) = e)_{i \in I, j \in J}$ to $(x_i ; r(x) = e)_{i \in I}$ is constructive.

Lemma 13.1 is already enough to prove an analogue of a famous theorem in group theory due to G. Baumslag and D. Solitar (see [Lyndon and Schupp, 1977, Theorem IV.4.9]):

THEOREM 13.3 [Glass, 1983]. *There is a two-generator one-relator ℓ-group which is isomorphic to a proper quotient of itself.*

Proof. Let $G = (x_0, x_1 ; x_1^{-1} x_0^2 x_1 = x_0^3)$ and $\theta : G \to G$ be the homomorphism given by: $x_0 \theta = x_0^2$ and $x_1 \theta = x_1$. Since $(x_1^{-1} x_0^2 x_1) \theta = x_1^{-1} x_0^4 x_1 = (x_1^{-1} x_0^2 x_1)^2 = x_0^6 = x_0^3 \theta$, θ is well defined. Moreover, θ is onto since $x_0 = x_1^{-1} x_0^2 x_1 x_0^{-2} = [x_1, x_0^{-1}] \theta$ and $x_1 = x_1 \theta$. Hence $G/\ker(\theta) \cong G\theta = G$. To complete the proof we must show that $\ker(\theta) \neq \{e\}$. Note that $[x_0, x_1^{-1} x_0 x_1] \theta = [x_0^2, x_1^{-1} x_0^2 x_1] = [x_0^2, x_0^3] = e$ so $[x_0, x_1^{-1} x_0 x_1] \in \ker(\theta)$. For groups, it is immediate from the Nielson-Schreier normal form that $[x_0, x_1^{-1} x_0 x_1] \neq e$ in the group generated by $\{x_0, x_1\}$ subject to $x_1^{-1} x_0^2 x_1 = x_0^3$; but since the amalgamation and HNN properties fail for ℓ-groups, we must try something else. Let $e < f \in B(\mathbf{R})$ be a bump. Now $\text{supp}(f^2) = \text{supp}(f^3)$ and letting $\alpha \in \text{supp}(f)$, it follows by Lemma 13.1 that any isomorphism between $[\alpha, \alpha f^2]$ and $[\alpha, \alpha f^3]$ can be extended to an element $h \in A(\mathbf{R})$ with $h^{-1} f^2 h = f^3$. In particular, we may assume $\alpha h = \alpha$, $\alpha f h = \alpha f$ and $\alpha f^2 h = \alpha f^3$. Let H be the ℓ-subgroup of $A(\mathbf{R})$ generated by f and h. Since $h^{-1} f^2 h = f^3$, the map $x_0 \mapsto f$, $x_1 \mapsto h$ induces a homomorphism ϕ of G onto H. Now $\alpha f h^{-1} f h = \alpha f^3 \neq \alpha f^2 = \alpha h^{-1} f h f$ so $\alpha \in \text{supp}([f^{-1}, h^{-1} f^{-1} h])$. Hence $[f^{-1}, h^{-1} f^{-1} h] \neq e$ and, consequently, $[f, h^{-1} f h] \neq e$. But $[x_0, x_1^{-1} x_0 x_1] \phi = [f, h^{-1} f h]$, so $[x_0, x_1^{-1} x_0 x_1] \neq e$. Therefore $\ker(\theta) \neq \{e\}$ and the proof is complete.

The crucial point to note here is that although the amalgamation and HNN properties fail, it is possible to build a homomorphic image of the (finitely) presented ℓ-group in which some word does not collapse to e. Hence its preimage in our (finitely) presented ℓ-group is not e, though of course lots of other words which we don't care about may collapse to e (even though, if they're group words and the relators are group relations, they may not in the corresponding group presentation).

Two further ingredients will be needed to prove Theorems C and D. The first is of a similar flavour to the result in combinatorial group theory that every group can be embedded in one in which any two elements of the same order are conjugate. It is due to K. R. Pierce [1972a] and can also be found in [Glass, 1981b, Theorem 10B].

LEMMA 13.4. *Every ℓ-group can be embedded in one in which any two strictly positive elements are conjugate.*

Actually, we will need the following consequence of its proof.

COROLLARY 13.5. *Let (G, Ω) be an ℓ-permutation group and $\{f_n: n \in \mathbf{Z}\}$, $\{g_n: n \in \mathbf{Z}\}$ be subsets of G with $e < f_n \leq g_n$ and $\text{supp}(g_n) < \text{supp}(g_{n+1})$ for all $n \in \mathbf{Z}$. Then G can be embedded in an ℓ-group H such that for some $h \in H$, $h^{-1} f_n h = g_n$ for all $n \in \mathbf{Z}$.*

The second idea concerns disjointness of conjugates. Its full strength will not be needed until we prove Theorem A. Consider the ℓ-group $G = (a,b \; ; \; b^{-n}ab^n \wedge a = e)_{n \in \mathbf{Z}^+}$. In G, $b^{-m}ab^m \wedge b^{-n}ab^n = e$ unless $m = n$. If G could be finitely presented, e. g., on two generators x,y, then there would be an element $w(x,y)$ of the free ℓ-group on $\{x,y\}$ such that $w(x,y) = e$ in G and in any ℓ-group generated by a and b, $w(a,b) = e$ implies $b^{-n}ab^n \wedge a = e$ for all $n \in \mathbf{Z}^+$. But as $w(a,b) = e$ in G, there must be a positive integer n_0 such that $w(a,b) = e$ in any ℓ-group generated by a and b in which $b^{-n}ab^n \wedge a = e$ for all $0 < n \leq n_0$. Hence $b^{-n}ab^n \wedge a = e$ for all $n \in \mathbf{Z}^+$ if $b^{-n}ab^n \wedge a = e$ for all $0 < n \leq n_0$. But this is palpably false: let $b \in A(\mathbf{R})$ be translation by $+1$ and $e < a \in A(\mathbf{R})$ with $\mathrm{supp}(a) = (0,1) \cup (n_0+1, n_0+2)$; then $b^{-n}ab^n \wedge a = e$ for all $0 < n \leq n_0$ but $b^{-(n_0+1)}ab^{n_0+1} \wedge a \neq e$. Can G be embedded in a finitely presented ℓ-group? The answer is yes and is instrumental in the proofs of Theorems A, C, and D(i).

We may view G as an ℓ-permutation group (G,Ω) by the Cayley-Holland Theorem (Theorem 2.2.5). Define c by

$$\alpha c = \begin{cases} \alpha b^{-n}ab^n & \text{if } \alpha \in \mathrm{supp}(b^{-n}ab^n) \text{ for some } n \in \mathbf{Z}^+ \\ \alpha & \text{otherwise.} \end{cases}$$

Then c is well defined, $b^{-1}ab \leq c$, $b^{-1}cb \leq c$ and $a \wedge c = e$ (since $a \wedge b^{-n}ab^n = e$ for all $n \in \mathbf{Z}^+$). Conversely, if $g^{-1}fg \leq h$, $g^{-1}hg \leq h$ and $f \wedge h = e$, then by induction on $m \in \omega$, $e \leq g^{-(m+1)}fg^{m+1} \leq g^{-m}hg^m \leq h$; thus $g^{-(m+1)}fg^{m+1} \wedge f = e$ for all $m \in \omega$.

Now let $\hat{G} = (x_i \; ; \; r_j(x) = e)_{i \in I, j \in J}$ and $\hat{H}$ be an ℓ-group with generators y_i $(i \in I)$ and z_k $(k \in K)$. If the map $\phi : x_i \mapsto y_i$ $(i \in I)$ is an embedding of $\hat{G}$ in $\hat{H}$ (so $r_j(y) = e$ for all $j \in J$) and $\hat{H}$ satisfies $s_m(y,z) = e$ $(m \in M)$-- and possibly other relations--then $\hat{G}$ can be embedded in

$$\hat{L} = (y_i, z_k \; ; \; r_j(y) = e, s_m(y,z) = e)_{i \in I, j \in J, k \in K, m \in M}$$

via $\theta : x_i \mapsto y_i$.. This is immediate since if ψ is the natural homomorphism of $\hat{L}$ onto $\hat{H}$, then the diagram

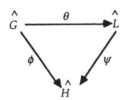

commutes.

Hence $G = (a,b \; ; \; b^{-n}ab^n \wedge a = e)_{n \in \mathbb{Z}^+}$ can be embedded in

$$L = (a,b,c \; ; \; b^{-n}ab^n \wedge a = e, \; b^{-1}ab \vee c = c, \; b^{-1}cb \vee c = c, \; a \wedge c = e)_{n \in \mathbb{Z}^+}.$$

But

$$L = (a,b,c \; ; \; b^{-1}ab \vee c = c, \; b^{-1}cb \vee c = c, \; a \wedge c = e)$$

by the "conversely" part. Hence G is embedded in a finitely presented ℓ-group.

We can also use the above to prove Theorem D(i) for three instead of two generators:

Since $x_i = (x_i \vee e)(x_i^{-1} \vee e)^{-1}$ (Lemma 0.1.8(b)), we may assume that each generator of $G = (x_1,...,x_m \; ; \; r_j(x) = e)_{j \in J}$ is positive. Now G is countable and hence may be embedded in $A(\mathbb{R})$ and hence in $A((0,1))$ [Holland,1963a] (see the proof of Lemma 2.3.2). Identify G with its image extended to $A(\mathbb{R})$; so $\mathrm{supp}(g) \subseteq (0,1) \subseteq \mathbb{R}$ for each $g \in G$. Let $a = \bigvee \{x_i : 1 \leq i \leq m\}$, let b be translation by $+1$, and let $c \in A(\mathbb{R})$ be defined by:

$$\alpha c = \begin{cases} \alpha b^{-n}ab^n & \text{if } \alpha \in \mathrm{supp}(b^{-n}ab^n) \text{ for some } n \in \mathbb{Z}^+ \\ \alpha & \text{otherwise.} \end{cases}$$

Since $\mathrm{supp}(b^{-n}ab^n) \subseteq (n,n+1)$, c is well defined and $\mathrm{supp}(b^{-n}ab^n) < \mathrm{supp}(b^{-m}ab^m)$ if $n < m$. By Corollary 13.5, the ℓ-subgroup of $A(\mathbb{R})$ generated by $G \cup \{a,b,c\}$ can be embedded in an ℓ-group H containing an element d such that $d^{-1}b^{-i}x_ib^id = b^{-i}ab^i$ $(1 \leq i \leq m)$. Hence H can be embedded in

$$L = (a,b,c,d,x_1,...,x_m \; : \; r_j(x) = e, \; a = \bigvee\{x_i \colon 1 \leq i \leq m\},$$

$$b^{-1}ab \vee c = c, \; b^{-1}cb \vee c = c, \; a \wedge c = e, \; d^{-1}b^{-i}x_ib^id = b^{-i}ab^i \;)_{1 \leq i \leq m, j \in J}.$$

But in L, $x_i = b^idb^{-i}ab^id^{-1}b^{-i}$ $(1 \leq i \leq m)$, so L is generated by $\{a,b,c,d\}$. Moreover, substituting for $x_1,...,x_m$ in $r_j(x)$ and $a = \bigvee\{x_i \colon 1 \leq i \leq m\}$ gives L generated by $\{a,b,c,d\}$ with relations indexed by J plus a finite number of others. Letting $y = a \vee c$, we see that L is generated by $\{y,b,d\}$ with $c = b^{-1}yb$ and $a = y(b^{-1}yb)^{-1}$. Hence every finitely generated ℓ-group is embeddable in a three-generator ℓ-group whose defining relations are finite, recursive, or recursively enumerable if those of G are. Since, as we saw, any finite number of relations are equivalent to a single relation, the modified Theorem D(i) follows.

To prove Theorem D(i), observe in the proof that every countable ℓ-group G can be embedded in a two-generator ℓ-group L, the generators x_i are mapped to $[b^{2i-1}ab^{-(2i-1)},a]$ $(1 \leq i \leq m)$, where a and b are the generators of L (see [Glass, 1981b,Theorem 10A]). Hence G can be embedded in $(a,b; \overline{w}(a,b) = e)$, where $\overline{w}(a,b)$ is obtained from the defining relation $w(x) = e$ of G by replacing each x_i occurring in $w(x)$ by $[b^{2i-1}ab^{-(2i-1)},a]$ $(1 \leq i \leq m)$.

I wish to thank Trevor Evans for kindly pointing out the general universal algebraic principle which strengthens the Neumanns' result to prove Theorem D(i). That one can get by with two instead of three generators as I had done, is entirely his.

As we saw (Theorem 11.4.4), the free product of any finitely generated ℓ-group with the free ℓ-group on two generators has trivial centre; i. e., if $G = (x_1,...,x_m \; ; \; r_j(x) = e)_{j \in J}$, then $H = (x_1,...,x_m,x_{m+1},x_{m+2} \; ; \; r_j(x) = e)_{j \in J}$ has trivial centre. But G is embeddable in its free product with the free ℓ-group on two generators, i. e., G is embeddable in H. Theorem D(ii) now follows.

To prove Theorem C we need the following technical result whose messy proof can be found in [Glass, 1984].

THE MESSUAGE LEMMA.[1] Let $G = (x_1,...,x_m; r_j(x) = e)_{j \in J}$ and $w(x)$ be an element of the free ℓ-group on $\{x_1,...,x_m\}$. Let

$$G(w) = (x_1,...,x_m,a_0,a_1,a_2,a_3,a_4; r_j(x) = e \, , \, a_0^{-1} \, | \, w(x) \, | \, a_0 = \bigvee\{|x_i| : 1 \leq i \leq m\},$$
$$a_2^{-1} \, | \, a_4 \, | \, a_2 = |a_0|, \; |w(x)a_4^{-1}| \wedge |a_4| = e, \; |a_2| \vee |a_3| \vee a_1^6 = a_1^6,$$
$$a_3^{-1}a_1^3a_3 = (|a_0| \vee a_1^{-1} \, | \, a_0 \, | \, a_1 \vee a_1^{-3} \, | \, a_0 \, | \, a_1^3 \vee a_1^{-4} \, | \, a_0 \, | \, a_1^4)^4)_{j \in J}.$$

(i)   If $w(x) = e$ in G, then $G(w) \cong \{e\}$.

(ii) If $w(x) \neq e$ in G, then G can be embedded in $G(w)$.

We now deduce Theorem C from the Messuage Lemma by a technique due to C. F. Miller III (see [Lyndon and Schupp; 1977, Theorem IV.3.5]); its analogue, that every countable group can be embedded in a three generator simple group, is due to Philip Hall [1974].

Let C be a countable ℓ-group. Let $(f_0,g_0),(f_1,g_1), ...$ be a list of all pairs of elements of C for which $f_n,g_n \neq e \; (n \in \omega)$. As in the proof of Theorem D(i) (for three instead of two generators), C can be embedded in

$$C^\dagger = (x_i,a,b,c,d \; ; \; r_j(x) = e, \; x_i \wedge a = x_i, \; a \wedge c = e, \; b^{-1}ab \vee c = c, \; b^{-1}cb \vee c = c,$$
$$d^{-1}b^{-i}x_ib^id = b^{-i}ab^i, \; d^{-1}b^n \, | \, f_n \, | \, b^{-n}d = b^n \, | \, g_n \, | \, b^{-n})_{i \in I,j \in J,n \in \omega}$$

where $C = (x_i \; ; \; r_j(x) = e)_{i \in I,j \in J}$, $J \subseteq \mathbb{Z}^+$ and each x_i is positive. Note that any two elements of C which strictly exceed e are conjugate in $C^\dagger$. Now let $C_0 = C$, $C_{m+1} = C_m^\dagger \; (m \in \omega)$ and $H = \bigcup\{C_n : n \in \omega\}$. By construction, H is a

---

[1]Messuage was "originally the portion of land intended to be occupied, or actually occupied, as a site for a dwelling house and its appurtenances " [Oxford English Dictionary].

countable simple ℓ-group. Hence we may assume that C itself is ℓ-simple.[1] By [Glass, 1981b, Theorem 10A] C can be embedded in a two-generator ℓ-group G. Let $w \in C \setminus \{e\}$. By the Messuage Lemma, G can be embedded in $G(w)$. It follows from Zorn's Lemma that there is an ℓ-ideal M of $G(w)$ maximal with respect to its intersection with C is $\{e\}$. Thus C is naturally embedded in $G(w)/M$ which has seven generators. Let N be an ℓ-ideal of $G(w)$. If $N \cap C \neq \{e\}$, then $N \cap C = C$ since C is ℓ-simple. Hence $w \in N$, so $Nw = N$ in $G(w)/N$. But in $G(w)/N$ all the defining relations of $G(w)$ hold. Therefore $G(w)/N \cong \{e\}$ by the Messuage Lemma. Consequently, $N = G(w)$. Thus M is a maximal ℓ-ideal of $G(w)$. It follows that $G(w)/M$ is ℓ-simple.

The Messuage Lemma together with Theorem A is also useful in proving a result related to Theorem B. We will not bother to distinguish between a finite presentation of an ℓ-group (i. e., a finite set of generators together with a finite set of relations on these generators) and the finitely presented ℓ-group given by this specific presentation. As in semigroups and groups (see [Lyndon and Schupp, 1977; §IV.4]), a property $\mathcal{P}$ of finitely presented ℓ-groups is said to be *Markov* if (i) $G \cong H$ and G satisfies $\mathcal{P}$ imply H satisfies $\mathcal{P}$, (ii) some finitely presented ℓ-group G_1 satisfies $\mathcal{P}$, and (iii) some finitely presented ℓ-group G_2 cannot be embedded in any finitely presented ℓ-group which satisfies $\mathcal{P}$.

THEOREM 13.6 [Glass, 1984]. *If $\mathcal{P}$ is any Markov property, there is no algorithm to determine for an arbitrary finitely presented ℓ-group whether or not it satisfies $\mathcal{P}$.*

Proof. Let G_1, G_2 be as in the definition of a Markov property $\mathcal{P}$ and G_0 be a finitely presented ℓ-group with insoluble word problem. We may assume that the generators of $G_1, G_2,$ and G_0 are disjoint. Let w be an element in the free ℓ-group on the generators of G_0, and H be the free product of G_0 and G_2 in the category of ℓ-groups. If $w = e$ in G_0, then $w = e$ in H and hence $H(w) \cong \{e\}$. Thus the free product L of $H(w)$ and G_1 in the category of ℓ-groups is isomorphic to G_1 and so satisfies $\mathcal{P}$. If $w \neq e$ in G_0 then $w \neq e$ in H and hence H (and therefore G_2) can be embedded in $H(w)$. It follows that G_2 can be embedded in L, a finitely presented ℓ-group. Consequently, L does not satisfy $\mathcal{P}$. We have therefore demonstrated that L satisfies $\mathcal{P}$ if and only if $w = e$ in G_0. Any algorithm which told us whether or not an arbitrary finitely presented ℓ-group satisfies $\mathcal{P}$ would therefore solve the word problem for G_0. This proves the theorem.

---

[1] Aliter: C can be embedded in $B(\mathbb{R})$ which is simple (see [Glass, 1981, Corollary 2L and Theorem 2G]. Hence C can be embedded in a countable simple ℓ-group.

Since the following are easily seen to be Markov properties, we obtain

COROLLARY 13.7 [Glass, 1984]. *For each of the following, there is no algorithm to determine whether or not an arbitrary finitely presented ℓ-group is*

(i) *abelian*

(ii) *archimedean*

(iii) *trivial*

(iv) *free*

(v) *free abelian*

(vi) *linearly ordered*

(vii) *linearly orderable*

(viii) *representable*

(ix) *normal valued*

(x) *with soluble word problem*

(xi) *with unique extraction of roots.*

The reader is encouraged to add his own favourite beast to the list.

An argument similar to that used in proving the Messuage Lemma can be used to establish (see [Glass, 1986a]):

LEMMA 13.8. *If $G = (a_1, a_2 ; r(a) = e)$ and*

$$G^* = (a_1, a_2, t, u, v, w, g, h ; r(a) = e,\ a_i \wedge h^{-1} a_j h = e,\ a_i \wedge h^{-2} a_j h^2 = e,$$

$$h^{-3} a_i h^3 = a_i,\ h^{-3} t h^3 = t,\ t^{-1} h^{-(i-1)} a_j h^{(i-1)} t = h^{-i} a_i^2 h^i,\ u^{-1} t u = h,$$

$$u \wedge e = e,\ u \wedge h^2 = u,\ v \wedge g^3 = g^3,\ v \wedge g^9 = v,\ w \wedge g^3 = g^3,\ w \wedge g^9 = w,$$

$$v^{-1} h v = h^2,\ (h \vee g^{-1} h g \vee g^{-3} h g^3 \vee g^{-4} h g^4)^4 = w^{-1} g^3 w)_{1 \leq i, j \leq 2},$$

then G can be embedded in G^* which is perfect.

Here *perfect* means that the convex ℓ-subgroup generated by the commutator subgroup is the entire ℓ-group.

Lemma 13.8 and Theorem D(i) yield Theorem D(iii).

To prove Theorem A we will show how to code recursive functions into finitely presented ℓ-groups.

A *recursive function* from ω into ω is one obtainable by an algorithm. Using Church's thesis, this can be shown to be equivalent to: a recursive function from ω into ω is one built up from the zero function ($\theta(n) \equiv 0$) and the successor function ($s(n) \equiv n+1$) using composition and induction on functions already known to be recursive (see, e. g., [Rogers, 1967]). Actually, using [J. Robinson, 1968], the key is composition.

A *recursive set* X is one for which there is an algorithm to determine for an arbitrary natural number, whether or not it belongs to X. A set Y for which there is only an algorithm to determine whether an arbitrary natural number belongs to Y is called a *recursively enumerable set*. (Caution: unlike usual English, "whether" does not mean "whether or not" here, as we now explain.) Intuitively, the algorithm associated with a recursive set X is a black box which when given an arbitrary natural number will play Orlando Gibbons' "Jubilate Deo" if the number belongs to X or Thomas Tallis' "Lamentations of Jeremiah" if it does not ("whether or not"). The algorithm associated with a recursively enumerable set Y is a black box equipped only with Gibbons' masterpiece; if it plays, the number is in Y ("whether") but there is no bound on how long we'll have to wait to know that silence will persist rather than Gibbons' piece will be played eons in the future (we don't know "or not"). If Y and $\omega \backslash Y$ are both recursively enumerable, then we can put the machines together and wait for which plays first to determine whether or not a natural number belongs to Y: i. e., if Y and $\omega \backslash Y$ are recursively enumerable, then Y is recursive (the converse is obvious). Note that a recursive (recursively enumerable) subset of a recursive set is also recursive (recursively enumerable).

Two important facts we will need are [Rogers,1967, §5.2]

(i) There exists a recursively enumerable set that is not recursive.

(ii) Every recursively enumerable set is the range of a recursive function (the converse is also true).

We wish to apply algorithms to words, not just sets of natural numbers. Given a countable language L (e. g., $L = \{\cdot,^{-1},e,\wedge,\vee\}$ for ℓ-groups), we can effectively enumerate (code) all the first order formulae in the language. If T is a theory in the language L such that the set of natural numbers inherited by a given set of axioms of T (as a subset of the set of formulae in L) is finite, recursive or recursively enumerable we say that the theory T is *finitely, recursively* or *recursively enumerably axiomatisable*. The crucial fact is that this definition turns out to be independent of the initial choice of the effective enumeration of the formulae of $\mathcal{L}$–see [Rogers, 1967]. Moreover, the set of sentences that are consequences of any finitely, recursively or recursively enumerably axiomatisable theory is a recursively enumerable set; i. e., the set of natural numbers attached to these sentences is a recursively enumerable set of natural numbers. Hence if $G = (x_1,...,x_m ; r(x) = e)$ is a finitely presented ℓ-group, $\{w(x) \in F : w(x) = e$ in $G\}$ is recursively enumerable, where F is the free ℓ-group on $\{x_1,...,x_m\}$; it has as axioms the finite set of axioms for ℓ-groups together with $\forall x(r(x) = e)$. To say that G has *soluble word problem* means $\{w(x) \in F : w(x) = e$ in $G\}$ is recursive. If a,d,a_X are generators of G, then $\{a_X^{-1}d^{-m}ad^m a_X d^{-m}a^{-1}d^m : m \in \omega\}$ is a recursive set of words and $\{a_X^{-1}d^{-m}ad^m a_X d^{-m}a^{-1}d^m : m \in X\}$ is recursively enumerable if $X \subseteq \omega$ is.

We can also effectively enumerate all finitely presented members of a recursively enumerably axiomatisable theory; e. g., if $r(x)$ is the nth word in the enumeration, $(x_1,...,x_m ; r(x) = e)$ gets $2^m 3^n$ attached--see [Rabin, 1958] for further details. Strictly, to say that the Markov property $\mathcal{P}$ is undecidable means that the set of numbers attached to the finitely presented ℓ-groups which satisfy $\mathcal{P}$ is not recursive.

To prove Theorem A we will require an amplification of the basic idea in the proof of Lemma 13.1 Let $e < a_0,b_0,c_0,d_0 \in B(\mathbf{R})$ each have one bump and let $a_0 \wedge b_0^{-1}a_0 b_0 = e = b_0 \wedge c_0^{-1}b_0 c_0 = c_0 \wedge d_0^{-1}c_0 d_0$. We wish to show the existence of an element $f_0 \in B(\mathbf{R})$ such that $f_0^{-1}a_0 f_0 = a_0$, $f_0^{-1}b_0 f_0 = c_0$, and $f_0^{-1}d_0 f_0 = d_0$. The conditions on $a_0,b_0,c_0,$ and d_0 imply that there are $\beta < \text{supp}(a_0) < \beta b_0$, $\gamma < \text{supp}(b_0) < \gamma c_0$ and $\delta < \text{supp}(c_0) < \delta d_0$. Any isomorphism of $[\beta,\beta b_0]$ onto $[\gamma,\gamma c_0]$ can be extended to an ordermorphism $\overline{f}_0$ of $\text{supp}(b_0)$ onto $\text{supp}(c_0)$ by Lemma 13.1. In particular, since $\gamma < \beta < \text{supp}(a_0) < \beta b_0 < \gamma c_0$, we may assume that $\overline{f}_0 | \text{supp}(a_0)$ is the identity. Also, since $\delta < \text{supp}(b_0) < \delta d_0$ and $\delta < \text{supp}(c_0) < \delta d_0$, we may extend $\overline{f}_0$ to $\hat{f}_0 : [\delta, \delta d_0] \cong [\delta, \delta d_0]$. By Lemma 13.1, there is $f_0 \in B(\mathbf{R})$ such that $f_0 | [\delta, \delta d_0] = \hat{f}_0$ and $f_0^{-1}d_0 f_0 = d_0$. By the choice of $\overline{f}_0$ and $\hat{f}_0$, $f_0^{-1}a_0 f_0 = \overline{f}_0^{-1}a_0\overline{f}_0 = a_0$ and $f_0^{-1}b_0 f_0 = \overline{f}_0^{-1}b_0\overline{f}_0 = c_0$; thus such a desired f_0 exists. Similarly, there exists $g_0 \in A(\mathbf{R})$ such that $g_0^{-1}a_0 g_0 = a_0$, $g_0^{-1}b_0 g_0 = c_0$ and $g_0^{-1}c_0 g_0 = d_0$.

So assume that a,b,c,d,f,g satisfy $a \wedge b^{-1}ab = e = b \wedge c^{-1}bc = c \wedge d^{-1}cd$, $f^{-1}af = a$, $f^{-1}bf = c$, $f^{-1}df = d$, $g^{-1}ag = a$, $g^{-1}bg = c$ and $g^{-1}cg = d$. Let G_0 be this finitely presented ℓ-group.

Let $\theta : \omega \to \omega$ be defined by: $\theta(n) = 0$ for all $n \in \omega$. Then let $G_0(\theta)$ be the ℓ-group G_0 with one extra generator a_θ and one extra relation $a_\theta = e$. Then $a_\theta^{-1} d^{-n} a d^n a_\theta = d^{-n} c^{-\theta(n)} a c^{\theta(n)} d^n$ for all natural numbers n.

Similarly, but with more work, it is possible to adjoin to G_0 a finite number of new generators and relations so that $a_s^{-1} d^{-n} a d^n a_s = d^{-n} c^{-s(n)} a c^{s(n)} d^n$ for all natural numbers n, where $s(n) \equiv n+1$.

Now suppose $\phi, \psi : \omega \to \omega$ are such that for some finitely presented ℓ-groups $G_0(\phi)$, $G_0(\psi)$ respectively, whose generators and relations contain those of G_0,

$$a_\phi^{-1} d^{-n} a d^n a_\phi = d^{-n} c^{-\phi(n)} a c^{\phi(n)} d^n \qquad \text{in } G_0(\phi),$$

$$\text{and} \quad a_\psi^{-1} d^{-n} a d^n a_\psi = d^{-n} c^{-\psi(n)} a c^{\psi(n)} d^n \qquad \text{in } G_0(\psi)$$

for all $n \in \omega$.

Now we wish to code in χ the composition of ϕ and ψ; i. e., $\chi(n) = \phi(\psi(n))$. So we take the generators of $G_0(\phi)$ and $G_0(\psi)$ together with their relations together with one new generator a_χ and a finite set of new relations to ensure that $a_\chi^{-1} d^{-n} a d^n a_\chi = d^{-n} c^{-\chi(n)} a c^{\chi(n)} d^n$ for all $n \in \omega$. To begin, let's try $a_\chi = a_\psi a_\phi$. This only gives

$$a_\phi^{-1} a_\psi^{-1} d^{-n} a d^n a_\psi a_\phi = a_\phi^{-1} d^{-n} c^{-\psi(n)} a c^{\psi(n)} d^n a_\phi$$

which in turn equals--even assuming that $c^{\psi(n)}$ does not cause problems--

$$a_\chi^{-1} d^{-1} a d^n a_\chi = d^{-n} c^{-\phi(n)} c^{-\psi(n)} a c^{\phi(n)} c^{\psi(n)} d^n = d^{-n} c^{-(\phi+\psi)(n)} a c^{(\phi+\psi)(n)} d^n$$

which is not what we want at all. What we need therefore is that when a_ϕ is applied, the $\psi(n)$ is the the exponent on the d instead of the c. But this is exactly what g does.

Let $a_\chi = f^{-1} a_\psi g a_\phi g^{-1} a_\psi^{-1} f$. Now
$$a_\chi^{-1} d^{-n} a d^n a_\chi = f^{-1} a_\psi g a_\phi^{-1} g^{-1} a_\psi^{-1} f d^{-n} a d^n f^{-1} a_\psi g a_\phi g^{-1} a_\psi^{-1} f$$

$$= f^{-1} a_\psi g a_\phi^{-1} g^{-1} a_\psi^{-1} d^{-n} a d^n a_\psi g^{-1} a_\psi^{-1} f$$

(since f commutes with d and a)

$$= f^{-1} a_\psi g a_\phi^{-1} g^{-1} d^{-n} c^{-\psi(n)} a c^{\psi(n)} d^n g a_\phi g^{-1} a_\psi^{-1} f$$

$$= f^{-1} a_\psi g a_\phi^{-1} (g^{-1} d g)^{-n} d^{-\psi(n)} a d^{\psi(n)} (g^{-1} d g)^n a_\phi g^{-1} a_\psi^{-1} f$$

$$= f^{-1} a_\psi g (g^{-1} d g)^{-n} d^{-\psi(n)} c^{-\phi\psi(n)} a c^{\phi\psi(n)} d^{\psi(n)} (g^{-1} d g)^n g^{-1} a_\psi^{-1} f$$

(assuming a_ϕ commutes with $g^{-1} d g$)

$$= f^{-1} a_\psi d^{-n} c^{-\psi(n)} b^{-\chi(n)} a b^{\chi(n)} c^{\psi(n)} d^n a_\psi^{-1} f$$

$$= f^{-1}d^{-n}b^{-\chi(n)}abx^{\chi(n)}d^n f$$

(assuming that b commutes with c,d and a_ψ)

$$= d^{-n}c^{-\chi(n)}acx^{\chi(n)}d^n, \text{ as desired.}$$

Notice that there are only a finite number of extra assumptions which can be incorporated in and taken along for the ride Of course, since $b\wedge c^{-1}bc = e$, b certainly does not commute with c so there are some changes that need to be made. These can be found in [Glass and Gurevich, 1983] or [Glass, 1981b, chapter 13], though in the latter some minor corrections are necessary (see Math. Rev. 85d 06015).

The above ideas are enough to show that if X is any subset of ω that is recursively enumerable, then there is a finitely presented ℓ-group $G(X)$ such that $a_\phi \in G(X)$ and $a_\phi^{-1}d^{-n}ad^na_\phi = d^{-n}c^{-\phi(n)}ac^{\phi(n)}d^n$ for all natural numbers n, where ϕ is a recursive function enumerating X.

Finally we define a finitely presented ℓ-group $H(X)$ which is just $G(X)$ with one new generator a_X and as extra relations has a_X commutes with $g^{-1}dg$ and $f^{-1}a_\phi ga_Xg^{-1}a_\phi^{-1}f$ commutes with each $d^{-n}ad^n$ $(n\in\omega)$--some finessing is again necessary to make this infinite set of relations a single relation but the reader should be able to manage this for himself using the disjointness of conjugates trick. Then, for each $n\in w$,

$$d^{-n}ad^n = f^{-1}a_\phi ga_X^{-1}g^{-1}a_\phi^{-1}fd^{-n}ad^nf^{-1}a_\phi ga_Xg^{-1}a_\phi^{-1}f.$$

So $g^{-1}a_\phi^{-1}fd^{-n}ad^nf^{-1}a_\phi g = a_X^{-1}g^{-1}a_\phi^{-1}fd^{-n}ad^nf^{-1}a_\phi ga_X$.

Hence $g^{-1}d^{-n}gd^{-\phi(n)}ad^{\phi(n)}g^{-1}d^ng = a_X^{-1}(g^{-1}d^{-n}g)d^{-\phi(n)}ad^{\phi(n)}(g^{-1}d^ng)a_X$.

Thus $d^{-\phi(n)}ad^{\phi(n)} = a_X^{-1}d^{-\phi(n)}ad^{\phi(n)}a_X$. Since $X = \{\phi(n) : n\in w\}$, $d^{-m}ad^m = a_X^{-1}d^{-m}ad^ma_X$ for all $m\in X$.

Now, in $A(R)$, we interpret each a_χ by $\hat{a}_\chi$ where $\hat{a}_\chi \mid \text{supp}(d_0^{-n}cod_0^n)$ is just $c_0^{\chi(n)}$ $(n\in w)$, and elsewhere $\hat{a}_\chi$ is the identity. Then each of the relations we have written down holds in $A(R)$ under this interpretation. Moreover, if $\hat{a}_X \mid \text{supp}(d_0^{-n}cod_0^n)$ is c_0 if $n\in w\setminus X$ and the identity elsewhere, the relations concerning a_X hold. Thus $H(X)$ has the ℓ-subgroup of $A(R)$ generated by the requisite elements as a homomorphic image. Since

$$\hat{a}_X^{-1}d_0^{-m}a_0d_0^m\hat{a}_X = d_0^{-m}c_0^{-1}a_0c_0d_0^m \neq d_0^{-m}a_0d_0^m$$

if $m\in \omega\backslash X$, $a_X^{-1}d^{-m}ad^m a_X \neq d^{-m}ad^m$ if $m\in \omega\backslash X$. Consequently, $a_X^{-1}d^{-m}ad^m a_X = d^{-m}ad^m$ if and only if $m\in X$ ($m\in \omega$). If X is not recursive, $H(X)$ cannot have soluble word problem. (If we could algorithmically determine if two arbitrary elements were equal or not, we'd be able to tell whether or not an arbitrary natural number belongs to X which is impossible since X is recursively enumerable but not recursive.)

I hope that the above oversimplification (with the accompanying lies) will help make a necessarily rather technical proof which already occurs in a book [Glass, 1981b] more readable by isolating the essential ingredients here.

Note that the above proof (and even the correct one) shows that the Turing degree of the word problem of $H(X)$ is at least that of X, but whether equality occurs, I do not know.

As a special example, it is possible to show that

$$L = (a,b,c \ ; a\wedge c = e, \ b\wedge c = c, \ cb^{-1}a^{-1}b\wedge b^{-1}ab = e, \ cb^{-1}c^{-1}b\wedge b^{-1}cb = e)$$

does have soluble word problem. The idea of the proof is as follows:

An easy induction shows that $[b^{-m}ab^m, b^{-n}ab^n] = e = [b^{-m}cb^m, b^{-n}cb^n]$ for all $m,n\in \mathbf{Z}$. Hence $[b^{-m}ab^m, b^{-n}cb^n] = e$ for all $m,n\in \mathbf{Z}$. Moreover, if N is the ℓ-ideal of L generated by a and c, then $b\notin N$ and b generates L/N; so L/N is abelian. By dividing c into those bumps $\hat{c}$ with $b^{-1}\hat{c}\, b\wedge\hat{c} = e$ and those bumps $c^{\dagger}$ with $b^{-1}c^{\dagger}b = c^{\dagger}$ (using the Cayley-Holland Theorem), an analysis of words is possible even though the $\hat{c}$'s and $c^{\dagger}$'s do not necessarily belong to L. (If $\hat{c}$ is a bump of a conjugate of a, then the word applied to any $\alpha\in \text{supp}(\hat{c})$ is obtained from just a's and b's and as $a << b$, it is a group word; it is easy to see if it is the identity on $\text{supp}(\hat{c})$ or not. If $\hat{c}$ is not a bump of a conjugate of a, each $b^{-m}ab^m$ acts as the identity and the same idea works with c in place of a. On $\text{supp}(c^{\dagger})$, a acts as the identity and we are in the abelian ℓ-group on two generators b^*,c^* subject to $c^*\wedge b^* = c^*$ and $c^*\wedge e = e$, which has soluble word problem (Corollary 3.2.9). The word is the identity if and only if it is in each of these cases.)

Note that the conjugacy problem for L can also be decided; i. e., there is an algorithm which when given two arbitrary elements of the free ℓ-group on (a,b,c) determines whether or not they are conjugate in L.

The proof can be extended to show that

THEOREM 13.9 [Glass, 1986]. *If a finitely presented ℓ-group G generates the ℓ-group variety $\mathcal{V}$, then one can uniformly construct an ℓ-group H which generates $\mathcal{V}\mathcal{A}$ and has the same Turing degree for its word and conjugacy problems as G does. In particular, it follows that $\mathcal{A}^2, \mathcal{A}^3, \mathcal{A}^4, \ldots$ are each generated by a single finitely presented ℓ-group.*

H is an ℓ-subgroup of $[(G \boxplus Z)\mathrm{WrZ}] \boxplus A_2$, where A_2 is a two-generator one-relator abelian ℓ-group (c.f., L is an ℓ-subgroup of $[(Z \boxplus Z)\mathrm{WrZ}] \boxplus A_2$, above). So, in some sense, H is a finitely presented "approximation" for $G\mathrm{WrZ}$ (it satisfies the same identities).

A more restricted case which is considerably easier was previously considered in [Glass, 1985]. If $G = (x_1, \ldots, x_m ; r(x) = e)$ is any finitely presented ℓ-group, then

$$G(p) = (a, x_1, \ldots, x_m ; r(x) = e, \; a \wedge e = e, \; a^{-p} x_i a^p = x_i,$$

$$a^{-k} | x_i | a^k \wedge | x_{i'} | = e)_{1 \leq i, i' \leq m, \, 1 \leq k < p}$$

has word and conjugacy problems of the same Turing degrees as those of G. This can be proved directly by spelling and shows that finitely presented ℓ-groups with soluble word and conjugacy problems abound. The existence of a finitely presented ℓ-group with soluble word problem by insoluble conjugacy problem remains open and is of central interest. For example, the conjugacy problems for free ℓ-groups and nilpotent ℓ-groups are still open.

THEOREM 13.10 [Kopytov, 1982]. *Any nilpotent class n ℓ-group that is finitely presented in the class of all nilpotent class n ℓ-groups has soluble word problem $(n \in \mathbf{Z}^+)$.*

Proof. Let G be a finitely presented nilpotent class n ℓ-group. So G is the quotient of the free nilpotent class n ℓ-group on $x_1, \ldots, x_m$ by the ℓ-ideal generated by $r(x)$, say. Let $w(x)$ be an ℓ-group word in $x_1, \ldots, x_m$. Then $w(x) = e$ if and only if $\forall x(r(x) = e \to w(x) = e)$ is a theorem of nilpotent class n ℓ-groups. Since there is a finite set of axioms for nilpotent class n ℓ-groups, the set of theorems for this variety is recursively enumerable. Hence $w(x) = e$ holds in G if $\forall x(r(x) = e \to w(x) = e)$ appears in the recursively enumerable list of theorems for nilpotent class n ℓ-groups.

It therefore remains to give an effective procedure to determine if $w(x) \neq e$ in G. Since any nilpotent class n ℓ-group is a subdirect product of linearly ordered groups (Theorem 10.10.4), if $w(x) \neq e$ in G there is a linearly ordered nilpotent class n group B generated by $b_1, \ldots, b_m$, say, such that $r(b) = e \neq w(b)$ in B. Without loss of generality, $w(b) > e$ in B. Since B is linearly ordered, every element of B can be expressed in terms of $b_1,...,b_m$ using only the group operations. Hence B, being nilpotent class n, has only a finite number of convex subgroups and we have $\{e\} = B_t \lhd \ ... \ \lhd B_1 \lhd B_0 = B$ where each B_i/B_{i+1} is a free abelian archimedean linearly ordered group of finite rank. Let $A_i = B_i/B_{i+1}$ and let $\{a_{ij} : j \in J_i\}$ be a basis for A_i with $a_{ij} < a_{ij+1}$. Choose $b_{ij} \in B_i \backslash B_{i+1}$ so that $b_{ij}B_{i+1} = a_{ij}$ ($0 \le i < t$; $j \in J_i$).Note that each element of B can be written uniquely in the form $c_0...c_{t-1}$, where

$$c_i = b_{i1}^{m_{i1}}......b_{ij_i}^{m_{ij_i}} \quad (0 \le i < t).$$

For example, if $B = \langle b_1, b_2, b_3 \ ; \ [b_1,b_2]^2 = [b_1,b_3]^3$ and $[b_i,b_j]$ central$\rangle$ is ordered so $b_3 >> b_2 >> b_1 >> [b_1,b_2] >> e$ and $b_1^{m_1}[b_2,b_3]^{m_4} > e$ if $m_1 + \pi m_4 > 0$, then we can write each element of B in the form $b_3^{m_3}b_2^{m_2}b_1^{m_1}[b_2,b_3]^{m_4}([b_1,b_2][b_1,b_3]^{-1})^{m_5}$ since $[b_1,b_2] = [b_1,b_2]^3[b_1,b_3]^{-3}$ and $[b_1,b_3] = [b_1,b_2]^2[b_1,b_3]^{-2}$. Here B_2/B_3 is free abelian of rank 2, all the other quotients being free abelian of rank 1.

Next observe that if $r(x) = \bigvee_p \bigwedge_q r_{pq}(x)$, where $r_{pq}(x)$ are group words, then $r(b) = e$ in B is the same as: for each p, there is q such that $r_{pq}(b) \le e$ and for some p_0, $r_{p_0 q}(b) \ge e$ for all q. Similarly $w(b) > e$ is the same as: for some p_0, $w_{p_0 q}(b) > e$ for all q (p and q running over suitable finite index sets).

Now let ψ_i be an embedding of A_i in $\mathbf{R}$. Since $\theta_i : A_i \to \mathbf{R}$ given by $\theta_i(a_{ij}) = \frac{1}{N}\psi_i(a_{ij})$ ($N \in \mathbf{Z}^+$) is also an embedding, we may assume that $\psi_i(a_{ij}) < \frac{1}{2}$ for all i,j, $0 \le i < t$, $j \in J_i$, and that $\{\psi_i(a_{ij}): 0 \le i \le t, j \in J_i\}$ is a set of algebraic numbers linearly independent over $\mathbf{Q}$. Let M be a positive integer very much bigger than $\sum_{i=0}^{t-1} \sum_{j \in J_i} k_{ij}$, where k_{ij} is 1 plus the maximum of the absolute value of the exponents of b_{ij} in the normal form for the various group words $r_{pq}(b)$ and $w_{pq}(b)$ occurring in $r(b)$ and $w(b)$, respectively. The map $\phi_i : A_i \to \mathbf{R}$ given by $\phi_i(a_{ij}) = M^i\psi_i(a_{ij})$ is an embedding of A_i in $\mathbf{R}$. Define $\phi : B \to \mathbf{R}$ by

$$\phi(c_0...c_{t-1}) = \sum_{i=0}^{t-1} \sum_{j \in J_i} m_{ij}\phi_i(a_{ij}) = \sum_{i=0}^{t-1} \sum_{j \in J_i} M^i m_{ij}\psi_i(a_{ij}), \text{ where } c_i = b_{i1}^{m_{i1}}...b_{ij_i}^{m_{ij_i}}..$$

Observe that $\phi(r(b)) = 0 < \phi(s(b))$, and that $\{\phi(b_{ij}): 0 \leq i \leq t\text{-}1, j \in J\}$ is a rationally independent set of algebraic numbers. By the above, this is just a finite system of inequalities and strict inequalities each of the form

$$\Sigma k_{ij}\phi(b_{ij}) \gtreqless 0 .$$

Conversely, if (*) is the system above with each $\phi(b_{ij})$ replaced by y_{ij}, then any solution in $\mathbf{R}$ of (*) with $\{y_{ij} : 0 \leq i \leq t\text{-}1, j \in J\}$ satisfying no non-trivial linear equation with integer coefficients in absolute value less than twice the sum of the absolute values of all the k_{ij}'s appearing in (*) will define a map from A_i into $\mathbf{R}$ and hence induce a possibly new order on each A_i and thus on B. Let B^* be this new linearly ordered group (isomorphic to B as a group). Then $r(b^*) = e < w(b^*)$ in B^*. Since B^* satisfies $r(b^*) = e$, it is a homomorphic image (as an ℓ-group) of G ; in B^*, $w(b^*) \neq e$. But the solution set in $\mathbf{R}$ of any finite set of inequalities and strict inequalities is a polyhedron in $\mathbf{R}^k$ and so contains points $\{y_{ij} : 0 \leq i < t, j \in J_i\}$ in the algebraic real numbers satisfying no non-trivial linear equation with integer coefficients in absolute value less than twice the sum of the absolute values of all the k_{ij}'s appearing in (*) (since $\{\phi(b_{ij}) : 0 \leq i < t, i \in J_i\}$ is a solution). Hence if $w(x) \neq e$ in G, there is an m-generator nilpotent class n group B^* with a finite system of quotients each of which is free Abelian of finite rank and isomorphic to a subgroup of the algebraic real numbers such that $r(b^*) = e \neq w(b^*)$ in B^*. Since the set of m-generator nilpotent class n groups with such quotients is recursively enumerable and the solution of a finite system of inequalities and strict inequalities in the algebraic real numbers is effective, we obtain an algorithm which will show if $w(x) \neq e$ in G; viz., enumerate all such groups and attempt to solve the corresponding system of inequalities and strict inequalities in each. If there is a solution in any of them, $w(x) \neq e$ in G.

   Since the set of words equal to e in G and the set of words not equal to e in G are both recursively enumerable, G has soluble word problem.

   Caution: An ℓ-group may well be finitely presented in a variety of ℓ-groups without being a finitely presented ℓ-group. For nilpotent ℓ-groups, I suspect the concepts are distinct, unlike groups; e. g., a group with two generators x and y in which $[x,[x,y]] = e = [y,[x,y]]$ is nilpotent, but unless an arbitrary ℓ-group whose generators x and y satisfy these equations is known, *a fortiori*, to be a subdirect product of o-groups, why do these imply $[x,[x^m \vee y^n, y]] = e$, say, for all $m,n \in \mathbf{Z}$?[1]

We next prove Theorem B. Recall (see Section 1.3) that if $G = (x_1,...,x_{10} ; u(x) = 0)$ and $H = (x_1,...,x_{10} ; v(x) = 0)$ are finitely presented abelian ℓ-groups, then they are isomorphic if and only if the polyhedral cones $Z(u)$ and $Z(v)$ with vertices $0 \in \mathbf{R}^{10}$ are piecewise homogeneous linear homeomorphic, where $x_1,...,x_{10}$ are identified with the projection maps of $\mathbf{R}^{10}$ to $\mathbf{R}$ and u and v are in normal form (see Section 1.3). Moreover, this correspondence is totally effective (Section 1.3). But the piecewise linear homeomorphism problem for 4-dimensional polyhedra is insoluble [Markov, 1960], and any 4-dimensional polyhedron can be embedded in $\mathbf{R}^9$. Now every polyhedron P in $\mathbf{R}^9$ corresponds to a polyhedral cone $\hat{P}$ in $\mathbf{R}^{10}$; viz.: $\hat{P} = \{\lambda(\alpha,1) : \alpha \in P, 0 \leq \lambda \in \mathbf{R}\}$. Hence the piecewise homogeneous linear homeomorphism problem for polyhedral cones in $\mathbf{R}^{10}$ with vertex the origin is insoluble. Consequently, the isomorphism problem for ten-generator one-relator abelian ℓ-groups is insoluble (every such polyhedral cone is $Z(w)$ for an appropriate $w(x)$ in the free abelian ℓ-group on ten generators). To complete the proof of Theorem B, let $\mathcal{V}$ be any recursively axiomatised variety containing an ℓ-group with more than one element. Then every abelian ℓ-group belongs to $\mathcal{V}$ by Theorem 10.2.2. Furthermore, every abelian ℓ-group $(x_1,...,x_{10} ; w(x) = 0)$ is finitely presented in $\mathcal{V}$ by

$$(x_1,...,x_{10} ; |w(x)| \vee \bigvee_{1 \leq i,j \leq 10} |[x_i, x_j]| = e),$$

where $w(x)$ is now written multiplicatively. Theorem B follows immediately.

It should be observed that in sharp contrast to Corollary 13.7, there is an algorithm to determine if an arbitrary finitely presented abelian ℓ-group has only one element since we can clearly determine whether or not $Z(w) = \{0\}$.

Finally, we close with a list of what we regard as the most challenging open problems in the subject.

1. Does there exist a finitely presented ℓ-group in $\mathcal{A}^3$ with insoluble word problem? This is the smallest power of $\mathcal{A}$ in the group case. Unfortunately, Harlompovich's example [Harlompovich, 1981] (or even Baumslag et al's [Baumslag, Gildenhuys and Strebel, 1985]) is not torsion free. Equally, replace $\mathcal{A}^3$ by normal valued.

---

[1]Given any finite set of words from the free ℓ-group on two generators, there is a two-generator metabelian ℓ-group G such that the commutator of any pair of these words belongs to the centre of G, yet G is not nilpotent of any class [Darnel and Glass, a].

2. Does there exist a finitely presented ℓ-group with soluble word problem but insoluble conjugacy problem? As we saw in Theorem 13.10, nilpotent class 2 ℓ-groups have soluble word problem; do they have soluble conjugacy problem? The answer is yes if it can be shown that whenever a and b belong to a finitely presented nilpotent class 2 ℓ-group in which they are not conjugate, then they fail to be conjugate in some totally ordered homomorphic image.

3. Given a recursively enumerable subset X of ω, does there exist a finitely presented ℓ-group whose word problem has Turing degree equal to that of X? Strengthenings, if 2 can be shewn, need to be examined.

4. Can every finitely generated ℓ-group with a recursively enumerable set of defining relations be embedded in a finitely presented ℓ-group? (We saw one example $(a,b ; b^{-n}ab^n \wedge a = e)_{n \in Z^+}$ where this was true. As we saw, $(a,d,a_X : a_X^{-1} d^{-m}ad^m a_X (d^{-m}ad^m)^{-1} = e)_{m \in X}$ can also be so embedded whenever $X \subseteq \omega$ is recursively enumerable.) It is the analogue of Graham Higman's famous theorem for groups (see [Lyndon and Schupp, 1977, §IV7]) and would show a beautiful connection between ℓ-groups and recursion theory, viz.: the finitely generated ℓ-subgroups of finitely presented ℓ-groups are precisely those defined by recursively enumerable sets of relations.

5. Does the analogue of the Boone-Higman Theorem hold (see [Lyndon and Schupp, 1977, §IV7])? That is, does a finitely generated ℓ-group have soluble word problem if and only if it can be embedded in an ℓ-simple ℓ-group which in turn can be embedded in a finitely presented ℓ-group?

6. Can every ℓ-group be embedded in one in which any two elements which are neither positive nor negative are conjugate?

A. M. W. Glass
Department of Mathematics and Statistics
Bowling Green State University
Bowling Green, Ohio 43403
U. S. A.

BIBLIOGRAPHY

M. E. Adams and D. Kelly

[1977]   *Disjointness conditions in free products of lattices*, Algebra Universalis 7(1977), 245-258.

S. Adyan

[1973]   *Periodic groups of odd exponent*, Proc. 2nd Internat. Conf. in the theory of groups, Canberra, 1973, 8-12.

N. L. Alling

[1960]   *On ordered divisible groups*, Trans. Amer. Math. Soc. 94(1960), 498-514.

[1962]   *On the existence of real-closed fields that are η_α-sets of power $\aleph_\alpha$*, Trans. Amer. Math. Soc. 103(1962), 341-351.

D. F. Anderson and J.Ohm

[1981]   *Valuations and semi-valuations of graded domains*, Math. Ann. 256(1981), 145-156.

M. Anderson and P. F. Conrad

[1981]   *Epicomplete ℓ-groups*, Algebra Universalis 12(1981), 224-241.

J. Arnold

[1929]   *Ideale in kommutative Halbgruppen*, Rec. Math. Soc. Moscow 36(1929), 401-407.

A. Kumar Arora and S. H. McCleary

[1986]   *Centralizers in free lattice-ordered groups*, Houston J. Math 12(1986), 455-482.

K. E. Aubert

[1979]   *Divisors of finite character*, Preprint No. 1, 1979, University of Oslo.

M. Auslander and D. A. Buchsbaum

[1959]   *Unique factorization in regular local rings*, Proc. Nat. Acad. Sci U. S. A. 45(1959), 733-734.

R. Baer and F. Levi

[1936]   *Freie Producte und ihre Untergruppen*, Comp, Math. 3(1936), 391-398.

K. Baker

[1968]   *Free vector lattices*, Canadian J. Math 20(1968), 58-66.

348

J. T. Baldwin, J. Berman, A. M. W. Glass, and W. Hodges

[1982]   *A combinatorial fact about free algebras,* Algebra Universalis 15(1982), 145-152.

R. Balbes

[1967]   *Projective and injective distributive lattices,* Pacific J. Math. 21(1967), 405-420.

R. Balbes and P. Dwinger

[1974]   *Distributive Lattices,* University of Missouri Press, 1974.

R. N. Ball

[1974]   *Full convex ℓ-subgroups of a lattice-ordered group,* Ph. D. Thesis, University of Wisconsin, 1974.

[1979]   *Topological lattice ordered groups,* Pacific J. Math. 83(1979), 1-26.

[1980a] *The distinguished completion of a lattice ordered group,* Algebra Carbondale 1980, Springer Lecture Notes 848.

[1980b] *Convergence and Cauchy structures on lattice ordered groups,* Trans. Amer. Math. Soc. 259(1980), 357-392.

[1982]   *The generalized orthocompletion and strongly projectable hull of a lattice ordered group,* Can. J. Math. 34(1982), 621-661.

[1984]   *Distributive Cauchy lattices,* Algebra Universalis 18(1984), 134-174.

[a]   *Distinguished extensions of a lattice ordered group,* submitted.

[b]   *The structure of the α-completion of a lattice ordered group,* Houston J. Math. (to appear).

[c]   *Cauchy completions are homomorphic images of submodels of ultrapowers,* Proc. Conference on Convergence Structures, Dept. Math., Cameron U., Lawton, Oklahoma.

[d]   *Paraprojectable lattice ordered groups,* in preparation.

R. N. Ball, P. F. Conrad, and M. R. Darnel

[1986]   *Above and below subgroups of a lattice ordered group,* Trans. Amer. Math. Soc. 271(1986), 1-40.

R. N. Ball and G. Davis

[1983]   *The α-completion of a lattice ordered group,* Czech. Math. J. 33(1983), 111-118.

R. N. Ball and M. Droste

[1985]   *Normal subgroups of doubly transitive automorphism groups of chains,* Trans. Amer. Math. Soc. 290(1985), 647-664.

R. N. Ball and A. W. Hager

[a]   *Epicomplete archimedean ℓ-groups and vector lattices,* Trans. Amer. Math. Soc. (to appear).

[b]   *Epimorphisms in archimedean ℓ-groups and vector lattices (and Baire functions),* J. Australian Math. Soc. (to appear).

[c]   *Archimedean-kernel-distinguishing extensions of archimedean ℓ-groups with weak unit,* Indian J. Math. (to appear).

[d]   *Algebraic extensions of archimedean ℓ-groups,* in preparation.

B. Banaschewski

[1957] *Über die Vervollständigung geordneter Gruppen,* Math. Nachr. 16(1957), 52-71.

M. A. Bardwell

[1980] *Lattice-ordered groups of automorphisms of partially ordered sets,* Houston J. Math. 6(1980), 191-225.

A. Baudisch, D. Seese, H.-P. Tuschik, and M. Weese

[1980] *Decidability and Generalized Quantifiers,* Akademie Verlag, Berlin, 1980.

G. Baumslag, D. Gildenhuys and R. Strebel

[1985] *Algorithmically insoluble problems for groups and Lie algebras, Part 2, (Isomorphism problem for solvable groups),* J. Algebra 97(1985), 278-285.

G. M. Bergman

[1984] *Specially ordered groups,* Comm. Algebra 12(1984), 2315-2333.

[a]   *Ordering free groups and coproducts of ordered groups,* unpublished preprint.

J. Berman

[1981] *Interval lattices and the amalgamation property,* Algebra Universalis 12(1981), 360-375.

S. J. Bernau

[1966] *Orthocompletions of lattice groups,* Proc. London Math. Soc. 16(1966), 107-130.

[1969] *Free abelian lattice groups,* Math. Ann. 180(1969), 48-59.

[1975] *The lateral completion of an arbitrary lattice group,* J. Austral. Math. Soc., Ser. A. 19(1975), 263-287.

[1977] *Varieties of lattice groups are closed under L-completion,* Instituto Nazionale Di Alta Matematica, Symposia Mathematica XXI(1977), 349-355.

W. M. Beynon

[1977] *Applications of duality in the theory of finitely generated lattice-ordered groups,* Canad. J. Math. 29(1977), 243-254.

A Bigard

[1968] *Sur les z-sous-groupes d'un groupe réticulé,* C. R. Acad. Sci. Paris Series A-b 266(1968), A261-A262.

[1969] *Contribution à la théorie des groupes réticulé,* Thesis, University of Paris, 1969.

A Bigard, K Keimel, and S Wolfenstein

[1977]   *Groupes et Anneaux Réticulés*, Lecture Notes in Math. **608**, Springer-Verlag, New York, 1977.

G. Birkhoff

[1942]   *Lattice-ordered groups*, Annals of Math. **43**(1942), 298-331.

[1960]   *Lattice Theory*, Amer. Math. Soc. Colloquium **XXV**(1960), Providence.

[1967]   *Lattice Theory*, Amer. Math. Soc. Colloquium **XXV**(1967), 3rd ed., Providence.

P. Bixler and M. R. Darnel

[1986]   *Special valued ℓ-groups*, Algebra Universalis **22**(1986), 172-191.

R. L. Blair and A. W. Hager

[1974]   *Extensions of zero-sets and of real-valued functions*, Math. Zeit. **136**(1974), 41-52.

R. D. Bleier

[1971]   *Minimal vector lattice covers*, Bull. Austral. Math. Soc. **5**(1971), 331-335.

[1975]   *Free ℓ-groups and vector lattices*, J. Austral. Math. Soc. **19**(1975), 337-342.

V. V. Bludov and N. Ya. Medvedev

[1974]   *On the completion of ordered metabelian groups*, Algebra and Logic **13**(1974), 369-373.

Z. I. Borevich and I. R. Shaferevich

[1966]   *Number Theory*, Academic Press, New York, 1966.

N. Bourbaki,

[1966]   *Elements of Mathematics, General Topology I*, Addison Wesley, Reading, 1966.

A. Bouvier and M. Zafrullah

[a]   On some class groups, preprint.

A. Brandis

[1965]   *Über die Multiplicative Struktur von Körpererweiter-ungen*, Math. Zeit. **87**(1965), 71-73.

D. Brignole and H. Ribeiro

[1965]   *On the universal equivalence of ordered abelian groups*, Algebra i Logika **4**,2(1965), 51-55.

S. Burris

[1985]   *A simple proof of the hereditary undecidability of the theory of lattice-ordered abelian groups*, Algebra Universalis **20**(1985), 400-401.

S. Burris and H. P. Sankappanavar

[1981]   *A Course in Universal Algebra*, Graduate Texts in Mathematics 78, Springer-Verlag, New York-Heidelberg-Berlin, 1981.

S. Burris and H. Werner

[1979]   *Sheaf constructions and their elementary properties*, Trans. Amer. Math. Soc. **248**(1979), 269-309.

R. Byrd, P. F. Conrad, and J. T. Lloyd

[1971]   *Characteristic subgroups of lattice-ordered groups*,Trans. Amer. Math. Soc. **158**(1971), 339-371.

R. Byrd and J. T. Lloyd

[1967]   *Closed subgroups and complete distributivity in lattice-ordered groups*, Math. Z. **101**(1967), 123-130.

C. C. Chang and A. Ehrenfeucht

[1962]   *A characterization of abelian groups of automorphisms of a simply ordering relation*, Fund. Math. **51**(1962), 141-147.

C. C. Chang and H. J. Keisler

[1973]   *Model Theory*, North-Holland, Amsterdam, 1973.

C. G. Chehata

[1952]   *An algebraically simple ordered group*, Proc. London Math. Soc. **2**(1952), 183-197.

D. J. Christensen and R. S. Pierce

[1959]   *Free products of α-distributive Boolean algebras*, Math. Scand. **7**(1959), 81-105.

A. H. Clifford

[1938]   *Arithmetic and ideal theory of a commutative semigroup*, Ann. Math. **39**(1938),594-610.

[1940]   *Partially ordered abelian groups*, Ann. Math. **41**(1940), 465-473.

I. S. Cohen and I. Kaplansky

[1946]   *Rings with a finite number of primes*, Trans Amer. Math. Soc. **60**(1946), 168-177.

P. M. Cohn

[1954]   *An invariant characterization of pseudovaluations of a field*, Proc. Cambridge Phil. Soc. **50**(1954), 159-171.

[1968]   *Bézout rings and their subrings*, Proc. Cambridge Phil. Soc. **64**(1968), 251-264.

[1973]   *Unique factorization domains*, Amer. Math. Monthly **80**(1973), 1-18.

P. M. Cohn and K. Mahler

[1953]   *On the composition of pseudo valuations*, Nieuw. Arch. Wisk. **3**(1953), 161-198.

352

W. W. Comfort and S. Negrepontis
[1974]   *The theory of ultrafilters*, Springer Verlag, Berlin, 1974.

P. F. Conrad
[1960]   *The structure of a lattice-ordered group with a finite number of disjoint elements*, Michigan Math. J. 7(1960), 171-180.
[1962]   *Regularly ordered groups*, Proc Amer. Math. Soc. 13(1962), 726-731.
[1964]   *The relationship between the radical of a lattice-ordered group and complete distributivity*, Pacific J. Math. 14(1964), 493-499.
[1966a] *Representations of partially ordered abelian groups as groups of real valued functions*, Acta Math. 116(1966), 199-221.
[1966b] *The lattice of all convex ℓ-subgroups of a lattice-ordered group*, Czech. Math. J. 15(90) (1965), 101-123.
[1968]   *Lex-subgroups of lattice-ordered groups*, Czech. Math. J. 18(93) (1968), 86-103.
[1970a] *Lattice-ordered Groups*, Tulane Lecture Notes, New Orleans, 1970.
[1970b] *Free lattice-ordered groups*, J. Algebra 16(1970), 191-203.
[1971a] *Minimal vector lattice covers*, Bull. Austral. Math. Soc. 4(1971), 35-39.
[1971b] *The essential closure of an archimedean lattice-ordered group*, Proc. London Math. Soc.38(1971), 151-160.
[1973]   *The hulls of representable ℓ-groups and f-rings*, J. Austral. Math. Soc. 16(1973), 385-415.
[1974a] *The topological completion and the linearly compact hull of an abelian ℓ-group*, Proc. London Math. Soc. 28(1974), 457-481.
[1974b] *Epi-archimedean groups*, Czech. Math. J. 24(99), (1974), 192-218.
[1974c] *The additive group of an f-ring*, Canad. J. Math. 26(1974), 1157-1168.

P. F. Conrad, J. G. Harvey, and W. C. Holland
[1963]   *The Hahn embedding theorem for lattice-ordered groups*, Trans. Amer. Math. Soc. 108(1963), 143-169.

P. F. Conrad and J. Martinez
[a]   *Locally finite conditions on lattice-ordered groups*, submitted.
[b]   *Signatures and discrete lattice-ordered groups*, submitted.
[c]   *Very large subgroups of lattice-ordered groups*, submitted.

P. F. Conrad and D. McAlister
[1969]   *The completion of a lattice-ordered group*, J. Australian Math. Soc. 9(1969), 182-208.

P. F. Conrad and J. R. Teller
[1970]   *Abelian pseudo lattice ordered groups*, Pub. Math. Debrecen 17(1970), 223-241.

M. R. Darnel

[1985]   *Epicomplete completely distributive ℓ-groups*,   Algebra Universalis **21**(1985), 123-132.

M. R. Darnel

[1987]   *Special valued ℓ-groups and abelian covers*, Order **4**(1987), 191-194.

M. R. Darnel and A. M. W. Glass

[a]   *Commutator relations and identities in lattice-ordered groups*, submitted.

F. Dashiell, A. Hager, and M. Henricksen

[1980]   *Order-Cauchy completions of rings and vector lattices of continuous functions*, Can. J. Math. **32**(1980), 657-685.

A. Day and J. Ježek

[1984]   *The amalgamation property for varieties of lattices*, Trans. Amer. Math. Soc. **286**(1984), 251-256.

R. Dedekind

[1871]   *Über die Theorie der ganzen algebraischen Zahlen*, Supplement XI to Dirichlet's "Vorlesungen über Zahlentheorie", second edition (1871), third edition (1879), fourth edition (1894) = Ges. Math. Werke, Wieweg 1932, III, 1-314, reprinted by Chelsea, New York, 1968.

R. DeMarr

[1965]   *Order convergence and topological convergence*, Proc. Amer. Math. Soc. **16**(1965), 588-590.

J. Dieudonné

[1941]   *Sur la theorie de la disibilité*, Bull. Soc. Math. France **49**(1941), 1-12.

M. Droste

[1985]   *The normal subgroup lattice of 2-transitive automorphism groups of linearly ordered sets*, Order **2**(1985), 291-319.

M. Droste and S. Shelah

[1985]   *A construction of all normal subgroup lattices of 2-transitive automorphism groups of linearly ordered sets*, Israel J. Math. **51**(1985), 223-261.

A. Ehrenfeucht

[1959]   *Decidability of the theory of linear ordering relation*, Notices Amer. Math Soc. **6**(1959), 268--269.

P. C. Eklof

[1977]   *Ultraproducts for algebraists*, in *Handbook of Mathematical Logic*, J. Barwise, ed., North-Holland, Amsterdam, 1977, 105-138.

354

J. Ellis

[1968]   *Group topological convergence in completely distributive lattice ordered groups*, Doctoral Dissertation, Tulane University, 1968.

R. Engelking

[1968]   *Outline of General Topology*, North Holland, Amsterdam, 1968.

Yu. L. Ershov

[1964]   *The decidability of the elementary theory of distributive, relatively complemented lattices and the theory of filters*, (in Russian), Algebra i Logika 3, 3(1964), 17-38.

C. J. Everett and S. Ulam

[1945]   *On ordered groups*, Trans. Amer. Math. Soc. 57(1945), 208-216.

T. Feil

[1980]   *Varieties of representable lattice ordered groups*, Doctoral Thesis, Bowling Green State University, 1980.

[1981]   *An uncountable tower of ℓ-group varieties*, Algebra Universalis, 14(1981), 129-131.

J. Ferrante and Ch. Rackoff

[1975]   *A decision procedure for the first-order theory of real addition with order*, SIAM J. Comp. 4(1975), 69-77.

[1979]   *The Computational Complexity of Logical Theories*, Lecture Notes in Math. 718, Springer Verlag, 1979.

I. Fleischer

[1976]   *Quantifier elimination for modules and ordered groups*, Bull. de l'Acad. Polon. Sci. 24(1976), 9-15.

C. D. Fox

[1983]   *On the Scrimger varieties of lattice ordered groups*, Algebra Universalis, 16(1983), 163-166.

R. Fraïssé

[1954]   *Sur l'extension aux relations de quelques proprietes des ordres*, Ann. Sci. Ecole Norm. Sup. (3) 71(1954), 363-388.

J. D. Franchello

[1978]   *Sublattices of free products of lattice ordered groups*, Algebra Universalis 8(1978), 101-110.

P. Freyd

[1964]   *Abelian Categories*, Harper and Row, New York, 1964.

L. Fuchs

[1951]   *The generalization of the valuation theory*, Duke Math. J. 18(1951), 19-56.

[1963]   *Partially Ordered Algebraic Systems*, Pergamon Press, New York, 1963.

[1965]   *Riesz groups*, Ann. Scuol. Norm. Sup. Pisa **19**(1965), 1-34.

M. R. Garey and D. S. Johnson

[1979]   *Computers and Intractability*, Freeman & Co., New York, 1979.

J. v. z. Gathen and M. Sieveking

[1978]   *A bound on solutions of linear integer equalities and inequalities*, Proc. Amer. Math. Soc. **72**(1978), 155-158.

C. F. Gauss

[1801]   *Disquisitiones Arithmeticus*, German edition, Springer-Verlag, Berlin, 1889; English translation, Yale University, New Haven, 1966.

L. Gillman and M. Jerison

[1976]   *Rings of Continuous Functions*, Van Nostrand, Princeton 1960; reprinted as Springer-Verlag Graduate Texts **43**, Berlin-Heidelberg-New York, 1976.

R. Gilmer

[1972]   *Multiplicative Ideal Theory*, Marcel Dekker, New York, 1972

A. M. W. Glass

[1974]   *ℓ-simple lattice-ordered groups*, Proc. Edinburgh Math. Soc. **19**(1974), 133-138.

[1976]   *Ordered Permutation Groups*, Bowling Green State University, Bowling Green, Ohio, 1976.

[1981a] *Elementary types of automorpphisms of linearly ordered sets -- a survey*, in Algebra Carbondale 1980, Proc., Springer Lecture Notes in Math. **848**, ed. R. K. Amayo, 218-229.

[1981b] *Ordered Permutation Groups*, London Math. Soc. Lecture Notes Series No.55, Cambridge University Press, 1981.

[1983]   *Countable lattice-ordered groups*, Math. Proc. Cambridge Phil. Soc. **94**(1983), 29-33.

[1984]   *The isomorphism problem and undecidable properties for finitely presented lattice-ordered groups*, in *Orders: description and roles*, Annals of Discrete Math. **23**, North Holland Math. Studies 99, ed. M. Pouzet and D. Richard, North Holland, Amsterdam (1984), 157-170.

[1985]   *Effective extensions of lattice-ordered groups that preserve the degrees of the conjugacy and word problems*, in *Ordered Algebraic Structures*, Lecture Notes in Pure and Applied Math. **99**, ed. W. B. Powell and C. Tsinakis, Marcel Dekker, New York, 1985, 89-98.

[1986a] *Effective embeddings of countable lattice-ordered groups*, in *Proc. 1st International Conference on Ordered Algebras*, Luminy 1984, ed. M.

Jambu-Giraudet and S. Wolfenstein, R&E **14**, Heldermann, Berlin, 1986, 63-69.

[1986b] *Generating varieties of lattice-ordered groups: approximating wreath products* , Illinois J. Math. **30**(1986), 214-221.

[1986c] *Lattice-ordered groups -- a very biased survey*, in *Proc. of Groups -- St. Andrews 1985*, ed. E.F. Robertson and C.M. Campbell, London Math. Soc. Lecture Notes Series No. **121**, Cambridge University Press, 1986, pp. 204-212.

[1987] *Free products of lattice-ordered groups*, Proc. Amer. Math. Soc **101**(1987), 11-16.

[a] *The universal theory of lattice-ordered abelian groups*, Proc. Le Mans Conference on Ordered Algebras 1987 (to appear).

A. M. W. Glass and Y. Gurevich

[1983] *The word problem for lattice-ordered groups*, Trans. Amer. Math. Soc. **280**(1983), 127-138.

A. M. W. Glass, Yu. Gurevich, W. C. Holland, and M. Jambu-Giraudet

[1981] *Elementary theory of automorphisms of doubly homogeneous chains*, in Logic Year 1979-80, The U. of Conn., Springer Lecture Notes in Math. **859**, ed. M. Lerman, J. H. Schmerl, and R. I. Soare, 1981, 67-82.

A. M. W. Glass, Yu. Gurevich, W. C. Holland, and S. Shelah

[1981] *Rigid homogeneous chains*, Math. Proc. Cambridge Phil. Soc. **89**(1981), 7-17.

A. M. W. Glass, W. C. Holland, and S. H. McCleary

[1980] *The structure of ℓ-group varieties*, Algebra Universalis **10**(1980), 1-20.

A. M. W. Glass and J. J. Madden

[1984] *The word problem versus the isomorphism problem*, J. London Math. Soc. **30**(1984) 53-61.

A. M. W. Glass and K. R. Pierce

[1980a] *Existentially complete abelian lattice-ordered groups*, Trans. Amer. Math. Soc. **261**(1980), 255-270.

[1980b] *Equations and inequations in lattice-ordered groups*, in *Ordered Groups*, Proc. Boise State Conf., Lecture Notes in Pure and Appl. Math. **62**(1980), J. E. Smith, G. O. Kenny, R. N. Ball, eds., Marcel Dekker, New York., 141-171.

[1980c] *Existentially complete lattice-ordered groups*, Israel J. Math. **36**(1980), 257-272.

A. M. W. Glass, D. Saracino, and C. Wood

[1984] *Non-amalgamation of ordered groups*, Math. Proc, Cambridge Phil. Soc. **95**(1984), 191-195.

A. M. W. Glass and J. S. Wilson
[a]   unpublished manuscript.

B. Glastad and J. L. Mott
[1982]   *Finitely generated groups of divisibility*, Contemporary Math. 8(1982), 231-247.

E. Golod and I. Safarevic
[1964]   *On class field towers*, (Russian), Izv. Akad. Nauk. SSSR 28(1964),261-272; English translation in Am. Math. Soc. Transl. (2) 48, 91-102.

G. Grätzer
[1971]   *Lattice Theory*, Freeman, San Francisco, 1971.
[1978]   *General Lattice Theory*, Academic Press, New York, 1978.
[1979]   *Universal Algebra*, 2nd ed., Springer-Verlag, New York, 1979.

G. Grätzer and H. Lakser
[1971]   *The structure of pseudocomplemented distributive lattices II; congruence extensions and amalgamations*, Trans. Amer. Math. Soc. 156(1971), 343-358.

G. Grätzer and E. T. Schmidt
[1957]   *On a problem of M. H. Stone*, Acta Math. Acad. Sci. Hungar. 8(1957),455-460.

[1963]   *Characterizations of congruence lattices of abstract algebras*, Acta Sci. Math. (Szeged) 24(1963), 34-59.

S. A. Gurchenkov
[1982a]   *Minimal varieties of ℓ-groups*, Algebra and Logic, 21, 83-87 (Algebra i Logika 21(1982), 131-137).
[1982b]   *Varieties of nilpotent lattice ordered groups*, Algebra and Logic 21, 331-339 (Algebra i Logika 21(1982), 499-510).
[1984a]   *Varieties of ℓ-groups with the identity $[x^p,y^p] = e$ have finite bases*, Algebra and Logic 23, 20-35 (Algebra i Logika 23(1984), 27-47).
[1984b]   *On covers in the lattice of ℓ-varieties*, Mat. Zametki 35(1984), 677-684.
[1984c]   *IXth All-Union Symposium on Group Theory*, Moscow, Theses Reports, 1984, p. 194.
[1985]   *About varieties of ℓ-groups with infinite axiomatic rank*, Siberian Math. J. 26(1985), 66-70.

Yu. Gurevich
[1965]   *Elementary properties of ordered abelian groups*, Transl. Amer. Math. Soc. 46(1965), 165-192.
[1967a]   *Hereditary undecidability of the theory of lattice-ordered abelian groups*, (in Russian), Algebra i Logika 6(1967), 45-62.

358

[1967b] *A contribution to the elementary theory of lattice-ordered abelian groups and K-lineals,* Soviet math. Dokl. **8**(1967), 987-989.

[1977] *Expanded theory of ordered abelian groups,* Annals of Math. Logic **12**(1977), 193-228.

[1980] *Ordered abelian groups,* in *Ordered Groups,* Proc. Boise State Conf., Lecture Notes in Pure and Appl. Math. **62**(1980), J. E. Smith, G. O. Kenny, R. N. Ball, eds., Marcel Dekker, New York., 173-174.

Yu. Gurevich and W. C. Holland

[1981] *Recognizing the real line,* Trans. Amer. Math. Soc. **265**(1981), 527-534.

Yu. S. Gurevich and A. I. Kokorin

[1963] *Universal equivalence of ordered abelian groups,* Algebra i Logika Sem. **2**, no. 1, 37-39.

A. W. Hager

[1969] *On inverse-closed subalgebras of C(X),* Proc. London Math. Soc. **19**(1969), 233-257.

[1985] *Algebraic closures of ℓ-groups of continuous functions,* in *Rings of Continuous Functions,* ed. C. E. Aull, Dekker Lecture Notes in Pure and Applied Mat. **95**, 165-169, Marcel Dekker, New York, 1985.

A. W. Hager and J. J. Madden

[1983] *Majorizing injectivity in abelian lattice-ordered groups,* Rend. Sem. Mat. Univ. Padova, **69**(1983), 181-194.

A. W. Hager and L. C. Robertson

[1977] *Representing and ringifying a Riesz space,* Symposia Math. **21**(1977), 411-431.

[1978] *Extremal units in an archimedean Riesz space,* Rend. Sem. Mat. Univ. Padova **59**(1978), 97-115.

[1979] *On the embedding into a ring of an archimedean ℓ-group,* Canad. J. Math.**71**(1979), 1-8.

H. Hahn

[1907] *Über die nichtarchimedische Grössensysteme,* S.-B. Kaiserlichen Akad. Wiss. Math. Nat. Kl. IIa, **116**(1907), 601-655.

P. Hall

[1974] *Embedding a group in a join of given groups,* J. Australian Math. Soc. **17**(1974), 434-495.

O. Harlompovich

[1981] *A finitely presented solvable group with unsolvable word problem,* Izvestia Akad. Nauk USSR, Ser. Mat. **45** No. 4, 1981, 852-873.

W. Heinzer

[1969]   *Some remarks on complete integral closure,* J. Austral. Math. Soc. 9(1969), 310-314.

H. Herrlich and G. Strecker

[1973]   *Category Theory, an Introduction,* Allyn and Bacon, Boston, 1973; 2nd edition, Heldermann Verlag, 1979.

G. Higman

[1954]   *On infinite simple groups,* Publ. Math. Debrecen 3(1954), 221-226.

H. Hironaka

[1964]   *Resolution of singularities of an algebraic variety over a field of characteristic zero,* Annals of Math. 79(1964), 109-326.

P. Hill

[1972]   *On the complete integral closure of a domain,* Proc. Amer. Math. Soc. 36(1972), 26-30.

O. Hölder

[1901]   *Die Axiome der Quantität und die Lehre vom Mass,* Leipzig Ber., Math.-Phys. Cl. 53(1901), 1-64.

W. C. Holland

[1963a]   *The lattice ordered group of automorphisms of an ordered set,* Michigan Math. J. 10(1963), 399-408.

[1963b]   *Extensions of ordered groups and sequence completion,* Trans Amer. Math. Soc. 107(1963), 71-82.

[1965]   *Transitive lattice-ordered permutation groups,* Math. Z. 87(1965), 420-433.

[1969]   *The characterization of generalized wreath products,* J. Algebra 13(1969), 155-172.

[1976]   *The largest proper variety of lattice-ordered groups,* Proc. Amer. Math. Soc. 57(1976), 25-28.

[1979]   *Varieties of ℓ-groups are torsion classes,* Czech. Math. J. 29(1979), 11-12.

[1985a]   *Varieties of automorphism groups of orders,* Trans. Amer. Math. Soc. 288(1985), 755-763.

[1985b]   *A note on lattice orderability of groups,* Algebra Universalis 20(1985), 130-131.

W. C. Holland and J. Martinez

[1979]   *Accessibility of torsion classes,* Algebra Universalis 9(1979), 199-206.

W. C. Holland and S. H. McCleary

[1969]   *Wreath products of ordered permutation groups,* Pacific J. Math. 31(1969), 703-716.

[1979] *The word problem for free lattice-ordered groups,* Houston J. Math. 5(1979), 99-105.

W. C. Holland, A. Mekler and N. R. Reilly
[1986] *Varieties of lattice-ordered groups in which prime powers commute,* Algebra Universalis 23(1986), 196-214.

W. C. Holland, A. Mekler, and S. Shelah
[1985] *Lawless order,* Order 1(1985), 383-397.

W. C. Holland and N. R. Reilly
[1986] *Structure and laws of the Scrimger varieties of lattice ordered groups,* Proc. First International Conference on Ordered Algebraic Systems, Marseilles (1984) (ed. S. Wolfenstein), R & E **14**, Heldermann Verlag, Berlin, pp.71-81.

[a] *Metabelian varieties of lattice ordered groups that contain only abelian o-groups.*

W. C. Holland and E. Scrimger
[1972] *Free products of lattice-ordered groups,* Algebra Universalis 2(1972), 247-254.

H. A. Hollister
[1978] *Nilpotent ℓ-groups are representable,* Algebra Universalis 8(1978), 65-71.

M. E. Huss
[1984] *Varieties of lattice ordered groups,* Ph.D. Dissertation, Simon Fraser University, 1984.

M. E. Huss and N. R. Reilly
[1984] *On reversing the order of a lattice ordered group,* J. of Algebra **91**, 176-191.

J. Isbell
[1960] *Uniform Spaces,* Mathematical Surveys, no. 12, Amer. Math. Soc., Providence, 1964.

[1966] *Epimorphisms and dominions,* in Proc. Conf. on Categorical Algebra, La Jolla 1965, Springer Notes (1966), 232-246.

P. Jaffard
[1950] *Corps demi-values,* C. R. Acad. Sci. Paris **231**(1950), Ser. A-B, 1401-1403.
[1953] *Contribution à la thèorie des groupes ordonnés,* J. Math. Pures Appl. **32**(1953), 203-280.
[1954] *Extensions des groupes réticulés et applications,* Publ. Sci. Univ. Alger. Ser. A-1(1954), 197-222.
[1955] *La notion de valuation,* Enseignment Math. **40**(1955), 5-26.

[1956] *Un contre-example concernant les groupes de divisibilité*, C. R. Acad. Sci. Paris **243**(1956), 1264-1268.

[1960] *Les Systémes d'Ideaux*, Dunod, Pris, 1960.

[1961] *Solution d'un probleme de Krull*, Bull. Soc. Math. France **85**(1961), 127-135.

J. Jakubik

[1970] *On subgroups of a pseudo lattice ordered group*, Pacific J. Math. **34**(1970), 109-115.

[1981a] *On value selectors and torsion classes of lattice ordered groups*, Czech. Math. J., **31**(1981), 306-313.

[1981b] *Prime selectors and torsion classes of lattice ordered groups*, Czech. Math. J., **31**(1981), 325-337.

[1981c] *On the lattice of torsion classes of lattice ordered groups*, Czech. Math. J., **31**(1981), 510-513.

[1982] *Torsion radicals of lattice ordered groups*, Czech. Math. J., **32**(1982), 347-36

M. Jambu-Giraudet

[1983] *Bi-interpretable groups and lattices*, Trans. Amer. Math. Soc. **278**(1983), 253-269.

R. E. Johnson

[1968] *Free products of ordered semigroups*, Proc. Amer. Math. Soc. **19**(1968), 697-700.

B. Jónsson

[1956] *Universal relational structures*, Math. Scand. **4**(1956), 193-208.

[1960] *Homogeneous universal relational structures*, Math. Scand. **8**(1960), 137-142.

[1961] *Sublattices of a free lattice*, Canadian J. Math. **13**(1961), 146-157.

[1962] *Algebraic extensions of relational systems*, Math. Scand. **11**(1962), 179-205.

[1965] *Extensions of relational structures*, Proc. International Symposium on the Theory of Models (Berkeley, 1963), North Holland, Amsterdam, 1965, 146-157.

[1967] *Algebras whose congruence lattices are distributive*, Math. Scand. **21**(1967), 110-121.

[1984] *Amalgamations of pure embeddings*, Algebra Universalis **19**(1984), 266-268.

I. Kaplansky

[1954] *Infinite abelian groups*, The University of Michigan Press, Ann Arbor, 1954.

[1970] *Commutative Rings*, Allyn and Bacon, 1970.

M. I. Kargapolov
[1963]   *Classification of ordered abelian groups by elementary properties*, (in Russian), Algebra i Logika 2(1963), 31-46.

M. I. Kargapolov and Yu. I. Merzlyakov
[1979]   *Fundamentals of the theory of groups*, Springer Verlag, Heidelberg, 1979.

H. J. Keisler
[1977]   *Fundamentals of model theory*, in *Handbook of Mathematical Logic*, J. Barwise, ed., North-Holland, Amsterdam, 1977, 48-103.

O. Kenny
[1975]   *The completion of an abelian ℓ-group*, Can. J. Math. **27**(1975), 980-985.

D. Kenoyer
[1984]   *Recognizability in the lattice of convex ℓ-subgroups of a lattice-ordered group*, Czech. Math. J. **34**(109) (1984), 411-416.

D. Kent
[1966]   *On the order topology in a lattice*, Illinois J. Math. **10**(1966), 90-96.

D. Kent and G. Richardson
[1974]   *Regular completions of Cauchy spaces*, Pacific J. Math. **51**(1974), 483-490.

D. Kent and R. Vainio
[1985]   *Ordered Cauchy spaces*, Internat. J. Math. and Math. Sci. **8**(3)(1985), 483-496.

N. G. Khisamiev
[1966]   *The universal theory of lattice-ordered abelian groups*, (in Russian), Algebra i Logika **5**, 3(1966), 71-76.

N. G. Khisamiev and A. I. Kokorin
[1966]   *Elementary classification of lattice-ordered abelian groups with a finite number of filaments*, Algebra i Logika **5**(1966), 41-50.

E. W. Kiss, P. Márki, P. Pröhle and W. Tholen
[1983]   *Categorical algebraic properties. A compendium on amalgamations, congruence extension, endomorphisms, residual smallness and injectivity*, Studia Sci. Math. Hungarica **18**(1983), 19-141.

V. M. Kopytov
[1975]   *Lattice-ordered locally nilotent groups*, Algebra and Logic **14**(1976), 249-251 (Algebra i Logika **14**(4), (1975), 407-413).

[1979]   *Free lattice ordered groups*, Algebra and Logic **18**(1979), 259-270 (English translation).

[1982]   *Nilpotent lattice-ordered groups*, Siberian Math. J. **23**(1982), 690-693.

[1983]   *Free lattice-ordered groups*, Siberian Math. J. **24**(1983), 98-101 (English translation).

[1985]   *A non-abelian variety of lattice ordered groups in which every solvable ℓ-group is abelian*, Mat. Sbornik **126/168**(1985), 247-266.

V. M. Kopytov and S. A. Gurchenkov

[1987]   *On covers of variety of abelian lattice ordered groups*, Siberian Math. J. **28**(1987), 406-408 (English translation).

V. M. Kopytov and N. Ya. Medvedev

[1975]   *Lattice-ordered locally nilpotent groups*, Algebra and Logic **14**(1976), 249-251 (Algebra i Logika **14**(1975), 407-413).

[1977]   *On varieties of lattice ordered groups*, Algebra and Logic **16**(1978), 281-285 (Algebra i Logika, **16**(1977), 417-423).

H. Kneser

[1924]   *Kurvenscharen auf Ringflächen*, Math. Ann. **91**(1924), 135-154.

W. Krull

[1931]   *Algemeine Bewertungstheorie*, J. reine angew. Math. **117**(1931),160-196.

F. Lacava

[1980]   *Osservazioni sulla teoria dei gruppi abeliani reticolari*, Boll. Un. Mat. Ital. A(5) **17**(1980), 319-322.

H. Lakser

[1973]   *Disjointness condition in free products of distributive lattices: an application of Ramsey's theorem*, Proc. U. Houston Lattice Theory Conference (1973), 156-168.

H. Lauchli and J. Leonard

[1966]   *On the elementary theory of linear order*, Fund. Math. **59**(1966), 109-116.

A. Lavis

[1963]   *Sur les quotients totalement ordonnes d'un groupe lineairment ordonne*, Bul. Soc. Royal Sci. Liege **32**(1963), 204-208.

W. Lenski

[1984]   *Elimination of quantifiers for the theory of archimedean ordered divisible groups in a logic with Ramsey quantifiers*, in *Models and Sets*, Proc. Logic Coll. '83, Aachen, Lecture Notes in Math. **1103**(1984), Springer, 261-280.

[1988]   *Ordered abelian groups in logics with Ramsey quantifiers*, Doctoral Dissertation, University of Heidelberg, 1988.

364

W. J. Lewis

[1973] *The spectrum of a ring as a partially ordered set,* J. Algebra **25**(1973), 419-434.

W. J. Lewis and J. Ohm

[1976] *The ordering of Spec R,* Can. J. Math. **28**(1976), 820-835.

J. T. Lloyd

[1964] *Lattice ordered groups and o-permutation groups,* Ph. D. Dissertation, Tulane University, 1964.

[1965] *Representations of lattice-ordered groups having a basis,* Pacific J. Math **15**(1965), 1313-1317.

P. Lorenzen

[1939] *Abstrakte Begrundung der multiplicative Idealtheorie,* Math. Z. **45**(1939), 533-553.

[1949] *Über halbgeordnete Gruppen,* Math. Zeit **52**(1949), 483-526.

R. J. Loy and J. B. Miller

[1972] *Tight Riesz groups,* J. Australian Math. Soc. **13**(1972), 224-240.

W. Luxemburg and A. Zaanen

[1971] *Riesz Spaces,* Vol. 1, North Holland, Amsterdam, 1971.

R. C. Lyndon and P. E. Schupp

[1977] *Combinatorial Group Theory,* Ergebnisse der Math. Grenzgeb. **89**, Springer, Heidelberg, 1977.

A. Macintyre

[1977] *Model completeness,* in *Handbook of Mathematical Logic,* J. Barwise, ed., North-Holland, Amsterdam, 1977, 139-180.

H. MacNeille

[1937] *Partially ordered sets,* Trans. Amer. Math. Soc. **42**(1937), 416-460.

R. Madell

[1980] *Complete distributivity and α-convergence,* Czech. Math. J. **30**(1980), 296-301.

J. J. Madden

[1983] *Two methods in the study of k-vector lattices,* Ph. D. Thesis, Wesleyan University, 1983.

J. Madden and J. Vermeer

[a] *Epicomplete archimedean ℓ-groups via a localic Yosida theorem,* (to appear).

A. I. Mal'cev

[1949] *Nilpotent torsion-free groups,* Izv. Akad. Nauk. SSSR Ser. Mat. **13**(1949), 201-212.

[1951] *On the full ordering of groups,* Trudy Mat. Inst. Steklov 38(1951), 173-175.

A. A. Markov

[1960] *Insolubility of the problem of homeomorphy,* in *Proc. International Congress of Math.,* 1958, ed. J. A. Todd FRS, University Press, Cambridge, 1960, 300-306.

J. Martinez

[1972] *Free products in varieties of lattice ordered groups,* Czech. Math. J. 22(97), (1972), 535-553.

[1973a] *Some pathology involving pseudo ℓ-groups of divisibility,* Proc. Amer. Math. Soc. 40(1973), 333-340.

[1973b] *Archimedean lattices,* Algebra Universalis 3(1973), 247-260.

[1973c] *Archimedean-like classes of lattice ordered groups,* Trans Amer. Math. Soc 186(1973), 33-49.

[1973d] *Free products of abelian ℓ-groups,* Czech. Math. J. 23(98), (1973), 349-361.

[1974a] *The hyperarchimedean kernel sequence of a lattice-ordered group,* Bull. Australian Math. Soc. 10(1974), 337-349.

[1974b] *Varieties of lattice ordered groups,* Math. Zeit. 137(1974), 265-284.

[1975a] *Torsion theory for lattice-ordered groups,* Czech. Math. J. 25(100), (1975), 284-299.

[1975b] *Doubling chains, singular elements and hyper-Z ℓ-groups,* Pacific J. Math. 61(2), (1975), 503-506.

[1976] *Torsion theory for lattice-ordered groups. Part II: Homogeneous ℓ-groups,* Czech. Math. J. 26(1976), 93-100.

[1977] *Pairwise-splitting lattice-ordered groups,* Czech Math. J. 27(102), (1977), 545-551.

[1979] *Nilpotent lattice-ordered groups,* Algebra Universalis 9(1979), 329-338.

[1980] *The fundamental theorem on torsion classes of lattice-ordered groups,* Trans. Amer. Math. Soc. 259(1980), 311-317.

[a] *Splits and local residues in lattice-ordered groups,* submitted.

[b] *A general theory of torsion classes for lattice-ordered groups,* University of Florida notes, 1977

S. H. McCleary

[1972a] *o-primitive ordered permutation groups I,* Pacific J. Math. 40(1972), 349-372.

[1972b] *Closed subgroups of lattice-ordered permutation groups,* Trans. Amer. Math. Soc. 173(1972), 303-314.

[1973a] *o-primitive ordered permutation groups II,* Pacific J. Math. 49(1973), 431-443.

366

[1973b] *o-2-transitive ordered permutation groups*, Pacific J. Math. **49**(1973), 425-429.

[1978] *Groups of homeomorphisms with manageable automorphism groups*, Comm. Algebra **6**(1978), 497-528.

[1985a] *Free lattice-ordered groups represented as o-2-transitive ℓ-permutation groups*, Trans Amer Math. Soc. **290**(1985), 69-79.

[1985b] *An even better representation for free lattice-ordered groups*, Trans. Amer. Math. Soc. **290**(1985), 81-100.

N. Ya. Medvedev

[1977] *The lattices of varieties of lattice-ordered groups and lie algebras*, Algebra and Logic **16**, 27-31 (Algebra i Logika **16**(1977), 40-45).

[1981] *Decomposition of free ℓ-groups into ℓ-direct products*, Siberian Math. J. **21**(1981), 691-696 (English translation).

[1982a] *ℓ-varieties without an independent basis of identities* (Russian), Math. Slovaca **32**(1982), 417-425.

[1982b] *On the theory of varieties of lattice-ordered groups*, Czech. Math. J. **32**(107), (1982), 364-372.

[1983a] *Coverings in the lattice of ℓ-varieties*, Algebra and Logic **22**, 39-44 (Algebra i Logika **22**(1983), 53-60).

[1983b] *Some questions of the theory of partially ordered groups*, Algebra and Logic **22**, 316-32 (Algebra i Logika **22**(1983), 435-442).

[1984] *Lattice of o-approximable ℓ-varieties*, Czech. Math. J. **34**(109), (1984) 6-17 (Russian).

J. Močkor

[1983] *Groups of Divisibility*, D. Reidel, Boston, 1983.

J. L. Mott

[1973] *The group of divisibility and its applications*, 1972 Conference on Commutative Algebra, Kansas, Springer, 1973.

[1974a] *Nonsplitting sequences of value groups*, Proc. Amer. Math. Soc. **44**(1974), 39-42.

[1974b] *Convex directed subgroups of a group of divisibility*, Canadian J. Math. **26**(1974), 532-542.

[1975] *The group of divisibility of Rees rings*, Math. Japonica **20**(1975), 85-87.

J. L. Mott and M. Schexnayder

[1976] *Exact sequences of semi-value groups*, J. reine angew. Math. **283/284**(1976), 388-401.

J. L. Mott and M. Zafrullah

[1981] *On Prüfer v-multiplication domains*, Manuscripta Math. **35**(1981), 1-26.

[a] *Factoriality in Riesz groups*, preprint.

[b] *Riesz groups as groups of divisibility*, preprint.

R. B. Mura and A. Rhemtulla

[1977]   *Orderable Groups*, Marcel Dekker, New York, 1977.

L. Nachbin

[1965]   *Topology and Order*,   Math. Studies No. 4, Van Nostrand, Princeton, 1965.

M. Nagata

[1957]   *A remark on the unique factorization theorem*, J. Math. Soc. Japan 9(1957), 143-145.

[1962]   *Local Rings*, Wiley (Interscience), New York, 1962.

T. Nakayama

[1942]   *Note on lattice-ordered groups*, Proc. Imp. Acad. Tokyo 18(1942), 1-4.

[1942-46] *On Krull conjecture concerning completely integrally closed integrity domains I,II,III*, Proc. Imp. Acad. Tokyo 18(1942),185-187; 233,236; Proc. Japan Acad. 22(1946), 249-250.

B. H. Neumann

[1943]   *Adjunction of elements to groups*, J. London Math. Soc. 18(1943), 4-11.

H. Neumann

[1967]   *Varieties of Groups*, Springer-Verlag, Berlin-Heidelberg-New York, 1967.

T. Ohkuma

[1954]   *Sur quelques ensembles ordonnés linéairment*, Fund. Math. 43(1954), 326-337.

J. Ohm

[1966]   *Some counterexamples related to integral closure in D[[X]]*, Trans. Amer. Math. Soc. 122(1966), 322-333.

[1969]   *Semi-valuations and groups of divisibility*, Canadian J. Math. 21(1969), 576-591.

A. Yu. Olshanskii

[1970]   *On the problem of a finite basis of identities in groups*,   Izv. Akad. Nauk. SSSR, Ser. Mat., 34, 316-384.

B. dePagter

[1981]   *f-algebras and orthomorphisms*, Doctoral Thesis, Leiden University, 1981.

P. Papangelou

[1965]   *Some considerations on convergence in abelian lattice groups*,   Pacific J. Math. 15(1965), 1347-1364.

K. R. Pierce

[1972a] *Amalgamations of lattice ordered groups*, Trans. Amer. Math. Soc. **172**(1972), 249-260.

[1972b] *Amalgamating abelian ordered groups*, Pacific J. Math. **43**(1972), 711-723.

[1976] *Amalgamated sums of abelian ℓ-groups*, Pacific J. Math. **65**(1976), 167-173.

R. S. Pierce

[1968] *Introduction to the Theory of Abstract Algebras*, Holt, Rinehart, and Winston, New York, 1968.

F. Point

[1983] *Quantifier elimination for projectable L-groups and linear elimination for rings*, Thèse, U. Mons, 1983.

W. B. Powell and C. Tsinakis

[1982] *Free products of abelian ℓ-groups are cardinally indecomposable*, Proc. Amer. Math. Soc. **86**(1982), 385-390.

[1983a] *Free products in the class of abelian ℓ-groups*, Pacific J. Math. **104**(1983),429-442.

[1983b] *The distributive lattice free product as a sublattice of the abelian ℓ-group free product*, J. Australian Math. Soc. **34**(1983), 92-100.

[1984] *Free products of lattice ordered groups*, Algebra Universalis **18**(1984), 178-198.

[1985] *Amalgamations of lattice ordered groups*, in Ordered Algebraic Structures, Lecture Notes in Pure and Applied Math. **99**, Marcel Dekker, New York, 1985; pages 171-178.

[1986] *Disjointness conditions for free products of lattice ordered groups*, Archiv der Math. **46**(1986), 491-498.

[a] *The failure of the amalgamation property for varieties of representable ℓ-groups*, Math. Proc. Cambridge Phil. Soc. (to appear).

[b] *Sets of disjoint elements in free products of lattice ordered groups*, Proc. Amer. Math. Soc. (to appear).

M. Presburger

[1929] *Über die Vollständigkeit eines gewissen Systems der Arithmetik...*, C. R. 1er Cong. des pays slaves, Warsaw, 1929, 92-101.

H. Prüfer

[1932] *Untersuchungen über Teilbarkeitseigenschaften in Körpern*, J. reine angew. Math. **168**(1932), 1-36.

M. O. Rabin

[1958] *Recursive unsolvability of group theoretic problems*, Ann. Math. **67**(1958), 172-194.

[1977] *Decidable theories*, in *Handbook of Mathematical Logic*, J. Barwise, ed., North-Holland, Amsterdam, 1977, 595-630.

J. A. Read

[1975] *Wreath products of non-overlapping lattice-ordered groups*, Canadian Math. Bull. **17**(1975), 713-722.

E. Reed

[1971] *Completions of uniform convergence spaces*, Math. Ann. **194**(1971), 83-108.

R. Redfield

[1974] *Ordering uniform completions of partially ordered sets*, Canadian J. Math. **26**(1974), 644-664.

N. R. Reilly

[1981] *A subsemilattice of the lattice of varieties of lattice-ordered groups*, Canadian J. Math. **XXXIII**(1981), 1309-1318.

[1983] *Nilpotent, weakly abelian and Hamiltonian lattice ordered groups*, Czech. Math. J. **33**(108), (1983), 348-353.

[1986] *Varieties of lattice ordered groups that contain no non-abelian o-groups are solvable*, Order **3**(1986), 287-297.

N. R. Reilly and R. Wroblewski

[1981] *Suprema of classes of generalized Scrimger varieties of lattice ordered groups*, Math. Zeit **176**(1981), 293-309.

P. Ribenboim

[1965] *On the existence of totally ordered abelian groups which are η_α-sets*, Bull. Acad. Polonaise Sci. **13**(1965), 545-548.

A. Robinson and E. Zakon

[1960] *Elementary properties of ordered abelian groups*, Trans. Amer. Math. Soc. **96**(1960), 222-236.

D. J. S. Robinson

[1982] *A Course in Group Theory*, Graduate Texts in Mathematics **80**, Springer, 1982.

J. Robinson

[1968] *Recursive functions of one variable*, Proc. Amer. Math. Soc. **19**(1968), 815-820.

H. Rogers, Jr.

[1967] *Theory of Recursive Functions and Effective Computability*, McGraw-Hill Series in Higher Math., McGraw-Hill, New York, 1967.

P. Samuel

[1971] *About Euclidean rings*, J. Algebra **19**(1971), 282-301.

370

D. Saracino and C. Wood

[1983]   *Finitely generic abelian lattice-ordered groups*, Trans. Amer. Math. Soc. **277**(1983), 113-123.

[1984]   *An example in the model theory of ℓ-groups*, Algebra Universalis **19**(1984), 34-37.

J. Schmid

[1979]   *Algebraically and existentially closed distributive lattices*, Zeitschr. math. Logik u. G. M. **25**(1979), 525-530.

P. H. Schmitt

[1982]   *Model theory of ordered abelian groups*, Habilitationsschrift, University of Heidelberg, 1982.

[1984]   *Model- and substructure complete theories of ordered abelian groups*, in Models and Sets, Proc. Logic Coll. '83, Aachen, Lecture Notes in Math. **1103**(1984), 389-418.

J. Schreier and S. Ulam

[1935]   *Eine Bemerkung über die Gruppe der topologischen Abbildungen der Kreislinie auf sich selbst*, Studia Math. **5**(1935), 155-159.

O. Schreier

[1927]   *Die untergruppen der freien Gruppen*, Abh. Math. Sem. Univ. Hamburg **5**(1927), 161-183.

E. B. Scrimger

[1975]   *A large class of small varieties of lattice ordered groups*, Proc. Amer. Math. Soc **51**(1975), 301-306.

P. Sheldon

[1971]   *How changing D[[X]] changes its quotient field*, Trans. Amer. Math. Soc. **159**(1971), 223-244.

[1973]   *Two counterexamples involving complete integral closure in finite dimensional Prüfer domains*, J. Algebra **27**(1973), 462-474.

[1974]   *Prime ideals in GCD-domains*, Canadian J. Math. **26**(1974), 98-107.

[a]   *A counterexample to a conjecture of Heinzer*, (preprint).

F. Šik

[1958]   *Automorphismen geordneter Mengen*, Casopis Pest. Mat. **83**(1958), 1-22.

[1965]   *Types spéciaux de réalisations des groupes réticulés*, C. R. Acad. Sc. Paris **261**(1965), 4948-4949.

R. Sikorski

[1948]   *A theorem on extensions of homomorphism*, Ann. Soc. Pol. Math. **21**(1948), 332-335.

[1952]   *Products of abstract algebras*, Fund. Math. **39**(1952), 211-228.

D. M. Smirnov

[1965]   *Generalized solvable groups and their group rings*, Math. Sbornik, **67**(1965), 366-383.

J. E. Smith

[1976]   *The lattice of ℓ-group varieties*, Ph.D. Thesis, Bowling Green State University, 1976.

[1980]   *The lattice of ℓ-group varieties*, Trans. Amer. Math. Soc., **257**(1980), 347-357.

[1981]   *A new family of ℓ-group varieties*, Houston J. of Math. 7(1981), 551-570.

[1984]   *Solvable and ℓ-solvable ℓ-groups*, Algebra Universalis, **18**(1984), 106-109.

D. Spikes

[1971]   *Semi-valuations and groups of divisibility*, Ph. D. dissertation, Louisiana State University, 1971.

W. Szmielew

[1955]   *Elementary properties of abelian groups*, Fund. Math. **41**(1955), 203-271.

A. Tarski

[1949]   *Arithmetical classes and types of Boolean algebras*, Bull. Amer. Math. Soc. **55**(1949), 64, 1192.

J. R. Teller

[1968]   *On abelian pseudo lattice ordered groups*, Pacific J. Math. **27**(1968), 411-419.

S. Thomas

[a]   Personal letter to A. M. W. Glass.

A. A. Vinogradov

[1949]   *On the free product of ordered groups*, (Russian), Math. Sbornik **25**(1949), 163-168.

[1971]   *Non-axiomatizability of lattice-ordered groups*, Siberian Math. J. **13**(1971), 331-332.

A. A. Vinogradov and A. A. Vinogradov

[1969]   *Non-locality of lattice-ordered groups*, Algebra and Logic 8(1969), 359-361.

S. M. Vosvi

[1983]   *On radical and coradical classes of ℓ-groups*, Algebra Universalis **16**(1983), 159-162.

W. Ward

[1940]   *Residuated lattices*, Duke Math. J. 6(1940), 641-651.

372

E. C. Weinberg
[1963] *Free lattice-ordered abelian groups*, Math. Ann. **151**(1963), 187-199.
[1965] *Free lattice-ordered abelian groups II*, Math. Ann. **159**(1965), 217-222.
[1967] Embedding in a divisible lattice-ordered group, J. London Math. Soc. **42**(1967),504-506.

V. Weispfenning
[1975] *Model-completeness and elimination of quantifiers in subdirect products of structures*, J. Algebra **36**(1975), 252-277.
[1976] *The model completion of a class of lattice-ordered abelian groups*, Notices Amer. Math. Soc. **23**(1976), A-349.
[1978] *Model theory of lattice products*, Habilitationsschrift, University of Heidelberg, 1978.
[1981] *Quantifier elimination for certain ordered and lattice-ordered abelian groups*, Bull. Soc. Math. Belg. **33**, ser. B, (1981), 131-156.
[1984] *Aspects of quantifier elimination in algebra*, in *Universal Algebra and its Links...*, Proc.**25**. Arbeitstagung, Darmstadt 1983, Heldermann V., Berlin, 1984, 85-105.
[1985] *The complexity of elementary problems in Archimedean ordered groups*, Proc. EUROCAL '85, Linz, Lecture Notes in Computer Science 204, Springer Verlag, 87-88
[1986a] *Existential equivalence of ordered abelian groups with parameters*, Tagungsbericht Oberwolfach 2, 1986.
[1986b] *The complexity of the word problem in abelian ℓ-groups*, Th. Comp. Sci. **48**(1986), 127-132.
[1986c] *Quantifier eliminable ordered abelian groups*, in *Algebra and Order*, R&E **14**, (S. Wolfenstein, ed.), Heldermann Verlag, Berlin, 1986, 113-126.
[1988] *The complexity of linear problems in fields*, J. Symb. Comput., **5**(1988), 3-27.
[a] *The complexity of almost linear Diophantine problems*, submitted.

S. Wolfenstein
[1968] *Valeurs normales dans un groupe réticulé*, Accademia Nazionale dei Lincei, **44**(1968), 337-342.

M. Yasuhara
[1974] *The amalgamation property, the universal-homogeneous models and the generic models*, Math. Scand. **34**(1974), 5-36.

K. Yosida
[1942] *On the representation of the vector lattice*, Proc. Imp. Acad. Tokyo **18**(1942), 339-343.

M. Zafrullah
[1978] *On finite conductor domains*, Manuscripta Math. **24**(1978), 191-204.

[a]   *Factoriality in partially ordered groups,* (preprint).

E. Zakon
   [1961]   *Generalized Archimedean groups,* Trans. Amer. Math. Soc. 99(1961),
            21-48.

O. Zariski
   [1947]   *The concept of a simple point of an abstract algebraic variety,* Trans.
            Amer. Math. Soc. 62(1947), 1-52.

INDEX

375